Highlights in
Condensed Matter Physics
and Future Prospects

NATO ASI Series

Advanced Science Institutes Series

A series presenting the results of activities sponsored by the NATO Science Committee, which aims at the dissemination of advanced scientific and technological knowledge, with a view to strengthening links between scientific communities.

The series is published by an international board of publishers in conjunction with the NATO Scientific Affairs Division

A	**Life Sciences**	Plenum Publishing Corporation
B	**Physics**	New York and London
C	**Mathematical and Physical Sciences**	Kluwer Academic Publishers
D	**Behavioral and Social Sciences**	Dordrecht, Boston, and London
E	**Applied Sciences**	
F	**Computer and Systems Sciences**	Springer-Verlag
G	**Ecological Sciences**	Berlin, Heidelberg, New York, London,
H	**Cell Biology**	Paris, Tokyo, Hong Kong, and Barcelona
I	**Global Environmental Change**	

Recent Volumes in this Series

Volume 278—Painlevé Transcendents
  edited by D. Levi and P. Winternitz

Volume 279—Fundamental Aspects of Inert Gases in Solids
  edited by S. E. Donnelly and J. H. Evans

Volume 280—Chaos, Order, and Patterns
  edited by Roberto Artuso, Predrag Cvitanović, and Giulio Casati

Volume 281—Low-Dimensional Structures in Semiconductors: From Basic
  Physics to Applications
  edited by A. R. Peaker and H. G. Grimmeiss

Volume 282—Quantum Measurements in Optics
  edited by Paolo Tombesi and Daniel F. Walls

Volume 283—Cluster Models for Surface and Bulk Phenomena
  edited by Gianfranco Pacchioni, Paul S. Bagus, and Fulvio Parmigiani

Volume 284—Asymptotics beyond All Orders
  edited by Harvey Segur, Saleh Tanveer, and Herbert Levine

Volume 285—Highlights in Condensed Matter Physics and Future Prospects
  edited by Leo Esaki

Series B: Physics

Highlights in Condensed Matter Physics and Future Prospects

Edited by
Leo Esaki

IBM Thomas J. Watson Research Center
Yorktown Heights, New York

Plenum Press
New York and London
Published in cooperation with NATO Scientific Affairs Division

Proceedings of NATO Science Forum '90 Highlights of the Eighties
and Future Prospects in Condensed Matter Physics,
held September 16 – 21, 1990,
in Biarritz, France

Library of Congress Cataloging-in-Publication Data

NATO Science Forum '90 Highlights of the Eighties and Future Prospects
  in Condensed Matter Physics (1990 : Biarritz, France)
   Highlights in condensed matter physics and future prospects /
edited by Leo Esaki.
       p.   cm. -- (NATO ASI series. Series B. Physics ; v. 285)
   "Proceedings of NATO Science Forum '90 Highlights of the Eighties
and Future Prospects in Condensed Matter Physics, held September
16-21, 1990, in Biarritz, France"--T.p. verso.
   "Published in cooperation with NATO Scientific Affairs Division."
   Includes bibliographical references and index.
   ISBN 0-306-44119-5
   1. Condensed matter--Congresses. 2. Superlattices as materials-
-Congresses. 3. Scanning tunneling microscopy--Congresses. 4. High
temperature superconductivity--Congresses. 5. Quantum Hall effect-
-Congresses.   I. Esaki, Leo, 1925-   . II. North Atlantic Treaty
Organization. Scientific Affairs Division. III. Title.
IV. Series.
QC173.4.C65N39   1990
530.4'1--dc20   91-44174
                                                          CIP

ISBN 0-306-44119-5

© 1991 Plenum Press, New York
A Division of Plenum Publishing Corporation
233 Spring Street, New York, N.Y. 10013

Printed in the United States of America

NATO Science Forum Series

Volume 285—Highlights in Condensed Matter Physics and Future Prospects
Edited by Leo Esaki

PREFACE

This volume contains the proceedings of the first NATO Science Forum "Highlights of the Eighties and Future Prospects in Condensed Matter Physics" (sponsored by the NATO Scientific Affairs Division), which took place in September, 1990, in the pleasant surroundings provided by the Hotel du Palais at Biarritz, France.

One hundred distinguished physicists from seventeen countries, including six Nobel laureates, were invited to participate in the four and a half day meeting. Focusing on three evolving frontiers: semiconductor quantum structures, including the subject of the quantum Hall effect (QHE), high temperature superconductivity (HiTc) and scanning tunneling microscopy (STM), the Forum provided an opportunity to evaluate, in depth, each of the frontiers, by reviewing the progress made during the last few years and, more importantly, exploring their implications for the future.

Though serious scientists are not "prophets," all of the participants showed a strong interest in this unique format and addressed the questions of future prospects, either by extrapolating from what has been known, or by a stretch of their "educated" imagination.

The areas dealt with in the Forum are indeed the outgrowths of three major milestones of condensed matter physics in the Eighties: The discovery of the QHE in the two-dimensional electron system by von Klitzing and the subsequent observation of the fractional QHE, including the remarkable advance in research on "engineered quantum structures with reduced dimensionality" such as superlattices, quantum wells, wires and dots; The development of the scanning tunneling microscope (STM) by Binnig and Rohrer made a profound impact on our understanding of surface structure by making possible the observation of atomic scale objects with unprecedented magnified power; The discovery of high temperature superconductivity (HiTc) in rare earth-copper oxides by Bednorz and Mueller was one of the greatest surprises in the Eighties; This has sparked intensive, world-wide investigations of these heavily-doped semiconductor-like materials, and yet no conclusive theory has emerged to account for the phenomenon.

It is noted that the three subject areas, independent as they may appear, have common threads. Electron tunneling, which constitutes the base of STM, has a rich history in solid-state physics, starting with Esaki tunnel diodes in semiconductors and Josephson junctions in superconductors, and leading to the observation of coherent tunneling in double barriers and superlattices. Also, reduced dimensionality is a common characteristic in both semiconductor quantum structures and HiTc materials. Furthermore, one of the conclusions in the Forum was the recognition of an "attraction" between semiconductor quantum structures and STM, namely, the potential for the control and manipulation of nanometer-scale structures by STM probes. The investigations of small structures with few electrons, impurities and spins will no doubt be expanded in the next decade. In the area of applied science, the integration of electronics and photonics in semiconductor quantum devices will also be accelerated. In the area of semiconductor integrated circuits, the ever decreasing size of individual devices will eventually approach nano-scale dimensions as the magnitude of integration advances toward the giga-level. One of the relevant evolving areas clearly identified in the Forum is the science and technology at the scale of nanometers.

I would like to take this opportunity to express my sincere gratitude to all the participants for contributing to the excitement and success of the Forum. I wish to thank Eli Burstein, Praveen Chaudhari and Erio Tosatti for their contributions to this forum as competent moderators for the Semiconductor, HiTc and STM panel sessions, respectively. A special debt of gratitude is acknowledged to Eli Burstein for his valuable advice in putting this forum together. I must also thank Leroy L. Chang and Marge Tumolo for their efforts in putting the manuscripts into final book form.

Preface

I am indeed deeply indebted to my co-directors, Klaus von Klitzing, K. Alex Müller and Heinrich Rohrer, and to the members of the organizing committee, and, last but not least, to the NATO officials, Jacques Ducuing and Giovanni Venturi, whose encouragement was indispensable in organizing this Forum.

July 1st, 1991   Leo Esaki

CONTENTS

OPENING

Welcoming Address
L. Esaki 1
Opening Address
J. Ducuing 3

QUANTUM HALL EFFECT

Quantum Hall Effect
K. von Klitzing...7
Fractional Quantum Hall Effect
D. C. Tsui...23
Theory and Implications of the Fractional Quantum Hall Effect
S. M. Girvin 35

SEMICONDUCTOR QUANTUM STRUCTURES

Implications of Semiconductor Superlattice Research
L. Esaki...55
Materials and Physics Aspects of Quantum Heterostructures
L. L. Chang..83
Electronic Energy Levels in Semiconductor Quantum Wells
and Superlattices
G. Bastard and R. Ferreira 117
III-V Versus II-VI Compound Superlattices
M. Voos 143
The First Principles View of Superlattices
H. Kamimura 161

Inelastic Light Scattering in Semiconductor Quantum Structures
    G. Abstreiter 191
Spectroscopic Investigations of Quantum Wires and Quantum Dots
    D. Heitmann, T. Demel, P. Grambow, M. Kohl
      and K. Ploog 209
Electron Optics in a Two-Dimensional Electron Gas
    H. van Houten 243
Some Recent Developments in the Physics of Resonant Tunneling
    L. Eaves 275
Nonlinear Optics and Optoelectronics in Quasi
      Two Dimensional Semiconductor Structures
    D. S. Chemla 293
Correlated Tunnel Events in Arrays of Ultrasmall Junctions
    T. Claeson and P. Delsing 333

HIGH TEMPERATURE SUPERCONDUCTIVITY

High Tc Superconductors: A Conservative View
    J. Friedel 365
High-Energy Spectroscopy Studies of High Tc Superconductors
    J. Fink, N. Nücker, H. Romberg, M. Alexander,
      P. Adelmann, R. Claessen, G. Mante, T. Buslaps,
      S. Harm, R. Manzke and M. Skibowski 377
The Role of the Short and Anisotropic Coherence Length
    G. Deutscher 399
Artificially Grown Superlattices of Cuprates
    O. Fischer, J-M. Triscone, O. Brunner,
      L. Antognazza, M. Affronte and L. Mieville 415
Excitations and Their Interactions in High Tc Materials
    J. R. Schrieffer 431
Some Relevant Properties of Cuprate Superconductors
    K. A. Müller 453

SCANNING TUNNELING MICROSCOPY

Local Probe Methods
    H. Rohrer 465
Ballistic Transport in Normal Metals
    P. Wyder, A. G. M. Jansen and H. van Kempen 495

Contents

Atom-Resolved Surface Chemistry With the STM:
    The Relation Between Reactivity and Electronic Structure
    P. Avouris 513
STM in Biology
    B. Michel...549
Manipulation and Modification of Nanometer Scale Objects
    with the STM
    C. F. Quate 573
The Physics of Tip-Surface Approaching:
    Speculations and Open Issues
    E. Tosatti 631
A New Spectroscopy of Carrier Scattering
    W. J. Kaiser, L. D. Bell, M. H. Hecht and L. C. Davis....................655

GENERAL

Role of Condensed Matter Physics in Information Management
    and Movement - Past, Present and Future
    C. K. N. Patel...667

Index 701

WELCOMING ADDRESS

Leo Esaki

IBM Thomas J. Watson Research Center
Yorktown Heights, New York 10598, USA

On behalf of the organizing committee, I am honored to welcome all of you attending this NATO Science Forum '90. I am happy to see so many distinguished participants at this beautiful site and also thankful for your positive response to and support of this unique forum which is focussed on the three important frontiers in condensed matter physics: semiconductor quantum structures, HiTc and STM.

Although this forum is our answer to a suggestion from the CSDL panel, headed by Dr. G. Venturi, under the NATO Science Committee, I have to acknowledge that the directors and committee members have all made a considerable effort to organize this unprecedented NATO event. Obviously, it is not an easy task to put together multiple subjects in a coherent and concise fashion. All of us in the organizing committee share responsibility for the selection of lecturers and invitees from various countries.

Our ultimate objective at this forum is very clear. It is the exploration of future prospects in the above-mentioned frontiers. You should all feel free to express your view on what would be of most significance in

Highlights in Condensed Matter Physics and Future Prospects
Edited by L. Esaki, Plenum Press, New York, 1991

the future course of research. History tells us that not only is individual creativity required, but that dynamical interactions among researchers are also essential for advancement in science. It is my hope that the collective deliberations that we are about to initiate here will provide a milestone for the shaping of the progress in condensed matter physics for the coming decade.

I would like to make a brief comment on our logo. This is the result of my request to my wife, Masako, to draw a picture of an experimentalist and a theorist in a serious discussion of an important research matter under the NATO banner. One of my friends whose research funds were perhaps squeezed, commented that the picture symbolizes the present hardship of researchers, with an illustration of a young man asking a senior official for money!

For the selection of this wonderful site, we owe thanks to the local arrangement chairman, Dr. V. Sadagopan and also to Mr. Pascal Stefanou of IBM France. It is not easy to find a hotel of just the right size, with 100 decent rooms, in a serene atmosphere. We chose this place because it meets these criteria, not because of the attraction of the superdelux facility of the Hotel du Palais. Nevertheless, this is a rather expensive venture. This is one reason why we have to make good use of our time in dealing with our busy schedule from 8:40 AM to 5:50 PM, every day. I am afraid I have already wasted too many minutes.

Now, I am delighted and honored to introduce Prof. Jacques Ducuing, Assistant Secretary General for NATO Scientific and Environmental Affairs. As the chairman of the NATO Science Committee, he was indeed instrumental in sponsoring and supporting this forum. I have known his name since 1962, because he was one of the co-authors of the seminal paper entitled "Interactions between light waves in a nonlinear dielectric," published in the Physical Review. The medium of this forum may be considered to be a highly non-linear interactive one because of the diverse backgrounds of the participants and we will hopefully generate intellectual higher harmonics in a most creative way throughout our meeting. No one is more suitable than Prof. Ducuing to present the opening address in our forum. I know everyone here is eager to hear from him.

OPENING ADDRESS

Jacques Ducuing

NATO Assistant Secretary General for Scientific and
Environmental Affairs

It is an honour and a great satisfaction for me to open this first
NATO Forum and to do it in this imperial decor. Satisfaction on
several counts. First, as a physicist, I am very pleased to see that
the first of what I hope will be a long and successful series of
meetings is devoted to an area of physics. I am also delighted to see
that this is a truly international meeting. In spite of the distance,
many of you have come from North America. This is very grati-
fying to us: it is one of the primary objectives of our Programme to
promote and sustain the dialogue between the scientific communities on both sides of the Atlantic Ocean. I am also happy to know
that there are a number of participants among us who come from
countries outside NATO. They can be sure that they are very
welcome here and my hope is that they will be more numerous in
the future.

Last but not least, this first NATO Forum corresponds to an important new endeavour of the NATO Science Programme. NATO has
a long tradition of supporting Science. During the past 30 years
more than 250,000 scientists have been supported directly in one
way or another. Although the international political context has
considerably changed and improved since 1958, the main objective

Highlights in Condensed Matter Physics and Future Prospects
Edited by L. Esaki, Plenum Press, New York, 1991

of our Programme, which is to stimulate cooperation between the scientists of our countries, remains the same. You can see it behind all the various features of our Science Fellowships, our Collaborative Research Grants, our Advanced Study Institutes, etc.

At this stage it might be appropriate to give you a rapid overview of the NATO Science Programme. It consists of three parts: Fellowships ... 400 MBF, Support of Science ... 417 MBF and Science for Stability ... 153 MBF. The well-known Fellowships, which are at the origin of the Programme and which are nationally administered so as to fit the specific needs of countries, have a budget of 400 MBF for 1990. A similar amount is devoted to the core of our activity to which I will return shortly (Support of Science), and around 150 MBF (about 5 M$) go annually to a scheme designed to help Greece, Portugal and Turkey to develop their scientific and technological base (Science for Stability). This is implemented through the cooperation of university, government and industry teams, with a strong involvement of the latter.

Support of Science includes our funding of Advanced Study Institutes, Advanced Research Workshops and Cooperative Research Grants, which take care of the travel and living expenses of scientists from different countries collaborating on a research project. Most of the funding goes to basic science without any preference for a specific domain. For a given type of activity, projects are selected by a multidisciplinary panel comprising a dozen scientists all from different countries and disciplines. In the present situation, only a small fraction of this core has a more "focused" character. This is what we call "Special Programmes" where we use the 3 different types of activity previously mentioned (ASI, ARW and CRGs) in a given area on which we want to put more emphasis. At the present time we have 3 Special Programmes. One is on Advanced Educational Technology, i.e. the study of the use of new technologies such as microcomputers, interactive disk systems and electronic communication for the improvement of the learning process. Another one, called Global Environmental Change, related to topics of strong current interest such as the ozone hole, the greenhouse effect and its impact on climatic change. Finally, a particularly successful one, called Chaos, Order and Patterns:

Aspects of Non-linearity, is devoted to the dynamics of nonlinear systems, a topic which is progressively pervading all areas of science from Physics to Biology.

Although we strongly believe that the Management of Science should follow the free flow of scientific thought and discovery, we also think that we ought to increase the part of our resources which goes to areas of strong current or potential interest. Special Programmes, guided by a panel of excellent and highly-motivated scientists, allow us to deal efficiently with these "cutting edge" areas as the panel can induce high-quality proposals and bring coherence to the set of various activities.

To go in this direction we must follow carefully the evolution of Science, have an accurate picture of trends, and be aware of the thinking of the scientific community. We hope that the NATO Forums will help us to do that. During the next few days you will review the current situation of a large part of Solid State Physics and in doing so you will point to directions which might be explored and offer new developments. This forward look is of great importance to us. For instance it will be particularly interesting to see which new perspectives will be offered by quantum structures and nanometer mesoscopic scale devices, or by the new advances in the probing and manipulation of atoms connected with tunneling microscopies. We intend to conduct Forums in other promising areas of Physics, preferably with a multidisciplinary character such as Clusters and beyond Physics in other fields of Science. In Chemistry an area like Supermolecular Chemistry would be a natural candidate with topics such as self-assembly and self-replication, the chemistry of monolayers, membranes, micelles,... In Life Sciences Genetic Research and Brain Research are also areas on which we would like to focus. More applied areas of Science will also come under scrutiny. NATO had for many years a strong interest in environmental matters and we are trying to relate it more closely to Science. This will induce us to explore more systematically domains like Environmental Chemistry.

Based on these Forums and Surveys we will reinforce and expand our Programmes in areas offering new prospects and developments and hopefully by doing so we will enhance the stimulation resulting from our activities. Thus this Forum assumes great significance as it is the first of a series of events which could profoundly change the NATO Science Programme. It is a spin-off of our Special Programme on Condensed Systems of Low Dimensionality and we are grateful to the Programme Director, Dr. Venturi, and the members of the Panel who supported the idea. We are also indebted to the co-directors and the members of the Committee for their splendid work. However, as everybody knows, this meeting would not have taken place without the personal involvement and leadership of Dr. Leo Esaki and I want to offer him the warmest thanks of the NATO Science Committee.

Let me wish us all a very successful meeting.

QUANTUM HALL EFFECT

K. von Klitzing

Max Planck Institut für Festkörperforschung
7000 Stuttgart 80, Germany

1. INTRODUCTION

The discovery of the quantum Hall effect in 1980 resulted from basic research on the building blocks of very large scale integrated, micro-electronic circuits - the silicon field effect transistor (Si-MOSFET). The influence of quantum confinement on the electronic properties of these devices have been a topic of much interest over the last twenty years and the QHE is only one of many results which demonstrate that quantum phenomena can dominate the behavior of microelectronic devices. The rapid advancement in miniaturization points to a greater role for quantum effects in future devices. Current research on quantum wires and quantum dots, as summarized by D. Heitmann in a later chapter, indicates that these may become the microelectronic structures of the future.

The optical properties of reduced dimensional devices reflect structure in the electronic density of states while the electrical properties of one-dimensional (1D) and weakly-coupled, zero-dimensional (OD) structures show quantum effects such as resistance variations in units of

$h/e^2 = 25812$ Ω, as in the QHE, and single electron transport. The quantum phenomena which result in the QHE, the AC Josephson effect and single electron transport, may be used to interconnect measurements of the voltage, U, the frequency, ν, and the current, I, as shown in Figure 1. The elementary relationships between U, ν and I for these effects are given below:

Josephson Effect $\qquad\qquad\qquad\qquad \nu = (2e/h)U$ $\qquad\qquad$ (1)

QHE and 1D ballistic transport $\quad$ U = (h/e^2)I $\qquad\qquad$ (2)

Single electron transistor $\qquad\qquad$ I = eν $\qquad\qquad$ (3)

The relations for the Josephson effect and the QHE, eqn. 1 and 2, seem to hold to a much higher level of accuracy than the best known values for the proportionality constants 2e/h and h/e^2. Thus, the Comité International des Poids et Measures has adopted the recommendation of the Comité Consultatif d'Electricité that fixed values for the Josephson constant, K_J = (483597.9 $\pm$ 0.2) GHz/V, and the von Klitzing constant, R_k = (25812.807 $\pm$ 0.005) Ohm, be used for all voltage and resistance calibrations:[1]

$K_{J\text{-}90}$ = 483597.9 GHz/V $\qquad\qquad\qquad$ $R_{K\text{-}90}$ = 25812.807 Ohm

These fixed values for K_J and R_K guarantee world-wide uniformity in all calibrations of electrical units with better reproducibilities than previous realizations of these units based on the definitions of the SI units.

In metrological applications of the QHE, it is not necessary to equate R_K with h/e^2, which is essentially the fine structure constant, α. However, since all theoretical studies of the QHE indicate that the quantized Hall resistance should be identical with h/e^2, we assume that the quantized resistance yields a direct measurement of the fine structure constant. Thus, it is possible to test quantum electrodynamics (QED) theory which expresses a measurable quantity, the anomalous magnetic moment of the electron, a_e, as a function of the fine structure constant:

$$a_e = \alpha/2\pi + C_4(\alpha/\pi)^2 + C_6(\alpha/\pi)^3 + C_8(\alpha/\pi)^4 + \ldots$$

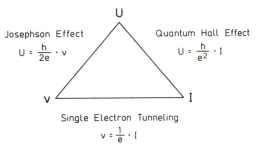

FIG. 1. The connection between the voltage, U, the current, I, and the frequency, ν, on the basis of quantum phenomena.

The coefficients C_4, C_6, C_8 have been calculated using QED theory under the assumption that electrons are point particles without internal structure. Based on an analysis of all available data related to the fine structure constant, B. N. Taylor and E. R. Cohen have concluded the following (unpublished, 1990):

If one assumes that the relation $R_K = h/e^2$ is exact, a small disagreement at the level of 10^{-8} between the QED-independent value of α and the α value determined from the anomalous magnetic moment exists. Such a disagreement may indicate that the electrons have a finite radius R and a rough estimate gives a value $R \cong 10^{-20}$ cm.

On the other hand, if one assumes *a priori* the validity of QED and the correctness of the resulting value of $\alpha(a_e)$, then the following result is obtained: $R_K = (1 + (60\pm22)\text{x}10^{-9})$ h/e^2. If the contribution of the anomalous magnetic moment for the least square adjustment of h/e^2 is deleted, the following relation between R_K and h/e^2 is found: $R_K = (1+(178\pm118)\text{x}10^{-9})h/e^2$. This result does not show a significant difference between R_K and h/e^2 but one may ask the question whether the equation $R_K = h/e^2$ is strictly correct.

Subsequent chapters of the text will attempt to answer this question and explain the exact quantization and the origin of the Hall plateaus. The connection between the quantized resistance observed in 1D structures for ballistic electronic transport in the absence of a magnetic field, and the quantization of the Hall resistance at high magnetic fields in 2D structures, will be stressed. The edge channel picture which results from the comparison of transport in these two different systems will be used to analyze recent experimental results. For a more general discussion of the QHE, the reader is referred to review articles[2,3] and a tutorial introduction into the physics and application of the QHE which had been presented at a NATO ASI.[4]

2. QUANTUM HALL PLATEAUS

Several approaches may be used to relate the quantized Hall resistance with the fundamental constant h/e^2. The simplest approach begins with the classical expression for the Hall resistance of a 2D system

$$R_H = U_H/I = B/n_s e \tag{4}$$

where U_H is the Hall voltage, I is the current, and n_s is the areal electron density. As the application of a strong transverse magnetic field B results in discrete Landau subbands with degeneracy[5]

$$N_L = eB/h, \tag{5}$$

one immediately obtains the quantized resistance

$$R_H = h/ie^2 \ (i = \text{integer}) \tag{6}$$

if the ratio B/n_s is adjusted such that an integer number of Landau levels is fully occupied with electrons, i.e., $n_s/N_L = i$. However, this picture fails to explain the observed plateaus in the Hall resistance vs B, for fixed n_s, since the quantization occurs only at singular values of the magnetic field given by

$$B_i = (h \, n_s)/(i \, e). \tag{7}$$

In contrast, experimental studies of the Hall effect vs B in typical 2D semiconductor systems, e.g., GaAs-AlGaAs heterostructures with $\mu(4.2K) = 10^5 cm^2/Vs$, exhibit wide Hall plateaus and vanishing longitudinal resistance R_{xx} as shown in Fig. 2. The data also indicate that small energy gaps due to spin splitting (B < 4T, Fig. 2) are not as well resolved as the cyclotron splitting. Additional complications include a decrease in the plateau width with increasing mobility and the formation of plateaus at fractional filling factors in the highest mobility samples at the lowest temperatures. The fractional Quantum Hall Effect (FQHE),

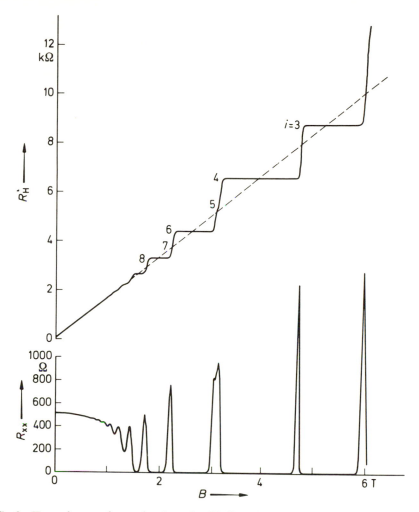

FIG. 2. Experimental results for the Hall resistance, R_H, and the longitudinal resistance, R_{xx}, of a GaAs/AlGaAs heterostructure, at T ≈ 10mK, as a function of the magnetic field, B. The Hall plateaus for filling factors i=3, 4, 6, and 8 are visible in the data.

which results from electron-electron interactions, will be discussed in the two contributions of D. C. Tsui and S. M. Girvin in this book.

An early model for the Hall plateaus relied upon the existence of electronic states which did not belong to the 2D system, e.g., interface states and impurity states in the depletion layer or in the doped AlGaAs layer. These localized states served to pin the Fermi level in the gap between Landau levels and maintain the integer filling factor condition over a finite range of fields. Although the dynamic exchange of electrons between the 2DEG and the extrinsic reservoir can produce small Hall plateaus, a quantitative explanation of the experimental situation requires the existence of localized states in the 2DEG. The dependence of the plateau width upon the electronic mobility suggests that the Hall plateaus originate from carrier localization in strong magnetic fields. Thus, it is generally accepted that the plateaus originate from a mobility gap, and not a density of states gap, in between Landau levels, i.e. the Landau levels are broadened, due to disorder, with localized states in the tails of these levels.[6] Calculations have shown that the Hall resistance remains constant if the occupation of localized states is varied. The semi-classical percolation picture for electronic motion in long range potential fluctuations indicates that electronic states lie on equipotential contours of the random potential at high magnetic fields. Electronic states which lie on the edge of the broadened Landau level are closed equipotential contours or localized states. The size of the equipotential contours tend to diverge upon approach of the Landau level center, resulting in extended states at the Landau level center. In the vicinity of the Landau level center, E_N, the localization length is predicted to vary as $\zeta \sim | E - E_N |^{-s}$. The peaks in the R_{xx} curves of Fig. 2 correspond to maximal scattering which occurs when the Fermi level lies among extended states in the Landau level center. For a symmetric density of states, the maxima in R_{xx} and the steps in R_{xy} correspond to half-integral filling factors. Thus, these structures should be spaced equidistantly on a $1/B$ plot if the broadened Landau subbands do not overlap. However, the lack of agreement between the calculated positions and the experimentally observed values demonstrates the existence of a strong overlap between the spin split Landau subbands (structures at 4.8T and 6.0T in Fig. 2). The overlap between the spin split levels is important for the interpretation of the nuclear spin lattice relaxation time where Korringa relaxation, i.e., the interaction between the nuclear spins and the 2D electronic spins, dominates the spin relaxation time.[7]

The detailed structure of the Landau level density-of-states has been the topic of many experimental investigations. All these studies indicate that a nearly constant background density of states is present in between the Landau levels, contrary to the predictions of the self consistent Born approximation (SCBA) theory. Calculations have shown[8] that the experimental results can be understood quantitatively by including only a few percent statistical variation in the carrier density. These fluctuations are strongly screened at half-integral filling factors when there are a large number of states at the Fermi level, but the screening vanishes as the number of mobile states at the Fermi level decreases with approach to the integral filling factors. This effect leads to a strong fluctuation in the Fermi level with respect to the Landau levels, which results in an increased thermodynamic density of states. Recent published measurements confirm the picture of an impurity induced increase in the density of localized states in the mobility gap.[9]

The sharp, narrow steps which connect the Hall plateaus at the lowest temperatures are smeared out with increasing temperatures. A systematic study of this effect on InGaAs/InP heterostructures[10] shows that the maximum slope, $d\rho_{xy}/dB$, varies with temperature as $T^{-\chi}$ with χ = 0.42 $\pm$ 0.04. (Very often, the measured quantities R_{xy} and R_{xx} are used to calculate the tensor components ρ_{xy}, ρ_{xx} and σ_{xx}. This procedure seems to be correct if identical results are obtained for the Hall bar and the Corbino geometries. However, inhomogeneous current distribution due to edge channels may lead to nonlocal resistances[11]). The universal temperature dependence of R_{xx} and R_{xy} at Fermi energies close to the Landau level center has been explained on the basis of the two parameter scaling theory which predicts delocalization of states due to the temperature dependence of the inelastic scattering rate[12.] Similar studies on GaAs/AlGaAs heterostructures with different impurities show that the exponent χ is not universal (S. Koch, unpublished). This behavior is demonstrated in Fig. 3 where the inverse half-width of the R_{xx} peaks, $(\Delta B)^{-1}$, and the maximum slope, $d\rho_{xy}/dB$, in the lowest Landau level are shown as a function of temperature for three samples. The slope of the curves increases with increasing impurity concentration for sheet doping with up to $2 \times 10^{10} cm^{-2}$ beryllium acceptors. These experimental results demonstrate the non-universal nature of the critical exponents associated with the localization to delocalization transition.

Observation of wide Hall plateaus does not necessarily imply perfect quantization of the Hall resistance in units of h/e^2. However, if the conditions summarized in the technical guidelines for the reliable measurement of the quantized Hall resistance are fulfilled,[13] one finds that the same value is measured for the quantized Hall resistance world-wide, independent of the material (Si, GaAs, InGaAs) and the geometry of the sample. Recent high precision comparisons of the Hall resistances in Si-MOSFETS and GaAs/AlGaAs heterostructures show a material-

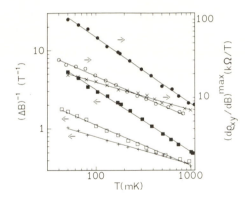

FIG. 3. The figure shows the reciprocal half-width $(\Delta B)^{-1}$ (left scale) and the maxima of $d\rho_{xy}/dB$ (right scale) for different GaAs/AlGaAs devices with controlled δ-doping of acceptor impurities in the vicinity of the maximum of the 2DEG wavefunction. The slopes of the curves increase from 0.36 ± 0.04 for devices without δ-doping, to 0.56 ± 0.05 for a doping level of $0.5\times10^{10}\text{cm}^{-2}$, and 0.81 ± 0.04 for a doping level of $2\times10^{10}\text{cm}^{-2}$

independent value for the quantized Hall resistance to 4×10^{-10} (A. Hartland, unpublished 1990)

It is surprising that a simple derivation of the Hall effect (eqns. 4-6) yields the correct value for the quantized resistance, especially since the assumptions used in the derivation of the classical Hall effect, i.e. constant Hall field, no impurities, and homogeneous current distribution, are not fulfilled in these systems. A microscopic picture of the QHE which includes the impurities, the boundary of the sample with electrical contacts, the finite current density and possible inhomogeneous current distribution, and electron-electron interactions appears too complicated for theoretical analysis. The observation of exact quantization, in spite of these complications, has led to the conclusion that the microscopic details of the system are irrelevant variables. Thus, Laughlin has tried to deduce the result on the basis of general gauge invariance arguments.[14] This approach provides no information regarding microscopic details like the current distribution, the influence of finite resistivity, $R_{xx} \neq 0$, on the value of the quantized Hall resistance, and energy dissipation ($\sim U_H I$) at the source and drain contacts resulting from the shorting of the Hall field in the vicinity of the metallic contacts. Experimental investigations of these effects have yielded intriguing results which are yet to be understood on the basis of Laughlin's theory.

Recently, the role of contacts as phase randomizing reservoirs has been demonstrated experimentally through transport studies of ballistic point contacts. The quantization of the resistance in ballistic point contacts[15,16], in units of h/e^2 as in the quantum Hall effect, has led to the edge channel picture of the QHE, which relates the electronic properties of a 2D system at high magnetic fields to ballistic transport in 1D systems in the absence of a magnetic field.

3. THE EDGE CHANNEL PICTURE

The edge channel picture of the QHE is based on a model, originally introduced by Landauer,[17] for electronic transport in an ideal 1D system. He pointed out that the electrical resistance of a system consisting of two reservoirs connected by an electron waveguide can be related to the transmission probability through the conductor. In the absence of backscattering, the characteristic feature of the 1D channel is a quantization of the resistance $R_o = h/2e^2$ which results from the cancellation of the energy dispersion relation $E(k)$ in the product of the density of states, $D(E) = (\pi dE/dk)^{-1}$, and the group velocity $v = \hbar^{-1}dE/dk$. Here, the factor of two reflects spin degeneracy in the absence of a magnetic field. Such a quantization of the channel resistance has been experimentally observed in transport studies of quantum point contacts. In these experiments, split gates are electrically biased to deplete the 2DEG of carriers except in a narrow constriction (point contact) of width comparable to the Fermi wavelength.[15,16] Quantum confinement in the constriction leads to the formation of 1D subbands and the electrical resistance of the constriction or point contact is simply related to the number, n, of occupied subbands in the point contact:

$$R = h/2e^2 n \qquad (8)$$

As the connection between 1D conducting channels and the quantum Hall effect has been discussed in much detail by M. Buttiker,[18] the following discussion will consider only a few aspects of the problem. The similarity between the quantized Hall resistance and the resistance of point contact is evident upon comparing eq. 6 and eq. 8. The point contact resistance in the absence of a magnetic field is identical to the quantized Hall resistance if 2n is replaced by the filling factor $i = n_s/N_L$. The factor of two in eq. 8 reflects the degeneracy of the spin split levels in the absence of a magnetic field. The similarities between 1D ballistic transport and the QHE imply that the 2D electronic system develops 1D-like characteristics at high magnetic fields. This connection is further investigated in the following discussion: The application of a strong magnetic field perpendicular to the plane of the 2DEG results in classical skipping orbits at the boundary of the 2D system, which are equivalent to 1D channels. The number of channels, N, is given by the number of occupied Landau levels which cross the Fermi level in the vicinity of the boundary of the sample and N is equal to i, the filling factor, since the spin degeneracy is removed by the strong magnetic field. Electrons occupying edge channels at opposite boundaries of the sample drift in opposite directions. Thus, carrier backscattering (dissipation) occurs only when carriers are scattered from one boundary to another and the vanishing conductivity, σ_{xx}, in the quantum Hall regime implies the absence of backscattering for integer filling factors. A simplified version

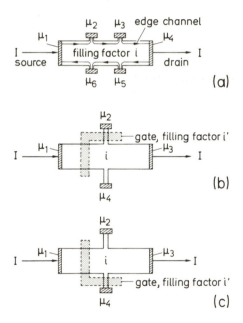

FIG. 4. (a) A standard Hall device with chemical potentials μ_i for the current and potential contacts.

(b) A Hall bar with two gate fingers. The filling factor i' below the gates may be reduced with a negative gate voltage. Measurements of the Hall resistance in this configuration are shown in Fig. 5a for the case where the filling factor outside the gate regions is i=4.

(c) The gate configuration shown in this figure leads to a Hall resistance R_{24} independent of the filling factor below the gated region (see the data of Fig. 5b).

of this formalism relates the transport current injected from a reservoir j into N channels to its chemical potential, μ_j, as follows:

$$I_{out}(j) = Ne\mu_j/h \tag{9}$$

The total current I_{tot} of a contact is the difference in the currents carried by outgoing and the incoming channels. Consider the case where the direction of the magnetic field is chosen such that the carriers which are confined to edge channels drift in the clockwise direction for the Hall bar geometry shown in Fig. 4. Here, the direction of the net transport current depends on the contact chemical potentials, μ_j, and it is independent of the edge current direction. On the basis of this formalism, the measured longitudinal and transverse resistances may be calculated even if the filling factor varies over the sample. We illustrate these points by calculating the resistances for the model systems shown in Fig. 4. For the ideal Hall device shown in Fig. 4a, the net current due to each reservoir is as follows:

Source reservoir $\quad\quad \mu_1 : \quad I_{tot} = I = (\mu_1 \text{-} \mu_6)$ ie/h

Potential reservoir $\quad\quad \mu_2 : \quad I_{tot} = 0 = (\mu_2 \text{-} \mu_1)$ ie/h

Potential reservoir $\quad\quad \mu_3 : \quad I_{tot} = 0 = (\mu_3 \text{-} \mu_2)$ ie/h

Drain reservoir $\quad\quad\quad \mu_4 : \quad I_{tot} = \text{-}I = (\mu_4 \text{-} \mu_3)$ ie/h

Potential reservoir $\quad\quad \mu_5 : \quad I_{tot} = 0 = (\mu_5 \text{-} \mu_4)$ ie/h

Potential reservoir $\quad\quad \mu_6 : \quad I_{tot} = 0 = (\mu_6 \text{-} \mu_5)$ ie/h

From these equations, it is possible to calculate measured quantities such as the two terminal resistance,

$$R_{14} = (\mu_1 - \mu_4)/(e\,I) = h/\left(i\,e^2\right) \tag{10}$$

the Hall resistance,

$$R_{26} = R_{35} = h/\left(i\,e^2\right) \tag{11}$$

17

and the longitudinal resistances which vanish in the QHE regime

$$R_{23} = R_{56} = 0. \tag{12}$$

A more complicated situation is considered in Fig. 4b. Here, the filling factor, i, of the ideal device (Fig. 4a) is reduced to i' < i, below the two gate-fingers which span the current path and one arm of the potential probe (shaded regions in Fig. 4b). Under this condition, only i' channels remain undisturbed along the sample boundary and i-i' channels are reflected at the barriers produced by the negative gate voltage. If we assume that adjacent edges with different chemical potentials do not equilibrate, the following results are obtained:

Source reservoir μ_1 : $I_{tot} = I = [i\mu_1 - i'\mu_4 - (i-i')\mu_1]\ e/h$
$= (\mu_1 - \mu_4)\ i'e/h$

Potential reservoir μ_2 : $I_{tot} = 0 = [i\mu_2 - i'\mu_1 - (i-i')\mu_2]\ e/h$
$= (\mu_2 - \mu_1)\ i'e/h$

Drain reservoir μ_3 : $I_{tot} = -I = [i\mu_3 = i'\mu_2 - (i-i')\ \mu_4]\ e/h$

Potential reservoir μ_4 : $I_{tot} = 0 = [i\mu_4 - i\mu_3]\ e/h$
$= [\mu_4 - \mu_3]\ ie/h$

From these equations, the Hall resistance R_{24} can be deduced:

$$R_{24} = (\mu_2 - \mu_4) / (e\ I) = h / (i'\ e^2), \tag{13}$$

this expression indicates that the Hall resistance reflects the filling factor in the gated regions. A measurement of the Hall resistance with the opposite field direction is equivalent to the gate configuration shown in Fig. 4c. In this case, the Hall resistance is found to be independent of the filling factor, i', in the gated regions and therefore the gate voltage. Thus,

$$R_{24} = h / (i\ e^2). \tag{14}$$

Experimental studies have investigated gate configurations of the type shown in Fig. 4b and they have confirmed the theoretical predictions.

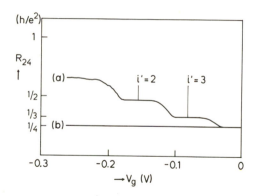

FIG. 5. (a) Experimental data for the Hall resistance R_{24} as a function of the gate voltage V_g for the gate configuration shown in Fig. 4b.
(b) Experimental data for the Hall resistance R_{24} as a function of the gate voltage V_g for the gate configuration shown in Fig. 4c.

Figure 5a shows the typical measured Hall resistance R_{24} as a function of the gate voltage for the sample configuration of Fig. 4b. Here, the magnetic field, B, has been chosen to obtain a filling factor i=4 in the ungated regions of the sample. Starting from the quantized resistance, $R_{24} = h/4e^2$, a stepwise increase in the Hall resistance is observed as the filling factor is reduced in the gated regions. The slight deviations in the measured resistances from the calculated values suggests that the assumption of adiabatic edge channel transport, i.e., no equilibriation between adjacent edge channels with different edge channels, is not strictly fulfilled. Thus, the deviations of the plateau values in Fig. 5a from the quantized values, $R_{24} = h/i'e^2$, is a measure of the interaction between the edge channels. If the edge channels are in equilibrium, then the measured Hall resistance is independent of the filling factor i'. The gate configuration of Fig. 4c yields a constant Hall resistance as observed in the data of Fig. 5b, in agreement with theoretical predictions. These experiments demonstrate that the electrical current in the quantum Hall regime flows along the boundary of the sample.

Recent experiments have also demonstrated that metallic contacts such as potential probes behave as reservoirs which enforce equilibrium between the edge channels[19]. Two gate barriers in series are used in these experiments and metallic contacts, i.e. Hall probes, between the barriers are electrically connected or disconnected using additional gates across the arms of the Hall probe. Thus, transport may be studied with

19

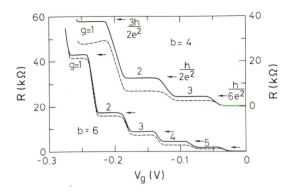

FIG. 6. The measured resistances, R, across a double barrier formed by two gate fingers across the width of a Hall bar, for filling factors b=4 and b=6 outside the barrier regions, and a gate voltage dependent filling factor g below the gated regions, (1 < g < b). The solid curves are observed if metallic contacts (Hall probes) are connected to the system in between the gated regions. The arrows indicate the expected plateau values for equilibrated transport. The dashed curves are obtained if the metallic contacts (Hall probes) are disconnected from the 2DEG.

or without a reservoir in between the barriers. Fig. 6 shows the experimental results. Here, the magnetic field was fixed to obtain filling factors b=6 (left scale) and b=4 (right scale) in the ungated regions of the sample. Application of a gate voltage, V_g, results in a reduced filling factor below the gate. If the reservoirs are connected to the channel, which corresponds to the full lines in the figure, the voltage drop across the double barrier is identical with two independent barriers in series. In the case of adiabatic transport where the reservoirs are disconnected from the channel, the two barriers behave as a single barrier with the overall resistance reduced by a factor of two, in an ideal case approximately realized for b=4 (dotted line in Fig. 6).

All these experiments demonstrate that the electronic properties of a 2 DEG in strong magnetic fields cannot be described by the components of the resistivity or the conductivity tensor. The electronic properties are influenced by nonlocal phenomena and depend upon the geometry of the whole device. This aspect of the quantum Hall effect will remain a topic of research in the future.

REFERENCES

1.   T. Quinn, Metrologia, **26** 69 (1989).

2. R. E. Prange, S. M. Girvin (eds.) The Quantum Hall Effect, (Springer, New York, Berlin, Heidelberg (1987).

3. K. v. Klitzing, Rev. Mod. Phys. **58**, 519 (1986).

4. K. v. Klitzing, NATO ASI Series, Series B Physics **170**, 229 (1987).

5. T. Ando, A. B. Fowler and F. Stern, Rev. Mod. Phys. **54**, 437 (1982).

6. T. Ando, J. Phys. Soc. Jpn. **52**, 1740 (1983).

7. A. Berg, M. Dobers, R. R. Gerhardts and K. v. Klitzing, Phys. Rev. Lett. **64**, 2563 (1990).

8. V. Gudmundsson and R. R. Gerhardts in: "High Magnetic Fields in Semiconductor Physics," ed. G. Landwehr, Springer Series in Solid State Sciences (Berlin), **71**, p. 67, **87**, p. 14.

9. R. J. Haug, K. v. Klitzing and K. Ploog, Proc. 19th Int. Conf. on the Physics of Semiconductors 1988, published by Institute of Physics, Polish Academy of Sciences, p. 307 (1988).

10. H. P. Wei, D. C. Tsui, M. A. Paalanen and A. M. M. Pruisken, Phys. Rev. Lett. **61**, 1294 (1988).

11. R. J. Haug and K. v. Klitzing, Europhys. Lett. **10**, 489 (1989).

12. A. M. M. Pruisken, Phys. Rev. Lett. **61**, 1297 (1988).

13. F. Delahaye, Metrologia **26**, 63 (1989).

14. R. B. Laughlin, in <u>Two-Dimensional Systems, Heterostructures, and Superlattices</u>, ed. G. Bauer, F. Kuchar and H. Heinrich, Springer Series in Solid State Sciences (Berlin/New York), Vol 53, p. 272.

15. B. J. van Wees, H. van Houten, C. W. J. Beenakker, J. G. Williamson, L. P. Kouwenhoven, D. van der Marel and C. T. Foxon, Phys. Rev. Lett. **60**, 848 (1988).

16. D. A. Wharam, T. J. Thornton, R. Newbury, M. Pepper, H. Ahmed, J. E. F. Frost, D. G. Hasko, D. C. Peacock, D. A. Ritchie and G. A. C. Jones, J. Phys. **C21**, L209 (1988).

17. R. Landauer, Philos. Mag. **21**, 863 (1970).

18. M. Büttiker in: <u>Festkorperprobleme/Advances in Solid State Physics</u>, **30**, ed. by U. Rössler (Vieweg, Braunschweig 1990), p. 40.

19. G. Müller, D. Weiss, S. Koch, K. v. Klitzing, H. Nickel, W. Schlapp and R. Lösch, Phys. Rev. **B42**, 7633 (1990).

FRACTIONAL QUANTUM HALL EFFECT

D. C. Tsui

Department of Electrical Engineering
Princeton University
Princeton, N.J.

Abstract - An overview is given of the fractional quantum Hall (FQH) effect. Emphasized are recent experiments on the anomaly in transport at $\nu = 1/2$, suggestive of a new correlated ground state, the fractional charge of quasiparticles/quasiholes of the 1/3 FQH liquid, and the long anticipated transition from the incompressible FQH liquids to deformable solids.

1. INTRODUCTION

The fractional quantum Hall effect (FQHE) is manifestation of a new state of electron matter - a peculiar, uniform density incompressible liquid. It is observed in highly perfect two-dimensional electron systems (2DES) realized in semiconductor heterostructures at low temperatures and in the presence of a strong magnetic field (B) perpendicular to the 2D plane. Prof. Girvin has talked in his lecture about some of the new physics that have evolved from the FQHE and their theoretical implications. In my talk, I have chosen an overview of the phenomenon

Highlights in Condensed Matter Physics and Future Prospects
Edited by L. Esaki, Plenum Press, New York, 1991

together with a discussion of two recent experiments. One is on the fractional charge carried by the quasiparticle/quasihole excitations and the other on the transition from the FQH liquids to electron solids. It is my hope that the discussion of these more recent developments will also give us some sense of the future prospects that the conference has attempted to bring out.

2. OVERVIEW OF THE PHENOMENON

The FQHE is characterized by the vanishing of the dissipative component of the transport coefficient ρ_{xx} and the quantization of the Hall component ρ_{xy} of the 2DES at Landau level filling factors of rational fractions, $v = p/q$.[1,2] It differs from the integer quantum Hall effect (IQHE) in several respects. First, the quantum numbers are non-integer and, therefore, it cannot be identified with the singularities associated with the complete filling of a finite number of single particle Landau levels. Moreover, the phenomenon is observable at much lower T. These two experimental facts by themselves suggest that the underlying physics is many-body physics and the relevant energy gap is much smaller than the cyclotron energy. Another important feature of the experimental data is that higher the mobility (μ) of the 2DES, more prominent is the FQHE. This point is very clear from the data shown in Fig. 1. Since μ indicates the perfection of the 2DES, we can conclude from these data that the FQHE is a many-body phenomenon resulting from the electron-electron interactions of 2DES in the clean limit in a strong magnetic field.

To date, the fractions observed in the lowest energy, N=0 and spin ↑ Landau level include:

$v = 1/3, 2/3,$

$1/5, 2/5, 3/5, 4/5,$

$1/7, 2/7, 3/7, 4/7, 5/7,$

$2/9, 4/9, 5/9, 7/9,$

$2/11, 3/11, 4/11, 5/11, 6/11, 7/11,$

$3/13, 4/13, 6/13, 7/13, 8/13, 9/13,$

$4/15.$

All are odd denominator fractions. The only well established even denominator FQHE is at $\nu = 5/2$, corresponding to fractional filling of the higher energy, spin unpolarized N = 1 Landau level.[3] Our current understanding of the FQHE is based on Laughlin's theory[4] that at $\nu = 1/q$ the 2DES is condensed into a series of especially stable, uniform density, incompressible liquids. The quasiparticle/quasihole excitations of the liquids are fractionally charged and are separated from the liquids by finite energy gaps. While Laughlin's wave function describes the $\nu = 1/q$ fundamental liquids and their electron-hole symmetric liquids at $\nu = (1-1/q)$, other liquids of odd denominator fractions are generated by using a hierarchy scheme,[5,6] which condenses the quasiparticles or quasiholes of the liquids to form their next higher order liquids.

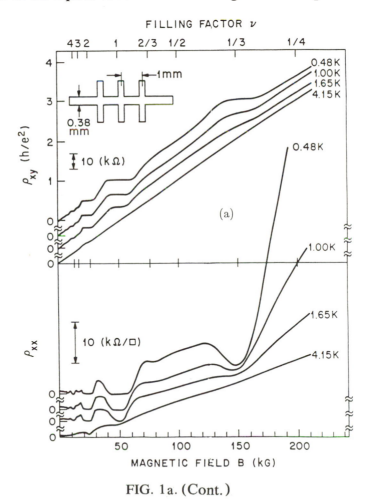

FIG. 1a. (Cont.)

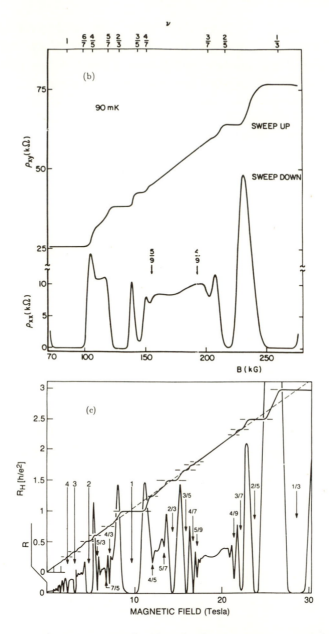

FIG. 1. Fractional quantum Hall effect in the transport coefficients ρ_{xx} and ρ_{xy} (or R and R_H) versus B. The samples are GaAs/Al_xGa_{1-x}As heterostructures and their two-dimensional electron densities and mobilities are
(a) n = 1.23×10^{11}/cm/^{2} and μ = 9×10^4 cm^2/Vsec (from Ref. 1),
(b) n = 2.13×10^{11}/cm^2 and μ = 7.1×10^5 cm^2/Vsec (from Ref. 2), and
(c) n = 2.35×10^{11}/cm^2 and μ = 1.3×10^6 cm^2/Vsec.

The transport at $\nu = 1/2$ is anomalous.[7] The broad ρ_{xx} minimum is signature that the 2DES at this filling cannot be a simple Fermi gas or Fermi liquid. This anomaly is more apparent in the more recent data from a higher quality sample (Fig. 2) in that, in spite of the deep ρ_{xx} minimum, ρ_{xy} shows no concomitant Hall plateau development. The temperature development of the minimum commences at $\sim 10K$, where all FQHE features are still absent, and saturates at a nonzero value below ~ 300 mK down to the lowest T ~ 25 mK of the experiment. We should recall from Prof. Girvin's lecture that our system of 2D electrons in a strong B field can be mapped into a system of composite objects which are bosons in a zero B field.[8] The FQH liquids result from condensation of these Bose particles. At $\nu = 1/2$, however, the same mapping maps our 2DES into a system of composite objects which are fermions. Since fermions even weakly interacting at B $= 0$ may be unstable, the ground state at $\nu = 1/2$ can be expected to be a new correlated state, distinctly different from the FQH liquids.

3. FRACTIONAL CHARGE

Several recent experiments have attempted to determine the charge carried by the quasiparticle/quasihole excitations of the FQH liquids.[9-11]

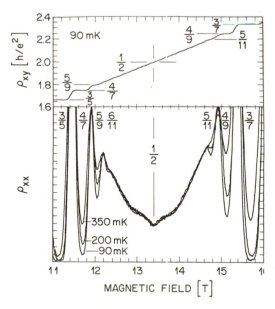

FIG. 2. ρ_{xx} and ρ_{xy} versus B around the $\nu = 1/2$ anomaly (from Ref. 7).

The one I want to mention is a study of the resistance fluctuations in a narrow sample, due to breakdown of the dissipationless current flow by resonant tunneling via states magnetically bound to the potential hills and valleys in the 2DES. In the semiclassical limit[12] when the cyclotron diameter is much smaller than the size of the potential, the nth energy level of the bound states is determined by Bohr-Sommerfeld quantization of the action integral:

$$2\pi e^* BA = nh, \tag{1}$$

where e^* is the charge of the quasiparticle, A is the area enclosed by the semiclassical orbit, and h is planck's constant. In the experiment of Simmons et al.,[11] quasi-periodic noise structures are observed in ρ_{xx} for B adjacent to the quantized Hall plateaus in narrow samples of a high mobility 2DES. Fig. 3 is an example of their data showing that, while the period is $\sim$ 150 G for the structures adjacent to all the IQHE plateaus (shown in the top two panels for $\nu = 2$ and $\nu = 1$, respectively), an approximately three times larger period ($\sim$ 500 G) is observed for the structures adjacent to the $\nu = 1/3$ FQHE plateau. They have demonstrated that the effect is local and close to a particular set of potential probes and that it can be attributed to the tunneling via magnetic bound states submicron in size. The quasi-periodicity is the minimum change in B to satisfy the quantization condition of Eq. (1) and the striking difference of the factor 3 is evidence that the charge carried by a quasiparticle of the $\nu = 1/3$ FQH liquid is $e^* = e/3$.

The quasiparticles and quasiholes are anyons and there is a great deal of current theoretical interest in the physical properties of many-anyon systems. In a recent paper, Kivelson[13] has emphasized the importance of the anyon statistics and discussed the conditions under which the experiment can measure the statistics of the quasiparticles. Since this is a relatively simple physical system where anyons are known to exist, research activities in this interesting area should see some increase in the near future.

4. FROM INCOMPRESSIBLE LIQUIDS TO DEFORMABLE SOLIDS

It has long been anticipated that in the sufficiently small ν limit, a triangular Wigner crystal[14] is favored over the FQH liquids. Moreover, random impurities are expected to pin the electron crystal to the host semiconductor and, as a result, the 2DES should be an insulator. Exper-

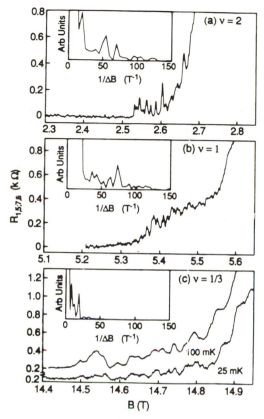

FIG. 3. The resistance along current flow, $R_{1,5,7,8}$, near the high field sides of resistance minima for (a) $\nu = 2$ at 25mK, (b) $\nu = 1$ at 25 mK, and (c) $\nu = 1/3$ at 25 and 100mK, all plotted with the same field scale. Insets show the Fourier power spectra of the region of fluctuations for each respective ν (from Ref. 11).

imentally, thermally activated conduction, indicative of an insulating state, is apparent for $\nu < 1/3$ even in the earliest FQHE data shown in Fig. 1(a), but it has been extremely difficult to tell if the insulating state is indeed due to the pinning of a solid and not Anderson localization of single particles in the extreme quantum limit.[15,16] Subsequent experiments on samples with constantly improved quality and reduced amount of random impurities (e.g. the data in Fig. 4) have made it very clear that better the quality of the 2DES smaller is the filling factor, below which the 2DES becomes insulating. The ρ_{xx} minimum, which is clear signature for a FQH liquid, has since become observable in $\nu < 1/3$ at 2/7, 3/11, 4/13, 2/9, 1/5, 2/11, and 1/7. However, only very recently Jiang et al.[17] has obtained data showing clear evidence that the ground state at ν = 1/5 is indeed a FQH liquid and, therefore, the insulating state observed for filling slightly above $\nu = 1/5$ cannot be due to single particle localization. It must be the result of electron-electron correlation.

Fig. 4(c) shows the data from Jiang et al. At $\nu = 1/5$, ρ_{xx} decreases exponentially with decreasing T down to the lowest available T of 80 mK and more recently to $\sim$ 30 mK. The data demonstrate for the first time that in a sufficiently clean 2DES the 1/5 state is a true FQH liquid in that $\rho_{xx} \rightarrow 0$ at $T \rightarrow 0$. Surprisingly, in a small region for $\nu > 1/5$ (e.g. ν = 0.21), $\rho_{xx} \rightarrow \infty$ as $T \rightarrow 0$, indicative of an insulator at T = 0, and is similar to that for ν immediately below 1/5. The fact that this insulating state is observed at ν above the 1/5 FQH liquid suggests that it must result from electron-electron correlation and a pinned Wigner solid is a distinct possibility. The experiment further suggests a phase diagram in

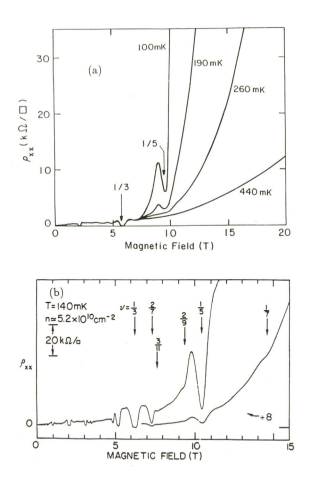

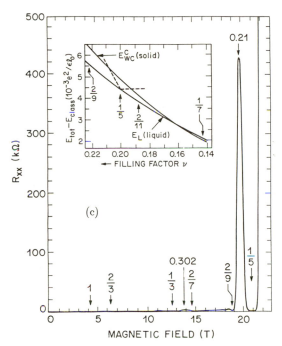

FIG. 4. Data from three different samples showing the transport features at $\nu <$ 1/3. (a) ρ_{xx} at several T from a n = 4.8x10^{10}/cm^2 and $\mu \approx$ 1.7x10^{6}cm^2/Vsec samples. The 1/5 minimum is clearly resolved for T < 260 mK, but the value of ρ_{xx} at the 1/5 minimum increases with decreasing T (from Ref. 15). (b) ρ_{xx} from a sample with lower disorder at 140 mK, showing a very weak minimum at ν = 1/7. The minimum value ρ_{xx} at ν = 1/5 in this sample still increases with decreasing T (from Ref. 16). (c) ρ_{xx} at 90 mK from a sample with n = 1/x10^{11}/cm^2 and $\mu \simeq$ 7.5x10^{6}cm^2/Vsec. At ν = 1/5, $\rho_{xx} \rightarrow$ 0 as T $\rightarrow$ 0, indicating that the ground state of the 2DES is a FQH liquid. At $\nu \sim$ 0.21 and $\nu <$ 1/5, ρ_{xx} increases exponentially with decreasing T, indicating an insulator at T = 0. The insert shows the result of a calculation for the total energy per flux quantum of the solid (E_{wc}^c) and the interpolated 1/m quantum liquids (E^L) as a function of filling factor.[18] A classical energy (E_{class} = 0.782133 $\nu^{-1/2}$) is subtracted for clarity. The dashed lines represent the cusp in the total energy[6] of the liquid at ν = 1/5. Its extrapolation intersects the solid at $\nu \simeq$ 0.21 adn $\nu \simeq$ 0.19 suggesting two phase transitions from quantum liquid to solid around ν = 1/5 (from Ref. 17).

which the solid phase below the $\nu = 1/5$ liquid is re-entrant above $\nu = 1/5$.

The insert in Fig. 4(c) shows the result of a calculation combining the theoretical results of Halperin[6] and of Lam and Girvin.[18] The dashed line represents the cusp in energy of the FQH liquid at $\nu = 1/5$ and the two intersects with the curve for the energy of the solid suggest the two phase transitions. It should also be emphasized that even though there is still no direct evidence that the observed insulating phase is a pinned solid, this notion is consistent with the nonlinear conduction threshold recently observed by Goldman et al.[19] Finally, a particularly exciting recent development, which I have not been able to go into, is the experimental use of powerful optical tools to probe into this interesting, but difficult regime.[20-22] We should expect to hear a great deal from these optical experiments in the very near future.

ACKNOWLEDGEMENTS

This work is a result of a long collaboration with my colleagues R. L. Willett, H. P. Wei, H. L. Stormer, J. A. Simmons, M. Shayegan, L. N. Pfeiffer, H. W. Jiang, A. C. Gossard, V. J. Goldman, L. Engel, J. P. Eisenstein, A. M. Chang and G. S. Boebinger, to whom I express my gratitude. It is supported by the NSF, ONR, AFOSR and a grant from the NEC Corporation.

REFERENCES

1. D. C. Tsui, H. L. Stormer and A. C. Gossard, Phys. Rev. Lett. **48**, 1559 (1982)

2. A. M. Chang, P. Berglund, D. C. Tsui, H. L. Stormer and J. C. Hwang, Phys. Rev. Lett. **53**, 997 (1984).

3. R. L. Willett, J. P. Eisenstein, H. L. Stormer, D. C. Tsui, A. C. Gossard and J. H. English, Phys. Rev. Lett. **59**, 1776 (1987).

4. R. B. Laughlin, Phys. Rev. Lett. **50**, 1395 (1983).

5. F. D. M. Haldane, Phys. Rev. Lett. **51**, 605 (1983).

6. B. I. Halperin, Phys. Rev. Lett. **52**, 1583 (1984).

7. H. W. Jiang, H. L. Stormer, D. C. Tsui, L. N. Pfeiffer and K. W. West, Phys. Rev. **B40**, 12013 (1989).

8. S. M. Girvin (paper in this volume).

9. R. G. Clark, J. R. Mallett, S. R. Haynes, J. J. Harris and C. T. Foxon, Phys. Rev. Lett. **60**, 1747 (1988).

10. A. M. Chang and J. E. Cunningham, Solid State Commun. **72**, 651 (1990).

11. J. A. Simmons, H. P. Wei, L. W. Engel, D. C. Tsui and M. Shayegan, Phys. Rev. Lett. **63**, 1731 (1989).

12. S. A. Kivelson and V. L. Pokrovsky, Phys. Rev. **B40**, 1373 (1989).

13. S. A. Kivelson, Phys. Rev. Lett. (rejected).

14. H. Fukuyama, P. M. Platzman and P. W. Anderson, Phys. Rev. **B19**, 5211 (1979).

15. R. L. Willett, H. L. Stormer, D. C. Tsui, L. N. Pfeiffer, K. W. West and K. Baldwin, Phys. Rev. **B38**, 7881 (1988).

16. V. J. Goldman, M. Shayegan and D. C. Tsui, Phys. Rev. Lett. **61**, 881 (1988).

17. H. W. Jiang, R. L. Willett, H. L. Stormer, D. C. Tsui, L. N. Pfeiffer and K. W. West, Phys. Rev. Lett. **65**, 633 (1990).

18. P. K. Lam and S. M. Girvin, Phys. Rev. **B33**, 473 (1984).

19. V. J. Goldman, J. E. Cunningham, M. Shayegan and M. Santos, Phys. Rev. Lett. (to be published).

20. A. J. Turberfield, S. R. Haynes, P. A. Wright, R. A. Ford, R. G. Clark, and J. F. Ryan, Phys. Rev. Lett. **65**, 637 (1990).

21. B. B. Goldberg, D. Heiman, A. Pinczuk, C. Pfeiffer and K. West, Phys. Rev. Lett. **65**, 641 (1990).

22. H. Buhmann, W. Joss, K. von Klitzing, I. V. Kukushkin, G. Martinez, A. S. Plaut, K. Ploog and V. B. Timofeef, Phys. Rev. Lett. **65**, 1056 (1990).

THEORY AND IMPLICATIONS OF THE FRACTIONAL QUANTUM HALL EFFECT

S. M. Girvin

Department of Physics
Indiana University
Bloomington, IN 47405, USA

Abstract - The fractional quantum Hall effect (FQHE) is a remarkable many-body phenomenon which is important both in its own right and for its connections to other problems in condensed matter physics and quantum field theory. The essence of the effect is a new and unprecedented type of condensation of the electrons into a highly correlated ground state.

In the 19th century, J. C. Maxwell realized that $4\pi/c \approx 377\Omega$ was the impedance of the vacuum. In 1915 A. Sommerfeld realized that $\alpha \equiv e^2/\hbar c \approx 1/137$ was an important dimensionless constant. It follows that e^2/h has units of conductance and this quantity rose to importance in the 1970's in the study of the localization problem by Mott, Anderson, Thouless, and many others. The discovery of the integer quantum Hall effect[1] by von Klitzing in 1980 gave us a rather accurate value of the quantum of resistance $h/e^2 \approx 25,812.807\Omega$ through quantization of the Hall conductivity in a 2d electron gas showing Hall "plateaus" with universal values

Highlights in Condensed Matter Physics and Future Prospects
Edited by L. Esaki, Plenum Press, New York, 1991

$$\sigma_{xy} = \nu \frac{e^2}{h}, \tag{1}$$

where δ is an integer quantum number. In 1983 Störmer, Tsui, and Gossard discovered that the quantum number ν could take on rational fractional values.[1]

In a translationally invariant system it is easy to show that the Hall plateaus can *not* be observed - indeed the Hall effect gives us very little information in this case. Consider the following gedanken experiment. We have in the frame of the laboratory a 2d electron gas at rest in the presence of a perpendicular uniform magnetic field. The current and electric fields are both zero. Since we have a translationally invariant system, we are permitted to jump into a frame of reference moving with speed v parallel to the 2d plane. The current density observed in this frame is (to lowest order is v/c)

$$J = nev, \tag{2}$$

where n is the density of electrons. The Lorentz transformation properties of the B field yield an electric field at right angles to the current which (to lowest order is v/c) has magnitude

$$E = \frac{v}{c} B. \tag{3}$$

The Hall conductivity is thus

$$\sigma_{xy} = \frac{J}{E} = \frac{nec}{B}. \tag{4}$$

This shows that in a translationally-invariant system, the Hall effect is purely kinematical in origin and gives no information about the dynamics of the system. In particular Eq. (4) holds for both classical and quantum liquids and solids.

It follows from Eq. (4) that there exist values of the density n and field B such that Eq. (1) is satisfied, however there is nothing special about these particular points. What is remarkable is that in real *disordered* systems there exist rather wide "plateau" regions of n and B in which Eq. (1) is essentially exactly satisfied. This follows for rather

complicated reasons[1] associated with the breaking of translation symmetry by the disorder potential.

Associated with (and indeed a prerequisite for) the ideal
quantization of σ_{xy} is an ideal vanishing of the dissipation at zero temperature

$$\sigma_{xx} = 0. \tag{5}$$

In the presence of disorder this can only occur if there exists an
excitation gap. To see this, consider a disordered system carrying a
current. Now jump into a frame comoving with the electrons. There one
sees zero current but finds instead that the impurities move through the
fluid with speed v. This motion can be adiabatic (i.e. not shaking off a
wake of excitations) only if there exists an excitation gap.[2]

The origin of the gap in the integer case is the discrete spacing $\hbar\omega_c$
of the kinetic energy states (Landau levels), where $\omega_c \equiv eB/mc$ is the
cyclotron frequency. In the fractional case the gap is much smaller $\Delta \sim$
e^2/ℓ where $\ell \equiv (\hbar c/eB)^{1/2}$ is the magnetic length and is due to the existence of highly non-trivial Coulomb correlations which are well-
represented by Laughlin's variational wave function,[1] Ψ_L. The theme of
this paper is that electrons in a magnetic field can act like bosons in zero
field and that hence the Laughlin state represents a special, new and
unprecedented kind of quantum *condensation*.

Despite the Fermi statistics of the electrons, the FQHE system acts
very much like superfluid helium. Recall that the excitation spectrum of
a Fermi system is characterized by the existence of a high density of
single-particle excitations which extend all the way down to zero energy
and out to wavevector $2k_F$ due to the existence of a sharp Fermi surface.
These are illustrated in Fig. (1a) along with the collective mode which
either has linear dispersion (neutral system) or a finite gap (Coulomb
system). The finite "phase space" for low-energy single-particle
excitations is what gives (disordered) metals finite dissipation.

In a Bose system such as ^{4}He on the other hand, the symmetry of
the wave function under particle exchange prevents the existence of low-
lying single-particle excitations. There is only a linearly dispersing collective (phonon) density-wave mode which is gapless at small wavevector
and which is shown in Fig. (1b) exhibits the so-called roton minimum at
large wavevector.[3]

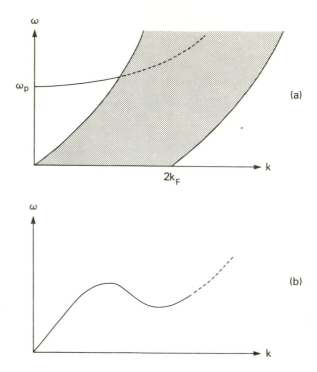

FIG. 1(a). Excitation dispersion for the collective plasmon mode and the particle-hole continuum in a charged Fermi system.

FIG. 1(b). Collective mode dispersion for superfluid helium showing the gapless phonon mode and the roton minimum. Note the absence of a particle-hole continuum in this Bose system.

The FQHE system is qualitatively like the ^{4}He in the sense that there is no single-particle continuum. This is because the kinetic energy has been quenched and the Fermi surface destroyed by the large applied magnetic field. There exists a phonon-like collective density-wave mode which, like ^{4}He exhibits a roton minimum, but unlike ^{4}He, has a large gap at small wavevector as shown in Fig. (2). This gap is associated with the incompressibility of the FQHE state and is the main focus of the present paper. In analogy with the Bijl-Feynman theory of the collective excitation in superfluid helium[3] we can define a variational wave function for the density-wave excited state of the form

$$\Psi_k = \frac{1}{\sqrt{N}} \bar{\rho}_k \Psi_L \qquad (6)$$

where Ψ_L is the Laughlin ground state[4] which will be described in detail below and ρ is the Fourier transform of the particle density

$$\rho_k = \sum_{j=1}^{N} e^{i\vec{k}\cdot\vec{r}_j}. \tag{7}$$

The overbar on ρ in Eq. (6) indicates that we are to project the operator onto the lowest Landau level.[5] This is a technical detail which will be ignored in the present discussion. The point is simply that the density couples to both the inter-Landau-level (cyclotron or magnetoplasmon) mode and the intra-Landau-level (magnetophonon/roton) mode and we are presently concerned only with the latter.

To see why Ψ_k represents a density wave, consider the following argument. If we evaluate Ψ_k for a configuration in which the particles are roughly uniformly distributed then ρ will be small (for finite k) and hence $|\Psi_k|^2$ will be small. This means that such configurations are unlikely to occur in this state. On the other hand, if we choose a non-uniform configuration of particles which has a density modulation at wavevector k, then ρ and hence $|\Psi_k|^2$ will be large in magnitude. If the amplitude of modulation is *too* large however, then $|\Psi_L|^2$ itself will decrease so there is some characteristic amplitude which is most likely to occur.

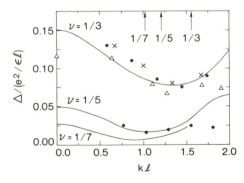

FIG. 2. Solid line shows the calculated dispersion curve for the collective density mode in the FQHE. Note the large gap and the magnetoroton minimum. The minimum occurs at a wave vector close to the reciprocal lattice vector of the crystal at the same density as indicated by the arrows. Triangles, dots and crosses are results from exact numerical calculations for small systems of particles.[6,7]

We can gain further insight into this by considering a simple harmonic oscillator analogy. Given the ground state of the oscillator, Ψ_0 we could guess as a variational excited state

$$\Psi_1(x) = x\Psi_0(x) \tag{8}$$

which is simply the coordinate times the ground state function and is of course the exact first excited state. We may interpret ρ as a collective coordinate describing one of the normal modes of oscillation of the fluid. Of course for an elastic continuum this is rigorously correct since one would have a harmonic medium. For a fluid, ρ_k is only a good approximation to a true collective coordinate. One would expect this continuum approximation to be good only at long wavelengths (relative to the particle spacing) but it works surprisingly well (especially in the FQHE) down to wavelengths on the same scale as the particle spacing.

To proceed, we compute the variational energy by writing

$$\Delta(k) = \frac{f(k)}{s(k)}, \tag{9}$$

where

$$f(k) \equiv \frac{1}{N} <\Psi_k|H - E_0|\Psi_k>, \tag{10}$$

with E_0 being the ground state energy and

$$s(k) \equiv \frac{1}{N} <\Psi_k|\Psi_k>. \tag{11}$$

Ignoring projection onto the lowest Landau level[5] the numerator becomes

$$f(k) = \frac{1}{N} <\Psi_L|\rho_k^+(H - E_0)\rho_k|\Psi_L>$$
$$= \frac{1}{N} <\Psi_L|\rho_k^+[H, \rho_k]|\Psi_L> \tag{12}$$

which by the oscillator strength sum rule[3,5] is given by the universal result

$$f(k) = \frac{\hbar^2 k^2}{2m} . \tag{13}$$

If we take the projection into account we replace $f(k)$ by $\bar{f}(k)$, the intra-Landau-level portion of the oscillator strength sum,[5] which is non-universal but still computable from Ψ_L. The denominator is the static structure factor

$$s(k) = \frac{1}{N} \langle \Psi_k | \Psi_k \rangle = \frac{1}{N} \langle \Psi_L | \rho_k^+ \rho_k | \Psi_L \rangle \tag{14}$$

which gives the mean square density fluctuation at wavevector k in the ground state. In the case of ^{4}He, Feynman was able to obtain $s(k)$ from neutron scattering data. In the FQHE we are forced to compute $s(k)$ by Monte Carlo simulation of the particle distribution $| \Psi_L |^2$. Again, the properly projected quantity $\bar{s}(k)$ is slightly more complicated but is readily found[5] from Ψ_L.

In ^{4}He $s(k)$ vanishes linearly for small k giving rise to a linearly dispersing collective mode

$$\Delta(k) = \frac{\hbar^2 k^2}{2m} \frac{1}{s(k)} . \tag{15}$$

At large k commensurate with the mean particle spacing there is a peak in $s(k)$ due to the existence of short-range solid-like correlations in the ground state particle distribution. This is the origin of the roton minimum in $\Delta(k)$.

The Bijl-Feynman expression nicely represents a dynamically quantity $\Delta(k)$ in terms of purely static properties of the ground state and works because in ^{4}He a single mode contains essentially all of the oscillator strength. This expression should not work as well for Fermi systems because of the particle-hole continuum. Nevertheless, for the 3D electron gas one finds that $s(k)$ vanishes quadratically (because of the long-range Coulomb force) and one obtains the *exact* collective mode gap energy

$$\lim_{k \to 0} \Delta(k) = \hbar \omega_p \tag{16}$$

where ω_p is the plasma frequency. At finite k the particle-hole continuum steals part of the oscillator strength from the plasmon but the approximation is still reasonable.[5]

In 2D the $1/r$ interaction is not long-ranged enough to produce a finite plasmon gap. It is necessary to have the 2D log r Coulomb interaction to produce a gap. In the FQHE, there are only short-ranged interactions, yet the structure factor vanishes sufficiently rapidly to produce a gap. The effect of the magnetic field is to produce a special rigid state which is incompressible and acts very much like a (2D) Coulomb-charged plasma (of bosons). The lack of long-range density fluctuations in the ground state is graphically illustrated in Fig. 7.7 in Chap. 7 of Ref.(1).

The predicted collective mode dispersion for the FQHE is shown in Fig. 2 which clearly illustrates the finite gap and the magnetoroton minimum. The single-mode approximation prediction is in excellent agreement with exact numerical calculations for small numbers of particles.[6,7] One of the experimental challenges of the coming years will be to directly observe this mode either by light scattering, or direct microwave or phonon absorption in ultra-high mobility samples. Since we have explicit wave functions it is straightforward to compute needed matrix elements.[8] One finds that the oscillator strength peaks at the roton minimum (ql $\sim$ 1) (as does the density of states[9] and that the oscillator strength is comparable to that of the cyclotron mode at that wavevector. To reach these high wavevectors optically unfortunately requires impurity or (ultra-small period) grating couplings.

One of the peculiar features of the magnetic field is that the intra-Landau level component of the density wave is largely transverse at long wavelengths and a mixture of transverse and longitudinal at the roton minimum. Thus one may be able to use either longitudinal or transverse phonons as a probe. Additionally, because the phonons are 3D and the electrons 2D, the phonon frequency and $k_{||}$ are independently adjustable.

To see why the Laughlin state acts like an incompressible Coulomb system we first consider the one-body eigenstates in the lowest Landau level. The many (degenerate) states are all of the form[4]

$$\phi(z) = f(z) \exp\left(-|z|^2/4\right), \qquad (17)$$

where $z \equiv (x + iy) / \ell$ is a complex dimensionless coordinate and $f(z)$ is an arbitrary polynomial. Up to an irrelevant normalization constant f can be expressed solely in terms of the location of its M zeros

$$\phi(z) = \prod_{j=1}^{M} (z - Z_j) \exp\left(-|z|^2/4\right). \qquad (18)$$

Using this the density distribution of the particle can be written

$$|\phi(z)|^2 = e^S \qquad (19)$$

where

$$S \equiv \sum_{j=1}^{M} 2 \ln |z - Z_j| - \frac{1}{2} |z|^2 \qquad (20)$$

which looks like the classical Boltzmann distribution for a 2D Coulomb charge located at z interacting with point charges at positions $\{Z_j\}$ and a neutralizing background. If the particle distribution is to be roughly uniform then clearly the density of point charges must, on average, cancel the background. That is, the magnetic field forces there to be a finite density of zeros in the wave function.[10] This density is B/Φ_o where $\Phi_o = hc/e$ is the flux quantum. Consider now the many-body ground state. Given that each particle needs to see a finite density of zeros, the Laughlin wave function takes advantage of this to place the zeros seen by any one particle at the position of the *other* particles[4,10]

$$\Psi_L(z_1, ..., z_n) = \prod_{i<j} (z_i - z_j)^m \exp\left(-\sum_{k=1}^{N} |z_k|^2\right), \qquad (21)$$

where m must be an odd integer to preserve analyticity and antisymmetry of the polynomial.[4] Thus each particle sees m zeros at the positions of the other particles, the particles avoid each other, and the state is highly correlated and has low energy.

The "binding" of the zeros to the positions of the particles is a kind of topological ordering which is the essence of the condensation in the

FQHE. Because particles see zeros as repulsive Coulomb charges, and because the zeros seen by any one particle are bound the positions of the *other* particles, the particles see each other as 2D Coulomb charges and hence the ground state distribution is that of a classical 2D one-component plasma. This causes the structure factor to vanish rapidly at long wavelengths and leads directly (within the single-mode-approximation) to the excitation gap central to the physics of the FQHE. Thus the gap exists if and only if topological order exists.

One must not be confused by the paradox that particles are repelled by zeros and yet zeros are bound to the particles. The reader is reminded that the zeros seen by any one particle are bound to the other particles. No particle sees its own zeros.

We turn now to the question of measuring the topological order represented by the binding of m zeros to the positions of the particles. To begin let us consider freezing the positions of all the particles except the first. The wave function then defines a complex function of a single argument z_1, the position of the first particle. If we integrate the gradient of the phase around some closed path Γ we obtain

$$\oint_\Gamma \vec{dr} \cdot \vec{\nabla} \text{Im} \, \log \Psi \;=\; 2\pi m N_0 \tag{22}$$

where N_0 is the number of other particles encircled by the path Γ. This phase integral is related to but not the same as the Berry's phase integral consider by Arovas, Wilczek, and Schrieffer.[11] They considered dragging a quasiparticle (vortex) excitation around a closed path, whereas here we are considering one of the original bare particles in the ground state.

We see that as the configuration of the particles is varied the number N_0 will fluctuate wildly (though not as strongly as for a compressible system). These phase fluctuations would seem to preclude the existence of an order parameter, however it turns out that we can take advantage of the simple *predictable* nature of the phase fluctuations to construct such an order parameter.

We take advantage at this point of the fact that the electrons in a magnetic field seem, as described above, to be acting like bosons. From modern notions of fractional statistics[12] we know that the statistics we assign to 2D particles is ambiguous. We can choose without loss of generality (or rigor) to represent an electron as a boson pierced by an

infinitesimal flux tube. The flux tube changes the Hamiltonian by introducing a non-trivial vector potential (to be described in detail below).

Upon adiabatic interchange of two of these composite particles, one
obtains an additional Berry's phase $\exp(i\theta)$ due to the Aharonov-Bohm
effect.[12] Choosing the strength of the flux tube such that $\theta = \pi$ gives a
factor of (-1) for exchange which precisely duplicates the minus sign that
one obtains from fermi statistics. Thus bosons with a certain peculiar
Hamiltonian (flux tubes) exactly mimic fermions. It is straight forward
to show that all observables are identical for the two systems.[12]

The point of making this transformation is that, since the system
seems to be behaving like a Bose system, it should be easiest to describe
its ordering within the "boson representation." It turns out to be convenient to choose the flux in the solenoids such that θ is not π but rather
$-m\pi$. Since m is an odd integer, the statistical transmutation from
fermions to bosons described above remains unaffected. The advantage
gained by this maneuver is that it turns out that the average density of
flux carried by the solenoids *precisely cancels* the externally applied field,
as illustrated in Fig. 3. Thus at the level of mean-field theory[13,14,12] one
is left with a model of bosons in zero (average) field which can then con-

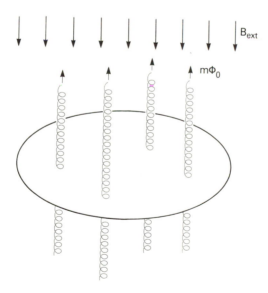

FIG. 3. Schematic illustration of composite bosons-plus-flux-tubes in an
external field. The average flux from the tubes cancels the external flux leaving
bosons in zero (mean) field which can then condense.

45

dense. The fact that it is composite objects described by statistical angle $\theta/\pi = $ -m which condense is closely related to the result to Arovas, Wilczek and Schrieffer[11] that vortices have statistics $\theta_v/\pi = 1/m$. It is a standard result that in a Chern-Simons field theory, the duality transformation[15,16] that interchange particles and vortices inverts the value of θ/π.

The possibility of a mean-field theory of fractional statistics was first suggested by Arovas, Wilczek, and Zee.[13] The first application of this notion was a demonstration of the existence of off-diagonal long-range order (ODLRO) in the FQHE.[14] Recently Laughlin has made use of this notion to study superconductivity of anyons.[17,18]

To make the condensation and ODLRO ideas more explicit, consider the Hamiltonian transformed to the Bose representation

$$H_B = \sum_{j=1}^{N} \frac{\hbar^2}{2m} \left(-i\vec{\nabla}_j + \vec{a}_j \right)^2 + V, \tag{23}$$

where V is the interaction potential and $\vec{a}_j$ is (proportional to) the vector potential seen by the jth particle due to the solenoids on the other particles. The only change from the fermion representation is the addition of the vector potential which obeys.

$$\vec{\nabla} \times \vec{a}_j = -2\theta \sum_{k \neq j} \delta^2 \left(\vec{r}_j - \vec{r}_k \right). \tag{24}$$

It follows from Eq. (24) that when one particle completely circles another, the Aharanov-Bohm phase is $\exp(-2i\theta)$ which is consistent with the fact that such a path is topologically equivalent to a pair of exchanges, each of which contributes $\exp(-i\theta)$.

With this in hand we can now reconsider the phase integral in Eq. (22) which becomes

$$\oint_\Gamma d\vec{r}_1 \cdot \left(\vec{\nabla} \text{Im} \log \Psi_B + \vec{a}_1 \right) = 2\pi \left(m + \frac{\theta}{\pi} \right) N_0, \tag{25}$$

which vanishes identically for the chosen value of $\theta/\pi = $ -m. Thus our transformation from fermions in a magnetic field to bosons in zero (average) field has completely eliminated the phase fluctuations. It is clear that the composite boson/flux tube object condenses. Indeed one can see this condensation explicitly in the traditional way by examining the one-body density matrix, which in the Bose representation becomes

$$\rho(z, z') = \prod_{j=2}^{N} \int d^2z_j \Psi_B^*(z', z_2, ..., z_N) \Psi_B(z, z_2, ..., z_N)$$

$$= \prod_{j=2}^{N} \int d^2z_j \Psi_F^*(z', z_2, ..., z_N) \Psi_F(z, z_2, ..., z_N) \exp\left(i \int_z^{z'} d\vec{r}_1 \cdot \vec{a}_1\right),$$

(26)

where the line integral of the vector potential is along some path connecting z and z'. This line integral is actually multiple-valued (path dependent) but because m is an integer, all paths are equivalent modulo 2π, which is all that counts in this case.

For large separations of the arguments z and z', $\rho(z,z')$ will go rapidly to zero unless Ψ_B contains no phase fluctuations, i.e., the bosons have condensed. This occurs here because the wild phase fluctuations in the Fermi wave function Ψ_F due to the zeros bound to the particles are precisely cancelled by the phase winding due to the solenoids.

If we view the zeros as vortices and the solenoids as anti-vortices then we see that in the Laughlin state the vortices and antivortices are on top of each other and cancel out. In this special case, it is possible to show that the boson density matrix decays algebraically asymptotically[14]

$$\rho(z, z') \sim |z - z'|^{-m/2}.$$

(27)

Thus there is not true ODLRO but only the algebraic ODLRO characteristic of 2D systems at finite temperature. A more precise analogy is to a 2D Coulomb charged (log r interaction) boson system which exhibits algebraic ODLRO even at zero temperature. This is because the system is incompressible - the particle number fluctuations are too small to allow the conjugate phase of the condensate to fully condense. The plasmon gap associated with this incompressibility is the analog of the magnetophonon gap and incompressibility in the FQHE. It is a remarkable feature of the FQHE that this structure arises from only short-range forces in the quantum Hamiltonian.

The exact result obtained in Eq. (27) holds only for the Laughlin state. The Laughlin state can be viewed as the unique ground state at

Landau level filling factor $\nu = 1/m$ in the presence of a very short range repulsion.[19,20] As the range of the repulsion is increased one finds that m - 1 of the zeros begin to fluctuate away from the positions of the particles. (One of the zeros must remain bound to preserve the antisymmetry.) As long as the zeros remain bound in the vicinity of the flux tubes, we anticipate that the boson density matrix will remain long-ranged just as it does for a superfluid at any temperature below the Kosterlitz-Thouless point. Here we have quantum rather than thermal fluctuations in the vortex positions.

For sufficiently long-ranged potentials one expects an unbinding transition in which the topological order is lost and the density matrix begins to decay exponentially.[14] This prediction has been confirmed in numerical work by Haldane and Rezayi[21] which clearly shows that at the unbinding transition, the ODLRO and the excitation gap simultaneously disappear, thus confirming our picture that ODLRO implies a gap and vice-versa. The Kosterlitz-Thouless analogy does not extend to the detailed nature of the transition since it is expected to be a first-order transition to a state of broken translation symmetry[8,14] (a Wigner crystal[22,23]).

Given the picture that it is a composite boson/flux-tube which condenses, one can attempt a Ginsburg-Landau description of the FQHE ordering.[14,15,24,25,26] Let Ψ be a complex field describing the condensate and consider the energy functional

$$H = \int d^2r \left| \left(\vec{p} + \vec{A} \right) \Psi \right|^2. \tag{28}$$

Let us first remind ourselves how this works for an ordinary superconductor. Consider a single vortex state

$$\Psi(r) = f(r)e^{i\vartheta} \tag{29}$$

where ϑ is the azimuthal angle and $f(0) = 0, f(\infty) = 1$. In the absence of a vector potential this state costs an infinite amount of energy because the current $\Psi^* p \to \Psi$ falls off only like r^{-1} at large distance. In the presence of a vector potential the velocity operator changes from $\vec{p}$ to $\vec{p} + \vec{A}$. This changes the current distribution which must be computed self-consistently from Maxwell's equations

$$\vec{\nabla} \times \vec{\nabla} \times \vec{A} = \vec{\nabla} \times \vec{B} = \frac{4\pi}{c} \vec{J}, \tag{30}$$

and

$$\vec{J} = \Psi^* \left(\vec{p} + \vec{A} \right) \Psi. \tag{31}$$

In the presence of a vortex the system can have a finite energy if the current falls to zero rapidly enough. Consider the line integral on a path Γ encircling the vortex at a large distance where the current vanishes

$$\oint_\Gamma \vec{dr} \cdot \Psi^* \left(\vec{p} + \vec{A} \right) \Psi = 0. \tag{32}$$

The single-valuedness of Ψ guarantees that

$$\oint_\Gamma \vec{dr} \cdot \Psi^* \vec{p} \Psi = -2\pi n \tag{33}$$

where $n = 0, \pm 1, \pm 2,...$, and without loss of generality, we assume $|\Psi| \rightarrow 1$ asymptotically. Hence we have

$$\int_\Gamma \vec{dr} \cdot \vec{A} = 2\pi n \tag{34}$$

or using Stoke's theorem

$$\int_\Gamma d^2 r \left(\vec{\nabla} \times \vec{A} \right)_z = 2\pi n. \tag{35}$$

Thus the total flux generated by the vortex must be quantized. The current at short distances near the vortex will distribute itself to fulfill this condition thereby removing the infinite energy.

We can now very easily modify this picture to accommodate the FQHE. The flux generated by the currents in the 2D electron gas is negligible. However each boson by construction, carries a solenoid with it. Hence we replace Eq. (30) by

$$\left(\vec{\nabla} \times \vec{A} \right)_z = \left(B_{ext} - 2\theta |\Psi|^2 \right) \tag{36}$$

where $|\Psi|^2$ represents the density of particles (and hence solenoids).

As we remarked earlier, in the boson representation of the Laughlin ground state (for which we take $\Psi = 1$ everywhere), the average flux vanishes so we have

$$\left(\vec{\nabla} \times \vec{A}\right)_z = -2\theta\left(|\Psi|^2 - 1\right). \tag{37}$$

The flux quantization condition becomes

$$\int_\Gamma d^2r\left(\vec{\nabla} \times \vec{A}\right)_z = 2\pi n = -2\theta \int_\Gamma d^2r\left(|\Psi|^2 - 1\right). \tag{38}$$

Thus flux quantization is replaced by fractional charge quantization

$$q^* = \int d^2r\left(|\Psi|^2 - 1\right) = n(\theta/\pi)^{-1}. \tag{39}$$

This is easily understood. Each particle carries θ/π units of flux in its solenoid so a single vortex must bind π/θ of a charge to have a finite energy. This allows us to identify Laughlin's fractionally-charged quasiparticles[4] as quantized vortices.[14,24]

In analogy with an extreme type-II superconductor, the system accommodates extra *charge* (away from the $\nu = \pi/\theta$ Landau level filling) in the form of quantized vortices. If these are localized by disorder then σ_{xx} remains zero and σ_{xy} remains quantized thus giving a plateau of finite width.

An additional analogy with type-II superconductivity is that, since the vortex energy is finite, one expects no Kosterlitz-Thouless phase transition at any finite temperature. (This is not to be confused with the quantum (zero-temperature) vortex-unbinding transition which occurs as a function of the range of the potential.) In principle, the charged vortices could interact sufficiently to undergo a glass transition at some finite temperature. Barring this however one expects thermally activated dissipation at any finite temperature due to thermally activated quasiparticle motion. This is the direct analog of flux-flow dissipation in a type-II superconductor.

SUMMARY

The last decade has seen remarkable experimental and theoretical progress since the discovery of the integer effect in 1980. We now have an excellent microscopic and quantitative understanding of the fractional effect.[1] This paper has focussed not on this but on recently developed macroscopic and qualitative understanding of the nature of the ordering in the FQHE.

Our theme has been that 2D electrons in a magnetic field act like bosons in zero field. The neutral collective excitations are density waves that have a gap. The charged excitations carry quantized fractional charge in a manner analogous to flux quantization and have finite energy. The object which condenses is not the original particle but a boson/flux-tube composite. A Ginsburg-Landau theory has been found which correctly reproduces the phenomenology of the condensation.

For the future we hope to see direct spectroscopic observation of the gap, and further elucidation of the recently discovered even denominator states discussed in D. Tsui's paper in this volume. It has also been proposed[27,28] that there may be interesting edge-state effects in the fractional regime. The edge-state approach is motivating a rapidly growing amount of interesting new work[29,30] in both the integer and fractional regimes which need to be further explored by theorists. In principle the edge-state picture is equivalent to the usual bulk Kubo formula for linear response. In practice, non-equilibrium effects at the edges may be crucially important.[30]

A case can be made that, while our quantitative and microscopic understanding of the simple fractions $\nu = 1/m$ is excellent, the more complex states further down in the hierarchy are less well understood. Further work here is certainly needed. One interesting new approach is that of Jain[31] who uses a modification of the ideas presented here. Instead of adding a flux tube to give statistical phase $\theta/\pi = m$, Jain uses $\theta/\pi = m - 1$ which maps the electrons into *fermions* in an effective magnetic field corresponding to a filled Landau level (as opposed to obtaining bosons in zero field as described above). Thus he makes a connection between the integer and fractional effects. He then generalizes this idea to hierarchical fractional states which are related back to higher integer states. This has the advantage of giving quite explicit wave functions, assuming that the Landau level mixing does not significantly spoil the kinetic energy. This picture is new and its implications are not yet fully clear, but the approach looks promising.

The Wigner crystal has long been sought and indeed the fractional Hall effect was discovered during a search for the crystal. There is increasing evidence (see D. Tsui's paper) that some sort of freeze-out of carriers into a crystalline or glassy state is occurring at low filling factors. It is hoped that the next decade will bring us something definitive in this area.

It seems clear that the fundamental issues in the quantum Hall effect have been fairly well settled but much interesting work remains to be done.

REFERENCES

1. The Quantum Hall Effect, 2nd ed., ed. by R. E. Prange and S. M. Girvin (Springer-Verlag, New York, Heidelberg, Tokyo, 1990).

2. Strictly speaking the system could be gapless as long as there existed a finite minimum group velocity for all excitations (as in a superfluid). For v less than this value the impurities will not "Cerenkov" radiate. It turns out however that all (known) FQHE states are incompressible and have a non-zero gap.

3. R. P. Feynman, Statistical Mechanics (Benjamin, Reading, 1972).

4. R. B. Laughlin in Chap. 7 of Ref. 1.

5. S. M. Girvin in Chap. 9 of Ref. 1.

6. F. D. M. Haldane and E. H. Rezayi, Phys. Rev. Lett. **54**, 237 (1985).

7. G. Fano, F. Ortolani and E. Colombo, Phys. Rev. **B34**, 2670 (1986).

8. S. M. Girvin, A. H. MacDonald and P. M. Platzman, Phys. Rev. **B33**, 2481 (1986).

9. P. M. Platzman, S. M. Girvin and A. H. MacDonald, Phys. Rev. **B32**, 8458 (1985).

10. B. I. Halperin, Helv. Phys. Acta. **56**, 75 (1983).

11. D. P. Arovas, J. R. Schrieffer and F. Wilczek, Phys. Rev. Lett. **53**, 722 (1984).

12. G. S. Canright and S. M. Girvin, Science **247**, 1197 (1990).

13. D. P. Arovas, J. R. Schrieffer and F. Wilczek, Nuc. Phys. **B251**, 117 (1985).

14. S. M. Girvin and A. H. MacDonald, Phys. Rev. Lett. **58**, 1252 (1987).

15. D. H. Lee and M. P. A. Fisher, Phys. Rev. Lett. **63**, 903 (1989).

16. X. G. Wen and A. Zee, Phys. Rev. Lett. **62**, 1937 (1989).

17. R. B. Laughlin, Phys. Rev. Lett. **60**, 2677 (1988).

18. R. B. Laughlin, Science **242**, 525, (1988).

19. F. D. M. Haldane in Chap. 8 of Ref. 1.

20. S. A. Trugman and S. Kivelson, Phys. Rev. **B26**, 3682 (1985); V. L. Pokrovksky and A. L. Tapalov, J. Phys. **C18**, L691 (1985).

21. E. H. Rezayi and F. D. M. Haldane, Phys. Rev. Lett. **61**, 1985 (1988).

22. P. K. Lam and S. M. Girvin, Phys. Rev. **B30**, 473 (1984); **31**, 613 (E) (1985).

23. D. Levesque, J. J. Weis and A. H. MacDonald, Phys. Rev. **B30**, 1056 (1984).

24. S. M. Girvin in Chap. 10 of Ref. 1.

25. N. Read, Phys. Rev. Lett. **62**, 86 (1989).

26. S. C. Zheng, T. H. Hansson and S. Kivelson, Phys. Rev. Lett. **62**, 82 (1989).

27. A. H. MacDonald, Phys. Rev. Lett. **64**, 220 (1990).

28. A. M. Chang, Sol. State Comm. **74**, 871 (1990).

29. D. A. Syphers and P. J. Stiles, Phys. Rev. **B32**, 6620 (1985); R. J. Haug, A. H. MacDonald, P. Streda and K. von Klitzing, Phys. Rev. Lett. **61**, 2797 (1988); Phys. Rev. Lett. **62**, 608(E) (1989); S. Washburn, A. B. Fowler, H. Schmid and O. Kem, Phys. Rev. Lett. **61**, 2801 (1988); S. Komijama, H. Hira, S. Sasoand and S. Yiyamizu, Phys. Rev. **B40**, 5176 (1989); A. M. Chang, J. E. Cunningham, Sol. State Comm. **72**, 651 (1989); L. I. Glazman and M. Jonson, Phys. Rev. Lett. **64**, 1186 (1990); B. J. van Wees, E. M. M. Willems, C. J. P. M. Harmans, C. W. J. Beenakker, H. van Houten, J. G. Williamson, C. T. Foxon and J. J. Harris, Phys. Rev.

Lett. **62**, 1181 (1989); C. W. J. Beenakker and H. van Houten, Phys. Rev. Lett. **60**, 2406 (1988).

30. L. P. Kouwenhoven, B. J. van Wees, N. C. van der Vaart, C. J. P. M. Harmans, C. E. Timmering and C. T. Foxon, Phys. Rev. Lett. **64**, 685 (1990).

31. J. K. Jain, Phys. Rev. **B41**, 7653 (1990).

IMPLICATIONS OF SEMICONDUCTOR SUPERLATTICE RESEARCH

L. Esaki

IBM Thomas J. Watson Research Center
Yorktown Heights, New York 10598, U.S.A.

Abstract - Following the past two decade evolutionary path in the inter-disciplinary research of semiconductor superlattices and quantum wells, significant milestones are presented with emphasis on electric field-induced effects in the frontier of semiconductor physics associated with technological advances.

1. INTRODUCTION

In the early twentieth century, the encounter with physical phenomena for which Newtonian mechanics could not possibly provide an adequate explanation, prompted the advent of quantum mechanics. Quantum mechanics, indeed, has been playing an indispensable role in the modern physics world, including our understanding of the behavior of semiconductors. In the late twentieth century, however, quantum mechanics has found an additional role, that is, its principles have been utilized in the design of novel semiconductor structures which exhibit

Highlights in Condensed Matter Physics and Future Prospects
Edited by L. Esaki, Plenum Press, New York, 1991

unprecedented properties. The advanced techniques of thin-film growth have facilitated engineering for *designed quantum structures*.

In 1969, research on such structures was initiated with a proposal of an *engineered* semiconductor superlattice by Esaki and Tsu (1) (2). In anticipation of advancement in epitaxy, two types of superlattices were envisioned, with alternative deposition of ultrathin layers: Doping and compositional, as shown at the top and bottom of Fig. 1, respectively.

Since the one-dimensional potential is introduced along with the superlattice (SL) axis (perpendicular to the deposited plane layers), we thought that, if our attempt was successful, elegantly simple examples in one-dimensional quantum physics, for instance, resonant electron tunneling (3), Kronig-Penney band model or Stark localization, which

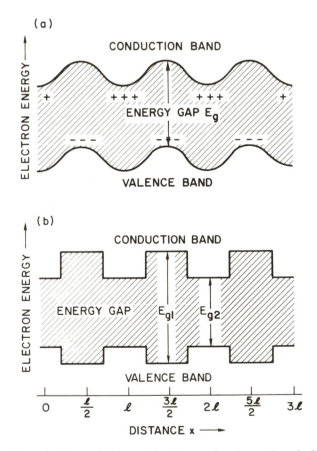

FIG. 1. Spacial variations of the conduction and valence bandedges in two types of superlattices: doping (top) and compositional.

had remained textbook exercises, could, for the first time, be practiced in a laboratory: *Do-it-yourself quantum mechanics* would be possible, since the principles of quantum theory dictate the design of semiconductor structures or devices.

At the inception of the SL idea, it was recognized that the long, tailored lattice period provided the unique opportunity to exploit electric field-induced effects. Our early analysis of electron dynamics in a modestly high electric field, along with the SL axis, led to the prediction of the occurrence of a negative differential resistance which could be a precursor of the Bloch oscillation. The introduction of the SL, apparently, allows us to explore the regime of electric field-induced quanitzation such as Stark ladders, which is not readily accessible in *natural* crystals.

Before reaching the SL concept, we had explored the feasibility of structural formation by epitaxy of ultra-thin barriers and wells which might exhibit resonant electron tunneling. It was thought that semiconductors and technologies developed with them, might be suitable for demonstration of the quantum wave nature of electrons associated with the interference phenomena, since their small Fermi energies due to low carrier densities help make the de Broglie wavelength relatively large. Namely, the Fermi wavelength $\lambda_f = 2\pi/k_f$, where k_f is the magnitude of the Fermi wavevector, is given as a function of the carrier density n, as follows: $\lambda_f = (8\pi/3n)^{1/3}$ for a three-dimensional (3D) system; $(2\pi/n)^{1/2}$ for a two-dimensional (2D) system; $(4/n)$ for a one-dimensional (1D) system.

The idea of the SL occurred to us as a natural extension of double- and multi-barrier structures: Namely, the SL is a series of quantum wells (QWs) coupled by resonant tunneling, where quantum effects are expected to prevail. An important parameter relevant to the observation of such effects is the phase-coherent length or, roughly, the electron inelastic mean free path, which depends heavily on bulk as well as interface quality of crystals and also on temperatures and values of the effective mass. As schematically illustrated in Fig. 2, if characteristic dimensions such as SL periods and QW widths are reduced to less than the phase-coherent length, the entire electron system will enter a *mesoscopic* quantum regime of reduced dimensionality, being placed in the scale between the macroscopic and the microscopic.

It was theoretically shown that the introduction of the SL potential perturbs the band structure of the host materials, yielding unprecedented

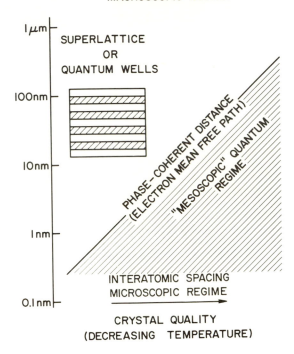

MACROSCOPIC REGIME

FIG. 2. Schematic illustration of a "mesoscopic" quantum regime (hatched) with a superlattice or quantum wells in the insert.

electronic properties of quasi-two-dimensional character (1) (2). Figure 3 shows the density of states $\rho(E)$ for electrons with $m^* = 0.067m_o$ in an SL with a well width of 100Å and the same barrier width, where the first three subbands are indicated with dashed curves. The figure also includes, for comparison, a parabolic curve $E^{1/2}$ for 3D, a steplike density of states for 2D (quantum well), a curve $\sum_{mn} > (E - E_m - E_n)^{-1/2}$ for 1D (quantum wire), and a delta function $\sum \delta(E - E_l - E_m - E_n)$ for 0D system (quantum dot) where the quantum unit is taken to be 100Å for all cases and the barrier height is assumed to be infinite in obtaining the quantized energy levels, E_l, E_m and E_n. Notice that the ground state energy increases with decrease in dimensionality if the quantum unit is kept constant. Each quantized energy level in 2D, 1D and 0D is identified with the one, two and three quantum numbers, respectively. The unit for the density of states here is normalized to $eV^{-1}cm^{-3}$ for all the dimensions, although $eV^{-1}cm^{-2}$, $eV^{-1}cm^{-1}$ and eV^{-1} may be commonly used for 2D, 1D and 0D, respectively.

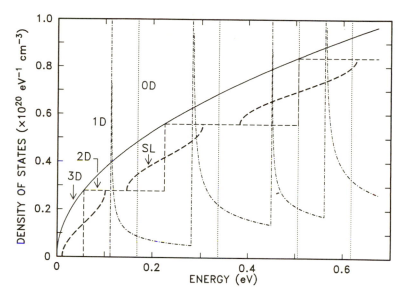

FIG. 3. Comparison of density of states in the three-dimensional (3D) electron system with those of a superlattice, and the two-dimensional (2D), one-dimensional (1D), and zero-dimensional (0D) electron systems.

2. EPITAXY AND SUPERLATTICE GROWTH

In the early 1970s, we initiated the seemingly formidable task of engineering nanostructures in the search for novel quantum phenomena (4). Heteroepitaxy is of fundamental interest for the SL and QW growth. Innovations and improvements in growth techniques such as MBE (5), MOCVD (metallo organic chemical vapor deposition) (6) and MOMBE (metallo organic molecular beam epitaxy) or CBE (chemical beam epitaxy) during the last decade have made possible high-quality heterostructures. Such structures possess predesigned potential profiles and impurity distributions with dimensional control close to interatomic spacing and with virtually defect-free interfaces, particularly, in a lattice-matched case such as $GaAs - Ga_{1-x}Al_xAs$. This great precision has cleared access to a *mesoscopic* quantum regime.

The semiconductor SL structures have been grown with III-V, II-VI and IV-VI compounds, as well as elemental semiconductors. Semiconductor hetero-interfaces exhibit a band offset, usually associated with a gradual band-bending in its neighborhood which reflects space-charge effects. According to the character of such discontinuity, known hetero-interfaces can be classified into four kinds: Type I, type II-staggered,

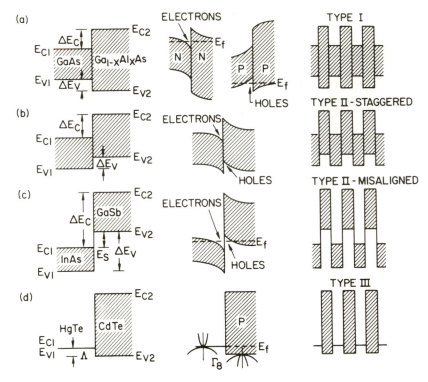

FIG. 4. Discontinuities of bandedge energies at four types of hetero-interfaces: band offsets (left), band bending and carrier confinement (middle), and superlattice (right).

type II-misaligned, and type III, as illustrated in Fig. 4(a)(b)(c)(d): Band offsets (left), band bending and carrier confinement (middle) and SLs (right).

3. RESONANT TUNNELING IN AN ELECTRIC FIELD

Resonant tunneling in multi-barrier structures can be observed with the application of an electric field perpendicular to the layer planes. In this context, Tsu and Esaki (7) derived the voltage dependence of the tunneling current in such structures with the transmission coefficient computed as a function of electron energy (Fig. 5), in assuming the conservation of the total electron energy as well as its momentum transverse to the direction of tunneling. Chang, Esaki and Tsu (8) measured the tunneling current and conductance as a function of applied voltage in GaAs-GaAlAs double barriers, as shown in Fig. 6, which exhibited a

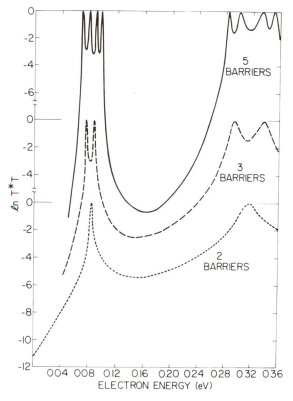

FIG. 5. Natural log of the transmission coefficient vs. electron energy in eV for cases of a double-, a triple-, and a quintuple- barrier. The peaks show resonance. The barrier and well widths are 20 and 50Å, respectively. The barrier height is 0.5 eV.

large component of the excess current primarily arising from imperfections in the structures, but agreed qualitatively with the Tsu-Esaki formula (7). The energy diagram is shown in the inset where resonance occurs at such applied voltages as to align the Fermi level of the electrode with the bound states, as shown in cases (a) and (c) of Fig. 6. This experimental result probably constitutes the first observation of man-made bound states in single QWs.

The technological advance in MBE has resulted in remarkably improved characteristics in resonant tunneling, which spurred renewed interest in such structures (9). Recently, high-quality resonant-tunneling double-barrier junctions, exhibiting not only large values of the peak-to-valley ratio but also fine structure, have been the focus of intense exper-

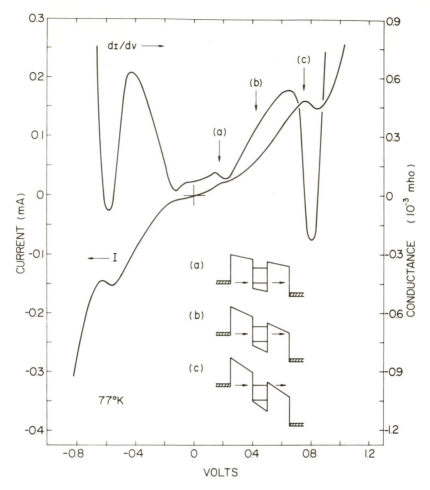

FIG. 6. Current- and conductance- voltage characteristics of a GaAs-GaAlAs double-barrier structure. Conditions at resonance (a) and (c), and at off-resonance (b), are indicated by arrows in the insert.

imental and theoretical investigations. The observation of resonant tunneling of holes reveals structure corresponding to respective bound states of both heavy and light holes (10). The existence of optical phonon-assisted resonant tunneling was found in the valley region (11) and analyzed (12). Resonant magneto-tunneling measurements manifested the energy levels quantized into Landau levels of index n (13). Later, in magneto-tunneling, LO phonon (magnetopolaron) - assisted process was clearly resolved (14) (15). Charge accumulation in a double-barrier resonant tunneling structure was studied by photoluminescence (16) and photoluminescence-excitation spectroscopy (17).

A variety of tunneling heterostructures have been developed with a combination of materials other than GaAs/AlAs, such as InGaAs/InAlAs (18) and InAs/AlSb (19). Research in polytype heterostructures of InAs/AlSb/GaSb (20), (21), has been rejuvenated with the observation of a large negative resistance based on resonant interband tunneling in GaSb/AlSb/InAs/AlSb/GaSb (22) and InAs/AlSb/GaSb/AlSb/InAs (23). Those multi-heterojunctions, indeed, correspond to a portion of the proposed polytype superlattices (20).

A great deal of engineering interest in resonant tunneling, obviously, has arisen from its potential application to high-speed semiconductor devices. Picosecond bistable operation has been observed in a double-barrier resonant tunneling diode, where a rise time of 2 ps is said

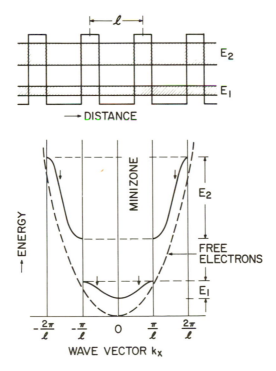

FIG. 7. (a) Potential profile of a superlattice, and (b) electron energy vs. wave vector in the minizones.

to be the fastest switching yet measured for an electronic device (24). Oscillations up to frequencies of 420 GHz were reported in a GaAs resonant tunneling diode at room temperature (25).

4. TRANSPORT AND OPTICAL PROPERTIES IN SUPERLATTICES AND QUANTUM WELLS SUBJECT TO AN APPLIED ELECTRIC FIELD

4.1. Theoretical Background

The introduction of the SL potential, as shown in Fig. 1, perturbs the band structure of the host materials: The degree of perturbation obviously depends on its potential profile. With the SL period, l, of the order of 50-100Å, the Brillouin zone is divided into a series of mini-

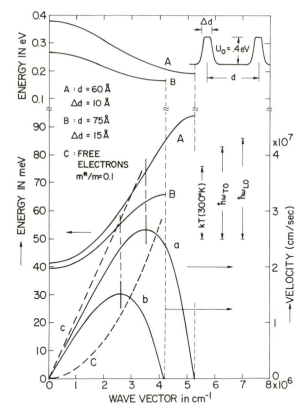

FIG. 8. E vs k in the first and second subbands and the group velocity vs. k in the first subband, for two superlattice structures and for free electrons (dashed curve).

zones, as shown schematically in the bottom of Fig. 7, giving rise to the first and second subbands, E_1 and E_2, separated by forbidden regions, as shown in the top. Actually, for two superlattice structures with 60Å (A) and 75Å (B) periods, we have made calculations for the electron energy E versus the wave vector k in the first and second subbands and also the group velocity, $\hbar^{-1}$ dE/dk, versus k in the first subband, as shown in Fig. 8, with the free electron case (C) for comparison.

With application of an electric field in the SL direction, x, the electron dynamics in a narrow subband has been analyzed in the framework of a simplified path integration method to obtain a relationship between the applied field F and the average drift velocity v_d. The equations of electron motion are

$$\hbar \partial k_x/\partial t = eF \quad \text{and} \quad v_x = \hbar^{-1}\partial E_x/\partial k_x, \tag{1}$$

then, the velocity increment in a time interval dt is

$$dv_x = eF\hbar^{-2}\left(\partial^2 E_x/\partial k_x^2\right)dt. \tag{2}$$

The average drift velocity, taking into account the electron scattering time τ, is written as

$$v_d = \int_0^\infty \exp(-t/\tau)dv_x = eF\hbar^{-2}\int_0^\infty \left(\partial^2 E_x/\partial k_x^2\right)\exp(-t/\tau)dt. \tag{3}$$

If a sinusoidal relationship, $E_x = \dfrac{E_1}{2}(1 - \cos k_x \ell)$, is assumed, we obtain

$$v_d = eF\tau/m^*(0)\left[1 + (eF\ell\tau/\hbar)^2\right]^{-1}, \tag{4}$$

where $m^*(0)$ is determined by $\hbar^2(\partial^2 E_x/2k_x^2)^{-1}$ at $k_x = 0$. Thus, the current - field curve is given by

$$J = env_d \tag{5}$$

where n is the electron concentration. The current J, plotted as a function of F in Fig. 9, has the maximum at $eF\ell\tau/\hbar = 1$ (indicated by an arrow) and thereafter decreases, giving rise to a differential negative resistance. The threshold field for the negative resistance is 10^3 V/cm if

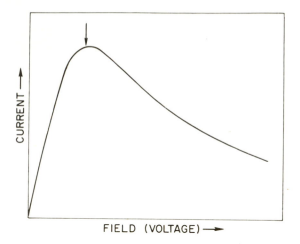

FIG. 9. Current versus electric field curve.

$\ell = 100\text{Å}$ and $\tau = 0.66$ ps. This result indicates that, if, in applying modestly high electric fields, electrons gain enough energy to reach beyond the inflection points (indicated by arrows in Fig. 7), they will be decelerated rather than accelerated by such electric fields.

If, in further increasing electric fields, electrons reach the mini-Brillouin zone boundary ($k_x = \pi/\ell$ in Fig. 7), an umklapp process (or Bragg scattering) will occur, that is, the electron wave vector, k_x, is changed by a reciprocal lattice vector, $2\pi/\ell$. If this happens, electrons reappear at the opposite point of the Brillouin zone, $k_x = -\pi/\ell$, and the process will be repeated. This is the physical picture of the so-called Bloch oscillation (the centroid of the electron wave packet moves back and forth, steadily, in the k-space), whose frequency is determined by the time period T required to cross the Brillouin zone, namely, $T = 2\pi\hbar/eF\ell$. Thus, the Bloch angular frequency ω_B is given by $\omega_B = 2\pi/T = eF\ell/\hbar$. In order to realize this oscillation, the scattering time τ should be greater than T, that is, $\omega_B\tau > 2\pi$. A precursor of this oscillation is the above-mentioned differential negative resistance which starts at $\omega_B\tau > 1$. The frequency of the Bloch oscillator is as high as 1 THz for $F = 4 \times 10^3$ V/cm and $\ell = 100\text{Å}$.

The Bloch oscillation involves resonating transitions between equally spaced and localized energy levels called Stark ladders which are the consequence of electric field-induced quantization. Fig. 10 shows schematically electronic wave functions of Stark ladders formed in a

LOW FIELD

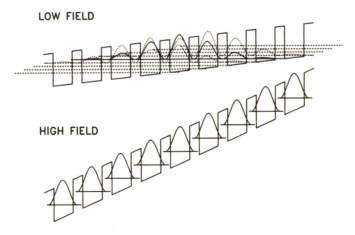

HIGH FIELD

FIG. 10. Schematic potential profiles with electronic wave functions of Stark ladders formed in a superlattice under low and high electric fields.

superlattice under low and high electric fields; at the low field, optical transitions may be possible because of sufficient overlap of wave functions between adjacent Stark states; the energy separation of these levels ΔE is equal to $\hbar\omega_B = eF\ell$. At the high field, field-induced electron hopping to non-overlapping adjacent Stark states may occur, leading again to a differential negative resistance (26).

Before the introduction by Wannier (27) of the term Stark ladder, James (28) heuristically reached the basic concept of Stark localization. The existence and real nature of these localized levels in bulk crystals, however, were quite controversial (29, 30). Experimental evidences supporting their existence are rather tenuous (31, 32, 33), because the observation, firstly, requires enough high fields to produce the energy separation greater than the collision broadening, that is,

$$\Delta E = eFa > \hbar/\tau, \quad \text{then} \quad \omega_B\tau > 1, \tag{6}$$

where a is the lattice constant which is typically 2 to 3 Å. If $\tau = 0.66$ ps, $\hbar/\tau = 1$ meV. Therefore, F should be of the order of 10^5 V/cm. On the other hand, the introduction of the SL structures with large lattice constants ℓ provide easy access to the electric field-induced quantization regime with a few tens of kV/cm.

Stark ladder states are localized over the length λ given by E_1/eF, where E_1 is the subband width. The k-space approach presented here is justified when $\Delta E = eF \ell \ll E_1$ or $\ell \ll \lambda$. If, however, the value of $eF\ell$ approaches to E_1 ($\lambda \to \ell$) because of high fields or narrow band-widths (tight binding SL), a hopping model by Tsu and Döhler (34) becomes valid. In this case, although a negative differential resistance may arise, the Bloch oscillation is unlikely to occur because of insufficient spacial overlap between adjacent Stark states.

Artaki and Hess (35) carried out Monte Carlo simulations for electron transport in SLs, taking into account scattering by acoustic and optical phonons and impurities; the collision broadening effect on SL transport associated with a negative resistance was assessed.

Incidentally, it should be appreciated that the quantum theory for the electric field effects in an SL presented here is quite different from the traditional approach to a bulk semiconductor with E_c and E_v band-

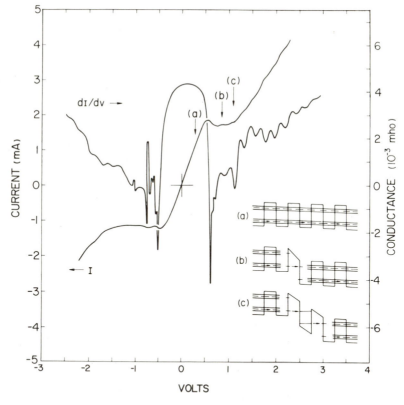

FIG. 11. Current- and conductance- voltage characteristics of a GaAs-AlAs superlattice. Band-type conduction (a) and resonant tunneling at an expanding high-field domain (b) and (c), are indicated in the insert.

widths of a few electron-volts: The electro-optical properties (a tail below the energy gap and an oscillatory behavior above it) have been treated with the effective mass approach, where the envelope eigenfunctions under an external electric field are Airy functions: The observation is consistent with the resultant Franz-Keldish effect.

4.2. Early Attempts in the Search for the Quantum Effects

Esaki et al. (36) found a weak negative resistance resembling Figure 9, starting at $F \sim 2 \times 10^4$ V/cm in an MBE-grown GaAs-GaAlAs SL with $E_1 \sim 40$ meV. Later, Esaki and Chang (37) studied transport properties in an SL of a tight-binding potential with $E_1 = 5$ meV, which exhibited an oscillatory behavior: The current and conductance as a function of applied voltage are shown in Fig. 11. A marginal band-conduction mechanism (case (a) in the inset) is broken down, as $eF\ell$ approaches to E_1, at relatively low voltages, resulting in a negative resistance with creation of a localized high-field domain. The current recovers due to resonant tunneling between the first miniband and the second miniband in the adjacent QWs (case (b) in the set) when their energies are aligned with each other. The oscillatory behavior in the conductance reflects the repetition of this resonance with increase in applied voltage, whose period, 0.24 V, agreed with the calculated value.

Tsu et al. (38) performed photocurrent measurements on GaAs-GaAlAs SLs subject to an electric field perpendicular to the plane layers with the use of a semitransparent Schottky contact. The photocurrents as a function of incident photon energy are shown in Fig. 12 for samples of three different configurations, designated A (35Å GaAs - 35Å GaAlAs), B (50Å GaAs - 50Å GaAlAs), and C (110Å GaAs - 110Å GaAlAs). The energy diagram is shown schematically in the upper part of the figure where minibands created by the SL potential are labeled E_1 and E_2 In sample C, the minibands are essentially discrete states. Calculated energy locations and bandwidths shown in the figure are found to be in satisfactory agreement with the observation. The photocurrent, as a function of applied voltage, exhibited a pronounded negative resistance which was interpreted with the hopping model (34) since it occured in the range of $eF\ell > E_1$.

Since the mid-80's, advanced growth facilities at many institutions have begun to provide high-quality SLs and QWs, meeting stringent requirements in the structural design. As a result, investigations into

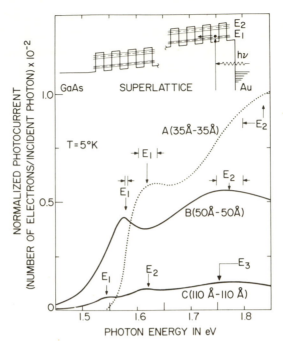

FIG. 12. Photocurrent versus photon energy for three superlattice samples A, B and C. Calculated energies and bandwidths are indicated. The energy diagram of a Schottky-barrier structure is shown in the upper part.

such quantum structures have been intensified and have proliferated extensively. We present here only a limited number of reports, selected rather arbitrarily by the author.

4.3. Recent Transport Experiments

Deveaud et al. (39) carried out transport measurements of 2D carriers in minibands using subpicosecond luminescence spectroscopy. Schneider et al. (40) studied the dynamics of electron transport along the growth direction of tight-binding GaAs-AlAs SLs by electrical and optical time-of-flight experiments. Helm et al. (41) observed infrared light (intersubband) emission from SLs excited by sequential resonant tunneling. Tatham et al. (42) made time-resolved Raman measurements of intersubband relaxation in GaAs QWs. Sibille et al. (43) studied negative differential conductance in transport properties of GaAs-AlAs SLs having relatively wide miniband-widths: The data were analyzed in terms of Esaki and Tsu's theory (1), (2). Hadjazi et al. (44) obtained preliminary results for microwave applications of superlattices. Grahn et al. (45) reported the observation of photoluminescence from the second and third subbands in GaAs-AlAs SLs with an applied electric field: The occupation of the higher subbands was achieved by sequential resonant tunneling of electrons. Brozak et al. (46) observed thermal saturation of transport in a narrow band of an SL: The conductivity was quenched as

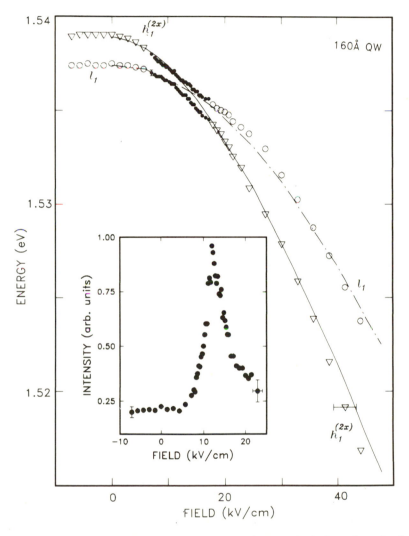

FIG. 13. Stark shifts of the light-hole exciton (open circles) and excited states of the heavy-hole exciton (open triangles), exhibiting strong coupling (solid circles) between them. The integrated intensity of $h_1^{(2x)}$ normalized to l_1 is shown as a function of electric field in the insert.

the temperature was raised. Beltram et al. (47) studied transport properties in a transistor with an SL structure built between the base and the collector, observing negative differential conductance and scattering-controlled transmission resonances. Vengurlekar et al. (48) reported miniband transport of minority electrons in a superlattice transistor. Lei et al. (49) provided an advanced theory of negative differential conductance in a superlattice miniband. Zaslavsky et al. (50) observed negative differential resistance in an edge-regrown superlattice (a modulated 2D electron system) at low temperatures.

4.4. Stark Shifts in QWs

Mendez et al. (51) studied the field-induced effect on the photoluminescence in GaAs-GaAlAs QWs, where the main peak is attributed to the excitonic transition between 2D electrons and heavy holes: When the electric field was applied perpendicular to the plane layers, pronounced field-effects, Stark shifts, were found. More recently, Vina et al. (52) studied the excitonic coupling as indicated in Fig. 13, which shows Stark shifts of the light-hole exciton (open circles) and excited states of the heavy-hole exciton (open triangles), exhibiting strong coupling (solid circles) between them. Chemla et al. (53) observed Stark shifts of the excitonic absorption peak, even at room temperature; Miller et al. (54) analyzed such electroabsorption spectra and demonstrated optical bistability, level shifting and modulation.

4.5. Formation of Stark Ladders in SLs with the Application of an Electric Field

Bleuse et al. (55) calculated field-induced Stark localization of the eigenstates in an SL, predicting a blue shift of the optical-absorption edge and the presence of oscillations in F^{-1}. Mendez et al. (56) observed the formation of Stark ladders in an SL by field-induced movement of the peak positions in photoluminescence and photocurrent spectra with applied electric fields. Fig. 14 illustrates the conduction- and valence-band potential profiles, including the Stark ladder formation, for a 40Å - 20Å GaAs-GaAlAs SL in the absence, (a), and in the presence of an electric field, (b) and (c). At F=0, the electron and hole wavefunctions are delocalized and interband transitions take place between SL minibands. With the application of an electric field, the localization of wavefunctions begin, forming the ladder whose energy spectrum is given by Eo + neFℓ (n = 0, $\pm$ 1, $\pm$ 2, ...), where Eo is the

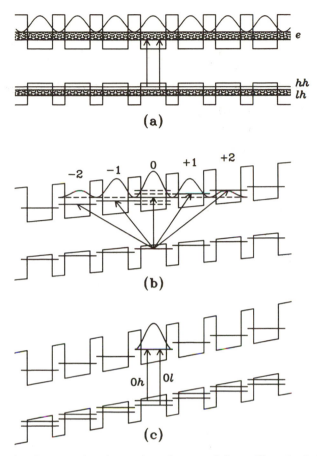

FIG. 14. Conduction- and valence-band potential profiles, including the Stark ladder formation, for a 40Å - 20Å GaAs-GaAlAs superlattice at zero field, (a), at a moderate field (2×10^4 V/cm), (b), and at a high field (10^5 V/cm), (c).

eigenenergy of an isolated QW. At a moderate field (2×10^4 V/cm), (b), electron and light-hole states are still extended over several SL periods whereas heavy holes are fully localized. Interband transitions (indicated by -2, -1, 0, +1, +2) involve Stark-ladder states like the ones represented here for the conduction band. For simplicity, the light-hole ladder states are not included. At a high field (10^5 V/cm), (c), all states are completely localized and transitions take place between levels corresponding to isolated QWs. Fig. 15 shows the interband transition energies as a function of field for the SL in Fig. 14, determined from the peak positions in photocurrent spectra. Each branch corresponds to a member

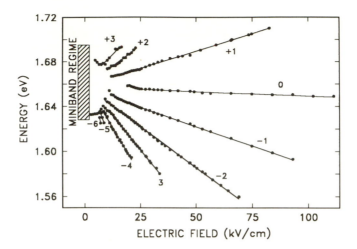

FIG. 15. Interband transition energies as a function of electric field for the superlattice in Fig. 14. Each branch corresponds to a member of the Stark ladder, labeled by an index, n.

of the Stark ladder, labeled by an index, n. With the same technique, Agullo-Rueda et al. (57) determined the coherence length of electrons from the number of observed interband transitions between the valence- and conduction-band Stark ladders, which was found at least ten periods when $\ell = 60$Å and increased with decreasing ℓ (or increasing the mini- band width). Later, it was shown that the coherence length did not depend strongly on temperature in the range 5 - 292K (58). Voisin et al. (59) confirmed the predicted electro-optical effect (55) by an electroreflectance study for an SL. Dignam and Sipe (60) presented cal- culations of the energies and absorption spectra of GaAs-GaAlAs SL excitons in the presence of an applied electric field, which were in good agreement with the experimental photocurrent spectra (58). Ohno et al. (61) observed localized surface states (Tamm states) intensionally intro- duced in GaAs-GaAlAs SLs by excitonic interband transitions in photoluminescence excitation spectra. Photocurrent experiments under an electric field showed additional transitions as well as anticrossing interactions between the Tamm states and the Stark-ladder states associ- ated with the SL.

5. CONCLUSION

Our original proposal and early works apparently triggered a wide spectrum of experimental and theoretical investigations in the new domain of semiconductor science and engineering. A variety of *engineered quantum structures* exhibited extraordinary transport and optical properties which may not even exist in any *natural* crystal. Thus, this new degree of freedom offered in semiconductor research *through advanced material engineering* has inspired many ingenious experiments, resulting in observations of not only predicted effects but also totally unknown phenomena. The growth of papers published on the subject

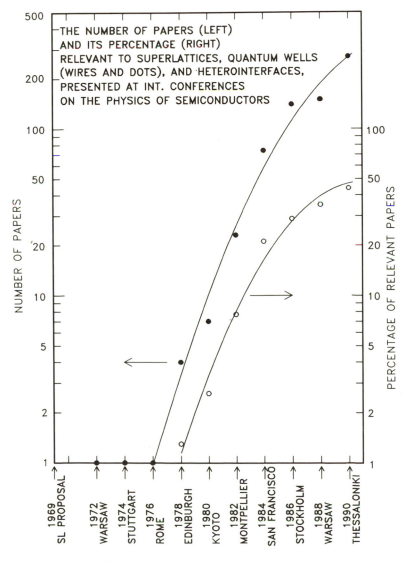

FIG. 16. Growth of papers at ICPS.

for the last two decades is indeed phenomenal. For instance, the number of papers and the percentage of those to the total presented at the International Conference on the Physics of Semiconductors (ICPS) being held every other year, increased as shown in Fig. 16. After the incubation period of several years, it really took off, with a rapid annual growth, around ten years ago. It says that, currently, nearly one half of semiconductor physicists in the world are working in this field. Activity in this new frontier of semiconductor physics has in turn given immeasurable stimulus to device physics, provoking new ideas for applications. Thus, unprecedented transport and opto-electronic devices have emerged. In this interdisciplinary research, beneficial cross-fertilizations are prevalent.

In this review, we have primarily focussed on intriguing phenomena which occur with the application of an electric field in the direction of the SL axis. The significance of such electric field-induced effects was recognized at the inception. The observation of Stark ladders probably consitutes one of the milestones in quantum structure research. It should be appreciated that an energy separation of 10 meV between adjacent Stark ladders was achieved only with 20 kV/cm, while bulk semiconductors would require far greater electric fields for such electric quantization. It is interesting to compare this with magnetic quantization of the orbital motion of electrons, in which an energy separation, $\hbar\omega_c$, of 10 meV between Landau levels for electrons with a small effective mass, is readily obtainable with a magnetic flux of 10T or so. The electric counterparts of the Landau level and cyclotron resonance are considered to be the Stark ladder and Bloch oscillation, respectively. "Whether the Bloch oscillation in superlattices is really observable or not" is, undoubtedly, one of the most fascinating unanswered questions.

REFERENCES

1. L. Esaki and R. Tsu, "Superlattice and negative conductivity in semiconductors," *IBM Research Note* RC-2418 (1969).

2. L. Esaki and R.Tsu, "Superlattice and negative differential conductivity in semiconductors," *IBM J. Res. Develop.* **14**: 61 (1970).

3. D. Bohm, "Quantum Theory:" (Prentice Hall, Englewood Cliffs, N.J. 1951), p.283.

4. L. Esaki, "A bird's-eye view on the evolution of semiconductor superlattices and quantum wells," *IEEE J. Quantum Electron.,* QE-**22**: 1611 (1986).

5. A. C. Gossard, "Growth of microstructures by molecular beam epitaxy," *IEEE J. Quantum Electron.*, QE-**22**: 1649 (1986).

6. M. Razeghi, "The MOCVD challenge," (Adam Hilger, Bristol and Philadelphia, 1989).

7. R. Tsu and L. Esaki, "Tunneling in a finite superlattice," *Appl. Phys. Lett.* **22**: 562 (1973).

8. L.L. Chang, L. Esaki, and R. Tsu, "Resonant tunneling in semiconductor double barriers," *Appl. Phys. Lett.* **24**: 593 (1974).

9. F. Capasso, K. Mohammed and A. Y. Cho, "Resonant tunneling through double barriers," *IEEE J. Quantum Electron.*, QE-**22**: 1853 (1986).

10. E.E. Mendez, W.I. Wang, B. Ricco, and L. Esaki, "Resonant tunneling of holes in AlAS-GaAs-AlAS heterostructures," *Appl. Phys. Lett.* **47**: 415 (1985).

11. V.J. Goldman, D.C. Tsui and J.E. Cunningham, "Evidence for LO-phonon-emission-assisted tunneling in double-barrier heterostructures," *Phys. Rev.* **B36**: 7635 (1987).

12. W. Cai, T.F. Zheng, P. Hu, B. Yudanin and M. Lax, "Model of phonon-associated electron tunneling through a semiconductor double barrier," *Phys. Rev. Lett.* **63**: 418 (1989).

13. E.E. Mendez, L. Esaki and W.I. Wang, "Resonant magneto-tunneling in GaAlAs-GaAs-GaAlAs heterostructures," *Phys. Rev.* **B33**: 2893 (1986).

14. M.L. Leadbeater, E.S. Alves, L. Eaves, M. Henini, O.H. Hughes, A. Celeste, J.C. Portal, G. Hill and M.A. Pate, "Magnetic field studies of elastic scattering and optic-phonon emission in resonant-tunneling devies," *Phys. Rev.* **B39**: 3438 (1989).

15. G.S. Boelinger, A.F. Levi, S. Schmitt-Rink, A. Passner, L.N. Pfeiffer and K.W. West, "Direct observation of two-dimensional magnetopolarons in a resonant tunnel junction" *Phys. Rev. Lett.* **65**: 235 (1990).

16. J.F. Young, B.M. Wood, G.C. Aers, R.L.. Devine, H.C. Liu, D. Landheer and M. Buchanan, "Determination of charge accumulation and its characteristic time in double-barrier resonant tunneling structures using steady-state photoluminescence," *Phys. Rev. Lett.* **60**: 2085 (1988).

17. H. Yoshimura, J.N. Schulman and H. Sakaki, "Charge accululation in a double-barrier resonant-tunneling structure studied by photoluminescence and photoluminescence-excitation spectroscopy," *Phys. Rev. Lett.* **64**: 2422 (1990).

18. A. Tackeuchi, T. Inata, S. Muto and E. Miyauchi, "Picosecond characterization of InGaAs/InAlAs resonant tunneling barrier diode by electro-optic sampling," *Jpn. J. Appl. Phys.* **28**: L750 (1989).

19. L.F. Luo, R. Beresford and W.I. Wang, "Resonant tunneling in AlSb/InAs/AlSb double-barrier heterostructures," *Appl. Phys. Lett.* **53**: 2320 (1988).

20. L. Esaki, L.L. Chang and E.E. Mendez, "Polytype superlattices and multi-heterojunctions," *Jpn. J. Appl. Phys.* **20**: L529 (1981).

21. H. Takaoka, Chin-An Chang, E.E. Mendez, L.L. Chang and L. Esaki, "GaSb-AlSb-InAs multi-heterojunctions," *Physica* **117B &118B**: 741 (1983).

22. L.F. Luo, R. Beresford and W.I. Wang, "Interband tunneling in polytype GaSb/AlSb/InAs heterostructures", *App. Phys. Lett* **55**: 2023 (1989); R. Beresford, L.F. Luo, K.F. Longenbach and W.I. Wang, "Resonant interband tunneling through a 110 nm InAs quantum well," *Appl. Phys. Lett.* **56**: 551 (1990); "Interband tunneling in single-barrier InAs/AlSb/GaSb heterostructures," *Appl. Phys. Lett.* **56**: 952 (1990); E.E. Mendez, H. Ohno, L. Esaki and W.I. Wang, "Resonant interband tunneling via Landau levels in polytype heterostructures," *Phys. Rev.* **B43**: 5196 (1991).

23. J.R. Soderstrom, D.H. Chow and T.C. McGill, "New negative differential resistance device based on resonant interband tunneling" *App. Phys. Lett.* **55**: 1094 (1989).

24. J.F. Whitaker, G.A. Mourou, T.C.L.G. Sollner and W.D. Goodhue, "Picosecond switching time measurement of a resonant tunnel diode," *Appl. Phys. Lett.* **53**: 385 (1988).

25. E.R. Brown, T.C.L.G. Sollner, C.D. Parker, W.D. Goodhue and C.L. Chen, "Oscillations up to 420 GHz in GaAs/AlAs resonant tunneling diodes," *Appl. Phys. Lett.* **55**: 1777 (1989).

26. R. Tsu and L. Esaki, "Stark quantization in superlattices," Phys. Rev. **B43**: 5204 (1991).

27. G.H. Wannier, "Wave functions and effective Hamiltonian for Bloch electrons in an electric field," *Phys. Rev.* **117**:432 (1960).

28. H.M. James, "Electronic states in perturbed periodic systems," *Phys. Rev.* **76**:1611 (1949).

29. A. Rabinovitch and J. Zac, "Electrons in crystals in a finite-range electric field," *Phys. Rev.* **B4**:2358 (1971).

30. W. Shockley, "Stark ladders for finite, one-dimensional models of crystals," *Phys. Rev. Lett.* **28**:349 (1972).

31. A.G. Chynoweth, G.H. Wannier, R.A. Logan and D.E. Thomas, "Observation of Stark splitting of energy bands by means of tunneling transitions," *Phys. Rev. Lett.* **5**:57 (1960).

32. S. Maekawa, "Nonlinear conduction of ZnS in strong electric fields," *Phys. Rev. Lett.* **24**:1175 (1970).

33. R.W. Koss, "Experimental observation of Wannier levels in semi-insulating GaAs," *Phys. Rev.* **B5**:1479 (1972).

34. R. Tsu and G. Döhler, "Hopping conduction in a superlattice," *Phys. Rev.* **B12**: 680 (1975).

35. M. Artaki and K. Hess, "Monte Carlo calculations of electron transport in GaAs/AlGaAs superlattices," *Superlatt. Microstruct.* **1**: 489 (1985).

36. L. Esaki, L.L. Chang, W.E. Howard, and V.L. Rideout, "Transport properties of a GaAs-GaAlAs superlattice," *Proc. 11th Int. Conf. on the Physics of Semiconductors*, Warsaw, Poland, 1972, (PWN-Polish Scientific Publishers, Warsaw, Poland), p. 431.

37. L. Esaki and L. L. Chang, "New transport phenomenon in a semiconductor superlattice," *Phys. Rev. Lett.* **33**: 495 (1974).

38. R. Tsu, L.L. Chang, G.A. Sai-Halasz, and L. Esaki, "Effects of quantum states on the photocurrent in a superlattice," *Phys. Rev. Lett.* **34**: 1509 (1975).

39. B. Deveaud, J. Shah, T. C. Damen, B. Lambert and A. Regreny, "Bloch transport of electrons and holes in superlattice minibands: direct measurement by subpicosecond luminescence spectroscopy," *Phys. Rev. Lett.* **58**: 2582 (1987).

40. H. Schneider, W.W. Rühle, K. v.Klitzing and K. Ploog, "Electrical and optical time-of-flight experiments in GaAs/AlAs superlattices," *Appl. Phys. Lett.* **54**: 2656 (1989).

41. M. Helm, P. England, E. Colas, F. DeRosa, and S.J. Allen, Jr., "Intersubband emission from semiconductor superlattices excited by sequential resonant tunneling," *Phys. Rev. Lett.* **63**: 74 (1989).

42. M.C. Tatham and J.F. Ryan, "Time-resolved raman measurements of intersubband relaxation in GaAs quantum wells," *Phys. Rev. Lett.* **63**: 1637 (1989).

43. A. Sibille, J.F. Palmier, H. Wang and F. Mollot, "Observation of Esaki-Tsu negative differential velocity in GaAs/AlAs superlattices, " *Phys. Rev. Lett.* **64**: 52 (1990); A. Sibille, J. F. Palmier, H. Wang, J. C. Esnaul and F. Mollot, "dc and microwave negative differential differential conductance in GaAs/AlAs superlattices," *Appl. Phys. Lett.* **56**: 256 (1990); A. Sibille, J. F. Palmier, F. Mott, H. Wang and J. C. Esnault, "Negative differential conductance in GaAs/AlAs superlattices," *Phys. Rev.* **B39**: 6272 (1989).

44. M. Hadjazi, A. Sibille, J. F. Palmier and F. Mollot, "Negative differential conductance in GaAs/AlAs superlattices," *Electronics Lett.* **27**: 1101 (1991).

45. H.T. Grahn, H. Schneider, W.W. Rühle, K. von Klitzing and K. Ploog, "Nonthermal occupation of higher subbands in semiconductor superlattices via sequential resonant tunneling," *Phys. Rev. Lett.* **64**: 2426 (1990).

46. G. Brozak, M. Helm, F. DeRosa, C.H. Perry, M. Koza, R. Bhat and S.J. Allen, Jr., "Thermal saturation of band transport in a superlattice," *Phys. Rev. Lett.* **64**: 3163 (1990).

47. F. Beltram, F. Capasso, D.L. Sivco, A.L. Hutchinson, S.N.G. Chu, and A. Cho, "Scattering-controlled transmission resonances and negative differential conductance by field-induced localization in superlattices," *Phys. Rev. Lett.* **64**: 3167 (1990).

48. A.S. Vengurlekar, F. Capasso, A.L. Hutchinson and W.T. Tsang, "Miniband conduction of minority electrons and negative transconductance by quantum reflection in a superlattice transistor," *Appl. Phys. Lett.* **56**: 262 (1990).

49. X.L. Lei, N.J.M. Horing and H.L. Cui, "Theory of negative differential conductivity in a superlattice miniband," *Phys. Rev. Lett.* **66**: 3277 (1991).

50. A. Zaslavsky, D.C. Tsui, M. Santos and M. Shayegan, "Observation of high-field negative differential resistance in an edge-regrown superlattice," to be published in *Appl. Phys. Lett.*

51. E.E. Mendez, G Bastard, L.L. Chang, and L. Esaki, "Effect of an electric field on the luminescence of GaAs quantum wells," *Phys. Rev.* **B26**: 7101 (1982).

52. L. Vina, R. T. Collins, E. E. Mendez and W. I. Wang, "Excitonic coupling in GaAs/GaAlAs quantum wells in an electric field," *Phys. Rev. Lett.* **58**: 832 (1987).

53. D.S. Chemla, T.C. Damen, D.A.B. Miller, A.C. Gossard, and W. Wiegmann, "Electroabsorption by Stark effect on room-temperature excitons in GaAs/GaAlAs multiple quantum well structures," *Appl. Phys. Lett.* **42**: 864 (1983).

54. D.A.B. Miller, J.S. Weiner, and D.S. Chemla, "Electric-field dependence of linear optical properties in quantum well structures," *IEEE J. Quantum Electron.*, QE-**22**: 1816 (1987).

55. J. Bleuse, G. Bastard and P. Voison, "Electric-field-induced localization and oscillatory electro-optical properties of semiconductor superlattices," *Phys. Rev. Lett.* **60**: 220 (1988).

56. E.E. Mendez, F. Agullo-Rueda and J.M. Hong, "Stark localization in GaAs-GaAlAs superlattices under an electric field,," *Phys. Rev. Lett.* **60**: 2426 (1988).

57. F. Agullo-Rueda, E.E. Mendez and J.M. Hong, "Quantum coherence in semiconductor superlattices," *Phys. Rev.* **B40**: 1357 (1989).

58. E.E. Mendez, F. Agullo-Rueda and J.M. Hong, "Temperature dependence of the electronic coherence of GaAs-GaAlAs superlattices," *Appl. Phys. Lett.* **56**: 2545 (1990).

59. P. Voisin and J. Bleuse, "Observation of the Wannier-Stark quantization in a semiconductor superlattice," *Phys. Rev. Lett.* **61**: 1639 (1988).

60. M.M. Dignam and J.E. Sipe, "Exciton stark ladder in GaAs/Ga$_{1-x}$Al$_x$As superlattices" *Phys. Rev. Lett.* **64**: 1797 (1990).

61. H. Ohno, E.E. Mendez, J.A. Brum, J.M. Hong, F. Agullo-Rueda, L.L. Chang and L. Esaki, "Observation of 'Tamm States' in superlattices," *Phys. Rev. Lett.* **64**: 2555 (1990).

MATERIALS AND PHYSICS ASPECTS OF QUANTUM HETEROSTRUCTURES

L. L. Chang

IBM Thomas J. Watson Research Center
Yorktown Heights, New York 10598, USA

Abstract - Quantum heterostructures have evolved over the last two decades to become a prominant, multi-disciplinary field in condensed matter science. We provide, in this work, an overview of both the materials and physics aspects. Structure processing in terms of deposition and characterizatiuon is described. The introduction of a heterostructure potential is shown to result in fundamental modifications of the materials, leading to prescribed electronic properties. The most extensively studied system of GaAs-AlAs is used for this illustration. Other heterostructure systems are discussed to the extent that they exhibit specific features.

1. INTRODUCTION

Throughout the history of semiconductor development, the strong coupling between materials and physics has time and again been demonstrated. Newly conceived ideas usually demand the ultimate in existing techniques in materials and processing. Inventions and innovations in the materials arena, on the other hand, invariably create opportunities for the search of new physical phenomena. This intimate relationship has recently been witnessed in quantum heterostructures, a field that has experienced enormous expansion since it was first proposed[1] and realized[2,3] about twenty years ago. It has evolved to become a predominant subject in condensed matter science, which is pursued earnestly worldwide in both academic, industrial and government laboratories. Throughout this period, we have made periodic reviews of the subject matter, focusing on both materials and physics aspects.[4-8] These articles, summarizing research activities at different times, also serve as a chronological record of the development of heterostructures themselves.

We show in Table I the multi-disciplinary nature of the field with a coverage beyond simple materials and physics. In addition to various growth methods for the fabrication of the structures, of which molecular beam epitaxy (MBE) and metalorganic chemical vapor deposition (MOCVD) are the most commonly used, a large number of characterizing techniques have been employed for their evaluation. The technique of x-ray diffraction remains the most dominant because of its capability of providing quantitative information of the structure. But both electron microscopy and diffraction are also popular; the former is particularly suitable for probing the interfaces on an atomic scale, and the latter is compatible with MBE growth conditions and is thus used as a powerful in situ evaluation tool. In the physics area, to which our efforts have

TABLE I. Multi-disciplinary studies in quantum heterostructures.

GROWTH	CHARACTERIZATION	PHYSICS	DEVICES
MBE	X-Ray Diffraction	Energy Quantization	Heterojunction FET
MOCVD	Electron Microscopy	DOS-Dimensionality	Heterojunction Bipolar
Liquid Phase	Electron Diffraction	Magnetic Quantization	Resonant-Tunneling Devices
Sputtering	Auger Spectroscopy	Lattice Vibrations	Hot-Electron Devices
Solid Phase	Secondary Ion Spectroscopy	Collective Excitations	Quantum-Well Lasers
Ion Beam	Nuclear Backscattering	Localization	Infrared Detectors
Lateral Epitaxy	EXAFS	Electron Dynamics	Optical Computation

been largely devoted, the presence of the heterostructure potential by ultrathin layers of dissimilar semiconductors creates quantum states with associated low-dimensional electron systems. Both the energy schemes and the density of states are drastically altered, giving rise to transport, optical and magnetic properties not present in bulk materials. Finally, device applications form an integral part of the studies in quantum heterostructures, including both new devices such as resonant-tunneling diodes and innovative structures of existing field-effect and bipolar transistors. Indeed, electronic devices making use of the new phenomena in quantum heterostructures provided one of the motivations for the development of the field from the very beginning, and they will no doubt receive more attention and efforts as the field becomes increasingly matured.

In terms of material systems, we plot in Fig. 1 energy gaps versus lattice constants for all the common semiconductors, including IV elements, III-V and II(IV)-VI compounds. Lines joining the materials indicate that, to our knowledge, the heterostructure systems have been investigated and quantum effects have been observed. The III-V's occupy the central part of the diagram; they remain the focus of heterostructure studies. However, great attention has recently been directed toward the other systems: the IV's because of the SiGe alloy and

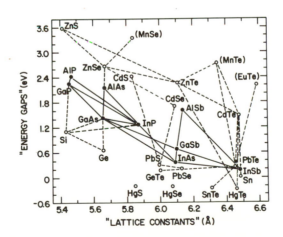

FIG. 1. Energy gaps vs lattice constants for semiconductors of IV elements, III-V and II(IV)-VI compounds and magnetic materials in parentheses. Lines connecting the semiconductors, solid for the III-V's and dotted for the others, indicate quantum heterostructures that have been investigated.

its desirable properties of carrier injection and transport in conjunction with Si-based devices; the II-VI's as a result of their relatively large energy gaps in the visible region suitable for light-emitting and display applications; and the IV-VI's whose long wavelengths have traditionally drawn interest in the infrared regime. Also included in Fig. 1 are magnetic semiconductors such as MnTe. Their alloys with ordinary semiconductors have opened up a new direction in quantum heterostructures, involving spin exchange and interaction in diluted magnetic semiconductors.

In this work, we will start with a description of the heterostructure deposition by MBE and its characterization by x-ray diffraction. This will be followed by a discussion of the key features of a quantum heterostructure, the underlying physics that makes it exhibit desirable and controllable properties as a new class of man-made electronic materials. The most extensively investigated GaAs-AlAs system will be used for the illustrative purpose. Other heterostructure systems will then be selectively described to the extent that they show new and unique properties.

2. STRUCTURE FORMATION

With the advance in semiconductor thin-film technology in recent years, a number of processes have been used for the deposition of quantum heterostructures, as can be seen in Table I. The process of MBE, basically an evaporation technique from molecular or atomic beams in an ultrahigh vacuum environment, remains the dominant. A schematic diagram of the system is shown in Fig. 2. The beams are usually generated thermally in effusion cells, travel in rectilinear paths in the vacuum space with guidance from orifices and shutters, and eventually condense on the substrate under kinetically favorable conditions. That MBE proceeds in an ultrahigh vacuum makes it compatible with modern analytic apparatuses, some of which are shown in Fig. 1: mass spectrometer, Auger analyzer and reflection high-energy electron diffractometer (RHEED). They can be used to monitor the beam and the background, detect surface environment and film composition and provide valuable information about crystallinity, surface smoothness and reconstruction. Additional features may be included in a commercial system today, such as individually cooled shrouds for the ovens to avoid cross heating and contamination, rotatable substrate holder during deposition to attain film uniformity, and interlock multi-chamber operation for clean and efficient processing.

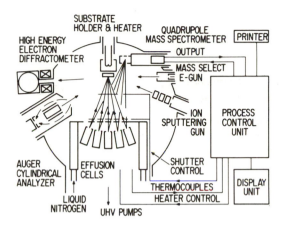

FIG. 2. A schematic diagram of the molecular beam epitaxy system with accessory apparatuses for the growth of quantum heterostructures.

Being an evaporation technique in nature, MBE is a deposition process, that deviates significantly from equilibrium. Evolved through a series of early developments, the foundation was laid by the kinetic studies of Ga and As on the surface of GaAs.[9,10] The results are summarized in Fig. 3 in terms of the sticking coefficients of the arsenic species as a function of the gallium flux. The situation is rather simple for As_2, involving a dissociative chemisorption of As_2 molecules on single Ga atoms. The sticking coefficient is proportional to the Ga flux and reaches unity in the limiting case. It is more complex for As_4, for which the growth process proceeds by both the adsorption and desorption of As_4 via a bimolecular interaction. Pairs of Ga surface atoms are involved, giving rise to a sticking coefficient of 0.5 in the limit. While both the As species are routinely used to deposit stoichiometric GaAs, the difference in detail in the growth mechanisms is expected to influence the film properties in terms of impurity incorporation and defect structures.[11]

One of the most important aspects of MBE deposition is its capability to produce extremely smooth films. Figure 4 shows the comparison between a GaAs substrate and a film of 1000 Å with both electron microscopy and RHEED pattern. The transmission like spotty pattern of the rough substrate undergoes a transition to become narrowly streaked

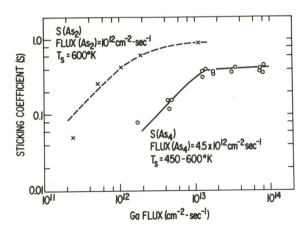

FIG. 3. Sticking coefficients of As on (100) GaAs to illustrate the kinetic behavior. Indicated are the fixed fluxes of As dimer and tetramer species and the substrate temperatures. (Ref. 10).

as the surface is smoothed out.[12] Nucleation is initiated at low-energy sites of surface steps and irregularities, and growth proceeds by a two-dimensional propagation to reduce surface roughness. The micrograph simultaneously exhibits a featureless finish, consistent with the RHEED observation. As can be seen in Fig. 4, additional streaks appear at fractional intervals, representing diffraction from a rearrangement of surface atoms into an orderly array. A variety of such surface structures or reconstructions have been observed under different experimental conditions. Investigations of this kind played a key role in the early development of MBE,[13] and they can be used to correlate with surface stoichiometry and growth conditions.[14] RHEED studies are also of use to assess lattice mismatch during heteroepitaxy,[15] and to provide a precise means for measuring the growth rate.[16] The latter was made possible by the observation that the diffraction intensity oscillates with time during deposition with the period corresponding to a monolayer growth. The RHEED oscillations have attracted great interest for studying surface diffusions and growth mechanisms.[17]

It is now clear that the popularity of MBE for growing quantum heterostructures is not accidental, for it offers characteristic features uniquely satisfying the stringent requirements that such a structure demands. The situation is summarized in Table II where the one-to-one matching is obvious: high purity films with extreme surface smoothness

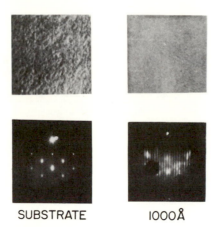

SUBSTRATE   1000Å

FIG. 4. Comparison of (100) GaAs surfaces between the substrate and a deposition of 1000 Å. Upper and lower pictures are transmission-electron micrographs with a magnification of 3 x 10⁴, and high-energy electron-diffraction patterns at [1$\bar{1}$0] azimuth, respectively.

are produced under precisely controlled conditions. Added to these are the typically low temperatures and rates of deposition, for example, 550° C and 1 μm / hr for GaAs. The slow growth rate, while undesirable for purposes such as thick coatings, is favorable for control of ultrathin layers. The relatively low growth temperature is necessary to minimize diffusion effects to achieve abrupt interfaces. The interdiffusion coefficient between GaAs and AlAs was found to be negligible under normal growth conditions.[18] We should also add to Table II the fact that the MBE process is quite flexible for use with a large number of materials.

While structural characterization of quantum heterostructures can be done in the MBE system, for example, compositional profiling by use of ion-sputtering with an Auger analyzer,[19] most evaluations are performed outside, after the sample is withdrawn from the chamber. The most commonly used is x-ray diffraction, including the large-angle Bragg reflection and the small-angle interference. They were employed at the early stage of superlattice development. The former is convenient to demonstrate the superlattice formation and measure its periodicity from the diffracted satellite peaks; the latter provides quantitative information about the layer thicknesses and interfaces. An example of the interference pattern of a GaAs-AlAs superlattice with only a limited number of six periods[20] is shown in Fig. 5. Both principal and secondary peaks are

TABLE II. Match between quantum heterostructures and molecular beam epitaxy.

MOLECULAR BEAM EPITAXY CHARACTERISTICS	HETEROSTRUCTURE REQUIREMENTS
Ultrahigh Vacuum Environment	High Purity Films
Slow Growth Rate	Ultrathin Layers
Planar Growth Process	Smooth Interfaces
Low Temperature Deposition	Abrupt Interfaces
Composition & Thickness Control	Stringent Structural Demand

observable, and they are in good agreement with theoretical calculations, not only in angular positions but also in intensities and linewidths. Of more importance is the fact that a change of one monolayer spacing of 2.8 Å is seen to result in a discernable variation. It was from these observations that the heterostructure layers were concluded to have a

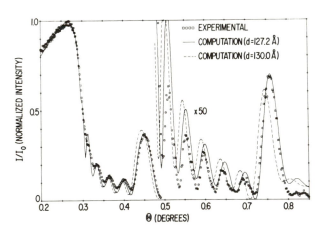

FIG. 5. X-ray interference pattern of a six-period GaAs-AlAs superlattice with a nominal layer thickness of 75 Å-50 Å. Calculated results are shown to demonstrate the agreement as well as the discrepancy with a one-monolayer variation.

degree of smoothness, abruptness and periodic coherency on the scale of atomic dimensions.

The other powerful technique for structural characterization is electron microscopy, including both scanning (SEM) and transmission (TEM) modes of operation. The former is easy to operate and has been used routinely for superlattices to reveal the periodicity for visual observation. The latter is more tedious for cross-sectional view but offers high resolution and defect information at the interfaces.[21] Under proper conditions, lattice-image TEM gives sufficient phase contrast to show atomic-level arrangement.[22] Figure 5 illustrates such an image with the electron beam incident on [110] of a GaAs-AlAs structure with 17-20 monolayers, respectively.[23] A close examination reveals the presence of a monolayer step at the A-A' interface.

The structural quality, while it is usually evaluated by metallurgical measurements, can also be assessed from the electronic properties of the resulting heterostructures. One method is to examine the linewidth of the optical transitions of the subbands, which is strongly influenced by the fluctuations of the confinement potential. It was shown earlier,[24] for example, that the photoluminescence linewidths were correlated with the interface roughness. This was used extensively, more recently, in connection with what is known as growth interruption.[25] The introduction of the interruption between heteroepitaxy greatly reduces the emission linewidths and sometimes splits the peaks, corresponding to monolayer variations of the quantum wells. Surface migration is believed to be responsible for this process: irregular monolayer islands are smoothed

FIG. 6. Lattice image at [110] incidence of a GaAs-AlAs superlattice with a layer thickness corresponding to 17-21 monolayers. A monolayer step, $a_0 = 2.83$ Å, is present at the A-A' interface. (Ref. 23).

out to a spatial extent greater than the exciton diameter. We should mention, in this connection, that it is usually desirable to enhance migration of the depositing species. This can be achieved by introducing the group III and V beams alternately during growth, a modified process of MBE known as migration enhanced epitaxy.[26]

As mentioned earlier, MBE is the most widely used but by no means the only process for the fabrication of quantum heterostructures. Great progress has recently been made for MOCVD,[27,28] a chemical-vapor deposition process using organometallic compounds, so that it becomes quite competitive to MBE. Satisfying adequately the requirements of heterostructures and, expectedly, being more adaptable for production, it has been used in a large number of materials. Various hybrid processes of MBE and MOCVD by use of gaseous sources in a vacuum chamber have also been developed;[29-32] they have been given different names, according to the details of the types of sources employed, such as gas-source MBE (GSMBE), metalorganic MBE (MOMBE) and chemical beam epitaxy (CBE). Other techniques, as listed in Table I, in comparison, have been used relatively rarely and mainly in special situations. Liquid phase epitaxy (LPE), because of its simplicity and low cost, is still being used for structures with somewhat crude definition or requiring a thick overgrowth. The sputtering process, not known for producing high-quality semiconductors, is widely applied to metallic heterostructures.

3. ENGINEERED ELECTRONIC PROPERTIES

The introduction of a heterostructure potential results in fundamental changes of the electronic structure of the semiconductor system. It alters both the electron, phonon and impurity energies, and modifies the excitation and scattering processes, leading to new transport, optical and magnetic properties. Furthermore, since the electronic structure can be prescribed from the configuration of the heterostructure --- the composition and thickness of the layers are precisely controllable on an atomic scale --- it is possible to design and fabricate such heterostructures to provide desired electronic properties. We list in Table III a few of the important phenomena that lead directly to changes in the key parameters of the heterostructure materials. It is in this sense that one speaks of tailored electronic properties or engineered electronic materials.

The most fundamental feature of a quantum heterostructure lies, of course, in the formation of quantum states or subbands from electric quantization. The first experiment for its observation was by resonant tunneling, in which resonance occurs when tunneling electrons through two barriers coincide in energy with the quantum states formed in the potential well between the two barriers.[3] A negative differential resistance was observed at the subband energy position, which could be varied by varying the layer thicknesses and compositions in a large number of heterostructure systems. Also extensively studied were issues related to the mechanism of resonant tunneling in terms of coherency, when the electron scattering time became comparable to the lifetime of the resonant state.[33] With recent interest in still lower dimensional electron systems, energy states in laterally confined structures,[34] as well as zero-dimensional dots[35] were evaluated and identified from tunneling experiments.

While transport measurements usually deal with only one type of carriers, optical experiments probe transitions involving states in both the conduction and valence bands. Among the different techniques, absorption gives the most direct information,[36] but luminescence is the most convenient and has been the most commonly used.[37] Figure 7 shows the luminescence excitation spectrum,[38] where the transitions are labeled with subband indices in numerals and the particles involved with e, h and l for electrons, and heavy and light holes. The spectrum exhibits sharp

TABLE III. Major phenomena in quantum heterostructures leading to new electronic properties.

PHENOMENA IN HETEROSTRUCTURES	CHANGES IN MATERIAL PARAMETERS
Electric Quantization	Quantized Subband Energies
Low-Dimensional Electron System	Modified Density of States
Impurity/Exciton Confinement	Enhanced Binding Energies
Modulated Lattice Vibrations	New Phonon Modes
Selective Doping	Reduced Impurity Scattering

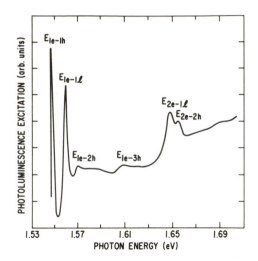

FIG. 7. Photoluminescence excitation spectrum of a GaAs-GaAlAs (Al = 0.3) multiple quantum well with a well thickness of 102 Å. The excitonic peaks associated with different inter-subband transitions are identified. (Ref. 37).

exciton features of ground and excited states transitions in good agreement with calculations including the effect of valence-band mixing. Spectra of this kind have been investigated with a wide range of layer thicknesses and materials to illustrate the central feature of subband control in quantum heterostructures. Other optical spectroscopies have also been used to probe the energy states of quantization. Two notable examples are electronic Raman scattering and reflectance modulation. The former measures the energy spacings of states within the same band by monitoring the energy shift of incident and inelastically scattered radiation,[39] while the latter can be used to access transitions at other symmetry points away from the Brillouin zone center.[40]

The controllability of the dimensionality of the electron system is one of the key features of heterostructures. The system is strictly two-dimensional in isolated quantum wells and deviates toward three-dimensional in superlattices with an increasing coupling between wells. The density of states, similarly, varies from a staircase-like to a parabolic behavior. Indeed, additional confinement can be applied, usually through lateral patterning as will be seen later, to further alter the density of states.

We show in Fig. 8 the Fermi surfaces of three structures. In one extreme, represented by bulk GaAs, the electron system is three-dimensional with a spherical Fermi surface. The other extreme has the cylindrical surface of the two-dimensional system, which is realized by a strong superlattice potential. Since the application of a magnetic field quantizes the energy surface perpendicular to the direction of the field, the angular dependence of the Shubnikov - de Haas oscillations, which arise as a series of Landau levels cross the Fermi energy, reflect directly the dimensionality of the electron system.[41] It varies from a cosine dependence through an intermediate case to an isotropic relation, as the system undergoes from two to three dimensionality. Experiments under a magnetic field have been very commonly used to probe low-dimensional systems, including both magneto-transport and magneto-absorption measurements. The interaction between electrons and phonons give rise to magneto-phonon oscillations.[42] Both interband magneto-absorption and cyclotron resonance have been extensively investigated.[43,44] The most significant observation, obviously, has been that of the quantum Hall effect,[45,46] resulting from electron localization and quasiparticle excitation.

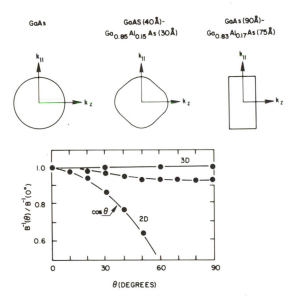

FIG. 8. Angular dependence of the period of Shubnikov-de Haas oscillations for three structures from the three-dimensional bulk, through an intermediate case, to the two-dimensional superlattice. The structure configurations and corresponding Fermi surfaces from calculations are also shown.

Impurities, which are the third item in Table III, play an essential role in determining both the transport and optical properties in a semi-conductor. In a quantum heterostructure, the impurity wavefunction is severely affected by the barrier potential. Consequently, the binding energy depends not only on the width of the quantum well but on the precise location of the impurity. The situation for an infinite barrier was first analyzed, indicating that the binding energy for an impurity at the well center is larger than that at the well edge, and that both increase with decreasing well width as the system effectively becomes two-dimensional.[47] The effect of a finite barrier height is to reduce the binding energy again at very narrow widths when the quantum state approaches the top of the barrier. Figure 9 illustrates this behavior from the results of the carbon acceptor for which systematic data were first obtained.[48,49] The binding energy of the silicon donor was also reported, consistent with the predictions of well-width and impurity-position dependence.[50]

Excitons are similar to impurities in that both involve coulombic interactions. Like impurities, their binding energy increases with decreasing well width. This has been demonstrated from both luminescence[51,52] and magneto-absorption[43] measurements for both heavy- and light-hole excitons. The binding energy of the former is in

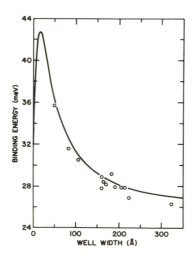

FIG. 9. Binding energies of the carbon impurity at the center of the well in GaAs-GaAlAs (Al = 0.3) multiple quantum wells. Shown are both the data points and the calculated curve. (Ref. 49).

fact lower because of their lighter mass in the plane of the layers. An interesting observation of late was the change of the exciton binding energy by the application of an electric field.[53] The field causes Stark localization, resulting eventually in isolated quantum wells. The electron system becomes strictly two-dimensional with a corresponding enhancement of the binding energy. In this connection, we should mention that the exciton energy can be further increased by additional confinement to one- and zero- dimensional systems. It can also be decreased by use of what is known as a type-II heterostructure in which the electrons and holes are spatially separated, as will be seen later. The ability to control the exciton binding energy by varying the configurations as well as external fields is an important aspect in engineering the electron properties of the heterostructure.

The introduction of the heterostructure potential modifies not only the electron subbands and dimensionality but also the vibrational modes of the lattice. In fact, they share a number of similarities and can be treated in an analogous manner in many aspects. Generally speaking, both propagating and confined modes exist, depending on whether or not their corresponding bulk dispersions in the component materials overlap. Since the early observation of folded acoustic phonons,[54] a large number of new modes have been reported, including both acoustic and optical modes as well as modes associated with the interface.[55,56] Figure 10 shows the folded acoustic phonons obtained from Raman scattering, the

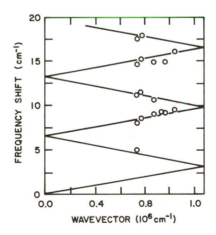

FIG. 10. Folded acoustic phonons from Raman Spectroscopy together with the calculated dispersions for a GaAs-AlAs superlattice with a layer thickness of 26 Å - 14 Å. (Ref. 57).

97

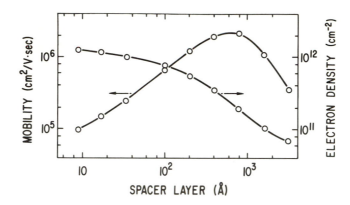

FIG. 11. Variations of electron density and mobility by controlling the spacer layer thickness in a selectively doped GaAs-GaAlAs (Al = 0.32) heterojunctions. (Ref. 63).

most commonly used technique to probe the phonon structures.[57] Theoretical calculations by use of the elastic-medium approximation are shown to compare quite favorably with the data points. For optical phonons, while the bulk modes in GaAs and AlAs do not overlap, the situation is different for the commonly used GaAlAs alloy, because of its two-mode behavior. Recently, to examine the deficiency of the dielectric continuum model, a microscopic model has been developed for confined phonon modes and their mixing with the interface modes.[58]

The last item in Table III deals with a special way of doping the heterostructure to lead to drastically reduced impurity scattering and thus enhanced mobility at low temperatures. Known as selective or modulation doping, impurities are introduced in the barriers while the electrons fall into adjacent wells where they are spatially separated from their parent impurities.[59] Additional increase in the mobility was achieved by leaving an undoped spacer layer in the barrier to further suppress the scattering by impurities.[60] Figure 11 shows the control of both the electron density and mobility by varying this spacer thickness.[61] The behavior in the region of thin spacers are readily understood. The eventual decrease in mobility results mainly from a reduction in electron screening when residual impurities in the well become dominant in the scattering process. This technique of selective doping has been widely used today to produce mobility values well beyond 10^6 cm^2/v.sec.[62] It has been applied to both electrons and holes, and to many other heterostructure systems.   This is an important achievement for both physics and devices, making possible, for example, the observation of the fractional quantum Hall effect on the one hand[46] and the realization of the high-electron-mobility field-effect transistor (HEMT) on the other.[63]

4. SPECIFIC HETEROSTRUCTURE SYSTEMS

It is clear from Fig. 1 that the GaAs-AlAs discussed thus far represents but one of a large number of heterostructure systems. Most of the others shown there are of interest in their own right, which obviously cannot be all described here. Instead, we limit ourselves in this section to a few examples with specially notable features. These are listed in Table IV, where we indicate in each case the specific physical effect involved and the resulting characteristics of the system; we also choose to use a different measurement technique for each illustration. Note that the last example is, again, GaAs-AlAs.   In this situation, the specific feature arises not from new semiconductor materials or alloy compositions as in the other cases, but from a change in configuration with additional spatial quantization to produce an electron system of one and zero dimensionality.

Heterostructures have usually been categorized into three types, based on the relative alignment of energy gaps or band edges of the component materials. The familiar GaAs-AlAs system is known as type I, where the gap of one semiconductor completely spans over that of the other; the same material, GaAs, provides potential wells for both electrons and holes. Staggered or, in the extreme situation, totally separated gaps constitute the type II system, of which the InAs-GaSb shown in Table IV is the most notable example.[64] This type differs from that of

TABLE IV. Specific quantum heterostructures of interest.

HETEROSTRUCTURE MATERIALS	PHYSICAL EFFECTS	SPECIFIC SYSTEM PROPERTIES	MEASUREMENT TECHNIQUES
InAs-GaSb	Band Structure	Electron-Hole System	Transport
GaSb-AlSb	Lattice Match	Strained System	Optical
CdTe-CdMnTe	Spin Exchange	Diluted Magnetic System	Magnetic
GaAs-GaAlAs	Further Confinement	One/Zero Dimensional Systems	Capacitance

GaAs-AlAs in two important aspects. One is the spatial separation of states and their associated carriers, electrons in InAs and holes in GaSb. The other is the creation of a semimetallic heterostructure with an increase in the layer thickness so that electrons and holes may coexist.[65] We should mention in this connection that there is also a type III structure, as exemplified by CdTe-HgTe, for which the band alignment appears similar to that of type II but with reversed band symmetry.[66] The band curvatures for the light particles in the two constituent materials are opposite, leading to unusual results in band mixing and interface states.[67]

The unique features of InAs-GaSb just mentioned have important consequences in both optical and conduction properties. The spatial separation of carriers gives rise to relatively weak absorption and emission, reduced exciton binding energy, and, more interestingly, a large number of optical transitions with different energy indices.[68] In transport, the increase in the InAs layer thickness results in a reversal of the ground states of electrons and holes and, consequently, a charge transfer and a sudden rise of the carrier density.[69] This sets the limit between the semiconductor and the semimetal regime, as shown in Fig. 12, where the

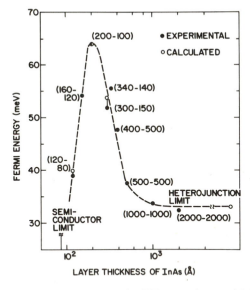

FIG. 12. InAs-GaSb superlattices with different layer thicknesses in angstroms in parentheses to illustrate the Fermi energy variation in the semimetallic regime between the two limits. Also shown are a few calculated points based on the assumption of equal numbers of electrons and holes.

Fermi level from extensive magneto-transport experiments is plotted.[70] The formation of the space-charge regions and multiple subbands eventually leads to a reduction and then a constant value of the Fermi level, as the superlattice approaches the heterojunction limit. Calculations based on the simple transfer process show good agreement with the experimental results, although band mixing[71] and interface states[72] may play a role for a quantitative understanding in detail. More recently, the introduction of Al in GaSb to tune the band edge alignment has shown that the heterostructures can be varied from a system of both carriers to that of electrons alone.[73]

The heterostructure of GaSb-AlSb was originally proposed in combination with InAs to form a polytype system.[74] It exhibits remarkable tunneling characteristics,[75] as demonstrated recently from resonant tunneling in a double-barrier structure.[76] We use here the GaSb-AlSb superlattice to illustrate the effect of strain, an important and inherent effect in heteroepitaxy. The layer grown on a substrate with a different lattice constant is strained, which begins to relax with the generation of dislocations when a critical thickness is reached. Most heterostructures, including GaAs-AlAs, are strained, as perfect lattice match is usually not possible. The explicit emphasis on this effect started with GaP-GaAsP,[77] leading to investigations in many other systems now referred to sometimes as strained-layer heterostructures. Figure 13 shows the transmission spectra.[78] The smaller lattice in GaSb introduces a biaxial dilation in the layer plane. This creates a hydrostatic tensile stress which shrinks the energy gap and, more importantly, a uniaxial compressive stress in the perpendicular direction which splits the valence states. The light holes are lowered in energy with respect to heavy holes, and this effect, in competition with that of quantization, gives rise to a reversal of their ground states under favorable conditions. As clearly shown in the figure, the fundamental gap of the superlattice is now defined by the light-hole state. Recent photoreflectance measurements have revealed a large number of transitions of both heavy and light holes with energies determined by the effect of strain.[79]

While the strain effect may be undesirable in many cases as it tends to introduce defects, the fact that it affects the band structure can also be used to our advantage. One notable example is to manipulate it in the Si-SiGe system to achieve selective doping, making use of the different, strain-induced valley splittings of the sixfold degenerate conduction band.[80] The other is to use this effect in the InAsSb system to lower the

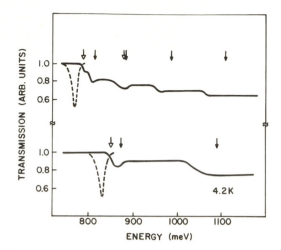

FIG. 13. Transmission spectra of GaSb-AlSb superlattices with layer thick-nesses of 181 Å - 452 Å (upper) and 84 Å - 419 Å (lower). Solid and open arrows indicate calculated transitions involving heavy and light holes, while the dotted curves represent luminescence.

effective energy gap of the superlattice in the long wavelength range for infrared applications.[81] In general, the relative band edges can be varied by proper strain to change the types of the heterostructure system. In addition, the strain also modifies the phonon energies,[82,83] which actually have been used to measure the strain itself. It should be pointed out that, although our interest in strain in the present context is focussed on its influence on heterostructure physics, the study of strain constitutes an important area in its own right. The subject is rather complex beyond what can be described by the simplified relaxation-dislocation model.

The next heterostructure system to be described involves what are not generally considered as ordinary semiconductors. These include alloys such as CdMnTe and ZnMnSe, as can be seen in Fig. 1, which are known as diluted magnetic semiconductors.[84] A variety of heterostructures have been fabricated with this class of materials of which CdTe-CdMnTe was the first achieved[85] and remained the most extensively studied.[86] Like ordinary semiconductor heterostructures, they exhibit energy quantization with associated optical properties involving excitons and phonons.[87,88] The central interest, unique to such materials and heterostructures, however, lies in the strong exchange interactions between the local moments of the magnetic ions and the spins of the conduction electrons, and among the magnetic ions themselves. The latter interaction results in different magnetic phases and their transitions as functions of temperature and the magnetic composition. The first magnetic measurement showed the important effect of dimensionality: the paramagnetic to spin-glass phase transition, which occurs in bulk materials, can not be sustained in a two-dimensional magnetic system.[89]

Of more relevance to the present work, however, is the magnetic spectroscopy, monitoring the electron-ion spin interaction and relaxation through photoexcitation. With polarized radiation, the magnetization shows peaks at energies corresponding to subband transitions in what amounts to a magnetic manifestation of electric quantization.[90] Time-resolved measurements at energies tuned at the transitions then provide dynamic information of the relaxation process. Figure 14 shows the spectra at different subband energies. The relaxation time shortens considerably at higher bands where the carriers penetrate deeper into the CdMnTe barrier regions and experience, on the average, more magnetic ions for accelerated magnetic relaxation. Similar observations were made by use of different heterostructures with varying well widths to change the subband energies. It is interesting to note that the magnetic relaxa-

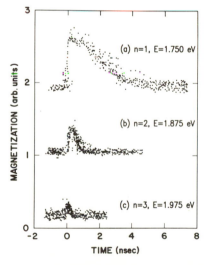

FIG. 14. Time-resolved magnetic spectroscopy of a CdMnTe superlattice with an Mn-composition of 0.065 - 0.38 and a layer thickness of 84 Å - 84 Å. Excitations were tuned at energies corresponding to subband transitions as indicated.

tion time can be much longer than the carrier spin relaxation time and also the carrier lifetime. The carriers, in other words, impart a magnetic imprint on the system which evolves long after the carriers themselves have relaxed and recombined.[86]

The last heterostructure system in Table IV deals with additional electric quantization in the lateral dimensions, instead of new component semiconductors. This was made possible with recent advances in lithography and etching techniques, and with proper use of gated structures on two-dimensional systems. The staircase density of states in two-dimensional systems becomes spiked and delta function-like as the dimensionality is further reduced to one and zero, known respectively as quantum wires and dots. Early evidence for the successful formation of such systems has been obtained from both transport and optical measurements.[91,92] More recent experiments have clearly shown the one- and zero dimensional states from resonant tunneling[34] and magneto-excitons.[93] The capacitance spectroscopy, which provides direct information of the density of states, is shown in Fig. 15 for quantum wires of different widths.[94] The oscillations reflect the passing of the Fermi level across the one-dimensional states as the gate voltage is varied. Similar experiments were extended to zero-dimensional states under both isolated and coupled quantum-dot conditions.

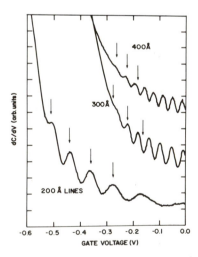

FIG. 15. Capacitance spectroscopy for quantum wires of different line widths, made on two-dimensional GaAs-GaAlAs electron gas. Arrows mark the maxima from theoretical calculations.

The field of small structures, as it is known today, represents now a large and growing field in condensed matter science, where two branches of physics studies merge, quantum heterostructures and mesoscopic systems. Investigations in heterostructures have traditionally emphasized energy quantization, when the size of the structure is comparable to the electron wavelength or the Fermi wavelength. Another relevant scale is the electron mean free path. Within this length, electrons transport ballistically. The most remarkable observation in this case is the quantized conductance through a narrow constriction defined by a pair of gates.[95] With varying gate voltages, the conductance shows plateaux at multiples of $e^2/\pi\hbar$, reflecting successive population of the one-dimensional states. For increasing sample size up to the scale of the inelastic diffusion length, phase coherence can still be maintained, as demonstrated by the Aharonov-Bohm oscillations[96] and universal conductance fluctuations[97] in this mesoscopic regime.

5. CONCLUDING REMARKS

It has been nearly twenty years since the first realization of the quantum heterostructure. This opened up a new field, drawing extensive interest in recent years and leading to a multi-disciplinary effort in research with a coverage of materials, physics and devices. In this work, we reviewed the materials and physics aspects with emphasis on the ability to control the structural configuration in heterostructures to change their electronic structure. In essence, new electronic materials can be designed and fabricated for both scientific and technological interest. In the latter case, for example, a large number of devices have been developed for applications in electronics as well as optoelectronics. In quantum heterostructures, we have realized exactly what Professor John Bardeen said more than thirty years ago (in Basic Research on Semiconductors, Proc. IEEE, 1959): "New materials and effects are not likely to be discovered empirically, but materials will be carefully tailored to produce the desired results."

At present, activities in quantum heterostructures remain strong, expanding and growing at a rapid pace, and the trend will continue in the coming period. Particular efforts will be focussed on new materials and configurations, dynamic transport and optical processes, and electron localization and condensation. Device interest will also accelerate, as the field becomes increasingly matured. But the area of the most thrust is likely to be that of small structures where studies in quantum heterostructures and mesoscopic systems meet, as just mentioned. From the processing viewpoint, innovative techniques of structure fabrication will be developed, including tilted-surface deposition, selective and edge regrowth, and in situ etching and re-evaporation. In terms of physics, studies in this area go beyond dimensionality effects. It offers the genuine opportunity to pursue a condensed system with only few electrons, impurities and spins.

We would like to observe, in passing, on the three topics covered in this conference: quantum heterostructures (QHS), high-temperature superconductivity (HiTc) and scanning tunneling microscopy (STM). Independent as they may appear, there are common threads. The process of electron tunneling, which constitutes the base of STM, was used as the first experiment both to detect the gap in superconductors and to demonstrate the creation of man-made quantum states in double-barrier heterostructures. For both QHS and HiTc, the materials consist of two-dimensional layered planes. In a way, the oxide and block layers in cuprate superconductors are the microscopic analogue to potential wells and barriers in QHS. On the other hand, QHS and STM share the common element of dealing with fine structures on the atomic scale. The combination of these two topics, in terms of STM control and manipulation of local atoms to influence the electronic properties of QHS, represents perhaps the most interesting and challenging endeavor of research in the coming period.

REFERENCES

1. L. Esaki and R. Tsu, "Superlattice and Negative Differential Conductivity in Semiconductors," IBM J. Res. Develop. **14**, 61 (1970).

2. L. L. Chang, L. Esaki, W. E. Howard and R. Ludeke, "The Growth of a GaAs-GaAlAs Superlattice," J. Vac. Sci. Technol. **10**, 11 (1973).

3. L. L. Chang, L. Esaki and R. Tsu, "Resonant Tunneling in Semiconductor Double Barriers," Appl. Phys. Lett. **24**, 593 (1974).

4. L. Esaki and L. L. Chang, "Semiconductor Superfine Structures by Computer Controlled Molecular Beam Epitaxy," Thin Solid Films **36**, 285 (1976).

5. L. L. Chang and L. Esaki, "Semiconductor Superlattices by MBE and Their Characterization," Prog. Crystal Growth and Characterization, **2**, 3 (1979).

6. L. L. Chang, "Recent Advances in Semiconductor Superlattices," J. Vac. Sci. Technol. **B1**, 152 (1983).

7. L. L. Chang and E. E. Mendez, "Compositionally Modulated Superlattices," in Synthetic Modulated Structures, ed. by L. L. Chang and B. C. Giessen (Academic Press, New York, 1985), p. 113.

8. L. L. Chang, "A Perspective of the Development of Semiconductor Superlattices and Quantum Wells," in Lattice Dynamics and Semiconductor Physics, ed. by J. B. Xia et al, (World Scientific, Singapore, 1990), p. 335.

9. J. R. Arthur, "Surface Stoichiometry and Structure of GaAs," Surf. Sci. **43**, 449 (1974).

10. C. T. Foxon and B. A. Joyce, "Interaction Kinetics of As_4 and Ga on (100) GaAs Surfaces Using a Modulated Molecular Beam Technique," Surf. Sci. **50**, 434 (1975).

11. B. A. Joyce, "Kinetic and Surface Aspects of MBE," in Molecular Beam Epitaxy and Heterostructures, ed. by L. L. Chang and K. Ploog (Martinus Nijhoff, Dordrecht, 1985), p. 37.

12. D. B. Dove, R. Ludeke and L. L. Chang, "Interpretation of Scanning High-Energy Electron Diffraction Measurements with Application to GaAs Surfaces," J. Appl. Phys. **44**, 1897 (1973).

13. A. Y. Cho, "Morphology of Epitaxial Growth of GaAs by a Molecular Beam Method:The Observation of Surface Structures," J. Appl. Phys. **41**, 2780 (1970).

14. S. M. Newstead, R. A. A. Kubiak and E. H. C. Parker, "On the Practical Applications of MBE Surface Phase Diagrams," J. Crystal Growth **81**, 49 (1987).

15. C. A. Chang, R. Ludeke, L. L. Chang and L. Esaki, "Molecular Beam Epitaxy of InGaAs and GaSbAs," Appl. Phys. Lett. **31**, 759 (1977).

16. J. J. Harris, B. A. Joyce and P. J. Dobson, "Oscillations in the Surface Structure of Sn-Doped GaAs during Growth by MBE," Surf. Sci. **103**, L90 (1981).

17. P. J. Dobson, B. A. Joyce, J. H. Neave and J. Zhang, "Current Understanding and Applications of the RHEED Intensity Oscillation Technique," J. Crys. Growth, **81**, 1 (1987).

18. L. L. Chang and A. Koma, "Interdiffusion Between GaAs and AlAs," Appl. Phys. Lett. **29**, 138 (1976).

19. R. Ludeke, L. Esaki and L. L. Chang, "GaAlAs Superlattices Profiled by Auger Electron Spectroscopy," Appl. Phys. Lett. **24**, 417 (1974).

20. L. L. Chang, A. Segmuller and L. Esaki, "Smooth and Coherent Layers of GaAs and AlAs Grown by Molecular Beam Epitaxy," Appl. Phys. Lett. **28**, 39 (1976).

21. P. M. Petroff, "Transmission Electron Microscopy of Interfaces in III-V Compound Semiconductors," J. Vac. Sci. Technol. **14**, 973 (1977).

22. H. Okamoto, M. Seki and Y. Horikoshi, "Direct Observation of Lattice Arrangement in MBE Grown GaAs-AlGaAs Superlattices," Jpn. J. Appl. Phys. **22**, L367 (1983).

23. M. Tanaka, H. Ichinose, T. Fututa, Y. Ishida and H. Sakaki, "Direct Observation of Atomic Step Structure at GaAs-AlAs Heterointerfaces in Transmission Electron Microscopy and Improved Lattice Image to Detect Interface by Material-Dependent Patterns," J. Physique **C5**, 101 (1987).

24. C. Weisbuch, R. Dingle, A. C. Gossard and W. Wiegmann "Optical Characterization of Interface Disorder in GaAs-GaAlAs Multi-Quantum Well Structures," Solid State Commun. **38**, 709 (1981).

25. H. Sakaki, M. Tanaka and J. Yoshino, "One Atomic Layer Heterointerface Fluctuations in GaAs-AlAs Quantum Well Structures and Their Suppression by Insertion of Smoothing Period in Molecular Beam Epitaxy," Jpn. J. Appl. Phys. **24**, L417 (1985).

26. Y. Horikoshi, M. Kawashima and H. Yamaguchi, "Low-Temperature Growth of GaAs and AlAs-GaAs Quantum-Well

Layers by Modulated Molecular Beam Epitaxy," Jpn. J. Appl. Phys. **25**, L868 (1986).

27. H. M. Manasevit, "Single-Crystal Gallium Arsenide on Insulating Substrates," Appl. Phys. Lett. **12**, 156 (1968).

28. J. P. Duchemin, S. Hersee, M. Razeghi and M. A. Poisson, "Metal Organic Chemical Vapor Deposition" in <u>Molecular Beam Epitaxy and Heterostructures</u>, ed. by L. L. Chang and K. Ploog, (Martinus Nihoff, Dordrecht, 1985) p. 677.

29. M. B. Panish, "Molecular Beam Epitaxy of GaAs and InP with Gas Source for As and P," J. Electrochem. Soc. **127**, 2729 (1980).

30. E. Venhoff, W. Pletschen, P. Balk and H. Luth, "Metalorganic CVD of GaAs in a Molecular Beam System," J. Crystal Growth **55**, 30 (1981).

31. E. Tokumitsu, Y. Kudou, M. Konagai and K. Takahashi, "Molecular Beam Epitaxial Growth of GaAs Using Trimethylgallium as a Ga Source," J. Appl. Phys. **55**, 3163 (1984).

32. W. T. Tsang, "Chemical Beam Epitaxy of InP and GaAs," Appl. Phys. Lett. **45**, 1234 (1984).

33. J. Luryi, "Frequency Limit of Double-Barrier Resonant-Tunneling Oscillators," Appl. Phys. Lett. **47**, 490 (1985).

34. S. Y. Chou, D. R. Allee, J. S. Harris and R. F. W. Pease, "Observation of Electron Resonant Tunneling in a Lateral Dual-Gate Resonant Tunneling Field-Effect Transistor," Appl. Phys. Lett. **55**, 176 (1989).

35. M. A. Reed, J. N. Randall, R. J. Aggarwal, R. J. Matyi, T. M. Moore and A. E. Wetsel, "Observation of Discrete Electronic States in a Zero-Dimensional Semiconductor Nanostructure" Phys. Rev. Lett. **60**, 535 (1988).

36. R. Dingle, W. Wiegmann and C. H. Henry, "Quantum States of Confined Carriers in Very Thin AlGaAs-GaAs-AlGaAs Heterostructures," Phys. Rev. Lett. **33**, 827 (1974).

37. C. Weisbuch, R. C. Miller, R. Dingle, A. C. Gossard and W. Wiegmann, "Intrinsic Radiative Recombination from Quantum

States in GaAs-AlGaAs Multiquantum Well Structures," Solid State Commun. **37**, 219 (1981).

38. R. C. Miller, A. C. Gossard, G. D. Sanders, Y. C. Chang and J. N. Schulman, "New Evidence of Extensive Valence-Band Mixing in GaAs Quantum Wells through Excitation Photoluminescence Studies," Phys. Rev. **B32**, 8452 (1985).

39. G. Abstreiter and K. Ploog, "Inelastic Light Scattering from a Quasi-Two-Dimensional Electron System in GaAs-AlGaAs Heterojunctions," Phys. Rev. Lett. **42**, 1308 (1979).

40. E. E. Mendez, L. L. Chang, G. Landgren, R. Ludeke, L. Esaki and F. H. Pollak, "Observation of Superlattice Effects on the Electronic Bands of Multilayer Heterostructures," Phys. Rev. Lett. **46**, 1230 (1981).

41. L. L. Chang, H. Sakaki, C. A. Chang and L. Esaki, "Shubnikov-de Haas Oscillations in a Semiconductor Superlattice," Phys. Rev. Lett. **38**, 1489 (1977).

42. D. C. Tsui, Th. Englert, A. Y. Cho and A. C. Gossard, "Observation of Magnetophonon Resonances in a Two-Dimensional Electronic System," Phys. Rev. Lett. **44**, 341 (1980).

43. J. C. Maan, G. Belle, A. Fasolino, M. Altarelli and K. Ploog, "Magneto-Optical Determination of Exciton Binding Energy in GaAs-Ga$_{1-x}$Al$_x$As Quantum Wells," Phys. Rev. **B30**, 2253 (1984).

44. H. L. Stormer, R. Dingle, A. C. Gossard, W. Wiegmann and M. D. Sturge, "Two-Dimensional Electron Gas at a Semiconductor-Semiconductor Interface," Solid State Commun. **29**, 705 (1979).

45. K. von Klitzing, G. Dorda and M. Pepper, "New Method for High-Accuracy Determination of the Fine-Structure Constant Based on Quantized Hall Resistance," Phys. Rev. Lett. **45**, 494 (1980).

46. D. C. Tsui, H. L. Stormer and A. C. Gossard, "Two-Dimensional Magnetotransport in the Extreme Quantum Limit," Phys. Rev. Lett. **48**, 1559 (1982).

47. G. Bastard, "Hydrogenic Impurity States in a Quantum Well: A Single Model," Phys. Rev. **B24**, 4714 (1981).

48. R. C. Miller, A. C. Gossard, W. T. Tsang and O. Munteanu, "Extrinsic Photoluminescence from GaAs Quantum Wells," Phys. Rev. **B25**, 3871 (1982).

49. W. T. Masselink, Y. C. Chang and H. Morkoc, "Binding Energies of Acceptors in GaAs-Al$_x$Ga$_{1-x}$As Quantum Wells," Phys. Rev. **B28**, 7373 (1983).

50. B. V. Shanabrook, "The Properties of Hydrogenic Donors Confined in GaAs-AlGaAs Multiple Quantum Wells," Surf. Sci. **170**, 449 (1986).

51. R. C. Miller, D. A. Kleinman, W. T. Tsang and A. C. Gossard, "Observation of the Excited Level of Excitons in GaAs Quantum Wells," Phys. Rev. **B24**, 1134 (1981).

52. E. S. Koteles and J. Y. Chi, "Experimental Exciton Binding Energies in GaAs/Al$_x$Ga$_{1-x}$As Quantum Wells as a Function of Well Width," Phys. Rev. **B37**, 6332 (1988).

53. F. Agullo-Rueda, J. A. Brum, E. E. Mendez and J. M. Hong, "Change in Dimensionality of Superlattice Excitons Induced by an Electric Field," Phys. Rev. **B41**, 1676 (1990).

54. C. Covard, R. Merlin, M. V. Klein and A. C. Gossard, "Observation of Folded Acoustic Phonons in a Semiconductor Superlattice," Phys. Rev. Lett. **45**, 298 (1980).

55. B. Jusserand, D. Paque and A. Regvency, "Folded Optical Phonons in GaAs/Ga$_{1-x}$Al$_x$As Superlattices," Phys. Rev. **B30**, 6245 (1984).

56. A. K. Sood, J. Menendez, M. Cardona and K. Ploog, "Interface Vibrational Modes in GaAs-AlAs Superlattices," Phys. Rev. Lett. **54**, 2115 (1985).

57. B. Jusserand and D. Paque, "Confined and Propagative Vibrations in Superlattices," in Heterojunctions and Semiconductor Superlattices, ed. by G. Allan, G. Bastard, N. Boccara, M. Lannoo and M. Voos (Springer-Verlag, Berlin, 1986). p. 108.

58. K. Huang and B. Z. Zhu, "Dielectric Continuum Model and Frohlich Interaction in Superlattices," Phys. Rev. **B38**, 13377 (1988).

59. R. Dingle, H. L. Stormer, A. C. Gossard and W. Wigmann, "Electron Mobilities in Modulation-Doped Semiconductor Heterojunction Superlattices," Appl. Phys. Lett. **33**, 665 (1978).

60. L. C. Witkowski, T. J. Drummond, C. M. Stanchak and H. Morkoc, "High Mobilities in $Al_xGa_{1-x}As$-GaAs Heterojunctions," Appl. Phys. Lett. **37**, 1033 (1980).

61. J. J. Harris, C. T. Foxon, K. W. Barnham, D. E. Lacklison, J. Hewett and C. White, "Two-Dimensional Electron Gas Structures With Mobilities in Excess of $3x10^6$ cm^2/v.sec," J. Appl. Phys. **61**, 1219 (1987).

62. L. Pfeiffer, K. W. West, H. L. Stormer and K. W. Baldwin, "Electron Mobilities Exceeding $10^7 cm^2$/v.sec in Modulation-Doped GaAs," Appl. Phys. Lett. **55**, 1888 (1989).

63. T. Mimura, S. Hiyamizu, T. Fujii and K. Nambu, "A New Field-Effect Transistor with Selectively Doped $GaAs/n-Al_xGa_{1-x}As$ Heterojunctions," Jpn. J. Appl. Phys. **19**, L225 (1980).

64. H. Sakaki, L. L. Chang, R. Ludeke, C. A. Chang, G. A. Sai-Halasz and L. Esaki, "InGaAs-GaSbAs Heterojunctions by Molecular Beam Epitaxy," Appl. Phys. Lett. **31**, 211 (1977).

65. L. L. Chang and L. Esaki, "Electronic Properties of InAs-GaSb Superlattices," Surf. Sci. **98**, 70 (1980).

66. J. N. Schulman and T. C. McGill, "The CdTe/HgTe Superlattice: Proposal for a New Infrared Material," Appl. Phys. Lett. **34**, 663 (1979).

67. M. Voos, "The Physics of Hg-Based Heterostructures" in Band Structure Engineering in Semiconductor Microstructures, ed. by R. A. Ahram and M. Jaros (Plenum Press, New York, 1988) p. 61.

68. L. L. Chang, G. A. Sai-Halasz, L. Esaki and R. L. Aggarwal, "Spatial Separation of Carriers in InAs-GaSb Superlattices," J. Vac. Sci. Technol. **19**, 589 (1981).

69. L. L. Chang, N. J. Kawai, G. A. Sai-Halasz, R. Ludeke and L. Esaki, "Observation of Semiconductor-Semimetal Transition in InAs-GaSb Superlattices," Appl. Phys. Lett. **35**, 939 (1979).

70. L. L. Chang, N. J. Kawai, E. E. Mendez, C. A. Chang and L. Esaki, "Semimetallic InAs-GaSb Superlattices to the Heterojunction Limit," Appl. Phys. Lett. **38**, 30 (1981).

71. M. Altarelli, "Electronic Structure and Semiconductor-Semimetal Transitions in InAs-GaSb Superlattices," Phys. Rev. **B28**, 842 (1983).

72. E. E. Mendez, L. Esaki and L. L. Chang, "Quantum Hall Effect in Two-Dimensional Electron-Hole Gas," Phys. Rev. Lett. **55**, 2216 (1985).

73. H. Munekata, T. P. Smith, F. F. Fang, L. Esaki and L. L. Chang, "Electrons and Holes in InAs-GaAlSb Quantum Wells," J. Physique **C5**, 151 (1987).

74. L. Esaki, L. L. Chang and E. E. Mendez, "Polytype Superlattices and Multi-Heterojunctions," Jpn. J. Appl. Phys. **20**, L529 (1981).

75. H. Takaoka, C. A. Chang, E. E. Mendez, L. L. Chang and L. Esaki, "GaSb-AlSb-InAs Multi-Heterojunctions," Physica **117-118B**, 741 (1983).

76. R. Beresford, L. F. Luo, K. F. Longenbach and W. I. Wang, "Resonant Interband Tunneling through a 110nm InAs Quantum Well," Appl. Phys. Lett. **56**, 551 (1990).

77. G. C. Osbourn, "Strained-Layer Superlattices from Lattice Mismatched Materials," J. Appl. Phys. **53**, 1586 (1982).

78. P. Voisin, C. Delalande, M. Voos, L. L. Chang, A. Segmuller, C. A. Chang and L. Esaki, "Light and Heavy Valence Subbands Reversal in GaSb-AlSb Superlattices," Phys. Rev. **B20**, 2276 (1984).

79. H. Shen, Z. Hang, J. Leng, F. H. Pollak, L. L. Chang, W. I. Wang and L. Esaki, "Interband Transitions from the Photoreflectance of GaSb/AlSb Multiple Quantum Wells," Superlattice and Microstructures, **5**, 591 (1989).

80. G. Abstreiter, H. Brugger, T. Wolf, H. Jorge and H. J. Herzog, "Strain-Induced Two-Dimensional Electron Gas in Selectively Doped Si/Si_xGe_{1-x} Superlattices," Phys. Rev. Lett. **54**, 2441 (1985).

81. G. C. Osburn, "InAsSb Strained-Layer Superlattices for Long Wavelength Detector Applications," J. Vac. Sci. Technol. **B2**, 176 (1984).

82. F. Cerdeira, A. Pinczuk, J. C. Bean, B. Batlogg and B. A. Wilson, "Raman Scattering from Ge_xSi_{1-x}/Si Strained-Layer Superlattices," Appl. Phys. Lett. **45**, 1138 (1984).

83. B. Jusserand, P. Voisin, M. Voos, L. L. Chang, E. E. Mendez and L. Esaki, "Raman Scattering in GaSb-AlSb Strained Layer Superlattices," Appl. Phys. Lett. **46**, 678 (1985).

84. J. K. Furdyna, "Magnetic Properties of Diluted Magnetic Semiconductors: A Review," J. Appl. Phys. **61**, 3526 (1987).

85. R. N. Bicknell, R. W. Yanka, N. C. Giles-Taylor, E. L. Buckland and J. F. Schetzina, "$Cd_{1-x}Mn_xTe$-CdTe Multilayers Grown by Molecular Beam Epitaxy," Appl. Phys. Lett., **45**, 92 (1984).

86. L. L. Chang, D. D. Awschalom, M. R. Freeman and L. Vina, "Optical and Magnetic Properties of Diluted Magnetic Semiconductor Heterostructures," in <u>Condensed Systems of Low Dimensionality</u>, ed. by J. L. Beeby (Plenum Press, New York, 1991). p. 165.

87. X. C. Zhang, S. K. Chang, A. V. Nurmikko, L. A. Kolodziejski, R. L. Gunshore and S. Datta, "Interface Localization of Excitons in $CdTe/Cd_{1-x}Mn_xTe$ Multiple Quantum Wells," Phys. Rev. **B31**, 4056 (1985).

88. E. K. Suh, D. V. Bartholomew, A. K. Ramdas, S. Rodriguez, S. Venogupalan, L. A. Kolodziejski and R. L. Gunshore, "Raman Scattering from Superlattices of Diluted Magnetic Semiconductors," Phys. Rev. **B36**, 4316 (1987).

89. D. D. Awschalom, J. M. Hong, L. L. Cheng and G. Grinstein, "Dimensional Crossover Studies of Magnetic Susceptibility in Diluted Magnetic Semiconductor Superlattices," Phys. Rev. Lett. **59**, 1733 (1987).

90. D. D. Awschalom, J. Warnock, J. M. Hong, L. L. Chang, M. B. Ketchen and W. J. Gallagher, "Magnetic Manifestation of Carrier Confinement in Quantum Wells," Phys. Rev. Lett **62**, 199 (1989).

91. K. F. Berggren, T. J. Thornton, D. J. Newson and M. Pepper, "Magnetic Depopulation of 1D Subbands in a Narrow 2D Electron Gas in a GaAs: AlGaAs Heterojunction," Phys. Rev. Lett. **5**, 1769 (1986).

92. W. Hansen, M. Horst, J. P. Kotthaus, U. Merkt, C. Sikorski and K. Ploog, "Intersubband Resonance in Quasi One-Dimensional Inversion Channels," Phys. Rev. Lett. **58**, 2586 (1987).

93. K. Kohl, D. Heitmann, P. Grambow and K. Ploog, "One-Dimensional Magnetoexcitons in $GaAs/Al_xGa_xAs$ Quantum Wires," Phys. Rev. Lett. **63**, 2124 (1989).

94. T. P. Smith, H. Arnot, J. M. Hong, C. M. Knoedler, S. E. Laux and H. Schmid, "Capacitance Oscillations in One-Dimensional Electron Systems," Phys. Rev. Lett. **59**, 2802 (1987).

95. B. J. van Wees, H. van Houten, C. W. J. Beenakker, J. G. Williamson, L. P. Kouwenhoven, D. van der Marel and C. T. Foxon, "Quantized Conductance of Point Contacts in a Two-Dimensional Electron Gas," Phys. Rev. Lett. **60**, 848 (1988).

96. R. A. Webb, S. Washburn, C. P. Umbach and R. B. Laibowitz, "Observation of h/e Aharonov-Bohm Oscillations in Narrow Metal Rings," Phys. Rev. Lett. **54**, 2696 (1985).

97. P. A. Lee and A. D. Stone, "Universal Conductance Fluctuations in Metals," Phys. Rev. Lett. **55**, 1622 (1985).

ELECTRONIC ENERGY LEVELS IN SEMICONDUCTOR QUANTUM WELLS AND SUPERLATTICES

G. Bastard and R. Ferreira

Laboratoire de Physique de la Matiére Condensée de
l'Ecole Normale Supérieure
24 rue Lhomond, F-75005 Paris

1. INTRODUCTION

The last ten years have witnessed an explosive increase of the fundamental and applied researches in quantum heterostructures obtained by refined growth techniques (such as Molecular Beam Epitaxy or Metal Organic Chemical Vapor Deposition). The reason for such a wide interest lays on the possibility of tailoring on the nanometer scale band edge profiles and thus ultimately specific electro-optical functions. The key feature of these man-made heterostructures is the preeminent part played by the quantum effects, such as size quantization, tunnelling phenomena, etc.. That quantum effects are so important arises from the smallness of the layer thicknesses. In most cases they are smaller than the de Broglie wavelength of either the electrons or the holes and thus the wave-like nature of the charge carriers necessarily comes into play. In the following, we present a brief survey of the results which have been achieved in the theoretical understanding of the electronic properties of semiconductor quantum wells and superlattices during the last decade.

Highlights in Condensed Matter Physics and Future Prospects
Edited by L. Esaki, Plenum Press, New York, 1991

In the last part of the present article, we shall attempt to predict future trends in the theoretical studies of these microstructures.

2. THE ENVELOPE FUNCTIONS APPROXIMATION

Besides ab initio methods which aim at evaluating the band offsets between two semiconductors when they are put in contact, as realized in the growth of heterolayers, people use empirical computation schemes when they need to evaluate the energy levels of actual heterostructures. It is almost necessary to assume the offsets are known to render the problem tractable, even for large scale computers. It is now recognized that among the empirical descriptions the envelope function framework[1-3] is a versatile and convenient method to calculate the electronic states of semiconductor heterostructures. In essence this scheme relies on the separation between the rapidly varying periodic parts of the Bloch functions, assumed to be the same in each kind of layer, and the more slowly varying envelope functions. The latter experience the potentials arising from the band bending due to charges and the band edge steps when one goes from one layer to the next. Thus, a great deal of the computational complexities is of bulk like origin, i.e. depends upon the fact that the extrema around which one is interested to build the heterolayer eigenstates is non degenerate or degenerate, the local effective mass tensor is scalar or not, etc.. In this respect, for III-V or II-VI heterolayers which retain the zinc blende lattice, there is a marked difference between the conduction and valence bands. The former is non degenerate (apart from spin) and displays small anisotropy and non parabolicity. The valence band edge instead is fourfold degenerate at the zone center (Γ_8 symmetry), which leads to complicated dispersion relations (although quadratic upon the wavevector $\mathbf{k}$).[4]

As a result of the simplicity of the host's conduction band edge one may readily calculate the conduction sub-bands by solving for each $k_\perp$ a scalar Schrödinger equation:

$$\left[P_z[1/2m(z)]P_z + \hbar^2 k_\perp^2/2m(z) + V_S(z) + V_{ext}(z)\right]\chi(z) = \varepsilon\chi(z) \qquad (1)$$

where $V_S(z)$ is the position dependent conduction band edge (0 in the well acting material, V_S in the barrier acting material), $m(z)$ the position dependent conduction band effective mass (m_w in the well, m_b in the barrier) $\mathbf{k}_\perp = (k_x, k_y)$ and $V_{ext}(z)$ an external potential (arising e.g. from charges or an external electric field). When $\mathbf{k}_\perp$ is small to allow the second term of eq.(1) to be treated perturbatively one finds the eigenenergies organize into two dimensional sub-bands:

$$\varepsilon_n(k_\perp) = E_n + \hbar^2 k_\perp^2 / 2m_n \tag{2}$$

where m_n is such that

$$1/m_n = P_w/m_w + P_b/m_b \tag{3}$$

and P_w, P_b are the integrated probabilities of finding the electron in the well and in the barrier respectively while in the nth state (confinement energy E_n). The complete wavefunction associated with eq.(2) is:

$$\Psi_n(\mathbf{r}) = u_{c0}(\mathbf{r})\chi_n(z) \exp(i k_\perp . r_\perp)/\sqrt{S} \tag{4}$$

where $r_\perp$ denotes the electron in-plane position, $u_{c0}(\mathbf{r})$ is the periodic part of the Bloch function at the zone center for the hosts' Γ_6 edge.

The dominant effect which arises from the realization of a quantum well structure is the energy shift of the ground eigenstate with respect to the bulk situation. It is of purely quantum origin and varies (roughly) like L^{-2} where L is the quantum well thickness. The confidence one might put in envelope type of calculations to evaluate the size quantization is illustrated in fig. 1a where the energy shift of the band-to-band emission (or absorption) edge of $Ga_{0.47}In_{0.53}As$ quantum wells is plotted against the well thickness. This shift is equal to the sum of the electron and hole confinement energies in the $Ga_{0.47}In_{0.53}As$ well (if one neglects the exciton binding energy). It is seen that an excellent agreement between the predicted and observed shifts is obtained over the investigated range of well thicknesses and also that it depends relatively little upon the band offsets between the barrier and the well: the InP and Al(In)As bandgaps are very similar: 1.42eV and 1.52eV respectively but the band offsets between both the barrier - acting materials and the Ga(In)As well are very uneven: 0.24 eV and 0.5 eV respectively. There has been recently a nice experimental demonstration of the consistency between the envelope functions type of calculations and the experimental results. By purposely inserting planes of isovalent (thus short range) impurities at various places of a GaAs-Ga(Al)As quantum well, Marzin and Gerard[34] were able through measurements of the shifts of the different optical transitions to map the quantum well envelope probability

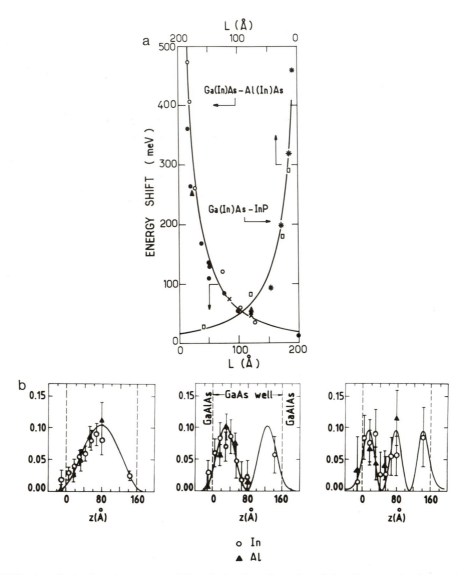

FIG. 1a. Calculated energy shift of the band-to-band fundamental absorption edge versus the Ga(In)As slab thickness in Ga(In)As - Al(In)As (lower horizontal scale) and Ga(In)As - InP (upper horizontal scale) quantum wells. The symbols correspond to a compilation of various experimental data. After ref. 9.
FIG. 1b. Calculated and measured probability densities versus the electron position for the three lowest bound states of a GaAs - Ga(Al)As quantum well. After ref. 34.

densities: These shifts are proportional at the lowest order of the perturbation theory to the probability densities of finding the electron or the
hole at the impurity plane. The outcome of their results are shown in fig.
1b where both repulsive (A1) and attractive (In) planar doping were
used.

3. VALENCE SUB-BANDS AT ZERO ELECTRIC FIELD

The bulk topmost valence hamiltonian is a 4×4 matrix, first
derived by Luttinger,[4] whose basis vectors can be identified with the four
components of a $J=3/2$ angular momentum. In the envelope function
approximation, the envelope eigenfunctions (which are 4×1 spinors on
the $|3/2, mj>$ basis where $-3/2 \leq mj \leq 3/2$) of the Luttinger hamiltonian
of a multiple quantum well structure are the solutions of:

$$\sum_{l'} H_{11'} \Psi_{l'}(z) = 0 \tag{5}$$

where $H_{11'}$ is a 4×4 matrix whose elements, as written on the basis
$|3/2,3/2>$; $|3/2,-1/2>$; $|3/2,+1/2>$; $|3/2,-3/2>$, are:

$$
\begin{pmatrix}
H_{hh} + V_p(z) & c & b & 0 \\
c^* & H_{lh} + V_p(z) & 0 & -b \\
b^* & 0 & H_{lh} + V_p(z) & c \\
0 & -b^* & c^* & H_{hh} + V_p(z)
\end{pmatrix} \tag{6}
$$

where:

$$\mathbf{k}_\perp = (k_\perp \cos\theta, k_\perp \sin\theta) \tag{7}$$

$$H_{hh} = -p_z(\gamma_1 - 2\gamma_2)p_z/2m_0 - \hbar^2 k_\perp^2 (\gamma_1 + \gamma_2)/2m_0 \tag{8}$$

$$H_{lh} = -p_z(\gamma_1 + 2\gamma_2)p_z/2m_0 - \hbar^2 k_\perp^2 (\gamma_1 - \gamma_2)/2m_0 \tag{9}$$

$$c = \hbar^2 \sqrt{3}\, \gamma_2 k_\perp^2 e^{-2i\theta}/2m_0 \tag{10}$$

$$b = -i\sqrt{3}\, \hbar^2 k_\perp e^{-i\theta}/2m_0 [\gamma_3 d/dz + d/dz\gamma_3] \tag{11}$$

while V_p (z) represents the position dependent valence band edge and $\gamma_1, \gamma_2, \gamma_3$ are the Luttinger parameters of the bulk materials. In eq. (10) the axial approximation[1] has been used. It renders the valence dispersions isotropic in the layer plane. It is interesting to note that the off diagonal elements of eq. (6) vanish at $k_\perp = 0$. Thus, a convenient method to find the $k_\perp$ dependence of the eigenstates consists of diagonalizing these off diagonal elements in a basis spanned by the $k_\perp = 0$ solutions of eq. (6). This basis includes the bound states and the continuum states of the problem. Since one is usually interested in knowing only the topmost dispersion relations, it is a sensible approximation to discard the continuum states and to retain only the $k_\perp = 0$ bound states. This incomplete basis has however the drawback of not complying with the conservation of the probability current at the interfaces. The latter can be expressed by integrating eq. (6) once with respect to z across an interface. Since the off diagonal terms of eq. (6) involve the γ_2, γ_3 parameters, which are in principle position dependent, the conservation of the probability current involves a 4x4 matrix $C_{11'}$ which is not diagonal. The use of a finite $K_\perp = 0$ basis does not allow to cope with the off diagonal terms of $C_{11'}$. In practice however, the penetration of the topmost valence levels outside the well is faint and the violation of the current conserving conditions is of minor importance. One noticeable exception to this rule of thumb is found in the HgTe-CdTe system where the differences in the γ_1, γ_2 parameters are large (γ_1 and γ_2 are of opposite signs in the two materials) and the penetration in the CdTe layers is non negligible due to the light conduction band mass in HgTe ($0.03\ m_0$).

At $k_\perp = 0$ the eigenstates of eq. (6) split into two decoupled sets of levels. Those corresponding to $m_J = \pm 3/2$ (respectively $\pm 1/2$) correspond to heavy (light) holes since the respective confinement energies involve a heavy $[m_0/(\gamma_1 - 2\gamma_2)]$ and a light $[m_0/(\gamma_1 + 2\gamma_2)]$ effective mass. We shall denote these $k_\perp = 0$ edges by HH_n (respectively LH_n) where $n \geq 1$. If the b and c terms were always discarded the in plane dispersions of the valence levels would exhibit the "mass reversal effect." Namely, the in plane effective masses would be lighter for the HH_n subbands $[m_0/(\gamma_1 + \gamma_2)]$ than for the LH_n ones $[m_0/(\gamma_1 - \gamma_2)]$. This would lead to the crossings of the heavy and light hole branches, for instance at $k_1 = [m_0(HH_1 - LH_1)/\gamma_2\hbar^2]^{1/2}$ for the topmost sub-bands. The non vanishing off diagonal terms replace these crossings by anti-crossings. The latter feature implies strongly non parabolic in plane dispersion relations. This is particularly noticeable for the LH_1 sub-bands which acquire a camel back shape near $\mathbf{k}_\perp = 0$. Large mixing of the heavy and light hole characters have to be expected when the anti-crossings take place and in fact the very notion of heavy and light hole becomes fuzzy at non vanishing $\mathbf{k}_\perp$. Such considerations can be quantified by evaluating the $\mathbf{k}_\perp$ dependence of the average value of J_z over the various eigenstates. As will be shown below these averages deviate quickly from the edge values ($\pm 3/2$ for HH_n, $\pm 1/2$ for LH_n).

It is actually possible to derive the in plane dispersion relations and the corresponding eigenfunctions of a single quantum well under flat band conditions ($F = 0$) almost analytically if one restricts the $\mathbf{k}_\perp = 0$ basis to the HH_1, HH_2 and LH_1 edges[5-7]. Under these assumptions the dispersion relations are the roots of:

$$(\varepsilon_{HH1} - \varepsilon[(\varepsilon_{LH1} - \varepsilon)(\varepsilon_{HH2} - \varepsilon) - |<\Phi_1|b|_{\chi 2}>|^2]$$
$$- |<\Phi_1|c|\chi_1>|^2(\varepsilon_{HH2} - \varepsilon) = 0 \tag{12}$$

where:

$$\varepsilon_{HHi} = HH_i - \hbar^2 k_\perp^2(\gamma_1 + \gamma_2)/2m_0 \tag{13}$$

$$\varepsilon_{LHi} = LH_i - \hbar^2 k_\perp^2(\gamma_1 - \gamma_2)/2m_0 \tag{14}$$

and θ_1, χ_1, χ_2 are the quantum well envelope functions of the LH_1, HH_1 and HH_2 states at $\mathbf{K}_\perp = 0$. Each of these eignevalues are twice degenerate (Kramers degeneracy) and to each of the twofold degenerate eigenergies one can associate the two orthogonal wavefunctions:

$$\Psi_{k\perp}\uparrow = 1/\sqrt{S} \left[1 + a^2 + \eta^2\right]^{-1/2} e^{ik_\perp} \bullet r_\perp$$
$$\times \left[a e^{-2i\theta}\chi_1(z), \Phi_1(z), 0, i\eta e^{i\theta}\chi_2(z)\right] \tag{15}$$

$$\Psi_{k\perp}\downarrow = 1/\sqrt{S} \left[1 + a^2 + \eta^2\right]^{-1/2} e^{ik_\perp} \bullet r_\perp$$
$$\times \left[-i\eta e^{-i\theta}\chi_2(z), 0, \Phi_1(z), a e^{2i\theta}\chi_1(z)\right] \tag{16}$$

where:

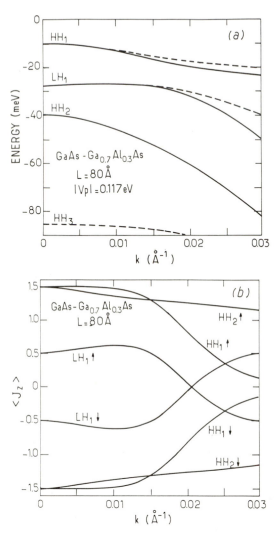

FIG. 2a. Calculated in-plane dispersion relations of the topmost valence sub-bands in a 80Å thick GaAs − Ga$_{0.7}$Al$_{0.3}$As single quantum well. Solid line: three level model. Dashed lines: inclusion of all the levels bound in the well at k = 0.

FIG. 2b. Calculated in-plane dependence of the J$_z$ averages over the energy eigenstates in the three level model. The notations ↑, ↓ refer here to the dominant character at k = 0. They agree with (are opposite to) those used in the text for the HH$_1$(LH$_1$, HH$_2$) sub-bands.

$$a = \hbar^2 \sqrt{3} \, k_\perp^2 \gamma_2 < \Phi_1 | \chi_1 > /[2m_0(\varepsilon - \varepsilon_{HH_1})] \qquad (17)$$

$$\eta = \hbar^2 \sqrt{3} \, k_\perp < \chi_2 | \gamma_3 d/dz + d/dz\gamma_3 | \phi_1 > /[2m_0(\varepsilon - \varepsilon_{HH2})] \qquad (18)$$

One readily checks that $< \Psi_{k_\perp}\uparrow | J_z | \Psi_{k_\perp}\uparrow >$ (respectively $< \Psi_{k_\perp}\downarrow | J_z | \Psi_{k_\perp}\downarrow >$) extrapolates to $+3/2$ (respectively $-3/2$) for the lowest lying hole states (the HH_1 branch). Moreover, J_z has no non-vanishing matrix element between any $\Psi_{k_\perp}\uparrow$. and $\Psi_{k_\perp}\downarrow$ corresponding to the same energy. This has led us to label the hole eigenstates according to the "spin" ($\downarrow$, $\uparrow$) even though neither $\Psi_{k_\perp} \uparrow$ nor $\Psi_{k_\perp} \downarrow$ are eigenstates of σ_z or J_z at finite $k_\perp$.

Figures 2a and 2b show the calculated dispersion relations and averages of J_z over the $\uparrow$ and $\downarrow$ eigenstates for a 80Å GaAs-Ga(Al)As quantum well (valence band offset $|V_p| = 117$ meV, $\gamma_1 = 6.85$, $\gamma_2 = 2.1$, $\gamma_3 = 2.9$ respectively) [7]. As mentioned previously, the J_z averages deviate from $\pm 3/2$ ($\pm 1/2$) for the heavy (light) hole branches when they anticross. We also show for comparison in fig. 2a the dispersion curves obtained by retaining more sub-bands in the $k_\perp = 0$ basis. It is seen that over significant $k_\perp$ range the three level model gives a fair account of the HH_1 dispersion.

Knowing the dispersion relations and the wavefunctions one can calculate a number of physical properties such as for instance the level lifetimes associated with defects, phonons etc...A particular class of relaxation effects are those which correspond to spin flips of the carrier. They are very fast for holes in bulk materials, the reason being that the spin-orbit effects are very large: a hole created at a finite **k** value with a given "spin," as obtained by quantizing **J** along **k**, will almost immediately lose its "spin" by any scattering mechanism, because a change in the hole wavevector also implies, in the final state, a different "spin" quantum number. Such a fast hole "spin" flip becomes inhibited in quantum wells, at least if $k_\perp$ is not too large.

The reason for this quenching is the lifting of the heavy and light hole degeneracy by the quantum well potential which makes the HH_1 states with a small $k_\perp$ to be nearly $\pm 3/2$ states, which cannot be coupled by any static potential. An example of such a behavior is shown in fig. 3 where the "spin"-flip and "spin"-conserving scattering times of holes due to ionized impurity or alloy fluctuations is plotted versus $k_\perp$ for bulk $Ga_{0.47}In_{0.53}As$ or $Ga_{0.47}In_{0.53}As$ - InP quantum wells[7].

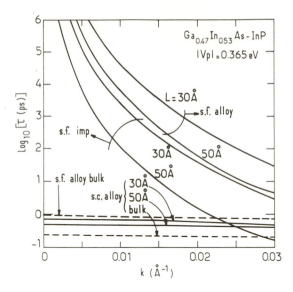

FIG. 3. Calculated dependence upon the in-plane wavevector of the impurity-
and alloy fluctuations-assisted "spin"-flip (s.f.) and "spin"-conserving (s.c.)
scattering times within the HH_1 sub-band of $Ga_{0.47}In_{0.53}As$ - InP quantum wells
and bulk $Ga_{0.47}In_{0.53}As$.

Two salient features are noticeable. Firstly, the "spin"-conserving
scattering is much faster than the "spin"-flip one in quantum wells.
Also, it is almost independent of the in-plane wavevector and of the
quantum well thickness: the bulk and quantum well figures are of comparable magnitudes. In contrast, the "spin"-flip relaxation times are
strongly $k_\perp$ dependent in quantum wells and over a significant $k_\perp$ range
may be longer than the recombination time (500 ps - 1 ns). The
"spin"-flip relaxation time of holes in bulk $Ga_{0.47}In_{0.53}$ As is instead very
short and $k_\perp$ independent. Thus, there is a clear effect associated with
the size quantization upon the strength of some scattering mechanisms.
The spin relaxation times play a key part in the spin orientation experiments where a circularly polarized light creates a preferential spin orientation of the photo-generated electrons and holes. This magnetization
decays due to the spin-flip relaxation and eventually the emitted light displays a residual degree of circular polarization. Recently, Uenoyama and
Sham[8] have succeeded in interpreting spin orientation measurements
performed in doped and undoped GaAs-Ga(Al)As quantum wells by
assuming that, instead of an immediate "spin"-relaxation, the holes keep
some memory of their initial polarization. The previous discussions have

explained why it should be so from considerations of the valence sub-band structure.

The band mixing effects at finite $k \perp$ has some important consequences upon the "vertical" transport of holes in a superlattice. Namely, it enhances the hole capability of hopping from one well to the other when $k \perp$, i.e. the in-plane velocity increases. The physics of this effect can be simply understood upon examination of eq. (6). At $k \perp = 0$ the heavy hole sub-band width for the z motion depends upon the transfer integral from one well to the other. The latter behaves like $\exp$-$\left[2m_{hh}/h^2 \left(|V_p| + \varepsilon_{HH1} \right) L_B \right]$ where L_B is the barrier thickness and $m_{hh}/m_0 = (\gamma_1 - 2\gamma_2)^{-1}$ is the heavy hole effective mass along the growth axis. The ground light hole sub-band width is given by a similar expression where $m_{lh} (= m_0/(\gamma_1 + 2\gamma_2))$ and ε_{LH1} replace m_{hh} and ε_{HH1} respectively, i.e. the tunneling length for the light hole is longer than that of the heavy hole. At finite $k \perp$ the heavy and light hole characters become mixed. Hence, one should expect the heavy hole motion along the growth axis to be characterized by a lighter effective mass when $k \perp$ increases. Therefore, the bandwidth for the ground superlattice sub-band should increase upon $k \perp$ This trend is demonstrated in fig. 4 where we show the $k \perp$ dependence of the ground valence sub-band width in GaAs $-$ Ga$_{0.7}$Al$_{0.3}$As superlattices with equal layer thicknesses and three different periods[9]. A twofold increase of the bandwidth is found upon the $k \perp$ increase from zero (no band mixing) to $3 \times 10^6 \text{cm}^{-1}$ and $d = 60\text{Å}$.

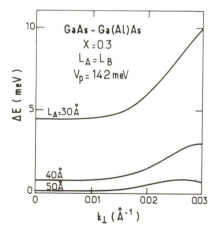

FIG. 4. Calculated $k_\perp$ dependence of the ground valence sub-band width in GaAs $-$ Ga$_{0.7}$Al$_{0.3}$As superlattices with equal layer thicknesses and $d = 60\text{Å}$, 80Å and 100Å.

4. ELECTRIC FIELD EFFECTS

The dominant effects on the valence eigenstates of single quantum wells which are associated with an electric field F applied parallel to the growth axis are[1-3,9] i) the spatial polarization of the eigenstates and ii) the lifting of the Kramers degeneracy at finite $K_\perp$.

The polarization effects also show up in energy shifts (due to the interaction of the induced dipole with the field) which are red shifts (quadratic upon F at low enough F) for the ground states HH_1 and LH_1 at $k_\perp = 0$. The lifting of the Kramers degeneracy occurs due to the combined effects of the non zero $k_\perp$, the finite spin-orbit energy and the non centro-symmetry of the tilted quantum well. Note that the field in principle prevents the existence of truly bound states. However, under most circumstances, the field induced ionization can be safely neglected (see fig. 5 for a relevant example[9]) as the finite lifetime of the virtual bound states remains much longer than the period of the carrier oscillations inside the tilted quantum well. In multiple well structures the electric field can cause either a spatial localization of states which were delocalized at $F = 0$ (the Wannier-Stark quantization[10]) or induces a partial delocalization when the potential energy drop over p periods peFd match the energy difference between two levels essentially localized in different

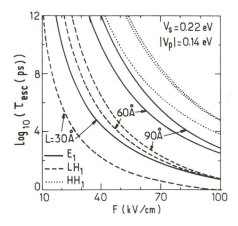

FIG. 5. Semiclassical estimates of the escape time of an electron (in the E_1 state), a heavy hole (in the HH_1 state) and a light hole (in the LH_1 state) at $k_\perp = 0$ out of quantum wells (L = 30Å, 60Å, 90Å) tilted by an electric field. For electron, light hole and heavy hole the escape time increases monotically with L.

wells (field induced resonant tunneling[11]). The latter mechanism is the one by which the holes coherently move along the growth axis of the structure while preserving their in plane momentum. This motion is actually negligible with respect to the one induced by the defects (or phonons): one may in fact show[12] that the coherent tunneling current in a finite and biased multiple quantum well structure collapses with the field, the reason being that the field in general misaligns the consecutive hole levels and only exceptionally aligns them. Once we abandon the coherent resonant tunneling picture and allow for the existence of relaxation mechanisms, the misalignment of consecutive eigenstates can be circumvented by external scatterers. Firstly the longitudinal motion no longer requires the in plane wavevector to be conserved. This means that the scatterers convert some part of the initial longitudinal carrier energy into transverse kinetic energy in the final states while conserving the overall carrier energy if they are elastic or allowing an energy relaxation of the carrier if they are inelastic. Thus, we can define an assisted time from one hole eigenstate $|\alpha, \mathbf{k}_\perp >$ to all other hole eigenstates $|\beta, \mathbf{k}'_\perp >$ by using the Fermi golden rule:

$$\hbar/2\pi\tau_{\alpha\mathbf{k}\perp} = \sum \mathbf{k}'_\perp | < \alpha\mathbf{k}_\perp | V_{def} | \beta\mathbf{k}'_\perp > |^2 \delta(\varepsilon_{\alpha\mathbf{k}\perp} - \varepsilon_{\beta\mathbf{k}'\perp}) \qquad (19)$$

It is clear from eq. (19) that the elastic assisted tunneling should exhibit resonances at the same electric field values as the purely coherent tunneling. In fact, when two eigenstates become aligned by the field their wavefunctions spatially delocalize. The assisted tunneling rate is proportional to the square of the matrix element of the scattering potential between the initial and final states of the transition. Clearly when these are delocalized the strength of the coupling is greatly enhanced compared with the non resonant case, hence the resonance. This qualitative discussion can be sustained by calculations as will be shown below.

In the context of coupled double barriers, it is now widely recognized that one deals with assisted (also called sequential[13]) tunneling from the emitter to the well and from the well to the collector. On the other hand, there have been attempts to interpret the experiments performed on coupled double wells in terms of the coherent (Rabi) oscillations that a carrier undergoes between the two wells when it is prepared in a quantum state which is not an eigenstate of the system. The observation of Rabi oscillations is easily achievable in the radio-frequency range where many quasi degenerate atomic states fall. On the other hand, in the context of laser excited double quantum wells, it seems to us hardly reasonable to imagine a smaller level splitting than the laser width (to obtain a coherent population of the two interacting states) yet larger than the broadening of the eigenstates (to be able to talk about distinct eigenstates of the two well system). If we thus abandon the idea of Rabi oscillations, we have to admit that the carrier will only be created in eigenstates of the double quantum well, in which they will stay for ever if there is no scattering to enable them to make transitions. Thus, we believe that the appropriate framework of interpretation of the tunneling in double quantum wells is at the present time the same as that prevailing in double barrier structures, i.e. that of the assisted tunneling.

The incoherent or assisted tunnelings of holes[14,15] nicely display the band mixing effects that were previously discussed: if heavy and light holes were always decoupled there would exist no possible conversion of a heavy to a light hole due to static scatters. Band mixing effects instead allow for such a coupling and thus for resonances in the assisted hole transfer when the electric field lines up heavy and light hole states. In general the band mixing effects accelerate the transfer over estimates

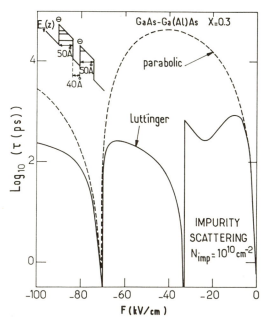

FIG. 6. The decimal logarithm of the impurity-assisted inter-well hole transfer time in a 50Å - 40Å - 50Å GaAs − $Ga_{0.7}Al_{0.3}As$ biased double well is plotted versus the electric field strength F. The initial state is HH'_1.

based on decoupled heavy and light hole models. We illustrate this point in fig. 6 where we show the calculated impurity assisted hole transfer time versus the electric field strength in the case of a symmetrical GaAs-Ga(Al)As double quantum well[14]. The initial hole state was taken as HH'_1, the edge of the topmost valence level in the right hand side well. The solid line corresponds to the full calculations while the dashed line is based on a model which does not take the band mixing into account (b = c = 0 in the Luttinger matrix).

The level scheme at $k_\perp = 0$ is shown in fig. 7 and allows to identify the resonance at -70 kV/cm as due to the anticrossing between HH'_1 and HH_2 and that near -35 kV/cm to the "frustrated" anticrossing between HH'_1 and LH_1. The anticrossing is "frustrated" because it does not occur at $k_\perp = 0$, but only for non vanishing $k_\perp$. It is striking that over a large electric field range, calculations which neglect the band mixing overestimate the transfer time by many decades.

Compared to electronic resonances the hole resonances are narrower and deeper around the field strength which realizes the anticrossing between the two interacting levels. This is due to the smaller hole energy splitting at resonance. The latter feature implies that, exactly

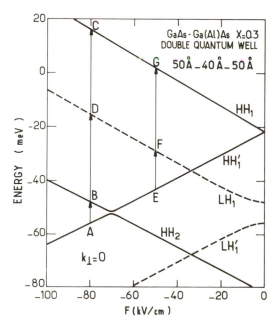

FIG. 7. Level scheme at $k_\perp = 0$ versus F for the heterostructure of fig. (6).

at resonance, the in-plane wave vector that the scatterer has to provide to the tunneling particle is smaller for a hole than for an electron. Since in general the scattering matrix element decreases with increasing in-plane wavevector transfer (see e.g. the case of coulombic scattering) the hole transition rate has to be larger at resonance than the electron one. This holds only at resonance where for both electrons and holes the initial and final states are delocalized over the whole heterostructure. On the other hand, off resonance the spatial localization of the initial and final states in different wells becomes much more pronounced for holes than for electrons when the energy detuning between the initial and final states increases, thereby decreasing sharply the transition rates. Thus, by varying F around a resonance one should find narrower resonance curves for the assisted transfer of holes than for that of the electrons if it were not for inhomogeneous broadenings. Actual quantum well samples display layer width fluctuations (islands). The sample can in the limit of large islands be envisioned as the superposition of many microsamples, each characterized by different resonance fields. It is clear that the individual narrow hole resonances are more liable to distortions by inhomogeneous broadening effects than the relatively broad electron ones.

Let δF be the width of an individual resonance occurring at F_0 and let ΔF be that of the resonance field distribution around F_1 due to the fluctuations in the layer width:

$$e\Delta F = |\, dE_1/dL\,|\,\Delta L/[h + (L + L')/2] \qquad (20)$$

where L and L′ (h) are the average thicknesses of the two quantum wells (barrier) and where we have assumed for simplicity that only one well width (L), and thus only one energy level (E_1), experiences fluctuations. Let us also empirically assume that the two functions describing the resonance versus $F(\tau^{-1}(F))$ and the resonance fields distribution ($P(F_0)$) are gaussians:

$$1/\tau(F) = 1/\tau_b + (1/\tau_0 - 1/\tau_b)\,\exp\!\left[-(F - F_0)^2/2(\delta F)^2\right] \qquad (21)$$

$$P(F_0) = \left[\Delta F\sqrt{2\,\pi}\,\right]^{-1}\exp\!\left[-(F_0 - F_1)^2/2(\Delta F)^2\right] \qquad (22)$$

where τ_0 and τ_b are the transfer times at resonance and far away from resonance respectively for a given resonance field F_0. If we assume these two times to be independent of F_0 the resulting resonance curve $< 1/\tau(F) >$ is the convolution of the two functions:

$$< 1/\tau(F) > = 1/\tau_b + (1/\tau_0 - 1/\tau_b) \, (\delta F/\sigma F)$$
$$\exp\left[- (F - F_1)^2/2(\sigma F)^2 \right] \tag{23}$$

where:

$$\sigma F = \left[(\delta F)^2 + (\Delta F)^2 \right]^{1/2} \tag{24}$$

More detailed numerical simulations have been undertaken to evaluate the effect of inhomogeneous broadening. We have found that the electronic resonances are relatively unaffected by fluctuations of one monolayer while the hole ones are more changed, as expected. The inhomogeneous broadening effects are more important in heterostructures containing narrow wells (because the $|dE_1/dL|$ are larger in narrow wells) or narrow barriers (because the matrix elements appearing in the assisted transfer rates depend exponentially on the barrier thickness and thus fluctuates widely in the case of moderately transparent barriers). Note that the heterostructures with narrow wells and/or barriers are exactly the ones used to detect the hole transfer. Besides the fluctuations in the well width discussed above there are also fluctuations in the electric field strength (and direction) due to the necessarily inhomogeneous doping of the n and p layers in the p-i-n structures. These fluctuations add to the well width ones and the actual ΔF can be quite large (and sample dependent).

Although deep and sharp hole resonances can be washed out by inhomogeneous broadening effects (and also by homogeneous ones), the background hole transfer time is a quantity which is very little affected by the layer fluctuations. It is remarkable that off resonance the assisted hole transfer time can be found in the range 100-500 ps for reasonable amount of defects. This means that the very idea of "immobile" holes is incorrect in truly quantum heterostructures such as double quantum wells.

Some words of caution have to be added regarding the order of magnitude of the assisted transfer times: in our model (Born approximation) the transfer time τ is inversely proportional to the areal concentration of defects. τ also depends upon the material under consideration, for instance upon the valence band offset $|V_p|$. Since the transfer is very much favored by the band mixing effects there exist situations where the band mixing in the final states is small. This is for example realized when in a double well under appropriate bias the ground hole level of the thick well is made to anti-cross the hole levels predominantly localized in a very narrow well. A long transfer time is predicted to occur under these circumstances. The assisted transfer from the ground hole state of the narrow well to the hole states of the wide well should be faster, because of the larger band mixing effects in the final states on the one hand and of the larger density of final states on the other hand.

Recent time-resolved measurements of the photoluminescence of double quantum wells have evidenced the hole escape outside a wide quantum well. Alexander et al[16] have reported strong evidences for resonances in the hole transfer time versus the electric field in biased 50Å- h -100Å GaAs-Ga$_{0.65}$Al$_{0.35}$As double quantum well structures and found short hole transfer times (some smaller than 20 ps). At resonance the hole transfer time was found to increase with increasing intermediate barrier thickness h while off resonance (transfer time of the order of a few hundreds picoseconds) the trend was not so clear. In any event, the hole transfer time has been demonstrated to be quite short, i.e. shorter than or equal to the recombination lifetime.

Qualitatively similar results on the fast hole escape have been reported by Vodjani et al[17] in their study of the photoluminescence associated with the electron-hole recombination in the central well of an operating double barrier structure. In these experiments, under resonant tunneling of electrons, the photoluminescence intensity is directly proportional to the number of minority holes injected in the well. Under illumination of the contact regions, at a smaller energy than the bandgap of the quantum well, the holes which are available for the recombination are those which have transferred from the contacts to the well. Thus, the photoluminescence intensity is a good indicator of the hole transfer.

5. EXCITONIC EFFECTS

The quasi bi-dimensional motions of electrons and holes in semiconductor quantum wells has a significant influence on the exciton binding energy in type I structures where the electron and the hole are essentially localized in the same layer[18-24]. This enhanced binding energy, which can be traced back to the fact that the hydrogen atom is four times more bound in two dimensions than in three dimensions, has a

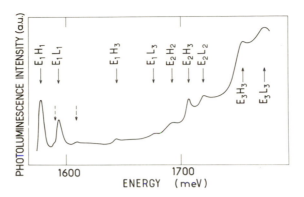

FIG. 8. An example of low temperature (T = 2K) photoluminescence excitation spectrum in a GaAs-Ga(Al)As two steps quantum well which evidences the excitonic peaks attached to the numerous optically allowed interband transitions. The dashed arrows point the shoulders due to the 2S state (or continuum onset) of the exciton formed between the electron and hole ground sub-bands. After ref. 41.

very profound influence on the quantum well absorption spectra, even at room temperature. Instead of displaying a series of staircases, characteristic of the two dimensional in-plane reduced motion, the absorption (and also the photoluminescence excitation) spectra display intense peaks which can be associated with the creation of 1S excitons (fig. 8). Smaller peaks, eventually shoulders, take place when the 2S (or dissociated) excitons are created.[18] A number of theoretical studies[9,21-24] have been devoted to the precise evaluation of the exciton binding energy in type I quantum wells, taking into account the complicated kinematics of the hole.

The electric field effects have also been calculated. In marked contrast with the bulk situation, where the excitons are readily field-ionized, it has been predicted and observed that the quantum well walls would largely hinder the electric field induced exciton dissociation. A summary of the agreement between an accurate theory[9] and observed[25] optical features is shown in fig. 9.

In type II quantum wells, such as realized in the GaSb-InAs or InP-Al(In)As systems, the electron and hole are spatially separated, a situation which also takes place in biased semiconductor superlattices where it is possible to form weakly bound excitons 1S between electrons and holes which are Wannier-Stark localized in different superlattice

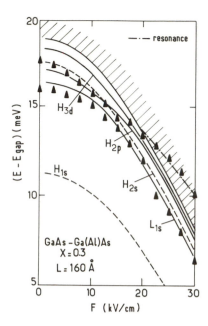

FIG. 9. Calculated and measured excitonic transitions in a 160Å thick GaAs - Ga(AlAs quantum well versus the electric field strength F. The notation H (L) and the subscripts 1S, 2S, 2P refer to heavy (light) main character and main radial component of the exciton wavefunction at F = 0. After ref. 9.

periods[26,27]. The spatial separation prevents the formation of tightly bound electron-hole pairs and for that reason the room temperature absorption spectra hardly show clear excitonic features. At low temperature instead, the spatially separated excitons have been observed[28,29] and it has been proposed that these weakly bound excitons could eventually condense into some new (albeit elusive) kind of correlated state[30].

Finally, it is worth pointing out that excitons in (true) superlattices have been observed,[31] with a binding energy which decreases with decreasing superlattice periods (keeping the hosts' layer thicknesses equal, see fig. 10). This feature nicely illustrates the progressive delocalization of the electron and hole eigenstates and the continuous passage from a quasi bi-dimensional to a three-dimensional situation.

There have also been interesting studies of the eventual formation of saddle point excitons, which arise from the pairing of electron-hole pair states located in the reciprocal space near the superlattice Brillouin zone ($q = \pi/d$). Detailed numerical simulations[32] show a remarkably

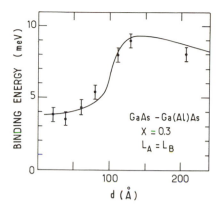

FIG. 10. Calculated and measured exciton binding energy versus the superlattice period d in GaAs-Ga(Al)As superlattices with equal layer thicknesses. After ref. 31.

good agreement with the observed spectra. It is however dubious that the extra optical features really arise from truly saddle point excitons. The extra oscillator strength are rather borrowed from the entire superlattice Brillouin zone.

6. FUTURE TRENDS

There is no doubt that the theoretical understanding of semiconductor heterolayers has a long way to go before a satisfactory description of the relaxation and transfer processes. The challenge is to obtain a consistent and reliable description of the carrier transport along the growth axis of, say a superlattice, depending on the superlattice quality (amount of residual doping, interface roughness, macroscopic smoothness, etc..). The experiments have already shown that several transport regime exist, ranging from the hopping conduction from one well to the other to the miniband regime, where a Boltzmann-like analysis may be relevant. One should also be aware that, in contrast with the regular bulk materials which are characterized by large bandwidths, the superlattices display narrow bands for the carrier motion along the growth axis. Their eigenstates are thus more sensitive to relatively small fluctuations and more prone to localization.

Applications of superlattice structures to efficient photo-detectors raise, on the theoretical side, the key question of the Wannier-Stark localization, the quantum counterpart of the semi-classical Bloch oscillator. High field transport may take place in a hopping-like fashion while under a vanishingly small electric field it would have been better described by means of the usual miniband conduction formalism. To which extent is the Bloch oscillator a valid concept, since our perception of this fascinating effect completely rests upon a semi-classical description of the carrier motion in these microstructures? We do know that the semi-classical picture, which requires the electric potential to vary infinitely slowly on the scale of the superlattice period, has no validity in actual cases (typically one applies a 40kV/cm field to a $1\mu m$ thick 30Å-30Å GaAs-Ga(Al)As superlattice). There are in fact ample experimental evidences that the miniband spectrum is destroyed and replaced by the evenly spaced Wannier-Stark spectrum[28,29]. Still, the calculation of the semi-classical and quantum extensions of the Bloch oscillator are very close to each other.[35] We need to understand on the theoretical side what kind of electron transport take place and between which states.

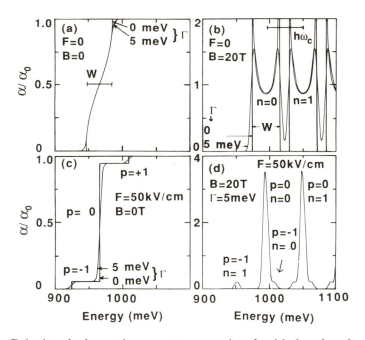

FIG. 11. Calculated absorption spectra associated with band-to-band transitions in a 40Å-45Å Ga(In)As - (Ga,Al,In)As superlattice for two values of the broadening parameter Γ: $\Gamma = 0$, 5 meV respectively. The spectra shown in figs. (10 a, b, c, d) correspond respectively to a three dimensional, quasi one dimensional, quasi bi-dimensional and quasi zero dimensional carrier motion. W is the sum of the ground electron and hole bandwidths. After ref. 33.

Finally, the improvements in the etching and patterning procedures of quasi bi-dimensional structures result in the creation of quantum wires and dots. The quality of these new objects has to improve greatly before one may undertake the finely detailed optical and electrical measurements which have already been carried out in the quasi bi-dimensional structures. There will certainly be a great deal of effort to quantify our understanding of the energy levels in quasi one and zero dimensional objects and then, but only then, their transport properties (if any). The recent experiments[36,37] on "electron" optics are fine demonstration of the wave-like nature of the electron, and probe by multi-channel quantized conductances[38–40]. For eventual optical applications (e.g. lasers) we really need to know to which extent the extra size quantization brought about by the patterning sensitively affects or not the wire or dot optical response. This is nowadays hard to assess due to the significant part played by the defects introduced during the etching and cutting fabrication stages. We suggest that a very good way to understand the effects of extra size quantization might be to study the optical properties of superlattices subjected to parallel electric and magnetic fields. Under

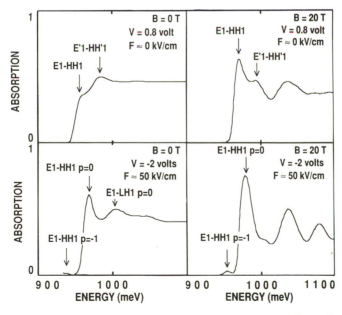

FIG. 12. Experimental absorption spectra for a 40Å-45Å Ga(In)As - (Ga,Al,In)As superlattice under different electric and magnetic field strengths. **F // B**. After ref. 33.

such circumstances it is possible, using the same structure and without introducing extra defects, to monitor transitions from a quasi 3D (B = F = 0) to a quasi 2D (B = 0, large F), or a quasi 1D (F = 0, large B) or a quasi 0D (F and B both large) situations[33] (see fig. 11). Preliminary experiments[33] have already shown that the optical absorption edge is radically altered at large F and B compared to the F = B = 0 situation (see fig. 12). A wealth of experiments and modelling are needed to fully quantify the effects of reduced dimensionalities in these flexible quantum wires and dots. Meanwhile the actual fabrication of laterally confined nanostructures will have certainly improved and the modelling will also be applicable to these structures.

ACKNOWLEDGEMENTS

We thank the CAPES (Brazil: R.F.) and CNRS (France: G.B.) for financial support. It is a pleasure to thank Drs. B. Soucail, J. A. Brum, L. L. Chang, A. Chomette, C. Delalande, B. Deveaud, L. Esaki, E. E. Mendez, M. H. Meynadier, P. Voisin and M. Voos for their active participation to the works reported here.

REFERENCES

1. M. Altarelli in Semiconductor Superlattices and Heterojunctions (Springer Verlag, Berlin 1986).

2. G. Bastard, Wave Mechanics Applied to Semiconductor Heterostructures (Les Editions de Physique, Les Ulis, 1988).

3. L. J. Sham, Superl. and Microstr. **5**, 335 (1989).

4. J. M. Luttinger, Phys. Rev. **102**, 1030 (1956).

5. A. Twardowski and C. Hermann, Phys. Rev. B **35**, 8144 (1987).

6. H. Chu and Y. C. Chang, Phys. Rev. B. **39**, 10861 (1989).

7. R. Ferreira and G. Bastard, submitted to Phys. Rev. B (1990).

8. T. Uenoyama and L. J. Sham, Phys. Rev. Lett. **64**, 3070 (1990).

9. G. Bastard, J. A. Brum and R. Ferreira (1989), to be published in Solid State Physics.

10. G. H. Wannier, Rev. Mod. Phys. **34**, 645 (1962).

11. R. Tsu and G. Döhler, Phys. Rev. B **12**, 680 (1975).

12. J. Bleuse and P. Voisin, 1988 (unpublished).

13. T. Weil and B. Vinter, App. Phys. Lett. **50**, 1281 (1987). See also B. Vinter and F. Chevoir, to be published in the Proceedings f the NATO Workshop on Resonant Tunnelling, El Escorial (1990).

14. R. Ferreira and G. Bastard, Europhys. Lett. **10**, 279 (1989).

15. K. Wassel and M. Altarelli, Phys. Rev. **B39**, 12803 (1989).

16. M. G. W. Alexander, M. Nido, W. W. Rühle and K. Köhler, to be published in the Proceedings of the NATO Workshop on Resonant Tunnelling. El Escorial (1990).

17. N. Vodjani, D. Côte, F. Chevoir, D. Thomas, E. Costard, J. Nagle and P. Bois, Semicond. Sci. Technol. **5**, 538 (1990).

18. R. C. Miller, D. A. Kleinman, W. T. Tsang and A. C. Gossard, Phys. Rev. **B24**, 1134 (1981).

19. G. Bastard, E. E. Mendez, L. L. Chang and L. Esaki, Phys. Rev. **B26**, 1974 (1982).

20. R. L. Greene, K. K. Bajaj and D. E. Phelps, Phys. Rev. **B29**, 1807 (1984).

21. D. A. Broido and L. J. Sham, Phys. Rev. **B34**, 3917 (1985).

22. G. E. W. Bauer and T. Ando, Phys. Rev. **B37**, 3130 (1988) and Phys. Rev. **B38**, 6015 (1988).

23. U. Ekenberg and M. Altarelli, Phys. Rev. **B35**, 7585 (1987).

24. L. C. Andreani and A. Pasquarello, Europhys. Lett. **6**, 259 (1988).

25. L. Vina, R. T. Collins, E. E. Mendez and W. T. Wang, Phys. Rev. Lett. **58**, 832 (1987).

26. J. A. Brum and F. Aguello-Rueda, Surf. Sci. **229**, 472 (1990).

27. M. M. Dignam and J. E. Sipe, Phys. Rev. Lett. **64**, 1797 (1990).

28. E. E. Mendez, F. Aguello-Rueda and J. M. Hong, Phys. Rev. Lett. **60**, 2426 (1988).

29. P. Voisin, J. Bleuse, C. Bouche, S. Gaillard, C. Allibert and A. Regreny, Phys. Rev. Lett. **61**, 1639 (1988).

30. T. Fukuzawa, E. E. Mendez and J. M. Hong, Phys. Rev. Lett. **64**, 3066 (1988).

31. A. Chomette, B. Lambert, B. Deveaud, F. Clerot, A. Regreny and G. Bastard, Europhys. Lett. **4**, 461 (1987).

32. B. Deveaud, A. Chomette, F. Clerot, A. Regreny, J. C. Maan, R. Romestain, G. Bastard, H. Chu and Y. C. Chang, Phys. Rev. **B40**, 5802 (1989).

33. R. Ferreira, B. Soucail, P. Voisin and G. Bastard, Phys. Rev. **B42**, 11404 (1990).

34. J. Y. Marzin and J. M. Gerard, Phys. Rev. Lett. **62** 2172 (1989).

35. R. Ferreira and G. Bastard, Surf. Sci. **229**, 424 (1990).

36. B. J. van Wees, H. van Houten, C. W. J. Beenarkker, J. G. Williamson, L. P. Kouwenhoven, D. van der Marel and C. T. Foxon, Phys. Rev. Let. **60**, 848 (1988).

37. D. A. Wharam, T. J. Thornton, R. Newbury, M. Pepper, H. Ahmed, J. E. F. Frost, D. Hasko, D. C. Peacock, D. A. Ritchie and J. A. C. Jones, J. Phys. **C21**, L209 (1988).

38. R. Landauer, IBM J. Res. Dev. **1**, 233 (1957); Philos. Mag. **21**, 863 (1970).

39. M. Büttiker, Y. Imry, R. Landauer and S. Pinhas, Phys. Rev. **B31**, 6207 (1985).

40. A. D. Stone and A. Szafer, IBM J. Res. Dev. **32**, 384 (1988).

41. M. H. Meynadier, C. Delalande, G. Bastard, M. Voos, F. Alaxandre, and J. L. Lievin, Phys. Rev. **B31**, 5539 (1985).

III-V VERSUS II-VI COMPOUND SUPERLATTICES

Michel Voos

Laboratoire de Physique de la Matiére Condensée de
l'ENS
24 rue Lhomond
75231 Paris Cedex 05
France

Abstract - At first, we describe an example of one of the important
effects recently observed in III-V compound superlattices, namely the
Wannier-Stark Quantization. We then describe original features of II-VI
compound systems which evidence some of the present differences
between III-V and II-VI compound heterostructures.

1. INTRODUCTION

Semiconductor superlattices (SL), which were first proposed by
Esaki and Tsu[1], consist of periodic stackings of thin alternate layers of
two different materials deposited on a suitable substrate by molecular
beam epitaxy (MBE) or metalorganic chemical vapor deposition. In first
approximation, they can be viewed as a series of quantum wells separated by potential barriers and coupled through resonant tunneling interaction. As a result of this coupling, the discrete energy levels, which
would correspond to isolated quantum wells, broaden in the z-direction
perpendicular to the plane of the layers and give rise to superlattice con-

duction and valence subbands or minibands, leading to band structures which are actually different from those of the corresponding host materials and can be tailored by changing parameters like the thickness of the layers for instance.

Superlattices involving III-V compound semiconductors have been extensively investigated for the last fifteen years about, and many new phenomena have been observed in these man-made heterostructures. Two striking examples are the quantum confined Stark effect [2] and the Wannier-Stark localization[3] which, depending on the layer thicknesses, result in a red and a blue shift of the band gap, respectively, when an electric field is applied in the z direction.

More recently, II-VI compound superlattices have been grown and additional aspects, due to particular features of the host materials, have been evidenced. In the case of Hg-based systems such as HgTe-CdTe or HgZnTe-CeTe superlattices, the unique inverted band structure of the zero-gap material (HgTe or HgZnTe) leads to a mass reversal for the light particles at each interface which gives rise to unusual band structures and properties. Another interesting situation is encountered in HgMnTe-CdTe and CdTe-CdMnTe superlattices which present original aspects due to the magnetic properties of the Mn ion. In addition, II-VI compound SL's involving wide gap semiconductors offer the possibility of obtaining light emission in the blue part of the spectrum.

After having presented an important effect recently observed in III-V compound SL's, we describe here some of the original properties of II-VI compound heterostructures which are also briefly compared to those of III-V compound systems.

2. III-V COMPOUND SUPERLATTICES: THE WANNIER-STARK QUANTIZATION

Many new effects have been observed in III-V compound heterostructures, and we describe here one of the most recent ones, namely the Wannier-Stark quantization (WSQ), which is interesting from the point of view of both basic and applied physics. This effect has been proposed recently[3-5] and observed independently in GaAs-AlGaAs SL's by Mendez et al[6] and Voisin et al[7], and also in GaInAs-AlGaInAs SL's by Bleuse et al[8].

The WSQ, which results from the application in the z direction of an electric field F to a semiconductor SL, has in fact been predicted and analyzed by Wannier[9] in the case of bulk materials, but never convincingly observed, probably because the required electric fields are much too large. The effect of an external electric field on the SL structure can be understood easily using a simple intuitive description. As already mentioned, a superlattice can be viewed as a series of quantum wells (QW) coupled through resonant tunneling which results in a broadening of the QW discrete energy levels into subbands of widths ΔE_1 and ΔHH_1 if we consider only the ground conduction (E_1) and heavy-hole (HH_1) subbands. Besides, the band gap E_g^{QW} of the isolated QW is[3] larger than that of the SL, E_g^{SL}, by about $(\Delta E_1 + \Delta HH_1)/2$. Under an electric field F along z, the energy levels in adjacent QW's are misaligned by eFd, where d is the SL period (Fig. 1). Resonant tunneling is thus inhibited and, for $eFd > \Delta E_1, \Delta HH_1$, the subband widths ΔE_1 and ΔHH_1 are equal to zero while the eigenfunctions are fully localized over each QW, leading to a two-dimensional (2D) situation. Since $E_g^{QW} = E_g^{SL} + (\Delta E_1 + \Delta HH_1)/2$, the fundamental optical transitions are, in these conditions, blue shifted by[3] $\Delta = (\Delta E_1 + \Delta HH_1)/2$ and, due to the 2D regime, the exciton binding energy and the optical excitonic transitions are enhanced. Besides, in an intermediate field regime, the eigenfunctions tend to localize over several adjacent wells so that the fundamental band to band transitions at E_g^{QW} are accompanied[3] by smaller ones occurring at $Eg^{QW} + peFd$ with $p = \pm 1, \pm 2...$, leading to an evenly spaced Wannier-Stark ladder. They correspond to "oblique" transitions in real space between an electron localized near the nth QW with a hole in the $(n + p)^{th}$ QW. These oblique transitions should disappear at high fields because, the localization becoming stronger, the overlap of the wavefunctions centered in adjacent QW's decreases rapidly.

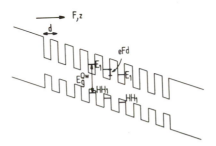

FIG. 1. Band structure of a usual semiconductor superlattice (SL) under an electric field F in the z direction perpendicular to the plane of the layers; d is the SL periodicity.

The WSQ has been observed in different SL's, as shown in Fig. 2 which presents[8] optical absorption data obtained at 2 K for different values of the voltage applied to a $Ga_{0.47}In_{0.53}As$-$Al_{0.24}Ga_{0.24}In_{0.52}As$ SL grown by MBE with respective layer thicknesses equal to 50 and 55 Å. The relatively small E_1-HH_1 excitonic peak occuring at 914 meV at zero bias (F = 0 Vcm^{-1}) shifts clearly towards high energies when F is increased. At the same time, satellite structures above and below the exciton peak appear, develop and then fade away are larger fields. These structures correspond to the predicted "oblique transitions" at E_g^{QW} + peFd, and their extinction at high electric field evidences the increasing field-induced localization. At a bias of 1.25 V (F ~ 20 kVcm^{-1}), the effective absorption edge is blue-shifted by about 5 meV, and the exciton line is clearly reaching saturation both in position and intensity, while the extinction of the oblique transition is not quite perfect yet. The observed blue shift, if we ignore linewidth effects, results from the competition of the blue shift due to the Wannier-Stark quantization, and of the red shift linked to the increase of the exciton binding energy. The later effect can be estimated reliably from earlier works[10], $E_b^{SL} \approx 2.5$ meV, and $E_b^{QW} \approx 8$ meV. It is thus deduced an experimental value of the bandgap blue shift Δ of 10 meV about, which compares[8] favorably with the calculated value $\Delta_{th} = E_g^{QW} - E_g^{SL} = 9$ meV.

The WSQ can be used, for instance, to make[11] original electro-optical modulators utilizing the shift of the observed optical transitions

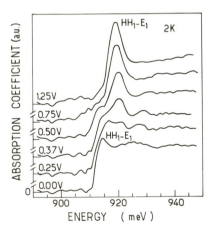

FIG. 2. Optical absorption coefficient at 2 K of a $Ga_{0.47}In_{0.53}As$-$Al_{0.24}Ga_{0.24}In_{0.52}As$ SL for different applied biases as a function of the incident photon energy.

under electric field. In addition to its interest in basic physics, the WSQ is thus potentially important for the development of electro-optical modulators for long-wavelength optical communications systems with an interesting characteristic, namely a very low driving voltage.

3. II-VI COMPOUND SUPERLATTICES

3.1 Hg-based superlattices

The band structure of bulk CdTe and HgTe is shown in Fig. 3 at 4 K. Besides, the band structure of bulk $Hg_{1-x}Zn_xTe$ and $Hg_{1-x}Mn_xTe$ is analogous to that of HgTe for Zn and Mn contents smaller than 12 and 7% respectively. CdTe is a usual wide gap semiconductor, but these Hg-based materials present two differences with respect to the band structure of III-V compounds. Indeed, they are zero-gap semiconductors or semimetals and, contrarily to what happens in the GaAs-AlGaAs

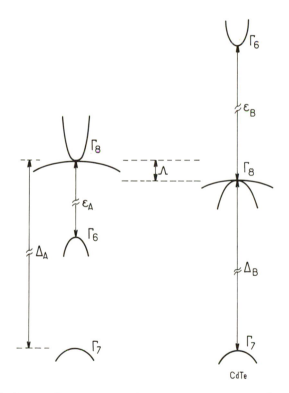

FIG. 3. Band structure of bulk HgTe and CdTe at 4 K.

system, the HgTe and CdTe conduction bands do not have the same symmetry. However the HgTe conduction band and the CdTe light hole band have the same Γ_8 symmetry, but not the same curvature. As a result, due to this unique inverted band structure, a mass reversal for light particles occurs at each interface and, for this reason. the corresponding Hg-based SL's have been called type III heterostructures.[12] The valence band offset, Λ, is shown in Fig. 3 and, from the point of view of the heavy-hole bands, the situation is comparable to that in the GaAs-AlGaAs system. Indeed, these heavy-hole bands have the same Γ_8 symmetry in HgTe and CdTe, and the HgTe layers are potential wells for heavy holes.

This mass reversal for light particles leads to interface states[13,14] and to unusual band structures. Figure 4 shows the calculated ground light-particle wavefunction for $\Lambda = 40$ meV in the case of a CdTe-HgTe-CdTe heterostructure.[14] It can be seen that the wavefunction peaks at each interface corresponding, for this reason, to an interface state. Besides, it is certainly worth pointing out that all the light-particle states are interface states.

Figure 5 gives[15,16] the band structure calculated at 2 K in the envelope function approximation in the case of a MBE-grown $Hg_{1-x}Zn_xTe$-CdTe SL with respective layer thicknesses equal to 105 and 20Å. Since x = 5.3%, this is necessarily a type III situation. As pointed out by Johnson et al.,[17] such a SL can be semiconducting or semimetallic depending, in particular, on the value of the valence band offset Λ. Figure 5(a) presents the calculated band structure of this SL along the growth axis (k_z) and in the plane of the layers (k_x) for $\Lambda = 40$ meV;

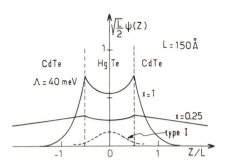

FIG. 4. Ground light-particle wavefunction of a CdTe-HgTe heterostructure (x = 1) calculated for $\Lambda = 40$ meV. Similar results for a $Hg_{0.75}Cd_{0.25}Te$-CdTe heterostructure (x = 0.25).

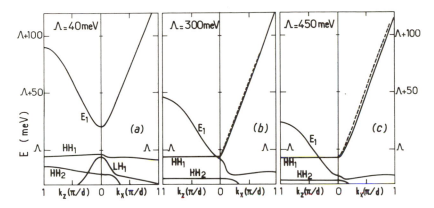

FIG. 5. Calculated band structure of the HgZnTe-CdTe SL described in the text at 2 K in the plane of the layers (k_x) and along the growth axis (k_z) for Λ = 40 meV (a), Λ = 300 meV (b) and Λ = 450 meV (c). The dashed lines in figures 5b and 5c represent the in-plane dispersion for $k_z = k_c$.

Figs. 5(b) and (c) show similar results obtained for Λ = 300 and 450 meV. The zero of energy is taken at the CdTe valence-band edge. E_1, LH_1 (HH_1, HH_2), respectively, denote the first light particle (heavy hole) bands for the motion along the growth axis z. For Λ = 40 meV (Fig. 5(a), the SL is a semiconductor, E_1 being the conduction band and HH_1 the highest valence band. For large valence band offsets, the E_1-HH_1 separation decreases until finally E_1 and HH_1 meet and then cross in the k_z direction for $k_z = k_c$. For Λ = 300 meV (Fig. 5(b)), HH_1 becomes the conduction band and the superlattice is semimetallic with a nearly zero gap and $K_c = 0.1 \, \pi/d$, where d is the SL period. For Λ = 450 meV (Fig. 5(c)), E_1 is at about 16 meV below HH_1 and $K_c = 0.3 \, \pi/d$. The most important change in the band structure when increasing Λ is the k_z dispersion of the SL conduction band. For Λ = 40 meV, this band is nearly isotropic around k = 0. However, for Λ = 300 meV and 450 meV, it is dispersionless for $0 < k_z < k_c$, corresponding to a strong mass anisotropy. In any case the in-plane mass m_x is extremely small near k_x = 0 ($m_x \approx 2.10^{-3} \, m_o$), as a result of the small HH_1-E_1 separation. The dashed line in Figure 5b and 5c gives the in-plane dispersion relation calculated for $k_z = k_c$. One can see that m_x depends on the growth-direction wavevector because $m_x \propto |HH_1$-$E_1|$ which depends on k_z. Figure 5 shows that the measurement of the conduction band anisotropy m_z/m_x, where m_x is the electron mass along k_z, can yield the nature, i.e. semiconductor or semimetallic, of the considered SL.

To obtain the electron mass anisotropy ratio in this SL, magneto-absorption experiments in the far infrared were done[15,16] at 1.7 K. From the data, the electron cyclotron mass m_c was obtained as a function of the angle θ between the z growth axis and the direction of the applied magnetic field. Figure 6 gives[16] m_c versus θ for an infrared photon wavelength equal to 118 μm. To interpret these data an ellipsoidal dispersion for the conduction band near k = 0 is assumed so that the cyclotron mass is given[18] at low magnetic field by $m_c^2 = (m_x^2 m_z)/(m_z \cos^2 \theta + m_x \sin^2 \theta)$, where m_x is here the in-plane mass at the photon energy and m_z the mass along the growth axis. The solid line in Figure 6, which is the calculated cyclotron mass using $m_z = 0.065\ m_0$, is in very good agreement with the experimental data. The measured mass-anisotropy ratio m_z/m_x is ~ 25 for $\lambda = 118\ \mu$m. Comparison to the band structure calculations shown in Figure 5 clearly supports the semimetallic nature of the superlattice. Indeed, calculations in the semiconductor configuration for $\Lambda = 40$ meV (Fig. 5a) lead to an anisotropy ratio of ≈ 1 at $\lambda = 118\ \mu$m, while the anisotropy is found to be ≥ 10 for $\Lambda \geq 300$ meV. In fact, from these investigations, Λ is found to be of the order of 350 meV.

It should also be pointed out that the investigated SL exhibits an unusual effect, namely a semimetal to semiconductor transition when the temperature is increased. This transition was evidenced[19] from far infrared magneto-absorption experiments, yielding a critical temperature of about 25 K, in good agreement with band structure calculations.

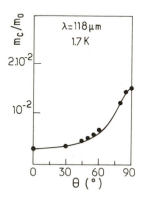

FIG. 6. Cyclotron mass m_c versus θ for $\lambda = 118\ \mu$m and T = 2 K. The dots correspond to the experimental points and the solid line to the calculated values in the ellipsoidal approximation described in the text.

Finally, it has been proposed[20] that Hg-based SL's could be interesting as infrared materials. Their optical absorption coefficient is comparable to its value in the corresponding bulk HgCdTe alloy and they present some advantages like a heavy mass along the z-direction which should result in a reduction of the tunneling and diffusion currents in a photodetector device. Calculations[21] of the band gap and of the corresponding cut-off wavelength λ_g have been done for HgTe-CdTe SL's as a function of temperature. For such SL's with equal layer thicknesses (d_1 and d_2) for both materials, these calculations yield $d_1 = d_2 = 35$ Å for $\lambda_g = 10$ μm and T = 77 K. Such thicknesses can be realized by molecular beam epitaxy, but some other technological problems should be solved if one wants really to use these SL's to make infrared detectors capable of competing with bulk HgCdTe.

3.2. Semimagnetic Superlattices

Many studies have been directed toward semimagnetic SL's (SMSL) in search of low-dimensionality magnetic effects.[22-26] The $Hg_{1-x}Mn_xTe$-CdTe type III system is interesting in the sense that the two-dimensional electron confinement occurs in the semimagnetic layers, and this system should be particularly attractive to probe the exchange interaction between the localized Mn magnetic moments and the SL band electrons. In HgMnTe-CdTe SL's, a strong effect of the exchange interaction on the Landau level energy is expected which can be evidenced from the temperature dependence of the electron cyclotron resonance.

Here are described some investigations done[27,28] in a $Hg_{0.96}Mn_{0.04}Te$-CdTe SL grown by molecular beam epitaxy on a (100) GaAs substrate with a 2 μm CdTe (111) buffer layer. The SL consists of 100 periods of 168 Å thick $Hg_{0.96}Mn_{0.04}$ Te layers interspaced by 22 Å thick CdTe barriers. It is n type with an electron density n = 6 x 10^{16} cm^{-3} and a mobility $\mu = 2.7$ x 10^4 cm^2/Vs at 25 K.

The SMSL band structure and Landau level calculations have been done using a 6 x 6 envelope function hamiltonian.[28,29] The Γ_6-Γ_8 energy separation is taken to be - 130 meV in $Hg_{0.96}Mn_{0.04}Te$ at low temperature. For the $Hg_{0.96}Mn_{0.04}Te$ layers, the theoretical model is modified to take into account the exchange interactions between localized d electrons bound to the Mn^{2+} ions and the Γ_6 and Γ_8 s- and p-band electrons. The s-d and p-d interactions are introduced in the molecular field approximation through two additional parameters, r and A, where $r = \alpha/\beta$ is the ratio between the Γ_6 and Γ_8 exchange integrals, α and β, respectively.[30,31] Here A is the normalized magnetization defined by $A = (1/6)\beta x N_0 <S_z>$, where N_0 is the number of unit cells per unit volume of the crystal, x is the Mn composition (x = 0.04), and $<S_z>$ is the thermal average of the spin operator along the direction of the applied magnetic field B. Extensive magneto-optical data[30,31] and direct magnetization measurements[32,33] performed on bulk alloys with dilute Mn concentration have shown that $<S_z>$ is well described by a modified spin - 5/2 Brillouin function, where the temperature T is replaced by an empirical effective termperature $(T + T_0)$. T_0 is found[33] to be ~ 5 K for x = 0.04. The exchange integral[31] $N_0\beta$ is known to be of the order of 1 eV and r ~ - 1. The resulting band structure along k_x and k_z, calculated for zero magnetic field using a valence band offset $\Lambda = 300$ meV, is shown in Figure 7 and one can note that this SMSL is semimetallic.

Far infrared magneto-absorption experiments were done[27-29] between 1.5 and 10 K in this SMSL. Figure 8 shows transmission spectra obtained at $\lambda = 41$ μm for different temperatures. The main absorption minimum corresponds to the ground electron cyclotron reso-

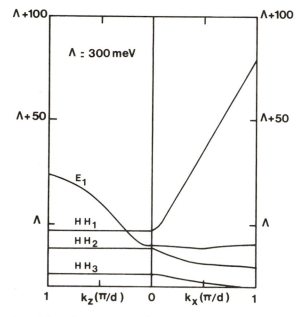

FIG. 7. Calculated band structure along k_x and k_z for a $Hg_{0.96}Mn_{0.04}Te$-CdTe superlattice at T = 2 K.

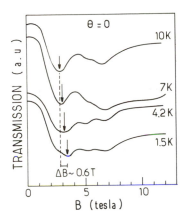

FIG. 8. Far-infrared magnetotransmission spectra at $\Lambda = 41$ μm for temperatures from 1.5 to 10 K. The arrows correspond to the calculated cyclotron resonance position. The marker above ΔB indicates the magnitude of the calculated shift for the observed cyclotron resonance transition.

nance. For each temperature, the calculated field position of the line is indicated by the arrows in Figure 8, and one can see that the agreement between experiments and theory is quite good. The most striking point is that this transition shifts to lower B with increasing temperature, the variation being $\Delta B \sim -0.6$ T between 1.5 and 10 K. There are two contributions to this temperature-dependent shift: the temperature dependence of the $Hg_{0.96}Mn_{0.04}Te$ energy gap and the temperature dependent Mn magnetization. In bulk HgTe, as T increases from 1.5 K to 10 K, the magnitude of the energy gap decreases by less than 3 meV. The effect in $Hg_{0.96}Mn_{0.04}Te$ is expected to be similar. Such a small shift in energy gap is calculated to have only a slight effect on the cyclotron resonance transition, at most -0.05 T. The observed shift of the observed transition is an order of magnitude larger than would imply the temperature dependence of the cyclotron resonance line, evidencing the effect of the exchange interaction in this semimagnetic semiconductor superlattice. The magnetization observed in the superlattice is consistent with the bulk $Hg_{0.96}Mn_{0.04}Te$ magnetization, as a result of the thick $Hg_{0.96}Mn_{0.04}Te$ layers and thin CdTe barriers in the investigated superlattice. To observe reduced-dimensionality magnetic effect would probably imply to be in the opposite situation, namely thin HgMnTe layers and thick CdTe barriers with respect to the distance between Mn ions.

Other SMSL's, such as the CdTe-CdMnTe system,[22-26,34,35] have been investigated. Such heterostructures are different from the previous one because they correspond to usual wide-gap semiconductors and also because the Mn ions are now in the potential barriers. One of the key differences that they present with the GaAs-AlGaAs system is the large magnetic field variation of the CdMnTe bandgap, and thus of the CdTe-CdMnTe conduction-band and valence-band offsets. These effects arise from the exchange interaction between the magnetic ions and the carriers.[36,37] Associated with the possible small value of the valence band offset Λ[34] a magnetic tuning of Λ leads to a type I $\rightarrow$ type II transition at a moderate magnetic field in the case of low Mn concentration in the barrier. This transition has been first observed very recently in another system, ZnSe-ZnFeSe,[38] and it is due to the very low Λ and the lattice mismatch.

Figure 9 describes[39] the effect of a magnetic field B on the conduction and valence bands in a CdTe-CdMnTe SL. Due to the exchange interaction, the conduction band gives rise to two bands with $m_j = \pm 3/2, \pm 1/2$. As a result, the valence band offset depends clearly on the magnetic field, so that the energy of the E_1 and HH_1 subbands is modified as shown in Fig. 9. Thus, as described in Fig. 10 (upper panel), a type I to type II transition can occur when the magnetic field is increased. In this case, the energy of the fundamental E_1-HH_1 band-to-band transition decreases[39,40] smoothly with increasing B, as schematized by the upper curve in the E vs B diagram (Fig. 10, lower panel). Also shown is the B-dependence[39,40] of the exciton binding energy R^* which exhibits a fast decrease in the vicinity of the type I - type II transition. As a result, the energy of the E_1-HH_1 excitonic optical transition as a function of B

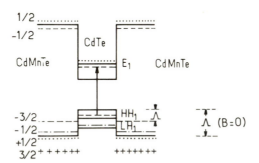

FIG. 9. Conduction and valence bands for a CdTe-CdMnTe heterostructure at zero magnetic field (solid line) and for a finite magnetic field, showing that the valence band offset Λ is field-dependent.

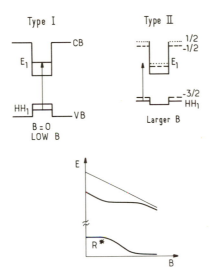

FIG. 10. Upper panel : schematization of a type I - type II transition induced by an applied magnetic field B.
Lower panel : typical variation versus B of the energy of the fundamental E_1-HH_1 band-to-band transition (upper curve); B-dependence of the exciton binding energy R^* (lower curve); resulting variation as a function of B of the energy of the E_1-HH_1 excitonic optical transition (middle curve).

should present a plateau (or perhaps a bump) when the type II situation is reached, as shown in Fig. 10.

Figure 11 presents typical photoluminescence data[39] (crosses, upper panel) obtained as a function of B at 1.7 K in a MBE-grown 86 Å - 86 Å CdTe-$Cd_{0.93}Mn_{0.07}$Te SL with 25 periods. These data correspond to the fundamental E_1-HH_1 excitonic luminescence transition, and the corresponding theoretical variation is given for three values of the parameter $Q_v^0 = \Lambda/\Delta E_g$, where Λ is the zero-magnetic-field strain-free value of the valence band offset and ΔE_g the difference between the band gaps of the host materials. The agreement between experiment and theory is quite good for $Q_v^0 \sim 15$-20%, yielding $\Lambda \sim 20$ meV, which is certainly the best value obtained up to now for this parameter. Besides, this evidences a magnetic-field-induced type I to type II transition around 1.9 T, which is supported by the decrease of the corresponding luminescence efficiency (Fig. 11, lower panel). The other curve shown in Fig. 11 (upper panel, circles) gives additional support to this type I - type II transition, and is discussed elsewhere.[39]

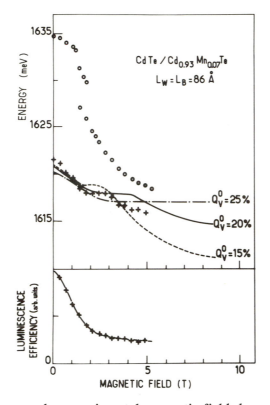

FIG. 11. Upper panel : experimental magnetic-field dependence of the energy of the E_1-HH_1 excitonic luminescence line at 1.7 K (crosses). The solid, dashed and dashed-dotted lines correspond to the theoretical variation of the energy of this excitonic optical transition for different values of the parameter Q_v^0 defined in the text. The other curve (circles) is discussed in details in Ref. 39.

Lower panel : experimental variation of the luminescence efficiency of the E_1-HH_1 line as a function of the magnetic field at 1.7 K.

3.3. Wide-gap Superlattices: Blue Light Emission

It has been shown recently that it is possible to grow wide-gap II-VI compound SL's whose quality is good and whose band gap lies in the blue part of the optical spectrum. For instance, Kolodziejski et al.[41] have grown ZnSe-ZnMnSe SL's exhibiting luminescence around 2.75 eV at low temperature. Besides, the observed luminescence efficiency was much larger than that of ZnSe epilayers, which is certainly an important result. Indeed, it is clear that blue-light emitting devices are quite important, but their fabrication raises problems which are not yet solved like, for instance, the doping of these II-VI materials.

4. CONCLUSION

It is well-known that III-V compound heterostructures have led to the observation of many new phenomena. They are still probably the best candidates for evidencing novel effects because the quality of the samples and of the often-required technology is high.

On the other hand, most of the effects detected in III-V compound SL's should be observable in II-VI compound heterostructures like, for example the Wannier-Stark quantization which has been reported[42] in CdTe-CdMnTe SL's. In addition, Hg-based systems, corresponding to type III structures, might lead to interesting phenomena such as tunneling which should give different results from those encountered in III-V compound heterostructures because of the unusual and unique interface wavefunction. Also, II-VI compound SL's involving Mn might allow the observation of new low-dimensionality magnetic effects. Finally, wide-gap II-VI compound heterostructures might give rise to blue-light emitting devices, provided some important problems, like doping for instance, are solved in the near future.

However, it should not be forgotten that they are attempts to make III-V compound semimagnetic SL's such as the InAs-InMnAs system,[43] and also that IV-VI compound semimagnetic SL's have been successfully grown and investigated, as exemplified[44] by the interesting studies performed on PbTe-PbMnTe heterostructures.

ACKNOWLEDGEMENTS

The author would like to thank G. Bastard, J. Bleuse, J. M. Berroir, C. Delalande, E. Deleporte, Y. Guldner, J. Manassés, B. Soucail, P. Voisin and J. P. Vieren from the Laboratoire de Physique de la Matiére Condensée de l'ENS; M. Allovon, E. Bigan and M. Quillec from CNET-Bagneux; A. Regreny from CNET-Lannion; C. Alibert, S. Bouchaud, S. Gaillard from EM2-Montpellier; L. L. Chang, L. Esaki and J. M. Hong from IBM; and J. P. Faurie from the University of Illinois at Chicago for their contribution to the investigations described here.

REFERENCES

1. L. Esaki and R. Tsu, IBM J. Res. Develop. **14**, 61 (1970).

2. G. Bastard, E. E. Mendez, L. L. Chang and L. Esaki, Phys. Rev. **B28**, 3241 (1983).

3. See, for instance, J. Bleuse, G. Bastard and P. Voisin, Phys. Rev. Lett. **60**, 220 (1988).

4. P. Voisin, A superlattice optical modulator, French patent (1986).

5. G. Bastard, NATO ASI on Interfaces, Quantum Wells and Superlattices, 1987 (Plenum, 1988).

6. E. E. Mendez, F. Agullo-Rueda and J. M. Hong, Phys. Rev. Lett. **60**, 2248 (1968).

7. P. Voisin, J. Bleuse, S. Bouche, S. Gaillard, C. Allibert and A. Regreny, Phys. Rev. Lett. **61**, 1639 (1988).

8. J. Bleuse, P. Voisin, M. Allovon and M. Quillec, Appl. Phys. Lett. **53**, 2632 (1988).

9. G. H. Wannier, Rev. Mod. Phys. **34**, 645 (1962).

10. A. Chomette, B. Lambert, B. Devaud, A. Regreny and G. Bastard, Europhys. Lett. **4**, 461 (1988).

11. E. Bigan, M. Allovon, M. Carré and P. Voisin, Appl. Phys. Lett. **57**, 327 (1990); E. Bigan, M. Allovon, M. Carré, A. Carenco and P. Voisin, IEEE LEOS Conference on Integrated Optoelectronics (1990), to be published; P. Voisin, B. Soucail, E. Bigan and M. Allovon, Device Research Conference, Santa Barbara (1990), to be published; K. K. Law, R. Yan, J. L. Merz and L. A. Coldren, ibid.

12. Y. Guldner, G. Bastard, J. P. Vieren, M. Voos, J. P. Faurie and A. Million, Phys. Rev. Lett. **51**, 907 (1983).

13. Y. C. Chang, J. N. Schulman, G. Bastard, Y. Guldner and M. Voos, Phys. Rev. **B31**, 2557 (1985).

14. G. Bastard, in Wave Mechanics Applied to Semiconductor Heterostructures, (Les Editions de Physique, Les Ulis, France, 1988).

15. J. M. Berroir, Y. Guldner, J. P. Vieren, M. Voos, X. Chu and J. P. Faurie, Phys. Rev. Lett. **62**, 2024 (1989).

16. M. Voos, J. Manassés, Y. Guldner, J. M. Berroir, J. P. Vieren and J. P. Faurie, to appear in Superlattices and Microstructures (1990).

17. N. F. Johnson, P. M. Hui and H. Ehrenreich, Phys. Rev. Lett. **61**, 1993 (1988).

18. N. W. Ashcroft and N. D. Mermin, in Solid State Physics, (Holt, Rinehard and Wilson, New York, 1976), p. 571.

19. M. Voos, J. Manassés, Y. Guldner, J. M. Berroir, J. P. Vieren and J. P. Faurie, to be published.

20. D. L. Smith, T. C. McGill and J. N. Schulman, Appl. Phys. Lett. **43**, 180 (1983)

21. Y. Guldner, unpublished.

22. R. N. Bicknell, R. W. Yanka, N. C. Giles-Taylor, E. L. Buckland, and J. F. Schetzina, Appl. Phys. Lett. **45**, 92 (1984).

23. L. A. Kolodziejski, T. C. Bonsett, R. L. Gunshor, S. Datta, R. B. Bylsma, W. M. Becker and N. Otsuka, Appl. Phys. Lett. **45**, 440 (1984).

24. K. A. Harris, S. Hwang, Y. Lansari, J. W. Cook Jr. and J. F. Schetzina, Appl. Phys. Lett. **49**, 713 (1986).

25. D. D. Awschalom, J. M. Hong, L. L. Chang and G. Grinstein, Phys. Rev. Lett. **59**, 1733 (1987).

26. L. L. Chang, this volume.

27. G. S. Boebinger, Y. Guldner, J. M. Berroir, M. Voos, J. P. Vieren and J. P. Faurie, Phys. Rev. **B36**, 7930 (1987).

28. Y. Guldner, J. Manassés, J. P. Vieren, M. Voos and J. P. Faurie, to be published.

29. J. M. Berroir, Y. Guldner, J. P. Vieren, M. Voos and J. P. Faurie, Phys. Rev. **B34**, 891 (1986).

30. G. Bastard, C. Rigaux, Y. Guldner, J. Mycielski and A. Mycielski, J. Phys. **39**, 87 (1978).

31. G. Bastard, C. Rigaux, Y. Guldner, A. Mycielski, J. K. Furdyna and D. P. Mullin, Phys. Rev. **B24**, 1961 (1981).

32. W. Dobrowolski, M. von Ortenberg, A. M. Sandauer, R. R. Galazka, A. Mycielski and R. Pauthenet, in Physics of Narrow Gap Semiconductors, Vol. 152 of Lecture Notes in Physics, edited by E. Gornik (Springer, Heidelberg, 1982), p.302.

33. J. R. Anderson, M. Gorska, L. J. Azevedo and E. L. Venturini, Phys. Rev. **B33**, 4706 (1986).

34. A. V. Nurmikko, R. L. Gunshor and L. A. Kolodziejski, IEEE J. Quant. Electro. QE **22**, 1785 (1986).

35. D. D. Awschalom, J. Warnock, J. M. Hong, L. L. Chang, M. B. Ketchen and W. J. Gallagher, Phys. Rev. Lett. **62**, 2309 (1989).

36. J. A. Gaj, R. Planel and G. Fishman, Solid State Commun. **29**, 435 (1979).

37. J. Wornock, A. Petrou, R. N. Bicknell, N. C. Giles-Taylor, O. K. Blanks and J. F. Schetznina, Phys. Rev. **B32**, 816 (1985).

38. X. Liu, A. Petrou, J. Warnock, B. T. Jonker, G. A. Prinz and J. J. Kretz, Phys. Rev. Lett. **63**, 2280 (1989).

39. E. Deleporte, J. M. Berroir, C. Delalande, G. Bastard, M. Hong and L. L. Chang, to appear in Phys. Rev. (1990).

40. G. Bastard and J. M. Berroir, unpublished.

41. L. A. Kolodziejski, R. L. Gunshor, T. C. Bonsett, R. Venkatasubramanian, S. Datta, R. B. Bylsma, W. M. Becker and N. Otsuka, Appl. Phys. Lett. **47**, 169 (1985).

42. A. Harwit, C. Hsu, F. Agullo-Rueda and L. L. Chang, to appear in Appl. Phys. Lett. (1990).

43. H. Munekata, H. Ohno, S. von Molnar, A. Segmuller, L. L. Chang and L. Esaki, Phys. Rev. Lett. **63**, 1849 (1989).

44. See, for instance, H. Clemens, H. Krenn, P. C. Weilguni, U. Stromberger, G. Bauer and H. Pascher, Surf. Sci. **228**, 236 (1990).

THE FIRST PRINCIPLES VIEW OF SUPERLATTICES

Hiroshi Kamimura

Department of Physics, University of Tokyo*
Bunkyo-ku, Tokyo, Japan 113

Abstract - In the present lecture very short period superlattices are chosen as an object of study. Very short period superlattices including strained superlattices have recently aroused keen interests as a new class of materials among semiconductor superlattices and have been extensively studies both theoretically and experimentally. In particular, very short period GaAs/AlAs superlattices with (001) interface was one of the highlight topics among various kinds of superlattices for recent several years. A number of first principles band structure and stability calculations have been performed for $(GaAs)_m(AlAs)_m$ superlattices with (001) interface, and a number of experimental groups have tested theoretical results. In the present lecture I will give an overview on the recent remarkable progress in research on the band structures and optical properties of this system, with strong emphasis on clarifying the characteristic features of the band structures and the optical properties, in particular by being compared with those of ordinary thick layer

---

* Present Address: Department of Applied Physics, Faculty of Science, Science University of Tokyo, 1-3 Kagurazaka, Shinjuku-ku, Tokyo Japan 162

Highlights in Condensed Matter Physics and Future Prospects
Edited by L. Esaki, Plenum Press, New York, 1991

superlattices. Finally, as a direction in the future prospect of research on this system, the improvement of the band gap problem in the local density approximation and the sample problem are discussed.

1. INTRODUCTION

One of the recent remarkable trends in condensed matter physics is a vigorous interaction among physics, materials science and technology. Let me express schematically such interactions by the correlation diagram shown in Fig. 1. By this correlation diagram I would like to indicate possibilities that the synthesis of new materials leads to the discovery of novel and exotic phenomena and new physical concepts, and new devices are invented from them.

We can find the examples of this diagram in the highlights of the eighties in the condensed matter physics. For example, the discovery of normal and fractional quantum Hall effects has been made on the striking developments of the growth techniques of semiconductor heterostructures and superlattices while the discovery of the high temperature superconductivity stands on the remarkable developments of the copper oxide ceramics technology. Research on the above two subjects has now accidentally opened a new frontier field of fractional statistics and anyon physics.

In this way the technology related to making artificial materials not found in nature has been playing an important role in the development of condensed matter physics. As a result the birth and rapid growth of new physical concepts and of new phenomena and the invention of new

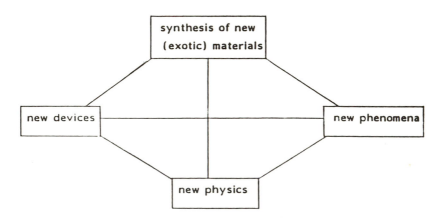

FIG. 1. The correlation diagram.

devices such as the scanning tunneling microscope have given tremendous impact not only to the other fields of physics, but also to other areas of science and technology. Thus, I believe, the condensed matter physics continues to be in a golden age in the nineties, with strong ties with various branches of physics, materials science and technology.

In this respect we might recall that, from as far back as the Stone Age, some 8,000 years ago, every significant period in the history of human civilization has been prompted and characterized by the discovery of a new material. The semiconductor superlattice, proposed by Esaki and Tsu in 1969[1,2], and recently made possible, might represent for us the new material for our new age, the "tailor-made material age." With this new tailor-made material, an important doorway to a new system of reduced dimensionality and a new semiconductor technology has been opened.

In this circumstance of rapidly growing semiconductor physics research, I would like to pay attention in this lecture to a new class of semiconductor materials, i.e. the very short period suprlattices, because these new semiconducting materials have recently stimulated a number of theoretical and experimental investigations as for electronic and structural properties and have opened possibilities of new devices. In this lecture I will first describe first-principles theoretical methods to calculate the electronic structure of a new material of a weakly correlated electronic system, and then try to elucidate the important features of the band structures and optical properties of very short period semiconductor superlattices.

2. VERY SHORT PERIOD SEMICONDUCTOR SUPERLATTICES

A superlattice is a one-dimensional periodic structure consisting of alternating thin layers. Recent remarkable progress in thin-film growth techniques such as the molecular beam epitaxy (MBE), the metal organic chemical vapor deposition (MOCVD) and the atomic layer epitaxy (ALE)[3,4] have made it possible to control the epitaxial growth to almost one monolayer at a heterointerface. In this way one can now make very short period superlattices with sharp interfaces.

In this lecture I choose a typical semiconductor superlattice of $(GaAs)_m(AlAs)_n$ as an object of study. Figure 2 illustrates a perfect $(GaAs)_m(AlAs)_n$ superlattice with (001) interface. Since GaAs/AlAs or $GaAs/Al_xGa_{1-x}As$ superlattice are very important as high speed elec-

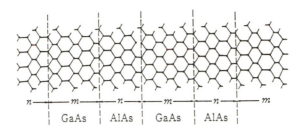

FIG. 2. The $(GaAs)_m(AlAs)_n$ superlattice with (001) interface.

tronic devices or quantum well lasers, a huge number of literatures on $GaAs/Al_xGa_{1-x}As$ systems have been published in the past decade, but these have been mostly concerned with superlattices of large layer thickness, and most of experimental results have been analyzed on the basis of the Kronig-Penney model.[2]

As regards very short period superlattices, however, the interaction between adjacent layers is so strong that the characters of constituent materials (e.g. GaAs and AlAs) are completely mixed. Thus the Kronig-Penney model does not hold for a system of very short period superlattices. In this context the first principles band structure calculations are essential for elucidating the electronic structures of the very short period superlattices.

In addition to elucidating the band structures and optical properties of $(GaAs)_m(AlAs)_m$ systems, I would like to investigate whether or not it is possible (1) to find in the first principles results the range of layer thickness from which the Kronig-Penny model begins to become effective and (2) to make the first principles derivation of the band offset ratio of conduction band to valence band.

3. CRYSTAL STRUCTURES AND A BRILLOUIN ZONE

Both GaAs and AlAs have a zinc-blende crystal structure. Because of their nearly equal bond lengths of about 2.452 A, GaAs and AlAs form a superlattice consisting of a sequence of alternate stacking of GaAs and AlAs layers without mismatch at interfaces. Figure 3 illustrates the unit cell of the $(GaAs)_2 \times (AlAs)_2$ superlattice with (001) interface. Hereafter we denote the $(GaAs)_m \times (AlAs)_n$ superlattice with (001) interface as "SL(m,n)" for simplicity. The primitive translational vectors of SL(n,n) are given by

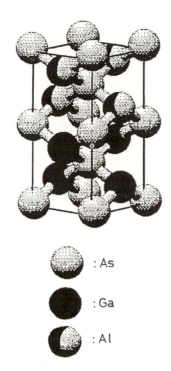

: As

: Ga

: Al

FIG. 3. The unit cell of $(GaAs)_2/(AlAs)_2$ superlattice.

$$a_1 = (1/2, 1/3, 0) \times a_0,$$
$$a_2 = (-1/2, 1/2, 0) \times a_0, \qquad (1)$$
$$a_3 = (0,0,0) \times a_0,$$

where we take the x and y axes parallel to the layer and the z axis along the direction of a superlattice. Further $a_0 = 10.683$ a.u. and then the period of SL(n,n) is $n \times a_0$.

The space group of SL(n,n) is D_{2d}^1. SL(n,n) contains $4 \times n$ atoms in a unit cell and their positions are, in case of SL(1,1), written as Ga:(0,0,0), Al:$(0,1/2,1/2) \times a_0$, and As:$(1/4,1/4,1/4) \times a_0$ and $(-1/4,1/4,3/4) \times a_0$. Those of SL(n,n)'s are obtained similarly. Thr Brillouin zone of SL(n,n) is shown in Fig. 4(a). This is obtained by folding the fcc Brillouin zone of bulk GaAs or AlAs shown in Fig. 4(b) along the z direction. The folded zone is also shown in Fig. 4(b) by dotted lines. Looking at this figure we can know that the symmetry points of the bulk crystal are folded into the points in the tetragonal Brillouin zone of SL(n,n) in the following way; $\Gamma \rightarrow \Gamma$, $X \rightarrow [\ \Gamma$ and X_{xy}], and $L \rightarrow [L$ or $L']$ etc.

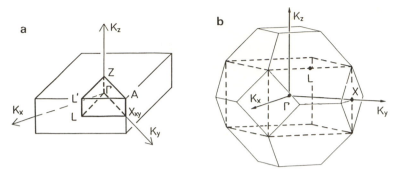

FIG. 4. (a) The Brillouin zone of SL(n,n). (b) The Brillouin zone of GaAs and AlAs.

4. METHOD OF CALCULATIONS

In order to calculate the band structures of new materials, each theory group can generally develop its own theoretical recipe. Our theory group have developed the total energy self-consistent pseudopotential method in the local density approximation (LDA) for weakly correlated electronic systems[5,6,7,8]. The psudopotential is the potential to which the outer or valence electrons in an atom or a solid are subject. The inner or core electrons are frozen in the pseudopotential. Two pseudopotential methods, the empirical pseudopotential and the density functional pseudopotential are often used. The former which is determined so as to reproduce atomic and solid properties is suited to analyzing excited state properties of an atom or a solid while the latter which is formed from the atomic wavefunctions of all electrons in an atom is suited to the determination of the ground state properties. The work mentioned in this lecture is concerned with the band structures and optical properties of SL(n,n) but not the stability and the ground state properties. Thus the empirical pseudopotential method is more appropriate.

In the density functional theory Kohn and Sham[9] reduced the problem of the many-electron Schrödinger equation to an effective one-electron problem. In the LDA the energy eigenvalues $\varepsilon_{\mu,k}$ and eigenfunctions $\Psi_{\mu,k}(r)$ are obtained by solving the following one-electron Schrödinger equation self-consistently,

$$
\left[-1/2 \, \nabla^2 + \sum_{\beta} V^{\beta}_{\text{ion}}(r - R_{\beta}) + V_{H}(r) + V_{xc}(r) \right] \Psi_{\mu,k}(r)
$$
$$
= \varepsilon_{\mu,k} \Psi_{\mu,k}(r),
$$
(2)

where $V_{ion}(r-R_\beta)$ is the pseudopotential of ion β positioned at R_β and $V_H(r)$ the Hartree potential due to valence electrons which is given from the total valence charge density $\rho_{tot}(r)$ by solving Poisson's equation

$$\nabla^2 V_H(r) = -4\pi\rho_{tot}(r). \tag{3}$$

Here

$$\rho_{tot}(r) = 2 \sum_{\mu, k}^{occ.} | \Psi_{\mu, k}(r)|^2, \tag{4}$$

where the summation is taken over the occupied states. In eq. (2) $V_{xc}(r)$ represents the exchange-correlation potential, for which the $X\alpha$ approximation is used, and thus V_{xc} is expressed as

$$V_{xc}(r) = -\frac{3\alpha}{2} \left[\frac{3}{\pi} \rho_{tot}(r) \right]^{1/3}. \tag{5}$$

Since it is known that $V_{xc}(r)$ determined by the LDA does not give a correct value of the band gap of semiconductor, Nakayama and Kamimura[5] have proposed to use α as a disposal parameter which is chosen so as to reproduce a band gap energy.

As for the model potentials of Al^{3+} and As^{5+} ions, the local pseudopotentials used by Pickett, Louie and Cohen[10] are adopted. The Fourier transform of these model potentials are given in the following form

$$V_{ion}^\beta(G) = \frac{1}{\Omega_\beta} \frac{a_1^\beta}{G^2} \left[\cos(a_2^\beta G) + a_3^\beta \right] \cdot \exp[a_4^\beta G^4], \tag{6}$$

where β denotes a kind of ion and Ω_β the atomic volume. As for Ga^{3+} the values adopted by Nakayama et al.[5] are used. The values of all parameters are given in Table I. The one-electron levels of Ga^{3+}, Al^{3+} and As^{5+} ions calculated using these ionic potentials $V_{ion}(G)$ are given in Table II, and compared with the one-electron levels obtained by solving all the electron levels self-consistently for Ga^{3+}, Al^{3+} and As^{5+} ions. It is seen from Table II that the energy levels of valence electrons are well described by the above pseudopotentials.

TABLE I. Parameters a_1^β and Ω_β for Ga^{3+} and As^{5+} pseudopotentials (in a.u. unit).

	a_1^β	a_2^β	a_3^β	a_4^β	Ω_β
Ga	-0.33845	1.27000	0.45660	0.00705	76.45
Al	-0.56918	1.04680	-0.13389	-0.02944	76.45
As	-0.70451	0.80323	0.16624	-0.01512	76.45

TABLE II. Energy levels of Ga^{3+}, Al^{3+} and As^{5+} (in ryd. unit). All values are obtained using the $X\alpha$ potential with $\alpha=0.7$.

		All-electron	Pseudopotential
Ga	4s	-0.5997	-0.6008
	4p	-0.1530	-0.1330
As	4s	-0.9867	-0.9594
	4p	-0.3365	-0.3308
Al	3s	-0.5193	-0.5392
	3p	-0.1553	-0.1521

5. PROCEDURE OF ADJUSTING A VALUE OF α

It is well-known that the band structure calculation within the LDA gives a lower value to the band gap of semiconductors, because this formalism does not include an accurate self-energy correction due to the non-local effects in the exchange-correlation functional. In order to remedy the above demerit, Nakayama and Kamimura[5] proposed the semiempirical treatment of taking account of the effect of self-energy correction in a potential, by adjusting a value of α in the $X\alpha$ approximation for an exchange-correlation potential. In this section, the effect of changing α on the band structures of GaAs and AlAs is investigated.

In Fig. 5(a), the calculated energy levels of GaAs is shown as a function of α Here the symbols Γ_p and X_q indicate the energy levels p and q at the symmetry points Γ and X in the fcc Brillouin zone of GaAs, respectively. From this figure we notice that all the valence band energy levels below the Fermi energy ε_F decrease with increasing a value of α. This is because a large exchange-correlation potential corresponding to a large value of α is effective in a region where a valence charge exists and as a result the energy gain of a valence band state due to the attractive potential becomes large. On the other hand, conduction band states seem less influenced by the local change of exchange-correlation potential. The same trend also appears in the case of AlAl. From the fact that the effect of large α is eminent in the valence band states, the present semiempirical treatment is just opposite to an idea developed in a first principles theory of the quasiparticle energies in semiconductors recently developed independently by Hybertsen and Louie[11] and Sham and Schlüter[12,13]. In this respect the present method to treat α as a disposal parameter is not sound physically from the first principles view, but is very successful for SL(n,n) as will be seen later.

Using the energy levels calculated for various values of α, Nakayama and Kamimura[5] have calculated the direct and indirect band gap energies in GaAs and AlAs as a function of α. The results are shown in Fig. 5(b). It is seen from Fig. 5(b) that both direct gap Γ_{6c} - Γ_{8v} and indirect gap X_{1c} - Γ_{8v} in GaAs and AlAs increase with increasing α, in particular the indirect gap increases sharply with α. It is also found that the experimentally observed values of band gap in both GaAs and AlAs, which are 1.52eV and 2.23eV respectively, are reproduced within an accuracy of 0.1eV by choosing $\alpha=1.15$. The reason why the band gap energies of bulk GaAs and AlAs are reproduced well by choosing the "same" value of α is that both materials are the semiconductors of type III-V with almost the same lattice constant and with similar charge distributions. Table III[5] shows the calculated energy levels of GaAs and

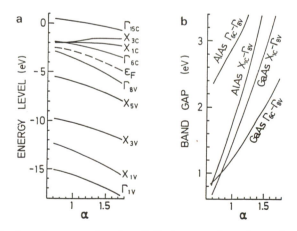

FIG. 5. (a) Calculated energy levels of GaAs as a function of α. ε_F denotes the Fermi energy. (b) Calculated direct gap (Γ_{6c}-Γ_{8v}) and indirect gap (X_{1c}-Γ_{8v}) energies of GaAs and AlAs as a function of α.

TABLE III. Calculated energy levels of GaAs and AlAs using $\alpha=1.15$, and the experimentally observed values (in eV unit).

	GaAs		AlAs	
	Present theory	Experiment	Present theory	Experiment
Γ_{1v}	−11.957	−13.1	−11.637	
Γ_{8v}	0	0	0	0
Γ_{6c}	1.419	1.52	3.184	3.13
Γ_{15c}	3.975	4.176	4.389	4.34
Γ_{1c}	8.376	8.33	8.715	
X_{1v}	−9.505	−10.75	−9.322	
X_{3v}	−6.535	−6.70	−5.659	
X_{5v}	−2.303	−2.80	−2.228	−2.356
X_{1c}	1.814	1.98	2.184	2.228
X_{3c}	2.259	2.58	2.537	2.428
L_{1v}	−10.266	−11.24	−9.97	
L_{1v}	−6.136	−6.70	−5.59	
L_{3v}	−0.944	−1.30	−0.85	
L_{1c}	1.615	1.85	2.78	

AlAs with $\alpha=1.15$ at the Γ-, X- and L-points, together with the experimentally determined values[14]. In this table the energy level of valence band maximum is taken to be zero. The agreement between theory and experiment is particularly good for the energy levels around the band gap energy region. Thus hereafter we adopt an optimized value of $\alpha=1.15$.

6. BAND STRUCTURES OF SL(n,n)

In this section I present the results of the band structures of SL(n,n) calculated by Nakayama and Kamimura[5,6,7,8], and discuss their characteristic features. Figure 6(a) shows the calculated band structure of SL(2,2) along the layered direction ΓZ, while Figs. 6(b) and (c) show the corresponding energy bands of GaAs and AlAs folded into the Brillouin zone of SL(2,2) along ΓZ. The energy bands corresponding to upper valence band states in SL(2,2) have an apparent one-to-one correspondence to those of GaAs or AlAs, while the energy bands corresponding to lower conduction band states in SL(2,2) are remarkably different from those of GaAs or AlAs. This indicates that there are strong inter- and intra-band mixings among the folded lower conduction bands of GaAs and those of AlAs when a superlattice is formed. In fact,

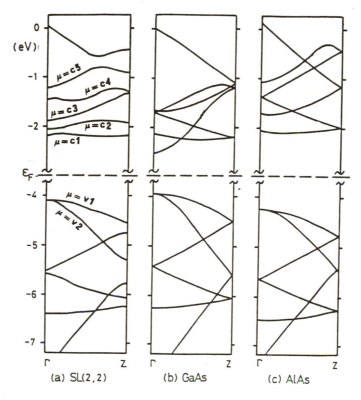

FIG. 6. (a) The calculated band structure of SL(2,2). (b) The folded energy band of GaAs. (c) The folded energy band of AlAs.

the valence band of SL(2,2) is the complete mixture of GaAs and AlAs characters while the lowest conduction band has an AlAs character and the second lowest conduction band has a GaAs character.

Figure 7 shows the band structures of SL(n,n) for n=4,78 and 10. As seen in this figure, the higher energy states of the valence band and the lower energy states of the conduction band become nearly flat as n increases from 4 to 10, reflecting the fact that the interaction between adjacent GaAs and AlAs layers becomes weaker as a value of n increases. If we would be able to find a trend that both the valence and conduction band states are nearly dispersionless along ΓZ in a certain energy region from the top of the valence band and the bottom of the conduction band, respectively, from a certain value of n, it may be said that the Kronig-Penny model becomes applicable from that value of n to higher values. As far as the present results up to n = 10 are concerned,

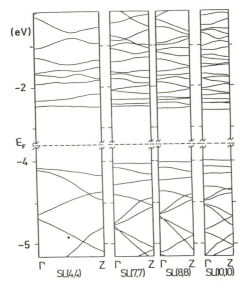

FIG. 7. The calculated band structures of SL(4,4), SL(7,7) SL(8,8), and SL(10,10).

we cannot see such a trend yet. Thus the Kronig-Penny model does not hold for the energy bands of $(GaAs)_n(AlAs)_n$ with n<10.

After Pickett et al.[10] for SL(9.9) and then Nakayama and Kamimura[5] reported the first self-consistent band structure calculation of SL(n,n) for n = 1 to 10, more than twenty theory groups have reported the band structure results of SL9n,n) for small values of n[15~37], among which the calculations of more than half groups are based on the LDA[15,19,23,24,25,26,28,30,34,35]. Most of the LDA band structures are similar to those of Nakayama and Kamimura[5,6,7,8].

7. THE OPTICAL PROPERTIES OF SL(n,n)

7.1. Dielectric Function and Absorption Spectra

The dielectric function $\varepsilon_2(\omega)$ of SL(n,n) is calculated by the following well-known formula as a function of the photon energy $\hbar\omega$ for polarization of light parallel (z) and perpendicular (xy) to the superlattice direction, using the energy dispersion and eigenfunction of each energy band:

The First Principles View of Superlattices

$$\varepsilon_{2,j}(\omega) \;=\;$$

$$\frac{e^2}{\pi m \omega^2} \sum_{\mu}^{unocc.} \sum_{\nu}^{occ.} \int_{\varepsilon_{\mu\nu}(k)\,=\,\hbar\omega} \frac{dS_{\mu\nu}}{|\,\nabla\varepsilon_{\mu\nu}\,|} \; \frac{|<\mu k\,|\,P_j\,|\,\nu k>|^2}{m} \qquad (7)$$

Here j denotes the polarization of light and p_j the momentum operator. The $dS_{\mu\,\nu}$ is the measure of integration in the equal energy surface which is defined by $\varepsilon_{\mu,\nu} = \varepsilon_{\mu,k} - \varepsilon_{\nu,k} = \hbar\omega$. This can be compared with the observed absorption spectra. Here the excitonic effect is not included. In Fig. 8 the calculated $\varepsilon_2(\omega)$ for xy and z polarizations of light for n=1,4,6,7 and 10 near the absorption edge are shown by solid line and dotted line, respectively. The following features emerge from this result:

1. For n=1, the fundamental absorption edge shows the feature of a three-dimensional van Hove singularity M_0.

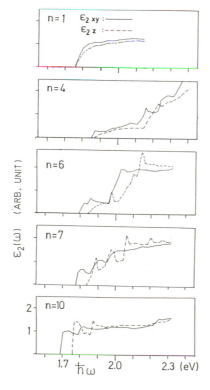

FIG. 8. The dielectric function $\varepsilon_2(\omega)$ of SL(n,n) with n=1, 4, 6, 7 and 10 as a function of photon energy $\hbar\omega$.

2. With increasing n the feature of the absorption edge changes from three-dimensional to two-dimensional nature.

3. The weak intensity near the absorption edge in SL(n,n) for 2 $\leq$n<7 is due to the dominance of AlAs character in the lowest conduction band. In Figs. 6 and 7, the lowest conduction band for 2$\leq$n<7 has the GaAs character only near Γ and the AlAs character in the most region of ΓZ, while in the second lowest conduction band the GaAs character predominates. Thus, although optical transitions in the k space are direct, some of transitions in SL(n,n) for 2$\leq$n< 7 occur from GaAs layers to AlAs layers in a real space. In this sense we call these transitions the cross-layer transition, while the transitions within the same layer the intra-layer transition.

4. There appears a large dichroism for n>6. In the spectra of z polarization for n>6 the high energy peaks correspond to transitions from the light holes to the low-lying conduction bands while in the spectra of xy polarization the low energy peaks correspond to transitions from the heavy hole states to the low-lying conduction bands. Since the energy difference between heavy hole and light hole states increases with increasing a value of n, the absorption edge for the xy polarization shifts to lower energy with increasing n.

Recently Anma et al.[38] measured the reflectance spectra by the ellipsometry technique and determined the shape of the absorption edge. According to their results, the E_0 edge exhibits a clear two-dimensional step for n$\geq$7 and the intensity is very weak for 2$\leq$n$\leq$6. These results are consistent with the present prediction.

7.2. The Band Gap Energy and Its Layer Thickness Dependence

The lowest conduction bands of both GaAs and AlAs have local minima (valleys) at the Γ-, X- and L-points in the fcc Brillouin zone. Then a question arises which minimum corresponds to the lowest energy state and which one the highest state, when a superlattice is formed. In order to respond to this question, Kamimura and Nakayama[7] calculated the energies of these minima in SL(n,n).

Figure 9 shows the calculated energies of the minima at Γ-, X- and L-points, where the energy is all measured from the top of the valence band at the Γ-point. In this figure the energies of various minima are plotted as a function of the period of a superlattice n. It is seen from this figure that, in the case of n=2 to 10 the conduction band state at Γ is the

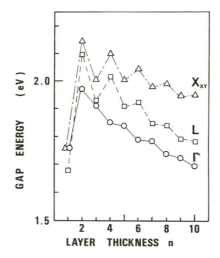

FIG. 9. Calculated band gap energies of SL(n,n) for three conduction band states referenced to the top of the valence band as a function of the layer thickness n.

lowest state and thus the band gap is direct, while that, in the case of n=1 the conduction band state at L is the lowest so that the band gap becomes indirect. This result is consistent with very recent band structure calculations of SL(1,1) by Christensen, Molinary and Bachelet[15],

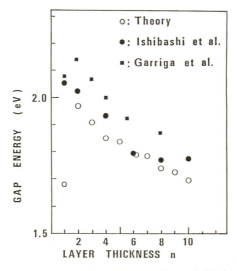

FIG. 10. Calculated minimum band gap energies of SL(n,n) as a function of layer thickness n, together with observed values.

Hamada and Ohnishi[26], Bylander and Kleinman[19], and Cardona, Suemoto, Christensen, Isu and Ploog[30].

Except n=1, the energies of all the conduction band minima decreases monotonously with increasing n for larger n (n$\geq$8), reflecting a confinement effect for the band edge states for n$\geq$2 which causes the larger band gap energy as the layer thickness decreases, while for smaller n (1$\leq$n$<$8) the energy change oscillates from the odd number of n to even number of n with increasing n, reflecting the matching condition of wave functions at the boundary of different layers. In Fig. 10 I replot only the calculated minimum band gap energy of SL(n,n) as a function of layer thickness n[5,6,7], together with the experimentally determined band gap energies.

Ishibashi, Mori, Itabashi and Watanabe[39] reported for the first time the photoluminescence measurements for $(GaAs)_n(AlAs)_n$ with n=1 to 24, and determined the band gap energies of $(GaAs)_n(AlAs)_n$ as a function of n. Their results for n=1 to 10 are also shown in Fig. 10. According to their result the band gap increases with decreasing n. As seen in Fig. 10, the magnitudes of the band gap energy coincide well with the theoretical calculations except n=1, where in experiment the band gap energy still increases for n=1 while in the theory it decreases considerably for n=1. As one of the reasons for this discrepancy I would like to point out that a sample of $(GaAs)_1(AlAs)_1$ might not form a superlattice but an alloying phase or a mixed phase, because the observed band gap energy of $(GaAs)_1(AlAs)_1$ is very close to that of the $Ga_{0.5}Al_{0.5}As$ alloy observed by Dingle, Logan and Arthur[40]. Calculations of the band gap energy of SL(n,n) by Gell, Ninno, Jaros and Herbert[16], and Hamada and Ohnishi[26] have also showed the sharp decrease of the band gap energy for n=1. Recently this tendency has been qualitatively confirmed by the photo-reflectance experiment by Garriga et al.[41], whose results for n$\geq$2 also coincide well with the theoretical results by Kamimura and Nakayama[7,8], as seen in Fig. 10.

7.3. Optical Spectra at a Higher Energy Region

As for the optical spectra in higher energy region, I would like to discuss here only the E_1 transition which corresponds to the M_1 saddle point type van-Hove singularity in GaAs. Figure 11 shows the calculated dielectric functions of GaAs and Sl(n,n) with n=1,2,4 and 7 as a function of photon energy. The first peak in the absorption spectra corresponds to the E_1 transition. In Fig. 12 the E_1 transition energy calculated by Kamimura and Nakayama[42] is shown as a function of layer thickness,

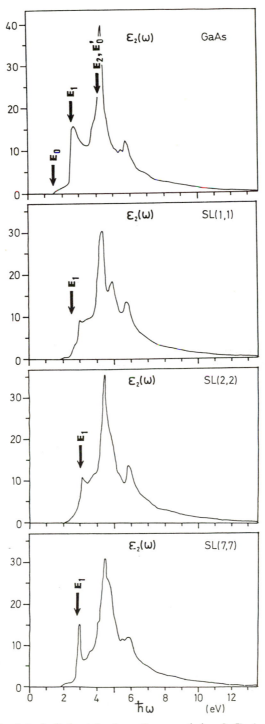

FIG. 11. The calculated dielectric function ε_2 (ω) of GaAs and SL(n,n) for n=1, 2 and 7 as a function of photon energy $\hbar\omega$.

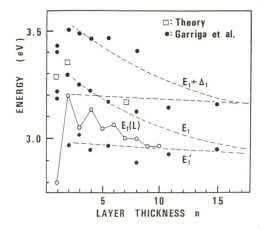

FIG. 12. Calculated E_1 transition energies of SL(n,n) and the experimentally observed ones as a function of layer thickness. $E_1(L)$'s represent interband transition energies at L-point, which are no longer a saddle point in SL(n,n).

which is obtained from the energy value of the peak in Fig. 11. (It should be noticed here that, because the present calculation gives a smaller value of E_1 transition energy than the experimentally obtained value in GaAs, the calculated E_1 transition energies are shifted by +0.29eV in order that the calculated E_1 transition energy of GaAs (2.56eV) coincides with the observed E_1 transition energy of GaAs (2.85eV)). As seen in this figure, the E_1 transition energy increases as the layer thickness n decreases from 7 to 2 while for n=1 the E_1 transition energy becomes lower than that of n=2. These trends reflect a certain kind of confinement effect.

Garriga et al.[41] estimated E_1 interband transition energies of SL(n,n) from the peaks of the observed dielectric function in addition to the above mentioned E_0 (fundamental band gap) energy. Their results of E_1 transition energies are also shown in Fig. 12. According to their assignment, the dotted lines are drawn for guide of eyes in this figure. According to their results, the E_1 transition energy increases with decreasing n from 8 to 2. They also find the sharp decrease of E_1 and $E_1+\Delta$ peaks as n decreases from 2 to 1, where $E_1+\Delta$ peaks correspond to the transition from lower valence band split by spin-orbit interaction to the conduction band at E_1 saddle point. These trends qualitatively coincide with the present theoretical calculation.

Next I would like to discuss the character of E_1 transition in SL(n,n); one may ask a question which kind of interband transitions contribute to E_1 transition of SL(n,n). In bulk GaAs, E_1 transition mainly corresponds to the interband transitions from the highest valence band to the lowest conduction band near L-point, which shows a strong spectral shape of a saddle point edge(M_1 type van-Hove singularity). On the other hand, in SL(n,n) the energy of the lowest conduction band state at L-point becomes lower in $(GaAs)_n(AlAs)_n$ with n=odd and thus the interband transition near L-point changes its singularity from M_1 to a weak M_0-type spectra. Instead of it, interband transitions at a k-value in the inner Brillouin zone which is off L-point give rise to a saddle point type (E_1 type) transition. This situation is shown in Fig. 13, by choosing SL(1,1) as an example. In the right side of this figure, a position of a saddle point in the transition energies in SL(1,1) is shown by a thick arrow. It is clearly seen that this position is shifted from the L-point saddle point in GaAs and that an L-point for SL(1,1) shown by a thin arrow in the left side of the figure is no longer a saddle point. In general, this new saddle point peak in SL(n,n) has been observed in experiments by Garriga et al.[41] In Fig. 12 I have also shown the calculated transition energies at L-point by $E_1(L)$. Observed peak energies must be larger than $E_1(L)$, as shown in Fig. 12.

7.4. Key Role of Photoluminescence (PL) in Determining a Conduction Band Character

Based on the band structures shown in Figs. 6 and 7, we can predict the appearance of at least two PL peaks in SL(n,n). From n = 2 to 6 the

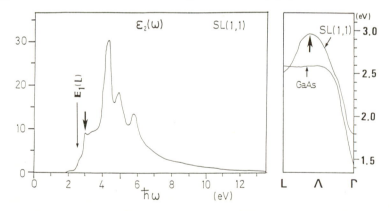

FIG. 13. (Right part): The position of a saddle point in SL(1,1) ("thick arrow"). It should be noticed that a saddle point shifts from L-point which is a saddle point in GaAs. (Left part): The optical spectrum of SL(1,1) in which a peak indicated by "thick arrow" corresponds to a saddle point transition. A transition at L-point shown by "thin arrow" does not create a peak.

intensity of a low energy PL peak is expected to be weak, because the lowest conduction band has mainly an AlAs character while that of a high energy PL peak may be strong, corresponding to a transition from the second lowest conduction band which has mainly a GaAs character. In fact, the observed PL decay time of the low energy peak is of order of 100 ns while that of the high energy peak is less than 2 ns[43].

As regards experiments, a number of PL measurements[43~48] have recently been reported for SL(n,n) in a wide range of n values, in particular under pressure. In these measurements at least two PL peaks have been observed. When pressure is applied, a lower energy peak shifts to lower energy while a higher energy peak shifts to higher energy in SL(n,n) for n≤10[43,44,45,46]. Since in GaAs and AlAs the conduction bands at Γ and X points in the Brillouin zone shift to higher and lower energies, respectively, by pressure, it has been concluded from the above experimental results that the lowest conduction band has mainly an AlAs character while the second lowest conduction band has a GaAs character for n≤10. The observed PL decay time has provided another support for this assignment[43,48,49].

In Fig. 14, the layer thickness dependence of two PL peaks reported by Kato et al.[43] for SL(n,n) with n = 4 to 18 (large circles) and

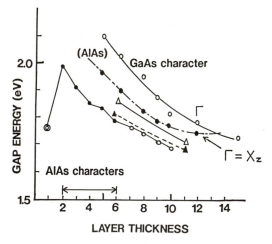

FIG. 14. PL peak energies as a function of the layer thickness. Large open and closed circles represent the observed values of two PL peak energies by Kato et al. (ref. 43) while open and closed triangles for n=6 and 11 represent those by Li et al. (ref. 46). Small open and closed circles together with a double circle for n=1 correspond to theoretical results by Kamimura and Nakayama (refs. 6, 7 and 8).

by Li et al[46] for n = 6, 11 and 17 (triangles) are shown for the range from n = 1 to 15, together with the theoretical result of transition energies corresponding to the transition from the lowest conduction band to the highest valence band[6,7], where closed and open symbols represent AlAs and GaAs characters, respectively.

It is seen from this figure that the agreement on the overall layer thickness dependence between theory and experiment is fairly good. However, it is noticed that, (1) the observed values of the lowest peak energy by Kato et al.[43] are a little larger than the calculated values while those by Li et al.[46] coincide well with the calculated values, and (2) the crossover from an AlAs to a GaAs character occurs at around n = 10 in experiments but at n = 6 in the theoretical result[6,7,8]. The following two problems may be considered as possible origins for the above discrepancy: (1) the band gap problem in the LDA and (2) the sample problem related to an atomic arrangement in an interface. These problems are discussed in the next section.

8. POSSIBLE ORIGINS FOR DISCREPANCY IN OPTICAL PROPERTIES AND FUTURE PROSPECT IN RESOLVING DISCREPANCY

In the previous section I have mentioned two possible origins for the discrepancy between theory and experiment as regards the photoluminescence. In this section I will explain these in detail.

8.1. The Band Gap Problem

The local density approximation (LDA) in the density functional theory has been very successful in describing many ground state properties of weakly correlated electron systems. However, problems occur when we apply the LDA to electronic excited states. For example, the one-electron energy eigenvalues in the LDA obtained by solving eq.(2) are not formally interpreted as quasiparticle energies. Thus the band gap energies in semiconductors and insulators calculated by the LDA are usually much smaller than the observed gap energies, often by 50% or more, as seen in Table IV.[11] This is called "the band gap problem." This band gap problem is due to the neglect of non-local effects in the exchange-correlation functional. Thus the non-local effect is essentially important in describing excited state properties.

Remarkable progress has recently been made in improving the band gap problem in the LDA. Hybertsen and Louis[11] and Sham and

TABLE IV. Band gaps of semiconductors and insulators (in eV).

Eg	LDA	QP	Expt
diamond	3.9	5.6	5.48
Si	0.52	1.29	1.17
Ge	< 0	0.75	0.744

Schluter[12,13] have independently proposed the first principles approaches to solve this problem. In general, the quasiparticle energies and wavefunctions are obtained by solving

$$\left[-\frac{1}{2}\nabla^2 + V_{ion}(r) + V_H(r) \right]\Psi(r)$$
$$+ \int dr'\Sigma(r, r'; E)\Psi(r') = E\Psi(r), \tag{8}$$

where Σ is the electron self-energy operator due to the non-local exchange correlation potential. The approach is based on calculating $\Sigma(r, r'; E)$ to the first order in the dressed Green's function G and the dynamically screened Coulomb interaction W, which is often called the GW approximation. In the GW approximation[50,11] Σ is expressed as

$$\Sigma(r, r'; E) = \frac{i}{2\pi} \int d\omega e^{-i\delta\omega} G(r, r'; E - \omega)W(r, r'; \omega), \tag{9}$$

where δ is a positive infinitesimal. The screened Coulomb interaction W is given by

$$W(r, r'; \omega) = \frac{1}{\Omega} \int dr'' \varepsilon^{1}(r, r''; \omega)V_c(r'' - r'), \tag{10}$$

where ε^{-1} is the time-ordered dielectric matrix and V_c is the bare Coulomb interaction. Hybertsen and Louie[11] have used a quasiparticle approximation for the Green's function G and calculated the dielectric matrices in two steps: (i) the ab initio static dielectric matrices are calculated with the use of the LDA, where the local field correction is taken into account, and (ii) the dielectric matrices are extended to finite frequency using a generalized plasmon-pole model employing exact sum rules.

In Table IV the band gap energies E_g thus calculated by Hybertsen and Louis[11] for diamond, Si and Ge are shown on the column "QP," together with observed values (column "Expt") and the calculated values with the LDA (column "LDA"). It is seen from this table that the excellent agreement has been achieved by considering the effect of the local fields and the dynamical screening in the LDA results.

In this context Zhang et al.[37] have extended the above-mentioned improvement method over the LDA to S(n,n). Because of very time-consuming calculations in ε, in particular, for a low-symmetry system, the results have been so far reported for n = 1 and 2. The band gap energies for local minima at L, Γ and X_{xy} thus calculated are shown in Table V[37]. As seen from this table, the band gap energies calculated in the quasiparticle approximation[37] become certainly larger than those calculated by Kamimura and Nakayama[7,8] in the LDA, but it is difficult, at present, to judge whether or not the agreement between the first principles values and observed ones has been improved by the quasiparticle approximation. For this purpose it will be necessary to extend the calculations by the quasiparticle approximation to SL(n,n) with higher values of n, on one hand, and to perform more accurate measurements for better samples in the future, on the other.

8.2. Sample Problems

Another possible origin for the discrepancy between theory and experiment in the PL energies is due to sample problems. A number of theoretical groups have discussed the stability of very short period superlattices, in particular for $(GaAs)_n(AlAs)_m$[22,25,51,52]. All of these results have shown that SL(1,1) is metastable, in the sense that a system consisting of pure bulk constituents would have a lower total energy. Christensen result[52] is very striking, i.e. SL(n,n) is metastable up to n = 7. If we follow these results, we may expect that an atomic arrangement in an interface region in SL(n,n) is not stoichiometric but alloying.

TABLE V. Quasiparticle energies (eV) for conduction-band states referenced to the top of the valence band for S(1,1) and S(2,2).

	Γ	L	X_{xy}
S(1,1)	2.11	1.85	2.13
S(2,2)	2.18	2.34	2.16

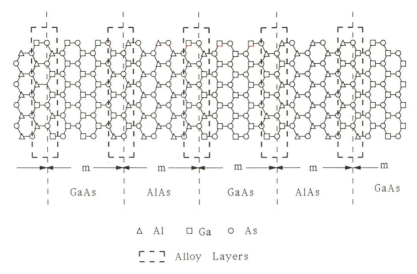

FIG. 15. A sketch of possible atomic arrangements in SL(4,4). Because of a metastable structure of SL(4,4), Ga and Al atoms near an interface form an alloying state.

In Fig. 15 I will sketch one of possible atomic arrangements which may occur, I imagine, in a real sample of a very short period superlattice, by choosing SL(4,4) as an example. In this system an As layer is taken as an interface, and on Ga and Al layers adjacent to the interface As layer Ga and Al atoms are randomly distributed, and as a result an alloying situation appears. Thus this system consists of 3 GaAs layers, 2 (Ga/Al)As layers and 3 AlAs layers, which form a supercell. If one will calculate the band gap of this system, one may expect that the calculated band gap energy of this system is much larger than that of SL(4,4), because the band gap of an alloy system of GaAs and AlAs is 2.05 eV[40]. I believe this kind of calculation will be very important in checking the sample characterization in the future research on superlattices. This kind of comparison between the first principles theory and sample characterization is possible only for very short period superlattices, because the first principles calculations can be made only for very short period superlattices. Hirose and Kamimura[53] are now performing such a calculation.

9. SUMMARY AND CONCLUDING REMARKS

In the present lecture I have shown the results of the first principles calculations of the electronic structures and optical properties of very short period superlattices of a GaAs/AlAs system, $(GaAs)_n(AlAs)_n$ with $1 \leq n \leq 10$, which have very recently aroused keen interests as a new class of semiconductor superlattice materials, and tried to clarify the characteristic features of its band structures and optical properties.

One of the outstanding features is the crossover effect in the character of the lowest conduction band from an AlAs character to a GaAs character with increasing a value of n although the value of n at which crossover occurs differs in theory (around n = 6) and experiment (around n = 10). This feature of long lifetime in a visible wavelength region in n = 2 to 6 will be very useful in the application of SL(n,n) to an optical device.

The second feature is that the conduction band structures of SL(n,n) with $n \leq 10$ are very unique and are different from those derived by the conventional phenomenological Kronig-Penny model. I have mentioned that the conduction band states are not treated accurately by the LDA so that the band offset ratios in SL(n,n) cannot be derived quantitatively. However, a band offset value in a valence band has been estimated from the first principles calculations. Although this value differs slightly, depending on a theoretical method adopted, it is in a range of 0.35 eV[5,6,7] to 0.5 eV[31].

In connection with discrepancy in the photoluminescence properties between first principles theories and experiments, I have suggested two directions as regards the future prospect of research on this system. Those are the improvement of the band gap problem in the local density approximation and the sample problem. Concerning the latter I have proposed a model calculation in order to check a sample characterization. In addition to the above two directions, an investigation with regard to an exciton effect in SL(n,n) which has been little done will be very interesting, in particular a relation between exciton effects and layer thickness.

Finally, I should like to emphasize that the theoretical recipe which I have mentioned for very short period $(GaAs)_n(AlAs)_n$ superlattices is useful not only for this system but also for strained short period superlattices. In fact, for example, Nakayama[54] has just recently applied the total-energy pseudopotential method with the LDA to a II-VI semiconductor strained superlattice system while Taguchi and Ohno[55] has applied it to $(InAs)_m(GaAs)_m$ strained superlattice systems with m = 1 to 7.

ACKNOWLEDGEMENTS

It is a great pleasure to acknowledge the collaboration of Takashi Nakayama on the work reported in this lecture, and stimulating conversations and discussions with him, David Ko, and Takahisa Ohno. I am also indebted to Mikio Eto, Kenji Hirose, Katsuyoshi Kobayashi, Shunichi Matsuno, Takashi Nakayama, Riichiro Saito and Shinji Tsuneyuki for their great assistance during the time of the preparation of the manuscript.

REFERENCES

1. L. Esaki and R. Tsu; IBM J. Research Develop. 14, 61 (1970).

2. For example, L. Esaki, in recent topics in semiconductor physics, edited by H. Kamimura and Y. Toyozawa (World Sc., Singapore, 1983) p. 1.

3. Y. Aoyagi, A. Doi, T. Meguro, S. Iwai and S. Namba, in Proc. 1st Int. Conf. on Electronic Materials, eds. T. Sugano et al., p. 242 (Materials Research Society, Pittsburgh, 1989).

4. H. Watanabe, in Proc. 20th Int. Conf. on Physics of Semiconductors, to be published (World Sc. Singapore, 1990).

5. T. Nakayama and H. Kamimura, J. Phys. Soc. Jpn. *54,* 4726 (1986).

6. H. Kamimura and T. Nakayama, in Proc. 18th Int. Conf. on Physics of Semiconductors, ed. B. Monemar, p. 643 (World Sc. Singapore, 1987).

7. H. Kamimura and T. Nakayama, Comments Cond. Mat. Phys. *13*, 143 (1987).

8. H. Kamimura and T. Nakayama, in Topics in Semiconductor Physics, eds. W. Stritrakol and V. Sa-Yakanit, p. 64 (World Sc. Singapore, 1988).

9. W. Kohn and L. J. Sham, Phys. Rev. *A140*, 1133 (1965).

10. W. E. Pickett, S. G. Louie and M. L. Cohen, Phys. Rev. *B17*, 815 (1978).

11. M. S. Hybertsen and S. G. Louie, Phys. Rev. Lett. *55*, 1418, Phys. Rev. *B34*, 5390 (1986), and related references therein.

12. L. J. Sham and M. Schlüter, Phys. Rev. *B32*, 3883 (1985).

13. L. J. Sham, in Topics in Semiconductor Physics, eds. W. Stritrakol and V. Sa-Yakanit, p. 1 (World Sc. Singapore, 1988), and related references therein.

14. Semiconductors, Physics of Group IV Elements and III-V Compounds, ed. O. Medelung, Landort-Bornstein, New Series, III /17a (Springer-Verlag, Berlin, Heidelberg, New York, 1982).

15. N. E. Christensen, E. Molinari and G. B. Bachelet, Solid State Commun. *56*, 125 (1985).

16. M. A. Gell, D. Ninno, M. Jaros, D. C. Herbert, Phys. Rev. *B34*, 2416 (1986).

17. M. A. Gell and D. C. Herbert, Phys. Rev. *B35*, 9591 (1987).

18. M. A. Gell, D. Ninno, M. Jaros, D. M. Wolford, T. K. Keuch and J. A. Bradley, Phys. Rev. *B35*, 1196 (1987).

19. D. M. Bylander and L. Kleinman, Phys. Rev. *B34*, 5280 (1986).

20. D. M. Bylander and L. Kleinman, Phys. Rev. *B36*, 3229 (1987).

21. C. G. Van de Walle and R. M. Martin, J. Vac. Sci. Technol. *B4*, 1055 (1986).

22. D. M. Wood, S. H. Wei and A. Zunger, Phys. Rev. Lett. *58*, 1123 (1987), Phys. Rev. *B37*, 1342 (1988).

23. S. H. Wei and A. Zunger, Phys. Rev. Lett. *59*, 144 (1987), J. Appl. Phys. *63*, 5794 (1988).

24. S. Ciraci and I. P. Batra, Phys. Rev. Lett. *58*, 2114 (1987), Phys. Rev. *B36*, 1225 (1987).

25. I. P. Batra, S. Ciraci and J. S. Nelson, J. Vac. Sci. Technol. *B5*, 1300 (1987).

26. N. Hamada and S. Ohnishi, Superlatt. Microst. *3*, 301 (1987).

27. J. Ihm, Appl. Phys. Lett. *50*, 1068 (1987).

28. A. Oshiyama and M. Saito, Phys. Rev. *B36*, 6156 (1987).

29. E. Yamaguchi, J. Phys. Soc. Jpn. *56*, 2835 (1987).

30. M. Cardona, T. Suemoto, N. E. Christensen, T. Isu and K. Ploog, Phys. Rev. *B36*, 5906 (1987).

31. S. Massidda, B. I. Min and A. J. Freeman, Phys. Rev. *B35*, 9871 (1988).

32. Jian-Bai Xia, Phys. Rev. *B38*, 8358 (1988).

33. M. Aloani, L. Brey and N. E. Christensen, Phys. Rev. Lett. *61*, 1643 (1988).

34. M. Aloani, L. Brey and N. E. Christensen, Phys. Rev. *B37*, 1167 (1988).

35. Y. Hatsugai and T. Fujiwara, Phys. Rev. *B37*, 1280 (1988).

36. S. B. Zhang, D. Tomanek, S. G. Louie, M. L. Cohen and M. S. Hybertsen, Solid State Commun. *66*, 585 (1988).

37. S. B. Zhang, M. S. Hybertsen, M. L. Cohen, S. G. Louis and D. Tomanek, Phys. Rev. Lett. *63*, 1495 (1989).

38. H. Anma, T. Yamaguchi, H. Okumura and S. Yoshida, Physica A *157*, 407 (1989).

39. A. Ishibashi, Y. Mori, M. Itabashi and N. Watanabe, J. Appl. Phys. *58*, 2691 (1985).

40. R. Dingle, R. A. Logan and J. R. Arthur Jr., Inst. Phys. Conf. Ser. *33a*, 210 (1977).

41. M. Garriga, M. Cardona, N. E. Christensen, P. Lautenschlager, T. Isu and L. Ploog, Phys. Rev. *B36*, 3254 (1987).

42. H. Kamimura and T. Nakayama, in Proc. 1st Int. Conf. on Electronic Materials, eds. T. Sugano et al. p. 173 (Materials Research Society, Pittsburgh, 1989).

43. H. Kato, Y. Okada, M. Nakayama and Y. Watanabe, Sol. St. Comm. *70*, 535 (1989).

44. K. Takarabe, S. Minomura, M. Nakayama and H. Kato, J. Phys. Soc. Jpn. *58*, 2242 (1989).

45. M. S. Skolnik, G. W. Smith, I. L. Spain, C. R. Whitehouse, D. C. Herbert, D. M. Whittaker and L. J. Reed, Phys. Rev. *B39*, 11191 (1989).

46. G. Li, D. Jiang, H. Han and Z. Wang, Phys. Rev. *B40*, 10430 (1989).

47. T. Nakazawa, H. Fujimoto, K. Imanishi, K. Taniguchi, C. Hamaguchi, S. Hiyamizu and S. Sasa, J. Phys. Soc. Jpn. *58*, 2192 (1989).

48. E. Finkman, M. D. Sturge and M. C. Tamargo, Appl. Phys. Lett. *49*, 1299 (1986).

49. M. H. Meynadier, R. E. Nahory, J. M. Worlock, M. C. Tamargo, J. L. de Miguel and M. D. Sturge, Phys. Rev. Lett. *60*, 1338 (1988).

50. L. Hedin, Phys. Rev. *139*, A796 (1965).

51. D. M. Bylander and L. Kleinman, Phys. Rev. Lett. *59*, 2091 (1987); Erratum* Phys. Rev. Lett. *60*, 472 (1988).

52. N. E. Christensen, Sol. St. Comm. *68*, 959 (1988).

53. K. Hirose and H. Kamimura, in preparation.

54. T. Nakayama, in Proc. 20th Int. Conf. on Physics of Semiconductors, eds. E. M. Anastassakis and J. D. Joannapoulos, p. 1182 (World Sc. Singapore, 1990).

55. A. Taguchi and T. Ohno, Phys. Rev. *B38*, 2038 (1988).

INELASTIC LIGHT SCATTERING IN SEMICONDUCTOR QUANTUM STRUCTURES

G. Abstreiter

Walter Schottky Institut
Technische Universität München
D-8046 Garching, Germany

Abstract - Raman scattering has been used successfully to study various properties of semiconductor heterostructures, superlattices and micro-structured devices. Light scattering by phonons leads to information on composition, strain, defects, temperature, superlattice period, layer thicknesses, interface roughness and many other important parameters. Folded acoustic modes and confined optical modes are the most widely studied lattice dynamical properties. Resonant electronic excitations on the other hand provide information on confinement energies, collective and single particle excitations in low dimensional systems, carrier density and velocity distribution. Selected examples are presented for Si/Ge and GaAs/AlAs systems.

1. INTRODUCTION

Inelastic light scattering is a versatile tool for the analysis of semi-conductor quantum structures. There appeared numerous publications in the past ten years which deal with various aspects of Raman scattering in

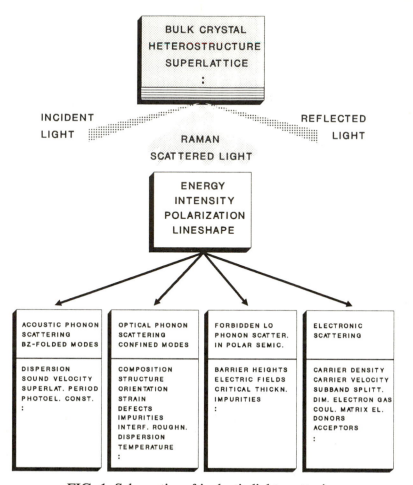

FIG. 1. Schematics of inelastic light scattering.

heterostructures, quantum wells, superlattices, surfaces or interfaces, and micro-structured devices. An overview of the many different aspects is given in the recent articles in Topics in Applied Physics, Light Scattering in Solids V, which concentrates on superlattices and other microstructures[1] and in the proceedings of a related NATO workshop held at Mont Tremblant in 1990.[2] Among the most intensively studied fundamental excitations are so-called zone-folded acoustic phonons in superlattices, confined optical modes in thin layers and resonant electronic light scattering in low dimensional carrier systems. In addition it has been shown that forbidden LO-phonon Raman scattering in polar semiconductors can be used also as a surface sensitive technique to study the formation of barrier heights and heterointerfaces. The basic principle of light scattering and the information which can be extracted from such measurements is summarized in Fig. 1. The semiconductors of interest are opaque in the visible spectral range. Therefore light scattering is performed usually in backscattering geometry as indicated in the figure. The relevant information on the elementary excitations is extracted from an analysis of the scattered light with respect to energetic position, intensity, lineshape and polarization of incident and scattered light. The basic scattering mechanisms are understood in detail by now. Therefore inelastic light scattering can be applied indeed as a versatile tool to investigate nearly all the essential properties of the semiconductor structures. Selected examples will be presented in this paper which includes lattice dynamical aspects and electronic excitations. Finally I will discuss some future prospects with respect to inelastic light scattering in microstructured and low-dimensional systems which includes also quantum wires and quantum dots.

2. LATTICE VIBRATIONS IN SUPERLATTICES

The lattice dynamical properties of thin layers and superlattices are changed drastically as compared to the bulk properties. Several review articles on phonon properties in superlattices have appeared in literature in recent years [3,4,5] The periodic layering of different materials gives rise to so-called folded acoustic modes and confined optical modes. This is shown schematically in Fig. 2 for the case Si/Ge. The dashed lines correspond to the bulk dispersion of the longitudinal acoustic (LA) and optic (LO) modes of Ge, the thin solid lines to those of Si within the first Brillouinzone $(2\pi/a)$. The superlattice period d causes a reduction of the Brillouinzone in growth direction to (π/d). As a consequence the acoustic phonon modes are "folded back" into the new smaller Brillouinzone giving rise to a series of modes close to the zone center and an opening of gaps in the phonon dispersion at the zone center and the new zone boundary. This behavior can be treated simply by an elastic continuum theory[6] where the two materials are characterized by their mass density and elastic stiffness constants. The resulting dispersion relation is then given by:

$$\cos(qd) = \cos(\omega d_A/v_A)\cos(\omega d_B/v_B)$$
$$- (1/2)(\rho_B v_B/\rho_A v_A + \rho_A v_A/\rho_B v_B)\sin(\omega d_A/v_A)\sin(\omega d_B/v_B)$$

where d_A and d_B are the individual thicknesses of materials A and B (period length of the superlattice $d = d_A + d_B$), ρ_A and ρ_B are the mass densities, and v_A and v_B the sound velocities, respectively. Eq. 1 leads to the "folding" of the acoustic branch with the average sound velocity weighted by the thickness and density of the materials. It gives rise also to the energy gaps at the zone center and the zone boundary. The elastic continuum theory is in surprisingly good agreement with the experimental results as long as the phonon branches of the bulk constituents have a linear dispersion, even if d_A and d_B correspond only to a few atomic planes. Folded acoustic modes have been observed first in GaAs/AlAs superlattices[7] and subsequently in many different systems such as various combinations of III-V semiconductors, II-VI compounds elemental group IV semiconductors and even amorphous layers (for details see the original references given in (3). An example for a Si/Ge superlattice grown on Ge substrates is shown in Fig. 3[8]. The sample consists of 20 periods of 4 atomic planes of Si and 12 atomic planes of Ge. A series of folded acoustic modes is observed in the Raman spectrum below 250 cm^{-1}. Their scattering intensity can be described with a photoelastic model. In superlattices with longer periods one can also study the dispersion throughout the whole superlattice Brillouinzone by changing the laser wavelength λ and consequently the scattering wave vector which is $q_s = 4\pi\eta/\lambda$ in backscattering where η is the refractive index. An example is shown in Fig. 4 for a Si/Si$_{0.5}$G$_{0.5}$ superlattice with $d=280\text{\AA}$[9]. The scattering wave vector, given in units of π/d, can be tuned even through the Brillouinzone boundary in such long period superlattices. The gaps in the phonon dispersion have been determined from such measurements. These few examples demonstrate the usefulness of Raman scattering by folded acoustic modes in superlattices for the investigation of various properties such as superlattice period, average sound velocity, phonon dispersion, photoelastic coefficients, compositional profile and so on.

The behavior of optical modes is somewhat different. In the situation shown in Fig. 2 the LO modes of the two materials don't overlap in energy. Consequently optical phonons cannot penetrate into the neighboring layers leading to phonon confinement. For Si/Ge this is especially pronounced for the Si optical modes because their energies are far above all the other modes. The confinement can be understood as

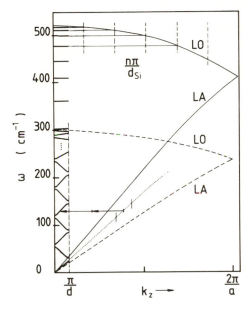

FIG. 2. Phonon dispersion of Si and Ge, Brillouinzone folding and phonon confinement in short period Si/Ge superlattices (schematically).

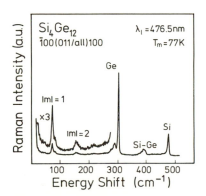

FIG. 3. Raman spectrum of a Si_4Ge_{12} superlattice grown on a (100) Ge substrate.

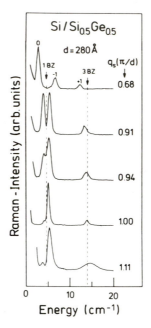

FIG. 4. Folded acoustic modes of a $Si/Si_{0.5}Ge_{0.5}$ multilayer structure. The spectra were obtained with different laser wavelengths which give rise to different scattering wave vectors q_s

standing waves which have to fit into the individual layers. In a simple approximation their energies can be determined from the bulk dispersion by fixing the wave vector to multiples of π/d_{Si}. This is shown also in Fig. 2. Details of the boundary conditions might lead to some deviations from this simple approximation. The optical branch of Ge on the other hand is overlapping with the LA modes of Si. Interaction leads to a finite dispersion of the "weaker confined" optical modes in Ge. This is also indicated in Fig. 2. Modes of this kind are clearly observed in the Raman spectrum shown in Fig. 3. The closely spaced peaks around 300 cm^{-1} correspond to the confined modes in the 12 monolayer thick Ge layers of order 1, 3 and 5[10]. The Si mode at about 480 cm^{-2} is the first order confined mode from the 4 monolayer Si. This mode, however, is shifted to lower energies due to the built-in biaxial tensile strain caused by the growth on Ge substrate.

Confined optical modes have been observed first and are so far most extensively studied in GaAs/AlAs superlattices[11,3]. They led to detailed information on the dispersion of bulk phonons, interface quality, boundary conditions, and strain. Under resonance conditions also new modes appeared in the spectra in between the LO and TO bulk phonon energies which were attributed to so-called interface modes with finite in-plane wave vector. This kind of breakdown in wave vector conservation under resonance conditions was first discussed by Merlin et al.[12] and studied in detail later for example by Sood et al.[13]. The dispersion of these modes have been studied recently away from resonance using micro-Raman spectroscopy[14]. An example is shown in Fig. 5. The laser is focussed on to the (110) cleavage plane of a GaAs/AlAs superlattice. The interface modes observed in the lowest spectrum are marked by arrows. Micro-Raman spectroscopy opens the possibility to study semiconductor heterostructures with completely different scattering geometries. It leads to new information and therefore will probably be used more extensively in the near future.

Another important feature of phonon Raman scattering is the study of resonance enhancements[15,16] which reveal detailed information on excitonic electronic excitations in quantum wells. Apart from the few selected examples the investigation of lattice dynamical properties of superlattices was also very helpful in analyzing the structures with respect to composition of alloys, crystalline nature, orientation, strain,

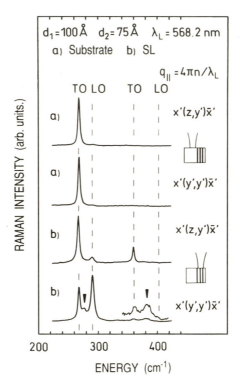

FIG. 5. Raman spectra from the substrate and the superlattice region of the (110) cleavage plane. The superlattice consists of 50 periods of 100Å GaAs and 75Å AlAs layers. The dashed lines indicate the bulk TO and LO phonon energies. The arrows mark the interface modes.

defects, impurities, local temperatures in devices and many other relevant properties. Phonon Raman scattering has indeed been established as one of the most versatile tools to study micro-structured semiconductors.

3. LIGHT SCATTERING BY FREE CARRIER EXCITATIONS

Fundamental electronic excitations in semiconductors have been studied extensively over the past twenty years[17,18,19]. Especially resonant inelastic light scattering experiments[20] showed that it is also possible to observe free carrier excitations in low dimensional systems with sheet densities of the order of 10^{11} cm^{-2}. Short after the proposal[21] the first observation of resonant light scattering by quasi-two-dimensional electron systems in GaAs/Al$_x$Ga$_{1-x}$As heterojunctions and quantum wells was reported[22,23]. Numerous publications have appeared in the past ten years on resonantly enhanced electron and hole excitations under various conditions (for reviews see[24,25]). Very recently also first electronic excitations in quantum wire systems have been reported[26,27].

Resonant inelastic light scattering by electronic excitations offers the unique possibility to separate single particle and collective excitations. Polarized spectra usually display plasmons which may be coupled to LO-phonons in polar semiconductors. Single particle excitations are observed in crossed polarizations (incident and scattered light polarized perpendicular to each other). The various possible excitations in 3,2,1, and 0-dimensional systems are shown schematically in Fig. 6. In 3 d the single particle excitations have a triangular lineshape. The plasmon frequency ω_p exhibits a weak dispersion: $\omega_p^2(q) = \omega_p^2(0) + 3q^2v_F^2/5$, where v_F is the Fermi velocity. In 2 d systems the motion of carriers is quantized normal to the layers. This leads to a subband splitting indicated by ω_{01} in Fig. 6b. Single particle (ω_{01}) and collective intersubband excitations (ω_{01}^{δ}) have been studied in many semiconductor systems and under various conditions[20,21]. In-plane single particle and plasmon excitations are only observed for a finite in-plane scattering wave vetor $q_\parallel$. Contrary to the 3 d case the plasmon frequency goes to zero for small wave vectors. The dispersion of a purely 2 d system is given by: $\omega_p^2(q_\parallel) = 2\pi n_{2d}e^2q_\parallel/m^x\varepsilon$, where n_{2d} is the 2-dimensional carrier concentration and ε the dielectric constant. This type of excitation was first reported by Olego et al.[28] in GaAs/Al$_x$Ga$_{1-x}$As for $q_\parallel \leq 1.7 \times 10^5$cm^{-1}. Recently the range of accessable wave vectors in light scattering experiments has been

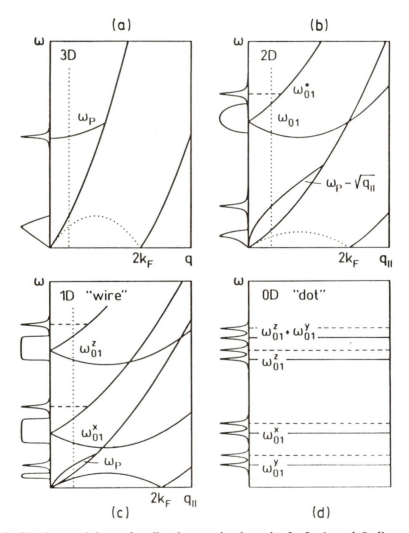

FIG. 6. Single particle and collective excitations in 3, 2, 1 and 0-dimensional carrier systems (schematically).

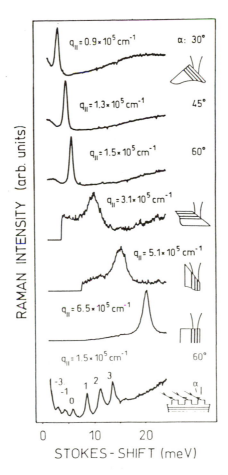

FIG. 7. Plasmon excitations in modulation doped GaAs/Al_xGa_{1-x}As multilayer structures measured with micro-Raman spectroscopy and grating coupler.

extended considerably by using grating coupler effects[29,30] or micro-Raman scattering[31] An example is shown in Fig. 7 for plasmon excitations in a layered two-dimensional electron gas in multiple GaAs quantum wells. The upper three spectra are obtained in the "usual" geometry, where α is the angle between the surface normal and the incident laser. The in-plane scattering wave vector is changed from $q_{\parallel} = 0.9 \times 10^5 \text{cm}^{-1}$ to $1.5 \times 10^5 \text{cm}^{-1}$ for α between $30°$ and $60°$. Micro-Raman spectroscopy allows backscattering from (110) cleavage planes or wedged surfaces. This technique extends the accessible wave vector range up to about $6.5 \times 10^5 \text{cm}^{-1}$. The plasmon is shifted up to more than 20 meV for the example shown in Fig. 7. The lower spectrum shows multiple plasmon excitations due to the grating coupler effect. A holographic grating with period length a was etched into the sample with the sheets of electrons left underneath. In this geometry the total wave vector is given by $q_{\parallel}^{\text{T}} = q_{\parallel} + mg$, where $m = m_i + m_s$ is the diffraction order of incident and scattered light and $g=2\pi/a$ is the reciprocal lattice constant of the periodic grating. Plasmon excitations in 2 d systems have been studied by various groups under different conditions (for details see original publications cited in Ref. 25).

Recently quantum wires and quantum dots, fabricated from layered semiconductor structures became very popular. Again inelastic light scattering is expected to provide detailed information on free carrier behavior in 1 and 0-dimensional systems. Possible excitations are shown schematically in Fig. 6 c and 6 d. So far experimental results exist only for wire structures[26,27]. It is assumed that carrier confinement is much stronger in growth direction (z) than in the lateral direction x where the quantizing potential is usually considerably weaker ($\omega_{01}^z > \omega_{01}^x$). The free motion in y-direction leads to a 1 d dispersion for the plasmons, which is very similar to the 2 d case. Plasmon excitations in a deep mesa-etched multi-quantum wire structure exhibit a strongly anistropic plasmon dispersion for $q_{\parallel}^y$ and $q_{\parallel}^x$. This is shown in Fig. 8 for periodic wires with period length a=8100Å and a geometrical wire width w=4100Å. Each wire contains five quantum wells with a sheet carrier concentration of $5 \times 10^{11} \text{cm}^{-2}$ and a spatial separation of 500Å. No significant dispersion of the plasmon peaks is observed with $q_{\parallel}^x$ perpendicular to the wires ($\gamma=0°$). This can be understood in terms of standing wave plasmons or confined plasmons normal to the wires with momentum $q_{\parallel}^x = m\pi/w_e$, where w_e is the effective width of the electron channels. Changing α along the wires ($\gamma=90°$), a dispersion of the

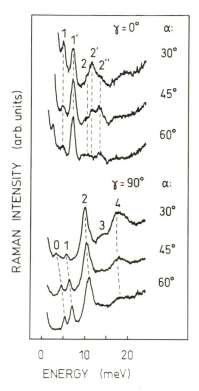

FIG. 8. Plasmon excitation in a wire structured $GaAs/Al_xGa_{1-x}As$ quantum well sample with excitation normal to the wires ($\gamma=0°$) and along to the wires ($\gamma=90°$).

plasmon is observed which agrees with the expected wave vector dependence with $q_{\parallel} = [(m\pi/w_e)^2 + (q_{\parallel}^{\circ}(\alpha))^2]^{1/2}$. An electron channel width $w_e \cong 2400\overset{\circ}{A}$ can be determined from such an analysis. These are the first experimental results of collective excitations in quantum wires studied by inelastic light scattering.

Apart from the behavior of free carrier excitations in reduced dimensions, electronic light scattering also provides information on various other properties like local carrier densities, velocity distribution or carrier temperature, subband splittings, donor and acceptor binding energies, Landau level spacings, magneto plasmons, spin splittings and so on. Therefore besides phonon Raman scattering also electronic light scattering has developed into a very important tool for the investigation of microstructured semiconductors.

4. FINAL REMARKS

It has been shown that many different aspects of low dimensional semiconductor systems are accessable to Raman scattering. With the use of optical multi channel detection it is nowadays possible to carry out such experiments at very low laser power densities or with high spatial resolution. This opens the possibility to study low temperature many body phenomena like fractional quantized Hall effect or Wigner crystallization. There are also first applications in the area of time resolved spectroscopy which enables us to study the dynamical behavior of hot carriers or hot phonons. There is also an increasing interest in lower dimensional systems like quantum wires and quantum dots. Inelastic light scattering will certainly play a prominent role in the elucidation of the various novel properties of these semiconductor microstructures including giant nonlinear optical effects which might also lead to new applications. This, however, requires better techniques in the preparation of quantum wires and quantum dots such that the high quality of heterostructures, quantum wells and superlattices is preserved. Micro Raman spectroscopy will be also used more and more for the analysis of semiconductor devices under operation. Finally, I expect that the concept of heterostructures and superlattices will be extended to many different material systems including strongly non lattice matched semiconductors and combinations with insulators, magnetic materials, metals and superconductors. This will provide new materials, which hopefully are also the basis for new effects. Inelastic light scattering will serve as one of the most versatile techniques to analyze the intriguing properties of these novel materials.

REFERENCES

1. Light Scattering in Solids V Springer Series: Topics in Applied Physics Vol. **66**, eds. M. Cardona and G. Güntherodt, Springer Verlag, Berlin, Heidelberg (1989).

2. NATO Advanced research workshop on Light Scattering in Semiconductor Structures and Superlattices, Mont Tremblant, Canada, eds. D. J. Lockwood and J. F. Young, to be published by Plenum (1990).

3. B. Jusserand and M. Cardona, Raman Spectroscopy of Vibrations in Superlattices in Ref. (1), p. 49.

4. M. V. Klein, JEEE J. QE **22**, 1760 (1986).

5. B. Jusserand and D. Paquet, in Semiconductor Heterojunctions and Superlattices, ed. by G. Allan, G. Bastard, N. Boccara, M. Lanoo, M. Voos (Springer, Berlin, Heidelberg 1986), p. 108.

6. S. M. Rytov, Akust. Zh.**2**, 71 (1956). (Sov. Phys. Acoust. **2**, 68 (1956)).

7. C. Colvard, R. Merlin, M. V. Klein and A. C. Gossard, Phys. Rev. Letters, **43**, 298 (1980).

8. G. Abstreiter, K. Eberl, E. Friess, W. Wegscheider and R. Zachai, Journal of Crystal Growth, **95**, 431 (1989).

9. H. Brugger, H. Reiner, G. Abstreiter, H. Jorke, H. J. Herzog and E. Kasper, Superlattices and Microstructures, **2**, 451 (1986).

10. E. Friess, K. Eberl, U. Menczigar and G. Abstreiter, Solid State Commun., **73**, 203 (1990).

11. B. Jusserand, D. Paquet and A. Regreny, Phys. Rev. **B30**, 6245 (1984).

12. R. Merlin, C. Colvard, M. V. Klein, H. Morkoc, A. Y. Cho, A. C. Gossard, Appl. Phys. Letters, **36**, 43 (1980).

13. A. K. Sood, J. Menendez, M. Cardona and K. Ploog, Phys. Rev. Letters, **54**, 2115 (1985).

14. A. Huber, T. Egeler, W. Ettmüller, H. Rothfritz, G. Tränkle and G. Abstreiter, Superlattices and Microstructures (1990) in press.

15. P. Manuel, G. A. Sai-Halasz, L. L. Chang, Chin-an Chang and L. Esaki, Phys. Rev. Letters **37**, 1701 (1976).

16. J. E. Zucker, A. Pinczuk, D. S. Chemla, A. Gossard and W. Wiegmann, Phys. Rev. Letters. **51**, 1293 (1983).

17. A. Mooradian, in Festkörperprobleme IV, ed. O. Madelung (Pergamon Vieweg, Braunchweig 1969) p. 74.

18. M. V. Klein, in Light Scattering in Solids IV, ed. M. Cardona, Topics Appl. Phys., Vol. 8, 2nd ed. (Springer, Berlin, Heidelberg 1984) p. 147.

19. G. Abstreiter, M. Cardona and A. Pinczuk, in Light Scattering in Solids IV, eds. M. Cardona and G. Güntherodt, Topics Appl. Phys., Vol. **54** (Springer, Berlin, Heidelberg 1984) p. 5.

20. A. Pinczuk, G. Abstreiter, R. Trommer and M. Cardona, Solid State Commun., **30**, 429 (1979).

21. E. Burstein, A. Pinczuk and S. Buchner, in Physics of Semiconductors, ed. B. L. H. Wilson (The Institute of Physics, London, 1979) p. 1231.

22. G. Abstreiter and K. Ploog, Phys. Rev. Letters, **42**, 1308 (1979).

23. A. Pinczuk, H. L. Störmer, R. Dingle, J. M. Worlock, W. Wiegmann and A. C. Gossard, Solid State Commun., **32**, 1001 (1979).

24. G. Abstreiter, R. Merlin and A. Pinczuk, IEEE J. QE, **22**, 1771 (1986).

25. A. Pinczuk and G. Abstreiter, in Ref. 1, p. 153.

26. J. S. Weiner, G. Danan, A. Pinczuk, J. Valladares, L. N. Pfeiffer and K. West, Phys. Rev. Letters, **63**, 1641 (1989).

27. T. Egeler, G. Abstreiter, G. Weimann, T. Demel, D. Heitmann, P. Grambow and W. Schlapp, Phys. Rev. Letters, **65**, 1804 (1990).

28. D. Olego, A. Pinczuk, A. C. Gossard and W. Wiegmann, Phys. Rev., **B25**, 7867 (1982).

29. T. Zettler, C. Peters, J. P. Kotthaus and K. Ploog, Phys. Rev., **B39**, 3991 (1989).

30. T. Egeler, G. Abstreiter, G. Weimann, T. Demel, D. Heitmann and W. Schlapp, Surface Science, **229**, 391 (1990).

31. T. Egeler, G. Abstreiter, G. Weimann and W. Schlapp, Superlattices and Microstructures, **5**, 123 (1989).

SPECTROSCOPIC INVESTIGATIONS OF QUANTUM WIRES AND QUANTUM DOTS

D. Heitmann, T. Demel, P. Grambow, M. Kohl and
K. Ploog

Max-Planck-Institut für Festkörperforschung
Heisenbergstr. 1, 7000 Stuttgart 80, Germany

Abstract - The remarkable progress of submicron technology in the eighties has made it possible to realize quantum wires and quantum dots. Starting from two-dimensional electronic systems (2DES) in semiconductor heterostructures electrons are further confined by lateral potentials acting on a submicron scale, which induce quantum confined energy states such that, for wires, a set of 1D subbands with free dispersion in only one direction is formed or, for dots, artificial "atoms" with a totally discrete energy spectrum are obtained. These low-dimensional systems exhibit unique properties. We give an introduction to this field of 1DES and 0DES by reviewing far-infrared excitations in quantum wires and dots and photoluminescence studies of quasi 1D excitons in quantum wires.

1. INTRODUCTION

Research on low-dimensional electronic systems in close interaction with technology and applications saw a remarkable increase in the last two decades. There was a great body of fundamental research in the seventies on the Si MOS systems.[1] Simultaneously, in a long and very hard effort layered epitaxial growth techniques have been developed and have been brought to such a high level of perfection that it was possible to realize, at will, semiconductor superlattices and heterostructures with tailored band-structures, designed and optimized for a special purpose. This has led to an enormous number of new fundamental results and applications with many unique or outstanding properties. Many of these properties arise from the two-dimensionality of the systems, two-dimensional in the sense that a strong confinement potential for electrons in the direction perpendicular to the layered structure leads to a quantization of the energy levels. Apart from the versatility of these systems we believe that their great attraction and the interest of so many researchers in this field is also due to the fact that with these novel growth techniques it was possible to design and realize simple quantum mechanical textbook concepts, i.e., the particle in a box. The high perfection, the versatility, and the beautiful simplicity of the 2DES has challenged many scientists to prepare and study systems with further reduced dimensionality, specifically quantum wires and quantum dots. Here, due to an ultrafine lateral confinement, the original free dispersion of the electrons in the lateral direction is also quantized. A potential acting in the x-direction creates a "quantum wire." (The growth direction is labeled z in the following.) Ultimately, with a confining potential in both the x- and y-directions, quantum dots, artificial "atoms" with a totally discrete energy spectrum are formed. It is hoped that the singularities in the density of states or the unique transport phenomena of these systems will result in further new and unexpected discoveries. There has already been a great amount of work in this field. Transport and capacitance in wires, narrow constrictions and dots has been explored,[1-19] vertical and lateral tunneling in 1DES and 0DES have been studied[20,21] and far infrared (FIR),[22-34] photoluminescence (PL),[35-57] and Raman spectroscopy[58,59] have been performed. In this paper we give an introduction to this field of 1DES and 0DES by reviewing recent investigations of quantum dots and quantum wires by FIR and PL spectroscopy.[11,25,29,47]

2. ONE- AND ZERO- DIMENSIONAL ELECTRONIC SYSTEMS

The term "1DES" is used here to denote electronic systems with an energy spectrum that consists of a set of quantum confined 1D subbands

$(i,j=0,1,2,3,..)$

$$E^{ij}(ky) = \frac{\hbar^2 k_y^2}{2m^*} + E_x^i + E_z^j, \tag{1}$$

where the electrons have a free dispersion only in the y-direction (k_y is the electron wave vector and we have assumed separability of the Hamiltonian for the different directions in space). Such systems are advantageously prepared by starting from a 2DES, e.g., in layered high-mobility modulation doped *AlGaAs/GaAs* heterostructures or quantum well systems or in metal-oxide-semiconductor devices. E_z^j, where z denotes the direction normal to the layered structures, are the quantized energy levels in the original 2DES with typical separations ranging from 10 to 100 meV. E_x^i are the quantum confined energy states due to the additional lateral confinement acting in x-direction. With a confining potential both in the *x*- and *y*-direction, quantum dots with a totally discrete energy spectrum are formed.

$$E^{ijk} = E_x^i + E_y^j + E_z^k. \tag{2}$$

3. PREPARATION OF QUANTUM WIRES AND QUANTUM DOTS BY HOLOGRAPHIC LITHOGRAPHY AND DRY ETCHING TECHNIQUES

To achieve for instance a lateral quantization of 2 meV in the GaAs system with $m^* = 0.065m_o$, one has to confine the electrons to a width w of some 100 nm. Different arrangements have been used to realize such systems, e.g., split-gate configurations,[8,9,22,24,28,30,31] "shallow"[3] and "deep"[11,25-27,29] -mesa etched structures, or isolation of electron channels by ion beam bombardment[6] through narrow masks or with focussed ion beams.[39,16] A very efficient way to produce narrow structures, in particular for large area samples that are required in optical experiments, is the method of holographic lithography. The superposition of two coherent laser beams results in a sinusoidally modulated intensity pattern which is used to expose a photoresist layer on top of the sample. After development, periodic photoresist stripes are obtained where the periodicity, *a*, can be controlled via the angle of incidence of the laser beams and the width of the stripes, *t*, can be varied in a certain range by the exposure and development time. This holographic method has the advantage that many parallel lines with excellent homogeneity over large areas, both in periodicity and width, can be prepared, which is advantageous for many experiments, e.g., optical experiments, FIR and microwave spectroscopy. Using a double exposure with an intermediate 90° rotation of the sample it is possible to prepare a cross-grating or a grid structure of dots. The photoresist stripes or dots are the starting point for an additional processing to prepare different configurations.

In Fig. 1 we have sketched some systems that we have realized by etching techniques. In Fig. 2 we show electron micrographs of quantum wire and quantum dot structures. In a technique, called "shallow"-mesa etching,[3] the doped AlGaAs is etched, leaving an 1DES below the remaining doped AlGaAs stripes (see Fig. 1a). "Deep"-mesa etched structures are shown in Fig. 1b and c. Here, starting from modulation doped AlGaAs/GaAs systems, or, for photoluminescence studies, from multi-layered undoped quantum wells, the etching is performed all the way down through the active GaAs layers. It is not easy to realize the

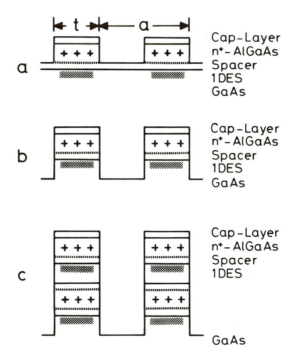

FIG. 1. Examples of etched quantum wire structures in the AlGaAs/GaAs system. (a) shows schematically a "shallow"-mesa etched structure, (b) a "deep"-mesa etched single-layer structure, (c) a "deep"-mesa etched double-layer structure.

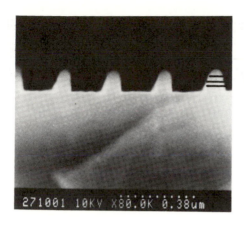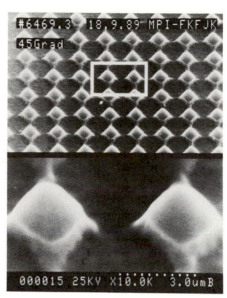

FIG. 2. Scanning electron micrograph of the profile of an AlGaAs/GaAs QW wire sample (left). The grating period is a=250nm, the wire width is 70nm. The approximate position of the three QW layers is indicated schematically in one of the stripes. The right side shows an array of 600 . 600nm^2 quantum dots.

latter structures. The etched surfaces create surface states which trap electrons and deplete the channel laterally or which quench the photoluminescence. With an optimized SiCl$_4$ plasma etching process it was possible to overcome these difficulties. We could fabricate narrow enough structures to achieve quantum confinement and nevertheless still had, in modulation doped systems, mobile electrons in the systems or strong enough photoluminescence. These methods have been discussed in Refs. 11-13. The deep-mesa etched structures have some unique advantages. Gated structures often do not work under illumination with bandgap radiation due to photoinduced leakage currents. Thus deep-mesa etched systems are ideal for photoluminescence and Raman spectroscopy. The deep-mesa etching technique also enables the fabrication of multi-layered quantum wire structures as sketched in Fig. 1c by starting from multi-quantum well systems. Such systems promise interesting technical applications for the integration of 1DES devices in multi-layered arrangements. These systems also show unique physical properties which allow one, e.g., to elucidate the dynamic response of 1DES.[25,26]

4. DETERMINATION OF THE 1D ENERGY SPECTRUM FROM MAGNETOTRANSPORT

A first approximation to calculate the energy levels in a 1DES, e.g., in a deep-mesa etched structure as sketched in Fig. 1b, is to use the particle in a box model with infinite potential barriers. However this is only a very rough approximation. The actual potential depends on charged surface states at the side walls, in other words, on the Fermi level pinning at the surfaces. It depends on the electric field of the remote ionized donors in the AlGaAs stripes and, in modulation doped samples, on the electrons in the channel itself due to self-consistent screening. The most sophisticated calculations for 1DES so far have been performed for the split-gate configuration by Laux et al.[60] It is found that for a small number of electrons in the channel the potential is nearly parabolic. (We will see later that this has important consequences for the FIR response.) However, with increasing 1D-charge density N_i, self-consistent effects become important. The potential flattens and the subband separation decreases drastically.

Also experimentally it is not easy to characterize the 1DES, i.e., to determine subband spacing, electrical wire width and number of electrons per 1D subband. In most studies so far 1DES are characterized by the magnetic depopulation of the 1D subbands which occurs if the 1DES is exposed to a magnetic field B oriented perpendicular to the original 2D plane.[2] The underlying physics can be explained without loss of generality if one assumes a parabolic confinement potential[2] $V(x) = 1/2\, m^* \Omega_0^2 x^2$. In this case the Schrödinger equation can be solved analytically. The magnetic field induces an additional potential $V_B(x) = 1/2\, m^* \omega_c^2 (x - x_0)^2$. In this case the Schrödinger equation can be solved analytically. The magnetic field induces an additional potential $V_B(x) = 1/2 m^* \omega_c^2 (x - x_0)^2$, where $\omega_c = eB/m^*$ is the cyclotron frequency. For this model the energy levels are given by

$$E^i(k_y, B) = \hbar\Omega\left(i + \frac{1}{2}\right) + \frac{\hbar^2 k_y^2}{2m_y^*(B)} \tag{3}$$

with $\Omega^2 = \omega_c^2 + \Omega_0^2$, and $m^*\Omega^2/\Omega_0^2$ This energy spectrum is shown in Fig. 3. Since the level spacing and the 1D density of states, $D_{1D}(E,B)$, increase with increasing B the 1D subbands become successively depopu-

214

lated, giving rise to oscillations of the Fermi energy. In a transport measurement this leads to Shubnikov-de Haas (SdH) type of oscillations. In a 2DES the number of occupied Landau levels increases with decreasing B, leading ideally to an infinite number of SdH oscillations periodic in $1/B$. In a 1DES, however, only a finite number of 1D subbands are occupied at $B = 0$, thus only a finite number of SdH oscillations occur, which are no longer linear in $1/B$.

We demonstrate this behavior for two types of our samples as shown in Fig. 1a and c. We performed magnetotransport measurements[11] at low temperatures ($T = 2.2K$) in perpendicular magnetic fields B. We defined on some of the one-layered quantum wire structures an active area of $2.5 \times 2.5mm^2$ by chemical mesa etching. Ohmic Au = Ge/Ni alloy contacts were aligned perpendicularly to the grating in order to measure the dc transport parallel to the stripes. On other samples a quasi-dc response was obtained by measuring the transmission of microwaves (30-40GHz) through the sample. Since for this measurement no contacts were needed, it was especially useful for

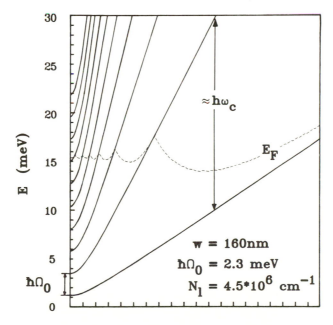

FIG. 3. Energy spectrum for electrons confined in a harmonic oscillator potential and in a magnetic field B. The dotted line shows the oscillations of the Fermi energy for the indicated linear charge density N_l. The hybrid 1D-subband- Landau-levels become successively depopulated with increasing B.

samples with multi-quantum well (MQW) wires. Alloyed contacts would
have short-circuited the channels of different layers.

In Fig. 4a we show measurements on a shallow-mesa etched sample
(Fig. 1c). The period of the wires was a = 500nm, the remaining width
of the n-doped AlGaAs was t = 250nm. Via Ohmic contacts the two-
terminal resistance was measured as a function of the magnetic field.
The dc conductivity shows well pronounced SdH-type oscillations. The

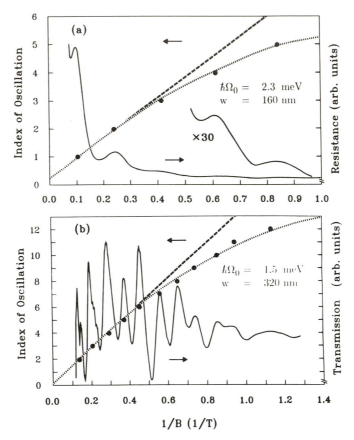

FIG. 4. Magnetotransport measurements (full lines, right scale) on "shallow"
mesa etched (a) (a = 550nm, t = 250nm) and in a "deep" mesa etched two-
layered (b) (a = 1000nm, t = 550nm) structure plotted versus 1/B. A fan
chart for the positions of the maxima in the magneto resistivity (full circles, left
scale) exhibits deviations from a linear 1/B dependence (dashed lines) and indicates a one-dimensional energy structure in our samples. The dotted lines show
the depopulation of 1D subbands within a harmonic oscillator model, calculated
for the indicated values of the confining potentials $\hbar\Omega_0$ and the wire width w.
(from Ref. 11).

important point is that the period of the oscillations is not constant in $1/B$, but exhibits distinct deviations at large values of $1/B$ and correspondingly small B. This clearly indicates the 1D character of our structure as was discussed above for the harmonic oscillator model. We calculated the depopulation of the 1D subbands by a magnetic field within this model using Ω_0 and the total 1D carrier density N_t as fitting parameters. The experimental fan chart in Fig. 4a is best described for $\hbar\Omega_0 = 2.3\text{meV}$ and $N_t = 4.5 \cdot 10^6 \text{cm}^{-1}$. For these values, six 1D subbands were occupied at $B = 0$. Defining the width w of the electron channel by the amplitude at the Fermi energy: $E_F (B = 0) + V (w/2) = 1/2\ m^* \Omega_0^2 (w/2)^2$, we found $w = 160\text{nm}$ which was smaller than the geometrical width $t = 250\text{nm}$. Therefore we defined a "lateral edge depletion region" on either side of width $w_{dl} = 1/2 (t - w) \approx 45\text{nm}$. As an example of deep mesa etched structures we show in Fig. 4b the quasi-dc conductivity of a two-layered quantum wire structure (Fig. 1c), measured in microwave transmission as described above. The structure has been prepared starting from a modulation doped, electronically decoupled MQW system with two periods where each period consisted of a 25nm undoped $Al_xGa_{1-x}As$ spacer, followed by a short-period AlAs/GaAs-superlattice (5 periods, each layer 22nm thick), a 50nm GaAs-QW, a 5nm AlAs spacer, and a 25nm n-doped $Al_xGa_{1-x}As$ layer ($N_d = 1.5 \cdot 10^{18}\text{cm}^{-3}$, $x = 0.3$). SdH oscillations can be clearly resolved and show again distinct deviations from a linear $1/B$ behavior at small B. An analysis within the harmonic oscillator potential model gives for the 1D confinement $\hbar\Omega_0 = 1.5\text{meV}$, $N_t = 15 \cdot 10^{16}\text{cm}^{-1}$ and 16 occupied 1D subbands. From the channel width $w = 320\text{nm}$ we deduced that there is a lateral depletion width $w_{dl} \approx 100\text{nm}$ on either side of the wire.

One of the most interesting aspects of 1DES is that the different 1D subbands behave in the ballistic regime in close analogy to waveguide modes in a microwave system, where the phase of the electron wave function is a well defined quantity. Accordingly novel device concepts based on quantum interference have been proposed (e.g. Ref. 17). For such devices it is highly desirable to work in the lowest mode, i.e., the 1D quantum limit. It was possible to realize the 1D quantum limit in narrow constrictions which lead to the beautiful observation of the quantized conductance in these systems.[8,9] However it still seems to be a problem to realize "long" quantum wires in the quantum limit. (In many experiments researchers apply magnetic fields, which effectively produce 1D edge channels, to demonstrate interference effects.[17,21]) We believe that the problem of long wires in the quantum limit is inherently connected with Coulomb effects which become important for the very low electron density under these conditions. This problem has also been addressed by Meirav et al.[18] The smallest number of occupied 1D subbands that we have reached so far in our deep mesa etched structures was four, similar to the situation in gated structures.[24] So it remains the challenge of the future to prepare long wires in the 1D quantum limit.

5. FIR SPECTROSCOPY OF QUANTUM DOTS

The analysis of the energy spectrum of quantum wires via the magnetic depopulation that we have discussed above, seems to be a very indirect way. It would appear to be much more direct to excite optical transitions between the subbands. However, it is found that the optical response exhibits a very complex behavior, where the subband separation can again be determined only indirectly. We shall demonstrate the FIR optical properties of low-DES using the example of quantum dots.[29]

Quantum dot samples were prepared by starting from modulation doped AlGaAs-GaAs heterostructures. An array of photoresist dots (with a period of 1000nm both in the x- and in the y-directions) was prepared by a holographic double exposure. Using an anisotropic plasma etching process, rectangular 200nm deep grooves were etched all the way through the 23nm thick undoped AlGaAs spacer layer into the active GaAs, leaving quadratic dots with rounded corners and geometrical dimensions of about 600 nm by 600 nm (see Fig. 2 and inset of Fig. 5). We have seen above that with the technique of "deep mesa etching" it was possible to realize 1DES in linear stripe systems with typical energy separations for the 1D subbands of about 2meV. The actual width of the electron channels was smaller than the geometrical width, indicating a lateral edge depletion of 100 to 120nm. For the dot structures here with increased etched surface area, this depletion is even more pronounced. Actually, we have prepared samples which had, in the dark, no mobile electrons. Via the persistent photoeffect we could then increase the number of electrons per dot, N, in steps up to $N = 210$. We determine N from the strength of the FIR absorption at high magnetic field. The potential that confines the electrons and thus determines the radius of the 2D disc, depends on the remote ionized donors and, in a self-consistent way, on N. We have estimated the radius from the observed resonance frequency and formula (8) which will be explained below. DC transport is inherently not possible in our quantum dot structures, thus

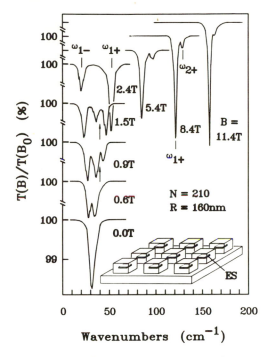

FIG. 5. Relative transmission of unpolarized FIR radiation for a dot structure with radius R = 160nm and N = 210 electrons per dot. The inset shows schematically the dot structure. (from Ref. 29).

we cannot, as in the case of 1DES, determine the level spacing directly from magnetic depopulation. However, since we have prepared the dots with exactly the same methods, we can assume that the energy levels in the dots have a similar separation as in the 1D wires. We expect a discrete energy spectrum with a typical energy separation of 1 to 2 meV, if we assume the $2(n+1)$ level degeneracy of the harmonic oscillator potential (n is the level index, see below).

The FIR experiments were performed in a superconducting magnet cryostat, which was connected via a waveguide system to a Fourier transform spectrometer. The transmission T(B) of unpolarized FIR radiation through the sample was measured at fixed magnetic fields, B, oriented normally to the surface of the sample. The spectra were normalized to a spectrum $T(B_0)$ with a flat response. The temperature was 2.2K. The active sample area was 3x3 mm^2 containing 10^7 dots. Experimental spectra for a sample with a dot radius R = 140nm and N = 210 electrons per dot are shown in Fig. 5. For B = 0 one resonance is observed at ω_{01} = 32cm^{-1}. With increasing B the resonance splits into two resonances, one, ω_{1-}, decreases in frequency, the other, ω_{1+}, increases. For B > 4T a second resonance, ω_{2+}, can be resolved which also increases with B. The most interesting observation is the obvious resonant coupling which occurs at a frequency of about 40cm^{-1}. Experimental resonance positions for two samples are depicted in Fig. 6. Figure 6a shows the resonance positions from a sample containing 210 electrons per dot, Fig. 6b a sample with only 25 electrons per dot. The interesting observation is the significant resonant anti-crossing at $\omega \approx 1.4\omega_{01}$, which was found for all our samples.

The FIR resonances observed here are, except for the resonant splitting and the higher modes, very similar to earlier observations on larger finite-sized 2DES in GaAs-heterostructures (dots with R $\approx$ 1.5μm).[61] Even for electron-systems on LHe with large radii R $\approx$ 1cm, a

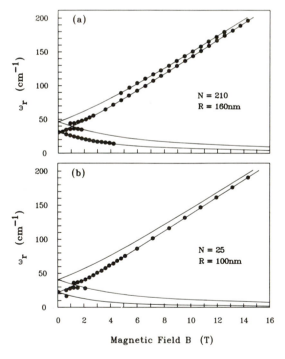

FIG. 6. Experimental B-dispersion of resonant absorption in quantum dot structures with R = 160nm and N = 210 (a), and with R = 100nm and N = 25 (b). The full lines are fits with the theoretical dispersion (equation (9)). All structures show an anti-crossing of the ω_{2-} with the ω_{1+} mode. (from Ref. 29).

similar dispersion and higher modes, but no anti-crossing was observed.[62,63] The experiments in Refs. 61-63 have been interpreted as depolarization, or equivalently, as edge magnetoplasmon modes.[64-68] However, recently Sikorski and Merkt observed for quantum dots in InSb systems containing only a very small number of electrons per dot (N < 50) resonances with the same dispersion as our ground modes $\omega_{1\pm}$ but without anti-crossing.[28] (An even more complex mode structure was found for coupled quantum dots by Lorke et al.[30]) For the quantum-confined systems one might expect, at first glance, that an adequate description of the FIR response can be given in terms of transitions between the discrete energy levels of the "atoms." However, it is known from FIR experiments on 1DES with quantum confined states that the optical response of such low dimensional systems is, with increasing number of electrons, strongly influenced by collective effects which increasingly give the excitations the character of local plasmon resonances. To elucidate this behavior for quantum dot structures and to explain the experimentally observed excitation spectrum we will approach this problem from two limits: (i) transition in an "atom" including collective effects and (ii) plasmon type of excitation in a 2DES of finite size.

The simplest way to treat a laterally confined 2D system in a magnetic field is to assume a parabolic confinement potential, which is actually not a bad assumption, as we will see later. We consider the one-electron Schrödinger equation:

$$H\left(\vec{r}, \vec{p}\right) = \frac{1}{2m^*} \left[\vec{p} + \frac{e}{c}\vec{A}\left(\vec{r}\right)\right]^2 + \frac{1}{2} m^* \Omega_0^2 r^2 \qquad (4)$$

where Ω_0 characterized the confining potential. The energy eigenvalues have been calculated by Fock[69]

$$E_{n,\ell} = \left(2n + |\ell| + 1\right) \cdot \sqrt{(\hbar\Omega_0)^2 + (\hbar\omega_c/2)^2} + \ell \cdot \hbar\omega_c/2 \quad (5)$$

where $n = 0,1,2\ldots$ and $\ell = 0,\pm1,\pm2$, are, respectively, the radial and azimuthal quantum numbers. This energy spectrum is shown in Fig. 7a.

Here we note that, whereas for the 1D wire structure (Fig. 3) there is a steady increase of all energy levels with B, the energy spectrum of the 0D dots is much more complex. In particular, high lying energy levels at small B decrease in energy with increasing B crossing several other energy levels. Thus in dots there is not a direct one-to-one correspondence between different energy levels at different fields. This makes an analysis of experimental data much more complex, see e.g., Ref. 14. Even more interesting is that a similar behavior also occurs, without any magnetic field, if we vary the electron-electron (ee) interaction in the system, e.g., via the number of electrons or the geometrical dimensions. For 1D wires e.g. the calculation of Laux et al[60] show a smooth lowering of all energy levels with increasing interaction. However for dots, Bryant[70] demonstrated very nicely that the energy levels in a dot can cross with increasing ee-interaction. So one has to be extremely careful in relating many-electron results or experimental data to a single-particle model.

Let us return to our one-particle spectrum (5). From calculations of the matrix elements one finds that dipole allowed transitions have transition energies

$$\Delta E^{\pm} = \sqrt{\hbar^2 \Omega_0^2 + (\hbar \omega_c / 2)^2} \pm \hbar \omega_c / 2. \tag{6}$$

This is, except for the anti-crossing, exactly the behavior that we observe in our experiments for the strongest mode in Fig. 6. What happens now if this external potential Ω_0 is "filled" with more than one electron? Recent self-consistent band structure calculations by Kumar et al.[71] show that ee-interaction, i.e., the self-consistent screening, strongly decreases the separation of the energy levels. For example, for the 300nm quantum dots in Ref. 71, the level spacing for one electron in a dot is 4 meV, while for $N = 10$ electrons it decreases to $\hbar \Omega_{01} = 2$ meV. Is it possible to observe this screened energy difference $\hbar \Omega_{01}$, the Hartree energy, in an optical dipole excitation? The answer for parabolic confinement is *No*! It has been shown for quantum wells[72] and quantum dots[73-75] with parabolic confinement, that the Hamiltonian for N electrons separates in a center of mass (CM) motion and into relative, internal motions. This CM motion solves exactly the one-electron Hamiltonian and is the only allowed optical dipole excitation. Thus, the optical response of quantum dots with parabolic confinement represents the collective CM excitation at the frequency of the unscreened external potential, i.e., for $B = 0$, at $\hbar \Omega_0$ and <u>not</u> at the screened one-particle energy separation $\hbar \Omega_{01}$. Thus, for parabolic confinement, dipole excitations are insensitive to ee-interactions, they are independent of the number of electrons in the dot (which has been observed very nicely in the experiments of Sikorski et al[28]) and reflect the energy spectrum of the "empty" atom. Shikin et al[67] have drawn the same conclusion from a classical description of an electron system in a parabolic confinement. The quantum mechanical CM motion corresponds exactly to the classical dipole plasma mode of the quantum dot. The result that the dipole excitation in a parabolic confinement is not affected by ee-interactions is a generalization of Kohn's famous theorem[76] which says that the cyclotron frequency and the related mass in a translationally invariant system is not affected by ee-interactions. The generalization has important consequences also in other fields. For example, it was hoped to detect the fractional quantum Hall effect (FQHE), which is a manifestation of ee-interactions, in an optical experiment. The generalization shows that a direct optical manifestation is not only forbidden in first order but also in second, i.e., a stronger than parabolic pertubative potential is needed to observe features of the FQHE in dipole excitation.

We also like to mention that the optical response is often calculated in RPA, where one starts from the screened one-particle spectrum with energy separation $\hbar\Omega$ (e.g. Ref. 77). The optical resonance frequency $\hbar\omega_r = \sqrt{\hbar^2\Omega^2 + \hbar^2\omega_d^2}$ is shifted with respect to $\hbar\Omega$ by the so-called depolarization shift, which is characterized by $\hbar\omega_d$. For the parabolic confinement, in this language, we thus find that with increasing number of electrons the decrease of the screened one-particle spacing $\hbar\Omega$ with respect to $\hbar\Omega_0$ is exactly cancelled by the increase of the depolarization shift. So it is very important to calculate the energy spectrum of the electrons self-consistently.[75]

The self-consistent band structure calculations, in particular for split-gate devices, show that the parabolic confinement is actually a very good approximation.[71] From this we can conclude that the unscreened potential $\hbar\Omega_0$ in our samples is approximately given by the optical response frequency of about 2 to 4meV. On the other hand, having demonstrated that it is possible to realize artificial atoms, one is naturally interested in studying next the fine structure associated with internal degrees of freedom and ee-interactions in the atom. The fact that we observe, weakly, higher order modes and resonant coupling, which is not allowed in a strictly parabolic confinement, shows that for our structures, weak, but definite nonparabolic terms exist in the potential. This makes our experiments sensitive to ee-interactions and the internal fine structure, which, however, enters in a very complex manner. A very interesting question is what is the origin of the resonant coupling in Figs. 5

and 6. We have argued in Ref. 29 (see also below) that the interaction between different dots is too weak to be a dominant process. Another possible effect is geometrical in nature, i.e., deviations from circular geometry, which, in principle, would produce such a coupling. To obtain at least a feeling for the influence of such effects we have calculated the one-electron energy spectrum for a potential V (x,y) = 1/2 $m^*\Omega_0^2(x^2 + y^2) + c(x^4 + y^4)$, where c was chosen such that the equipotential line at the highest occupied energy level for a system with N = 25 extents 10 percent more in [1,0]-direction then in [1,1]-direction. The results are shown in Fig. 7b. The degeneracy of crossing energy levels, whose azimuthal quantum numbers differ in the parabolic confinement (Fig. 7a) by 4, is now lifted. The experimentally observed anti-crossing corresponds to transitions between the ground state (0,0) and the split energy levels of the (0,1) state. The equipotential lines for the chosen value of c at the highest occupied energy level for a system look very similar to the geometrical shape of the dots, which is quadratic with rounded corners. Although it is expected that this shape of the confining potential overestimates the nonparabolicity deep inside the dot, the calculated anti-crossing of the (0,1) and the (0, $\bar{1}$) state is not strong enough to reproduce the experimentally observed splitting of 1meV. This seems to indicate that geometrical effects are not the dominant contributions to our observed anti-crossing (see also other arguments given below). However one has to be extremely careful in transferring single particle results to the optical response of a system with more electrons. In this case deviation from a parabolic confinement immediately lead to mixing of the CM motion and internal motions. In particular, higher modes do not occur at $2\Omega_0$, but, in general at significantly lower energies and approach with increasing N the classical plasmon type limit discussed below. Stimulated by our experimental findings several groups are now studying the anti-crossing starting for different models. Chakraborty et al[74] calculated the excitation spectrum for a small number of electrons in a parabolic confinement. For two interacting dots they found indeed an anti-crossing of modes, however, so far the splitting seems to be significantly smaller as compared to the experimental findings. Gudmundsson et al[75] have performed RPA calculations for 25 electrons in potentials of different shape. So far it is not clear how strong the anti-crossing is in this model. We will give a possible explanation starting from a classical plasmon type of excitation below.

Since the FIR response of the quantum dots reflects dominantly the

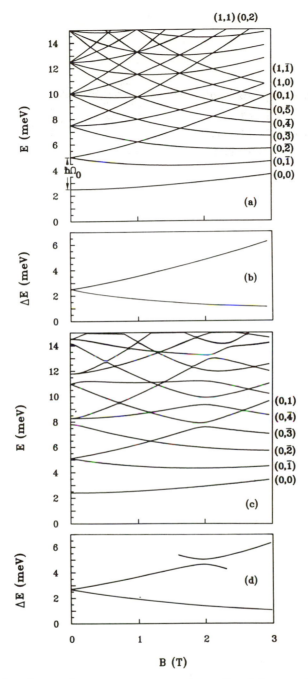

FIG. 7. Calculated one-electron-energy spectrum for (a) a parabolic confinement potential and (c) for a potential that includes a noncircular nonparabolicity term (see text). The lower parts show transition energies. (from Ref. 29).

rigid CM motion and the response frequency is, with increasing N, significantly larger than the one-particle energy level separation, it is also very elucidating to approach the dynamic excitations of dots from a classical plasmon type of excitation in a finite geometry. A simple way is to assume a disc-like 2DES with 2D density N_s and radius R and start from linear edge plasmons (e.g. Refs. 64-66), i.e., excitations which exist at an edge of a semi-infinite 2DES. These excitations have the dispersion $\omega_{ep}^2 = 0.81\omega_p^2(q)$ where

$$\omega_p^2(q) = N_s e^2 q / 2m^* \varepsilon_0 \varepsilon_{eff} \tag{7}$$

is the 2D plasmon frequency. For a disc the circumference quantizes the q-vectors in values $q = i/R$ ($i = 1,2, ...$). For $B = 0$ we thus have

$$\omega_{0i}^2 = 0.81 N_s e^2 i / 2m^* \varepsilon_0 \varepsilon_{eff} R. \tag{8}$$

In a magnetic field one calculates a set of double branches

$$\omega_{i\pm} = \sqrt{a_{i1}\omega_{0i}^2 + a_{i2}(\omega_c/2)^2} \pm a_{i3}\omega_c/2 \tag{9}$$

with $a_{i1} = a_{i2} = a_{i3} = 1$ (ω_c is the cyclotron frequency). This model already agrees very well with more sophisticated theories which determine more accurately the coefficients a_{ik} ($k = 1,2,3$) of ω_{0i} and ω_c. The values of the coefficients a_{ik} and the spacings of higher modes depend slightly on the way in which the 3D density profile for the electron distribution is modeled. For example, the squared frequency ratios at $B = 0$, $\omega_{02}^2/\omega_{01}^2$ ($\omega_{03}^2/\omega_{01}^2$), are 2 (3) for a step-like density profile[65] and 3/2 (15/8) for a density profile ($n(r) = n_0\sqrt{1 - r^2/R_0^2}$ ($r \leq R_0$), which is the classical self-consistent solution of a parabolic confinement potential (n is the density, r the radial distance).[68] Note that the B-dispersion of the lowest mode agrees exactly with the quantum mechanical result. These classical treatments also give the higher modes which are so far not available from quantum mechanical treatments in Refs. 72, 73.

We have fitted in Fig. 5 the dispersion with formula (9), using a_{21} as a fitting parameter and all other $a_{ik} = 1$. Taking into account the nonparabolicity of m^* in GaAs, we find a good agreement with the experimental dispersion. From the fit we obtain for the higher mode a frequency of $\approx 1.5\omega_{01}$. This is in between the expected frequency range of $\sqrt{15/8} \cdot \omega_{01}$ for the ω_3 mode of a parabolic confinement and the value of the ω_2 mode in the step-like density profile of $\omega_{02} = \sqrt{2} \cdot \omega_{01}$. From the fit it becomes clear that the splitting is caused by an anti-crossing of the ω_- branch of the higher mode and the ω_{1+} mode.

The obvious questions to ask are, also in this plasmon approach, which interaction causes the splitting and what determines its strength. An anti-crossing is not expected in the models discussed above (Eq. (9)) since the modes ω_{i+} and ω_{i-} represent, respectively, left and right circularly polarized eigenmodes, which are as such decoupled. Geometrical effects, i.e. deviations of the electronic system from a perfectly circular shape would, in principle, enable a mode coupling. However, we do not believe that this is a dominant effect since (i) such effects are independent of the dimensions and should thus be also present in the experiments of Ref. 61 where samples of a very similar shape show no splitting; (ii) for our samples one would expect that a splitting due to the geometrical shape effect would be more pronounced at large N and R where the electronic system extends closer to the geometrical edge. The surprising observation is that the experimental interaction strengths, expressed by the relative splitting $\Delta\omega/\omega$, increases with decreasing radius.

These observations, and the fact that this splitting is not observed on larger GaAs- systems with a very similar shape,[61] in particular not on the LHe-systems of Ref. 62 where beautifully sharp intersecting edge magnetoplasmons were found with, however, no interaction at all, leads to the conclusion that the smallness of our structures is the important parameter. This is why we think that a "nonlocal interaction" within the dot, which we will specify below, is important. Nonlocal effects are well-known for homogeneous 3DES and 2DES. They arise from the inherent finite compressibility of the Fermi gas, the "Fermi-pressure," and lead to corrections $f \cdot q^2 v_F^2$ (v_F = Fermi velocity) for the squared plasma frequency, where the prefactor f is $3/5$ in 3DES and $3/4$ in 2DES.[1] A microscopic description of the effects, which we here call "nonlocal," is that plasma oscillations are not only driven by Coulomb interaction, but also by the Fermi velocity of the individual electrons. This means that we introduce within a Thomas-Fermi approach quantum correction onto the so far classical electron gas. The Fermi pressure becomes important at small plasmon wavelengths and correspondingly high values of q. In homogeneous 2DES nonlocal effects are small under usual experimental conditions because the relevant parameter, a^*q, is small as compared to 1 (a^* is the effective Bohr radius). However, nonlocal interaction in 2DES can be clearly observed in resonant magnetic field experiments[78] where it leads to an anti-crossing of the 2D magnetoplasmon dispersion with harmonics n . ω_c of the cyclotron frequency.

For 0DES there is so far no theoretical treatment of the resonant nonlocal interaction. Only non-resonant nonlocal effects on the linear 1D edge magnetoplasmon dispersion have been considered until now.[64,65] Therefore, it is interesting to compare the 2D nonlocal interaction strength with the splitting that is observed here. The strength of the nonlocal resonant coupling in 2D, measured in terms of the frequency splitting, is $\Delta\omega/\omega_0 = 2.6\sqrt{q}$ (q in units of nm^{-1}).[78] If we use the same model for the interaction here and for q = 1/R we find for the sample of Fig. 6c with R = 90nm $\Delta\omega/\omega_{01} = 0.27$ which agrees surprisingly well with the experimental value of 0.30. This very close agreement might be accidentally to a certain degree and should be compared with a so far not available rigorous theory. In particular we expect that the nonlocal interaction also depends on the exact three dimensional density profile of the electron system. However, this close agreement demonstrates that the experimentally observed splitting is of the expected order for nonlocal interaction. Moreover, independent of the absolute value we expect within our simple analogy $\Delta\omega/\omega_{01} \propto \sqrt{1/R}$. This behavior is indeed approximately observed in our experiments and thus supports our interpretation.

6. FIR SPECTROSCOPY OF QUANTUM WIRES

As for the quantum dots, it is similarly found that the FIR response of quantum wires is strongly governed by collective effects.[22-25] We demonstrate this for a two-layered quantum wire structure as shown in Fig. 1c. In the experimental FIR spectra in Fig. 8 we observe at B = 0 two resonances at $\omega_{r1} = 62\text{cm}^{-1}$ and $\omega_2 = 35\text{cm}^{-1}$ if the incident electric field is polarized perpendicular to the wires. The resonance shifts with increasing B to higher frequencies. With increasing B resonances are also observed for parallel polarization with exactly the same resonance frequencies as for perpendicular polarization. Generally one finds for n-layered quantum wire structures n resonances; hence for a single-layered wire system only one resonance is observed. For the B-dependence one obtains experimentally that the resonances obey the relation $\omega_{ri}^2(B) = \omega_{ri}^2(B = 0) + \omega_c^2$.

Similar to the case of the quantum dots, one finds that the resonance frequencies ω_{ri} in the FIR spectra are significantly higher in energy

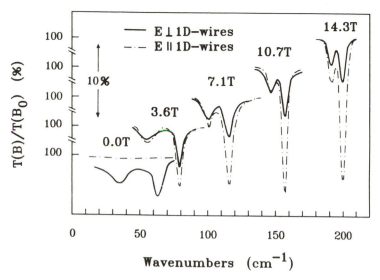

FIG. 8. Experimental FIR spectra, measured on a double-layered quantum wire structure at indicated magnetic fields B. Full lines and dash-dotted lines denote, respectively, polarization of the incident FIR radiation with the electric field vector perpendicular and parallel to the wires. (from Ref. 25).

than one would expect from the 1D subband separation $\hbar\Omega_0$, which, for the case of quantum wires, can be directly determined from the dc magneto-transport measurements. At B = 0, the two-layered system has: $\hbar\Omega_0 = 1.5\text{meV}$ (see Fig. 4b), $\hbar\omega_{r1} = 8\text{meV}$ and $\hbar\omega_{r2} = 4\text{meV}$ (see Fig. 8). This again reflects the fact that the optical response is a collective excitation, which is very interesting in this two-layered wire system. The higher mode is associated with a coupled in-phase CM motion in both quantum wires, the lower mode is an anti-phase motion of the electrons in the two coupled wires. These resonances resemble plasmon excitations in two (n)-layered, homogeneous 2DES where it is known that the collective excitation spectrum consists of two (n) branches.[79-81] The highest mode represents a so-called "optical" plasmon in contrast to the lower "acoustic" modes which have a linear q-dispersion at small q. The excitations in our quantum wires can be considered, if we neglect quantum effects, as localized optical and acoustic plasmon modes whose frequencies are approximately determined by a "plasmon in a box" model, in which the wavevector $q = \pi/w$ (w is the electronic width of the channel) is inserted into the 2D plasmon dispersion (6).[25] A more accurate description for a 1D wire with parabolic confinement is presented in Ref. 68.

7. 1D EXCITONS IN QUANTUM WIRES

So far we have treated modulation doped systems. There is also a great interest in the investigation of undoped systems with narrow quantum wells. The optical spectra of quantum-well (QW) systems in the visible and near infrared regime are governed by efficient intrinsic free exciton effects and exhibit a superior optical performance.[82,55] The important point here is that the confinement in a quantum well does not only lead to a - via the well width adjustable - blue shift of the exciton transition, but much more that the squeezing of the exciton wave functions in the z-direction also leads to a shrinkage of the exciton wave function in the lateral dimensions. This increases the binding energy, the effective Rydberg energy, and the oscillator strength, which makes it possible to observe QW excitons even at room temperature and which produces strong non-linear optical effects. These beautiful properties of the 2D excitons have challenged many scientists to study experimentally[35-56] and theoretically[83-87] systems with further reduced dimensionality, namely 1D quantum wires and 0D quantum dots. Besides the expectation of novel physical phenomena it is hoped that the optical properties, i.e. oscillator strengths, nonlinear coefficients and others, can be further improved with respect to the 2D systems to realize high performance optical devices.

We have prepared quantum wire structures[47] starting from MBE grown QW samples on semi-insulating (001) GaAs substrates. They consisted typically of three identical GaAs QW with well widths $L_z = 14$ nm separated by 10 nm thick $Al_xGa_{1-x}As$ barriers. A mask consisting of periodic photoresist lines was prepared by holographic lithography. With subsequent reactive ion etching nearly rectangular grooves of about 150 nm depth were etched into the samples (see Fig. 2). The resulting wire widths, w, ranged down to 60 nm where we still could detect a strong enough luminescence. On each QW wire sample a small part was left unpatterned to take reference spectra. Photoluminescence excitation (PLE) spectra obtained from QW wire samples with different widths, w, are shown in Fig. 9 together with the corresponding reference spectrum. For the quantum wire samples the electric field vector of the exciting light is polarized perpendicular to the wires. The two peaks in the reference spectrum at 808.93 nm and 805.48 nm are the e_1-hh_1 (heavy hole) and e_1-lh_1 (light hole) transitions, respectively. The PLE spectrum of the QW wire samples exhibit a different behavior. Three transitions can be

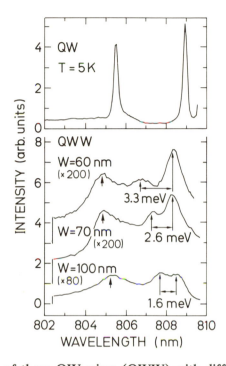

FIG. 9. PLE spectra of three QW wires (QWW) with different width w and of the corresponding reference QW. The two peaks in the reference spectrum at 808.93nm and 805.48nm are the e_1-hh_1 (heavy hole) and e_1-lh_1 (light hole) transitions, respectively. The PLE spectrum of the QW wire samples exhibit three transitions. Two transitions occur in the wavelength regime of the excitonic hh- and lh-ground-states of the reference sample. Note the shifts of these transitions to smaller wavelengths with respect to the corresponding reference transitions. A third additional transition can be observed on the smaller wavelength side of the hh-ground-state. These transitions represent quasi 1D exciton transitions related to quantum confined 1D subbands. The energy separation between the 1D exciton transitions increases with decreasing wire width w, which is a direct manifestation of the increasing lateral 1D confinement.

resolved. Two transitions occur in the wavelength regime of the excitonic hh- and lh-ground-states of the reference sample. Note the shifts of these transitions to smaller wavelengths with respect to the corresponding reference transitions (e.g., for w = 60nm, 0.4 meV). A third additional transition can be observed on the smaller wavelength side of the hh-ground-state. These transitions represent quasi 1D exciton transitions related to quantum confined 1D subbands. In particular, the third new transition, which will be labeled hh_{12} in the following, is due to the second 1D subbands in the conduction- and valence-band. The indices of hh_{ij} denote the quantum numbers for the z- and x-direction, respectively.

As a direct manifestation of the quantum confinement one expects that the energy separation between the hh_{11}- and hh_{12}- transition should increase with decreasing wire width w. We indeed find this wire-width dependence if we compare the PLE spectra of QW wire samples with different values of w in Fig. 9. For the case of w = 100 nm, the third transition, hh_{12}, is only separated by 0.75 nm (1.6meV) from the hh_{11} transition. With decreasing wire width this separation increases to 3.3 meV for w = 60 nm. We also would like to note that there is a strong characteristic dependence of the excitation strength for the two 1D excitons for different polarizations of the exciting light with respect to the direction of the wires, which we attribute to the intrinsic symmetry of the 1D exciton wave function.[47] In all our measurements, however, we were not able to resolve the lh_{12}, which should occur for t=70nm at about 803nm. The weakness of this transition was also observed in InP/InGaAs QW wires by Gershoni et al[40], who were the first to observe separated 1D subband excitons.

For a further confirmation of the 1D character and for additional quantitative information we have performed magnetooptical PLE experiments in magnetic fields B oriented perpendicularly to the layers. This method is well established for 2D QW excitons.[88-90] The exciton energies in a magnetic field are determined by an interplay between the excitonic Coulomb interaction, governed by the effective Rydberg energy and by the magnetic energy which is determined by the cyclotron resonance energy.[91] For low exciton states, in particular for the ground state and small magnetic fields, the Coulomb interaction dominates and the exciton energy is only affected by the diamagnetic shift. This diamagnetic shift is proportional to the square of the transversal extent of the exciton wave function and can thus be used to study the expected shrinkage of the 1D exciton wave function. For higher exciton states and large B the magnetic energy dominates and the transitions obtain the character of Inter-Landau-Level transitions with a linear B-dependence. These Inter-Landau-Level transitions can be easily observed with increasing magnetic field. Extrapolation to B=0 allows us then to determine the exciton binding energy.

The energy positions in the PLE-spectra of a QW wire with $L_x=70$ nm and of the corresponding reference-QW in a magnetic field are sum-

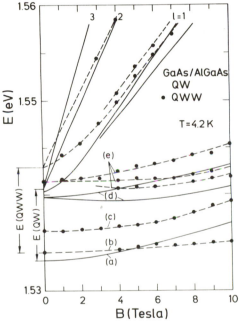

FIG. 10. Energy positions of the maxima in the magnetic-field dependent PLE-spectra of a $L_x \approx 70$ nm QW wire (QWW). The solid lines indicate the magnetic field dispersion for the transitions of the corresponding reference QW. (a), (b), (c), (d) and (e) denote the hh_1, hh_{11}, hh_{12}, lh_1 and lh_{11} transitions, respectively; the index ι labels the order of the Inter-Landau Level transitions. In addition, the exciton binding energies for the QW and the QW wire, E(QW) and E(QWW), respectively, which were deduced from zero-field extrapolations, are indicated. They show directly the increased binding energies of the 1D excitons. The reduced dimensionality is also reflected in the smaller diamagnetic shift of the QW wire sample. (from Ref. 47).

marized in Fig. 10. For clarity, we did not plot the magnetic field dispersion of the QW in discrete data points but as solid lines. The magneto dispersion of the reference sample is very similar to earlier measurements on QW systems, e.g. Refs. 88-90. The hh_1 and lh_1 transitions show a diamagnetic shift, which is stronger for the hh_1 transition reflecting the weaker Coulomb interaction of the hh_1 exciton. For the lh_1 transition the spin-splitting is resolved. The magnetic field dependence of the QW wires is strikingly different. The hh_{11} transition of QW wire shows a smaller diamagnetic shift as compared to the reference sample, indicating a stronger exciton binding energy. The diamagnetic shift of the hh_{12} transition demonstrates its excitonic character, which is less pronounced than that of the ground-state exciton. These features are qualitatively expected for 1D magneto-excitons and confirm our interpretation of the hh_{11} and hh_{12} transitions.

We analyzed the magnetic field dependence following similar evaluations for 2D systems, e.g. Refs. 89,90. The diamagnetic shift ΔE of the hh-ground state exciton of the reference sample was 0.51 meV at 3 Tesla, which is consistent with published values. The corresponding exciton binding energy is E_B of 7.5 meV, which was determined by extrapolating the Inter-Landau-Level transitions to B=0. The value of 7.5 meV agrees very well with reported binding energies of comparable QW samples.[92,93] For the hh_{11} transition of QW wire we could determine $E_B = 8.7$meV, which indicates an enhancement of the ground-state excitonic interaction in QW wire by about 15% with respect to the 2D QW reference sample. From the diamagnetic shift of the excitons we evaluate that the transverse extension of the hh_1 exciton of the reference QW was 1.09 a_B, if we follow the analysis of Ossau et al.[90] This value is in excellent agreement with the calculated value of Bastard et al[94] for a 14 nm QW. The corresponding value of the QW wire hh_{11} exciton (0.97 a_B) is by about 11 percent reduced, which directly indicates the shrinkage of the excitonic wavefunction due to the additional lateral confinement. It was also possible to determine the exciton binding energy of the hh_{12} transitions for the QW wire. It was found to be $E_B = 8.1$ meV.

8. CONCLUSIONS

Quantum wires and dots are beautiful systems. In these systems it is possible to realize, in a first approach, simple textbook quantum mechanical concepts. On a closer look they are much more complex and the interpretation of their physical properties involves elements from very different fields of physics which makes them extremely interesting objects of research. The fabrication of quantum wires and dots needs a very sophisticated technology which is at present well suited to prepare single samples for research. However, we think that there are still a great amount of additional research, hard work and good ideas necessary to implement these concepts in real devices. This remains the challenge of the nineties.

ACKNOWLEDGEMENT

We acknowledge financial support from the BMFT

REFERENCES

1. T. Ando, A. B. Fowler and F. Stern, Rev. Mod. Phys. **54**, 437 (1982).

2. K.-F. Berggren, T. J. Thornton, D. J. Newson and M. Pepper, Phys. Rev. Lett. **57**, 1769 (1986).

3. H. van Houten, B. J. van Wees, M. G. J. Heijman, J. P. André, D. Andrews and G. J. Davies, Appl. Phys. Lett. **49**, 1781 (1986).

4. T. P. Smith, III, H. Arnot, J. M. Hong, C. M. Knoedler, S. E. Laux and H. Schmid, Phys. Rev. Lett. **59**, 2802 (1987).

5. M. L. Roukes, A. Scherer, S. J. Allen, Jr., H. G. Craighead, R. M. Ruthen, E. D. Beebe and J. P. Harbison, Phys. Rev. Lett. **59**, 3011 (1987).

6. H. van Houten, B. J. van Wees, J. E. Mooij, G. Roos and K.-F. Berggren, Superlattices and Microstructure **3**, 497 (1987).

7. G. Timp, A. M. Chang, P. Mankiewich, R. Behringer, J. E. Cunningham, T. Y. Chang, and R. E. Howard, Phys. Rev. Lett. **59**, 732 (1987).

8. B. J. van Wees, H. van Houten, C. W. J. Beenakker, J. G. Williamson, L. P. Kouwenhoven, D. van der Marel and C. T. Foxon, Phys. Rev. Lett. **60**, 848 (1988).

9. D. A. Wharam, T. J. Thornton, R. Newbury, M. Pepper, J. E. F. Frost, D. G. Hasko, D. C. Peacock, D. A. Ritchie and G. A. C. Jones, J. Phys. C**21**, L209 (1988).

10. T. J. Thornton, M. Pepper, H. Ahmed, D. Andrews and G. J. Davies, Phys. Rev. Lett. **56**, 1189 (1986).

11. T. Demel, D. Heitmann, P. Grambow and K. Ploog, Appl. Phys. Lett. **53**, 2176 (1988).

12. P. Grambow, T. Demel, D. Heitmann, M. Kohl, R. Schüle and K. Ploog, Microelectronic Engineering **9**, 357 (1989).

13. P. Grambow, E. Vasiliadou, T. Demel, K. Kern, D. Heitmann and K. Ploog, Microelectronic Engineering **11**, 47 (1990).

14. W. Hansen, T. P. Smith, III, K. Y. Lee, J. A. Brum, C. M. Knoedler, J. M. Hong and D. P. Kern, Phys. Rev. Lett. **62**, 2168 (1989).

15. H. Fang, R. Zeller and P. J. Stiles, Appl. Phys. Lett. **55**, 1433 (1989).

16. A. D. Wieck and K. Ploog, Appl. Phys. Lett. **56**, 928 (1990).

17. S. Datta, M. R. Melloch, S. Bandyoadhyay, R. Noren, M. Vaziri, M. Miller and R. Reifenberger, Phys. Rev. Lett. **55**, 2344 (1985).

18. U. Meirav, M. A. Kastner, M. Heiblum and S. J. Wind, T. M. Moore and A. E. Wetsel, Phys. Rev. **B40**, 5871 (1989).

19. K. Ismail, T. P. Smith, III, and W. T. Masselink, Appl. Phys. Lett. **55**, 2766 (1989).

20. M. A. Reed, J. N. Randall, R. J. Aggarwal, R. J. Matyi, T. M. Moore and A. E. Wetsel, Phys. Rev. Lett. **60**, 535 (1988).

21. L. P. Kouwenhoven, F. W. J. Hekking, B. J. van Wees, C. J. P. M. Harmans, C. E. Timmering and C. T. Foxon, Phys. Rev. Lett. **65**, 685 (1990).

22. W. Hansen, M. Horst, J. P. Kotthaus, U. Merkt, Ch. Sikorski and K. Ploog, Phys. Rev. Lett. **58**, 2586 (1987).

23. J. Alsmeier, Ch. Sikorski and U. Merkt, Phys. Rev. **B37**, 4314 (1988).

24. F. Brinkop, W. Hansen, J. P. Kotthaus and K. Ploog, Phys. Rev. **B37**, 6547 (1988).

25. T. Demel, D. Heitmann, P. Grambow and K. Ploog, Phys. Rev. **B38**, 12732 (1988).

26. T. Demel, D. Heitmann, P. Grambow and K. Ploog, Superlattices and Microstructures **5**, 287 (1989).

27. K. Kern, T. Demel, D. Heitmann, P. Grambow, K. Ploog and M. Razeghi, Surface Sci. **229**, 256 (1990).

28. Ch. Sikorski and U. Merkt, Phys. Rev. Lett. **62**, 2164 (1989).

29. T. Demel, D. Heitmann, P. Grambow and K. Ploog, Phys. Rev. Lett. **64**, 788 (1990) and Proc. of the 6th Int. Winterschool on Localization and Confinement of Electrons in Semiconductors, Mauterndorf, Austria, (1990), Springer, Eds. F. Kuchar, H. Heinrich and G. Bauer, p. 51.

30. A. Lorke, J. P. Kotthaus and K. Ploog, Phys. Rev. Lett. **64**, 2559 (1990).

31. T. Demel, D. Heitmann and P. Grambow in "Spectroscopy of Semiconductor Microstructures," Eds. G. Fasol, A. Fasolino and P. Lugli, Plenum Press, p. 75 (1989).

32. U. Merkt, Ch. Sikorski and J. Alsmeier, Proceedings of NATO ARW on "Spectroscopy of Semiconductor Microstructures," Venice (1989), Eds. G. Fasol, A. Fasolino and P. Lugli, p. 89.

33. D. Heitmann, T. Demel, P. Grambow and K. Ploog, Advances in Solid State Physics **29**, Ed. U. Rössler (Vieweg, Braunschweig), p. 285, 1989.

34. U. Merkt, Advances in Solid State Physics **30**, Ed. U. Rössler (Vieweg, Braunschweig), p. 77, 1990.

35. K. Kash, A. Scherer, J. M. Worlock, H. G. Craighead and M. C. Tamargo, Appl. Phys. Lett. **49**, 1043 (1986).

36. J. Cibert, P. M. Petroff, G. J. Dolan, S. J. Pearton, A. C. Gossard and J. H. English, Appl. Phys. Lett. **49**, 1275 (1986).

37. M. A. Reed, R. T. Bate, K. Bradshaw, W. M. Duncan, W. R. Frensley, J. W. Lee and M. D. Shih, J. Vac. Sci. Technol. **B4**, 358 (1986).

38. H. Temkin, G. J. Dolan, M. B. Panish and S. N. G. Chu, Appl. Phys. Lett. **50**, 413 (1987).

39. Y. Hirayama, S. Tarucha, Y. Suzuki and H. Okamoto, Phys. Rev. **B37**, 2774 (1988).

40. D. Gershoni, H. Temkin, G. J. Dolan, J. Dunsmuir, S. N. G. Chu and M. B. Panish, Appl. Phys. Lett. **53**, 995 (1988).

41. H. E. G. Arnot, M. Watt, C. M. Sotomayor-Torres, R. Glew, R. Cusco, J. Bates and S. P. Beaumont, Superlattices and Microstr. **5**, 459 (1989).

42. M. Kohn, D. Heitmann, P. Grambow and K. Ploog, Phys. Rev. **B37**, 10927 (1988).

43. M. Kohl, D. Heitmann, P. Grambow and K. Ploog, Superlattices and Microstr. **5**, 235 (1989).

44. M. Tsuchinya, J. M. Gaines, R. H. Yan, R. J. Simes, P. O. Holtz, L. A. Coldren and P. M. Petroff, Phys. Rev. Lett. **62**, 4668 (1989).

45. K. Tsubaki, H. Sakaki, J. Yoshino and Y. Sekiguchi, Appl. Phys. Lett. **45**, 663 (1984).

46. A. Forchel, H. Leier, B. E. Maile and R. German, Advances in Solid State Physics, Ed. U. Rössler, Vieweg, Braunschweig (1988).

47. M. Kohl, D. Heitmann, P. Grambow and K. Ploog, Phys. Rev. Lett. **63**, 2124 (1989).

48. M. Kohl, D. Heitmann, P. Grambow and K. Ploog, Surf. Sci. **229**, 248 (1990).

49. K. Kash, J. M. Worlock, M. D. Sturge, P. Grabbe, J. P. Harbison, A. Scherer and P. S. D. Lin, Appl. Phys. Lett. **53**, 782 (1988).

50. K. Kash, J. M. Worlock, A. S. Gozdz, B. P. Van der Gaag, J. P. Harbison, P. S. D. Lin and L. T. Florez, Surf. Sci. **229**, 245 (1990).

51. M. Tanaka and H. Sakaki, Appl. Phys. Lett. **54**, 1326 (1989).

52. E. Kapon, D. M. Hwang and R. Bhat, Phys. Rev. Lett. **63**, 430 (1989).

53. M. Tsuchiya, J. M. Gaines, R. H. Yan, R. J. Simes, P. O. Holtz, L. A. Coldren and P. M. Petroff, Phys. Rev. Lett. **62**, 466 (1989).

54. M. Kohl, D. Heitmann, P. Grambow and K. Ploog, Phys. Rev. **B41**, 12338 (1990).

55. C. Weisbuch in "Physics and Application of Quantum Wells and Superlattices," Eds. E. E. Mendez and K. von Klitzing, Plenum Press, New York, p. 261 (1987).

56. K. Kash, J. Luminescene **46**, 69 (1990).

57. D. Heitmann, M. Kohl, P. Grambow and K. Ploog in "Physics and Engineering of 1- and 0-Dimensional Semiconductors," Eds. S. Beaumont and C. M. Sotomayor-Torres, Plenum Press, 1990, p. 255.

58. J. S. Weiner, G. Danan, A. Pinczuk, J. Valladaras, L. N. Pfeiffer and K. West, Phys. Rev. Lett. **63**, 1641 (1989).

59. T. Egeler, G. Abstreiter, G. Weimann, T. Demel, D. Heitmann and W. Schlapp, Surface Sci. **229**, 391 (1990).

60. S. Laux, D. J. Frank and F. Stern, Surf. Sci. **196**, 101 (1988).

61. S. J. Allen, Jr., H. L. Störmer and J. C. Hwang, Phys. Rev. **B28**, 4875 (1983).

62. D. C. Glattli, E. Y. Andrei, G. Deville, J. Poitrenaud and F. I. B. Williams, Phys. Rev. Lett. **54**, 1710 (1985).

63. D. B. Mast, A. J. Dahm and A. L. Fetter, Phys. Rev. Lett. **54**, 1706 (1985).

64. A. Fetter, Phys. Rev. **B32**, 7676 (1985).

65. A. Fetter, Phys. Rev. **B33**, 5221 (1986).

66. V. B. Sandomirskii, V. A. Volkov, G. R. Aizin and S. A. Mikhailov, Electrochimica Acta **34**, 3 (1989).

67. V. Shikin, T. Demel and D. Heitmann, Surf. Sci. **229**, 276 (1990).

68. V. Shikin, S. Nazin, D. Heitmann and T. Demel, Phys. Rev. **B43**, 11903 (1991).

69. V. Fock, Z. Phys. **47**, 446 (1928).

70. G. W. Bryant, Phys. Rev. Lett. **59**, 1140 (1987).

71. A. Kumar, S. E. Laux and F. Stern, Phys. Rev. **B42**, 5166 (1990).

72. L. Brey, N. Johnson and P. Halperin, Phys. Rev. **B40**, 10647 (1989).

73. P. Maksym and T. Chakraborty, Phys. Rev. Lett. **65**, 108 (1990).

74. T. Chakraborty, V. Halonen and P. Pietiläinen, Phys. Rev. B, in press.

75. V. Gudmundsson and R. R. Gerhardts, Proc. Intern. Conference on High Magnetic Fields in Semiconductor Physics, Würzburg, Germany, 1990.

76. W. Kohn, Phys. Rev. **123**, 1242 (1961).

77. W. Que and G. Kirczenow, Phys. Rev. **B37**, 7153 (1988) and, Phys. Rev. **B39**, 5998 (1989).

78. E. Batke, D. Heitmann, J. P. Kotthaus and K. Ploog, Phys. Rev. Lett. **54**, 2367 (1985).

79. G. Fasol, N. Mestres, H. P. Hughes, A. Fischer and K. Ploog, Phys. Rev. Lett. **56**, 2517 (1986).

80. A. Pinczuk, M. G. Lamont and A. C. Gossard, Phys. Rev. Lett. **56**, 2092 (1986).

81. J. K. Jain and P. B. Allen, Phys. Rev. Lett. **54**, 2437 (1985).

82. R. Dingle, W. Wiegmann and C. H. Henry, Phys. Rev. Lett. **33**, 827 (1974).

83. M. H. Degani and O. Hipolito, Phys. Rev. **B35**, 9345 (1987).

84. I. Suemune and L. A. Coldren, IEEE **QE 24**, 1778 (1988). I. Suemune, L. A. Coldren and S. W. Corzine, Superlattices and Microstructures **4**, 19 (1988).

85. G. Bastard in "Physics and Application of Quantum Wells and Superlattices," Eds. E. E. Mendez and K. von Klitzing, Plenum Press, New York, p. 21 (1987).

86. J. W. Brown and H. N. Spector, Phys. Rev. **B35**, 3009 (1987).

87. G. W. Bryant, Phys. Rev. **B37**, 8763 (1988).

88. J. C. Maan, G. Belle, A. Fasolino, M. Altarelli and K. Ploog, Phys. Rev. **B30**, 2253 (1984).

89. D. C. Rogers, J. Singleton, R. J. Nicholas, C. T. Foxon and K. Woodbridge, Phys. Rev. **B34**, 4002 (1986).

90. W. Ossau, B. Jäkel, E. Bangert, G. Landwehr and G. Weimann, Surf. Sci. **174**, 188 (1986).

91. O. Akimoto and H. Hasegawa, J. Phys. Soc. Japan **22**, 181 (1967).

92. R. C. Miller, D. A. Kleinman, W. T. Tsang and A. C. Gossard, Phys. Rev. **B24**, 1134 (1981).

93. U. Ekenberg and M. Altarelli, Phys. Rev. **B35**, 7585 (1987).

94. G. Bastard, E. E. Mendez, L. L. Chang and L. Esaki, Phys. Rev. **B26**, 1974 (1982).

ELECTRON OPTICS IN A TWO-DIMENSIONAL ELECTRON GAS

H. van Houten

Philips Research Laboratories
5600 JA Eindhoven
The Netherlands

Abstract - Quantum balistic transport in semiconductor nanostructures is discussed from the point of view of electron optics in the solid state. Several representative phenomena are introduced: 1. The conductance quantization of a quantum point contact in units of $2e^2/h$; 2. The generation and detection of collimated electron beams; 3. Coherent electron focusing, an interference experiment involving quantum point contacts as point source and detector; 4. Reduction of backscattering by a magnetic field, and the transition from ballistic to adiabatic transport in the quantum Hall effect regime. Ballistic and adiabatic transport are given a unified theoretical basis by the Landauer-Büttiker formalism for transport as a transmission problem. Finally, it is argued that the single-particle description (underlying the optical analogy) breaks down in zero-dimensional transport: tunneling through localized states in a quantum dot or electron cavity is regulated by the Coulomb-blockade of tunneling.

Highlights in Condensed Matter Physics and Future Prospects
Edited by L. Esaki, Plenum Press, New York, 1991

1. INTRODUCTION

The standard description of electronic transport in a metallic conductor in terms of a local resistivity might be expected to break down completely in systems small compared to the mean free path, where the electron motion is ballistic, rather than diffusive. As it turns out, classical size effects in thin metallic films or wires do not require a novel type of treatment, because the transport along the film or wire remains diffusive, even though the motion in the transverse directions is ballistic. A completely different situation arises only in the case of fully ballistic transport, when all dimensions of the sample are smaller than the mean free path. The classical ballistic transport regime in metals was pioneered by Sharvin's work on point contacts in metals.[1]

The *quantum ballistic transport* regime has only recently become accessible. The high mobility two-dimensional electron gas in GaAs-AlGaAs heterostructures offers a large mean free path (exceeding 10 μm), in combination with a large Fermi wavelength, because of the low electron density. Typically, the Fermi wavelength is 40 nm - two orders of magnitude larger than in metals. In the Eighties, several groups fabricated narrow "quantum wires" in GaAs-AlGaAs heterostructures, in search for clear manifestations of a quantum size effect.[2] Although a wealth of interesting results was obtained, this search was inconclusive (primarily because of elastic scattering by impurities, rough boundaries, and side-probes used for voltage measurements). The introduction in 1987 of unconventional transport geometries involving point contacts proved to be required to uncover the characteristic properties of quantum ballistic transport.[3,4,5] This new discipline may be characterized as *electron optics in a two-dimensional electron gas.*[6,7]

Several major topics will be discussed in this presentation, each of which centers around the *quantum point contact*, which is in essence a very short constriction or quantum wire of continuously adjustable width. The conductance of a quantum point contact was found to be quantized in units of $2e^2/h$.[3,5] This phenomenon is a consequence of the discrete number of 1D subbands in the constriction with a cut-off energy below the Fermi level. These subbands act as loss-less propagating modes in an "electron waveguide." This effect is the zero-field counterpart of the quantum Hall effect, another major topic of this Conference. As will be discussed, the *equipartitioning* of current among the waveguide modes provides the fundamental connection between the two phenomena. The current equipartitioning also underlies the Landauer-Büttiker formalism, which expresses conductances in terms of transmission probabilities between two reservoirs.[8,9] The experiments revealing the quantized point contact conductance have furnished the first real opportunity to apply this formalism to actual experimental situations. The subsequent experimental and theoretical work on 2D electron optics has significantly contributed to the present widespread use and acceptance of the Landauer-Büttiker formalism.

The electrons injected by a quantum point contact form a *beam*, which may be *collimated* due to the horn-like shape of the transition region between point contact and 2D electron gas.[10,11] This collimation effect has been demonstrated experimentally.[12] Part of the significance of the collimation effect is that it can account for a number of magnetoresistance anomalies in narrow multi-probe conductors (such as the quenching of the low-field Hall effect).[13,14] In addition, the collimated beam can and has been used to demonstrate a variety of geometrical electron optics effects (including beam focusing by means of a lens defined electrostatically in the 2D electron gas).[15,16] These effects are beyond the scope of the present article, however.

Coherent electron focusing experiments in a 2D electron gas[4,6,7] have provided a direct demonstration of the motion of electrons in skipping orbits along the boundary from point source to detector. In addition, large quantum interference oscillations in the detector signal were found at low temperatures, showing that a sufficiently narrow quantum point contact behaves as a spatially coherent and monochromatic point source. These results first demonstrated the feasibility of *coherent electron optics* in the solid state.

The resistance of a quantum point contact, measured with separate current and voltage probes, was found to be suppressed in a sufficiently strong magnetic field.[17] This is another consequence of the electron motion in skipping orbits along the sample edge: if the orbit diameter is smaller than the size of the point contact, the electrons are perfectly transmitted. This suppression of backscattering due to motion in skipping orbits is the weak-field precursor of the guiding along equipotential lines of electrons in strong magnetic fields. Quantum-mechanically, this type of motion corresponds to the formation of extended edge channels. It has been stressed by Büttiker[18] that the quantization of the Hall resistance requires the absence of back scattering, as well as the equipartitioning of current among the available edge channels at the

Fermi level. Independently, an anomalous quantum Hall effect was found in an experiment using quantum point contacts to *selectively* populate and detect edge channels.[6,19] *Adiabatic* transport, characterized by the absence of inter-edge channel mixing, was found to occur in these experiments for distances of at least a few μm, showing that the 2D electron gas boundary in a strong magnetic field can behave like an ideal multi-mode electron waveguide. The present understanding of the quantum Hall effect in terms of edge channel transport has been influenced in an essential way by these and subsequent investigations.

This paper concludes by a discussion of the limitations of the picture of quantum transport as electron optics. It is argued that the analogy breaks down in going from 1D transport through quantum wires to transport through "quantum dots." The reason is the importance of electron-electron interaction effects[20-21] in small capacitance systems. The Coulomb-blockade of tunneling, familiar from small metallic tunneling structures where quantum levels are not resolved, now *regulates* structure in the conductance due to resonant tunneling through zero-dimensional states.

The present article is intended as a pedestrian introduction to quantum transport in semiconductor nanostructures. The text is based in part on previous specialized reviews of our work on this subject, written in collaboration with C. W. J. Beenakker and B. J. van Wees,[7,22-24] to which we refer for details, and for more complete references to the literature.

2. QUANTUM BALLISTIC TRANSPORT

2.1. Conductance Quantization of a Quantum Point Contact

In the ballistic transport regime, it is the scattering of electrons at the sample boundaries which limits the current, rather than impurity scattering. Perhaps counterintuitively, this is true even in the case that boundary scattering is perfectly specular (as is very nearly the case in the systems considered here, as demonstrated by the electron focusing experiments discussed below). Consider as an example the canonical example of a ballistic conductor - a point contact. In the case of specular boundary scattering, this is equivalent to a channel of width W and length L much shorter than the transport mean free path l connecting two wider regions as illustrated in Fig. 1. The current I through the narrow constriction in response to a voltage difference V between the

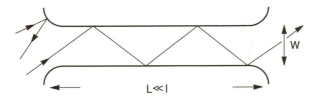

FIG. 1. Electron trajectories characteristic for the ballistic transport regime, for the case of specular boundary scattering. A non-zero resistance in the ballistic regime results from backscattering at the connection between the narrow channel and the wide 2D electron gas regions. (From van Houten et al.[25]).

wide regions to the left and right is *finite* even in the absence of impurities, because electrons are scattered back at the entrance of the constriction. The *contact conductance* $G = I/V$ is proportional to the constriction width - but independent of its length. One can not therefore describe the contact conductance in terms of a local conductivity, as one can do in the diffusive transport regime. This was already known to Sharvin,[1] who first treated transport through classical ballistic point contacts in metals.

Many of the principal phenomena in classical and quantum ballistic transport are exhibited in the most clean and extreme way by *quantum point contacts*. These are short and narrow constrictions in a two-dimensional (2D) electron gas, with a width of the order of the Fermi wave length. The starting point for the fabrication of a quantum point contact is the two-dimensional electron gas, present at the interface between two semiconductor layers in a heterostructure (GaAs and $Al_xGa_{1-x}As$). The electrons are confined in a potential well at the interface between the two materials. This well is formed because of the difference in bandgap, which leads to a potential step in the conduction band, and because of an attractive force due to the presence of positively charged ionized donor impurities in the AlGaAs layer. The electrons in the well are free to move along the interface, so that their motion is two-dimensional. Additional confinement occurs if one defines a narrow channel in the plane of the 2D electron gas. The lateral confining potential in such a channel has a set of discrete quantum levels, occupied by the electrons. These are now free to move in one direction only (along the channel) so that their motion is effectively one-dimensional. More precisely, the system is called quasi-one dimensional if several discrete levels are occupied. The discrete levels are labeled by a quantum number n. The one-dimensional subband corresponding to the energy levels with

247

the same value of n may be seen as a *mode in an electron waveguide*. Each mode corresponds to a standing wave in the lateral confining potential, similar to the transverse modes in an optical fiber (see Fig. 2a). The total energy of an electron in mode n is a sum of the discrete

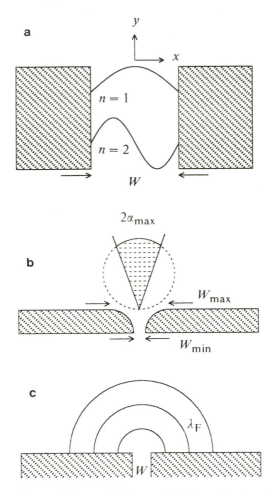

FIG. 2. The quantum point contact as a building block of solid state electron optics. a. If $2W \gtrsim \lambda_F$, it behaves like an electron waveguide with $2W/\lambda_F$ propagating modes (shown are two such modes propagating along the y-direction represented by their transverse wave-functions). b. A quantum point contact with a gradual connection to the wide 2D electron gas regions injects a collimated electron beam as indicated by the shaded collimation cone. The dotted circle is a plot of the classical $\cos\alpha$ distribution expected for an abrupt connection. The effects of diffraction are ignored here. c. A quantum point contact with $2W < \lambda_F$ acts as a spatially coherent point source. Consecutive wavefronts are separated by the Fermi wavelength λ_F. (From van Houten[26]).

energy E_n and the kinetic energy of the motion along the channel. For the electrons at the Fermi level, this is the Fermi energy $E_F = (\hbar k_F)^2/2m$. A limited number of modes are thus occupied, given by the largest integer N such that $E_N \gtrsim E_F$. For a square-well lateral confining potential of width W the number of occupied modes is N = Int[k_F W / π] (where "Int[]" denotes truncation to an integer). Note that N also equals the ratio of the channel width W to half the Fermi wavelength $\lambda_F/2$ (since $k_F \equiv 2\pi/\lambda_F$). Fig. 3 shows the characteristic peaks in the density of states $\rho(E)$ for a multi-mode (or quasi-one-dimensional) channel. Note that $\rho(E)$ tends to infinity at the population threshold of each mode. The density of states peaks should give rise to a quantum size effect on the conductance of a quantum wire, if the transport mean free path exceeds its width, and if the temperature is sufficiently low. Experiments on long and narrow channels failed to show such an effect convincingly, however.[2,24]

This situation changed in an unexpected manner with the advent of the quantum point contact. The number N of occupied modes in a quantum point contact is continuously adjustable by varying the negative gate voltage. The conductance G (the reciprocal of the resistance) decreases step-wise as the gate voltage is increased (thereby reducing the width of the point contact, and hence N).[3,5] This effect is shown in Fig. 4. The step height is given by $2e^2/h$ - a combination of fundamental constants familiar from the quantum Hall effect. As pointed out in the original papers,[3,5] the experimental data imply that

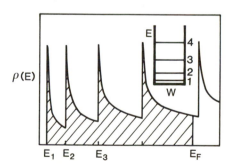

FIG. 3. Density of states $\rho(E)$ as a function of energy for a quasi-one dimensional channel. The shaded region indicates the occupied states at T = 0. The inset shows the lowest discrete energy levels in a square well confinement potential. (From Beenakker and van Houten[24]).

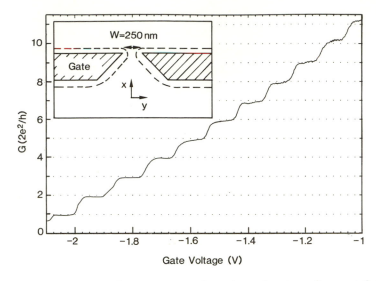

FIG. 4. Point contact conductance as a function of gate voltage at 0.6 K, demonstrating conductance quantization in units of $2e^2/h$. The constriction width increases as the negative gate voltage is reduced (see inset). (From van Wees et al.[3]).

- The conductance of a quantum point contact is approximately quantized in units of $2e^2/h$;

- A quantum point contact behaves like an ideal electron waveguide.

Whereas the effect was not anticipated theoretically, its explanation is relatively simple. The reason for the conductance quantization is that

- The current in an ideal electron waveguide is equally divided among the occupied modes.

At first sight this is rather surprising, since the propagation velocity of each mode at the Fermi energy depends on the mode index n (the available kinetic energy of each mode is $E_F - E_n$). The fundamental reason for this equipartitioning of the current in a quantum point contact is that the current carried by each mode is proportional to the product of its propagation velocity and of its contribution to the total density of states - a product which is independent of the mode index n, and of the details of the lateral confinement potential. The reader may convince himself of this fact by considering that the propagation velocity of mode n is proportional to the square root of the available kinetic energy, $(E_F - E_n)^{1/2}$, while the 1D-density of states of mode n is proportional to $(E_F - E_n)^{-1/2}$. This is the case of zero magnetic field. It is stressed, however, that the cancellation of group velocity and density of states, and with it the current equipartitioning, holds also for non-zero field. Each mode contributes $2e^2/h$ to the conductance, so that the conductance G of an ideal electron waveguide is quantized according to

$$G = \frac{2e^2}{h} N. \tag{1}$$

The deviations from exact quantization found in the experiments are estimated at 1%, which is not as good as the 0.1 part per million accuracy achieved in the quantum Hall effect, but still quite remarkable. There are several causes for these deviations. Firstly, quantum mechanical reflections of the modes may occur at the entrance and exit of the quantum point contact, causing transmission resonances in the conductance with a period which depends on the length of the point contact, and with an amplitude which depends on the shape of the transition region between the point contact and the wide two-dimensional electron gas regions. Secondly, if the point contact is very short, the transmission through evanescent modes (those which have $E_n > E_F$) may not be neglected. Further corrections are due to the finite probability of impurity scattering and to imperfections in the smoothness of the boundaries defining the point contact. For these reasons, quantization in zero magnetic field can not be used as a resistance standard.

Let us now briefly consider a non-ideal electron waveguide, one in which scattering occurs within the narrow channel. Forward scattering causes mixing between the modes, but does not lead to deviations from the quantized conductance values given in Eq. (1). The conductance is reduced below this value only in the case of backscattering in the channel, and at entrance or exit (as a consequence of imperfect matching of modes). In such a case only a fraction T_n of the current in the n-th mode occupied in the channel is transmitted, and the total conductance becomes

$$G = \frac{2e^2}{h} \sum_{n=1}^{N} T_n. \tag{2}$$

This relation between conductance and transmission probabilities of quantum channels or modes at the Fermi energy is referred to as the *Landauer formula* because of Landauer's pioneering 1957 paper.[8] The identification of the finite G in the limit of perfect transmission as a *contact* conductance is due to Imry.[27]

Eq. (2) refers to a *two-terminal* conductance measurement, in which the same two contacts (modelled as "reservoirs") are used both to drive a current through the system and to measure the voltage drop across it. More generally, one can consider a multi-reservoir conductor, to model e.g. *four-terminal* resistance measurements in which the current source and drain are distinct from the voltage probes. Traditionally, such measurements are intended to eliminate the ill-defined contact resistance associated with the current source and drain, so as to obtain the intrinsic resistivity of the conductor. Contact resistances may not be eliminated in the transport regimes considered here, however. A generalization of the Landauer formula (2) to multi-terminal conductances is due to Büttiker.[9] The theoretical framework referred to as the *Landauer-Büttiker formalism* provides a unified description of the transport experiments in the ballistic and adiabatic transport regimes.

2.2. Mode-Coupling and Electron Beam Collimation

It may be shown straightforwardly that the equipartitioning of current among the available modes in a wide electron waveguide (with a large number of occupied modes) implies an *isotropic velocity distribution* of the electrons - as should be expected on the basis of the classical correspondence. Consequently, the electron *flux* incident from a wide region on a point contact of *abrupt* shape has a cos α distribution (α denotes the angle with respect to the axis of the quantum point contact). Because of time reversal invariance the electrons emitted at the exit of an abrupt point contact must have the same angular distribution.

The angular distribution of electrons emitted by a quantum point contact of a *horn*-like shape deviates substantially from the cos α distribution of an abrupt point contact. A *collimated electron beam*[10,11] is injected instead by such a quantum point contact. The corresponding angular distribution is approximately a truncated cos α cone, of maximum angle α_{max}, as indicated schematically in Fig. 2b. This collimation effect may be understood by studying the mode-coupling between wide and narrow regions.

Let us first have a look at the classical correspondence between a mode and a trajectory in a channel of constant width W along the y-direction (see Fig. 2a). Each mode n corresponds to the condition that the channel width W is n times half the wavelength in the x-direction, λ_x. It is convenient to rewrite this condition in terms of discrete values for the transverse wavenumber $k_x = 2\pi/\lambda_x$, which is just the transverse electron momentum in units of $\hbar$. The quantization condition for each mode may then be written as

$$k_x = \pm n \frac{\pi}{W}, \quad n = 1, 2, 3, \dots, N. \tag{3}$$

with the total number of modes given by $N = \text{Int}[\, k_F W/\pi\,]$. The *total* momentum for each mode being fixed (at $mv_F = \hbar k_F$), one finds that a mode with index n corresponds to the electron trajectories inclined at an angle α_n with the channel axis (the y-direction). This angle is given by

$$\sin \alpha_n = \frac{k_x}{k_F} = \frac{n}{N}. \tag{4}$$

Consider now a quantum point contact with a gradual connection to the wide two-dimensional electron gas region. In the narrowest part of the quantum point contact $N_{min} = \text{Int}[k_F W_{min}/\pi]$ modes are occupied, while at its exit, where the width is W_{wide}, the number of available modes is $N_{wide} = \text{Int}[k_F W_{wide}/\pi]$. Provided the widening of the constriction from W_{min} to W_{max} is sufficiently gradual, the mode index n is a constant of the motion (this is called the *adiabatic* approximation). This implies that no transitions between modes of different index occur, so that a specific mode n in the narrowest part of the quantum point contact couples only to a single mode in the wide section at the exit - the one with the same index n. At the exit, the current is therefore carried by only N_{min} modes - a fraction W_{min}/W_{wide} of tahe number of available modes N_{wide}. In view of Eq. (4) this implies a maximum angle α_{max} for trajectories at the exit, given by

$$\sin \alpha_{max} = \frac{N_{min}}{N_{wide}} = \frac{W_{min}}{W_{wide}}. \tag{5}$$

(The latter equality holds true for a constriction of uniform density only.) According to this adiabatic approximation, the angular distribution would be sharply cut-off for angles larger than $\pm \alpha_{max}$ (as illustrated in Fig. 2b). A numerical simulation[12,14] of semi-classical trajectories has shown that the actual distribution is smooth, although its halfwidth is well described by the result (5) for α_{max}. Diffraction at the exit of the quantum point contact will lead to a broadening of the collimation cone.[28] Based on the analogy with optics, it may be estimated that diffraction leads to an angular uncertainty of $\pm \Delta \alpha_{diff}$ which may be approximated for $W_{wide} > \lambda_F$ by

$$\Delta \alpha_{diff} \approx \frac{\lambda_F}{2 W_{wide}} = \frac{1}{N_{wide}}, \tag{6}$$

so that the influence of diffraction on the width of the total collimation cone is relatively small if $N_{min} \gg 1$. (For the opposite case of a spatially coherent point source considered below, we typically have $N_{min} = 1$, and diffraction is predominant.) The collimation cone has an analogy in optics, where it arises in the conical reflector. The geometrical optics approximation would have been sufficient to arrive at Eq. (5), which is essentially a classical result. The adiabatic nature of the mode-coupling in the horn-like entrance and exit of the quantum point contact also plays an important role in the detailed theory of the conductance quantization:[29,30] it ensures an insensitivity to quantum mechanical reflection in the wide parts (since reflection occurs predominantly for high-index modes with a long longitudinal wavelength, which do not couple to the modes occupied in the narrow part of the point contact). This is why we have opted here for an explanation of the collimation effect in terms of modes.

The collimation effect at a quantum point contact was originally invoked[10] to explain the non-additivity of the resistance of point contacts in series.[31] The effect may more sensitively be detected by using a second quantum point contact as a collector (inset of Fig. 5), and using a weak magnetic field to deflect the collimated electron beam.[23] By this method, the result (5) for the collimation effect has been confirmed experimentally by Molenkamp et al..[12] The experimental results are reproduced in Fig. 5.

The implications of this study[10,12] are that

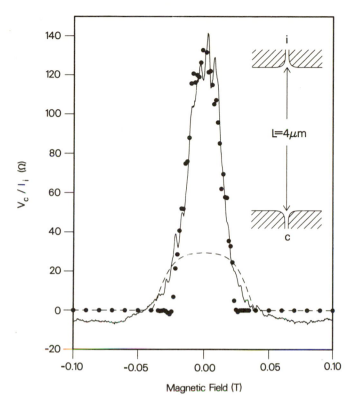

FIG. 5. Detection of a collimated electron beam over a distance of 4μm. In this four-terminal measurement, two ohmic contacts to the 2D electron gas region between the point contacts are used: One of these acts as a drain for the current I_i through the injector, and the other is used as a zero-reference for the voltage V_c on the collector. The drawn curve is the experimental data, at T = 1.8 K. The dots are the result of a semi-classical simulation, using a hard-wall potential with contours as shown in the inset. The dashed curve results from a simulation without collimation (corresponding to rectangular corners in the potential contour). (From Molenkamp et al.[12]).

- A quantum point contact can be used to inject a collimated beam of electrons in the 2D electron gas;

- The collimation effect is due to adiabatic transport in a horn-shaped potential over short distances. The collimated beam is formed as the point contact widens abruptly into the 2D electron gas, where adiabaticity breaks down.

A similar horn-shaped channel widening occurs at attachment sites of voltage probes in a narrow multi-probe conductor. This is why

- Collimation plays a central role in the theory of magnetoresistance anomalies in ultra-small Hall bar geometries.[13,14]

Adiabaticity also breaks down in long but smooth channels due to a small but non-negligible amount of diffuse boundary scattering.[32] The failure to account for the latter effect, as well as that of a finite phase coherence time, is apparently responsible for the poor agreement with experiment of purely quantum mechanical treatments of transport in multi-probe structures.[14,13]

2.3. Coherent Electron Focusing

As noted in the previous subsection, a narrow single-mode quantum point contact ($N = 1$) with an abrupt shape behaves like an optical point source, which may be used to perform the solid state electronics equivalent of quantum interference experiments known from optics (see Fig. 2c). We know this because of the experimental and theoretical study of coherent electron focusing.[4,6,7,33]

The geometry of the experiment in a 2DEG is similar to the transverse focusing geometry pioneered in metals by Tsoi.[34] It consists of two point contacts on the same boundary in a perpendicular magnetic field. Fig. 6 illustrates electron focusing in two dimensions as it follows from the classical mechanics of electrons at the Fermi level. The injector (i) injects a divergent beam of electrons ballistically into the 2DEG. Electrons are detected if they reach the adjacent collector (c), after one or more specular reflections at the boundary connecting i and c. The trajectories of the electrons at the boundary are referred to as *skipping orbits*. The radius of curvature of each of these trajectories is simply the cyclotron radius for an electron moving with the Fermi velocity $l_{cycl} = mv_F/eB$. The focusing action of the magnetic field is evident in Fig. 6 (top) from the black lines of high density of trajectories. These lines are known in optics as *caustics*, and are plotted separately in Fig. 6 (bottom). The caustics intersect the 2DEG boundary at multiples of the cyclotron diameter from the injector. As the magnetic field is increased, a series of these focal points shifts past the collector. The electron flux incident on the collector thus reaches a maximum whenever its separation L from the injector is an integer multiple of $2l_{cycl}$. This occurs when $B = pB_{focus}$, $p = 1, 2, ...$, with

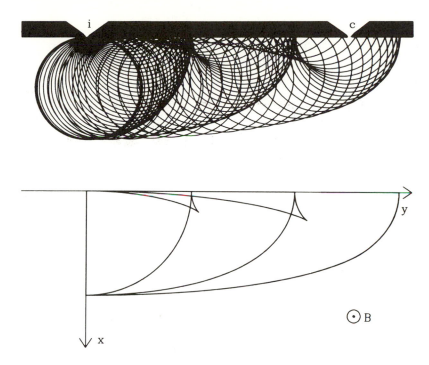

FIG. 6. Illustration of classical electron focusing by a magnetic field. Top: Skipping orbits along the 2DEG boundary. The trajectories are drawn up to the third specular reflection. Bottom: Plot of the caustics, which are the collection of focal points of the trajectories. (From van Houten et al.[6]).

$$B_{focus} = \frac{2mv_F}{eL}. \qquad (7)$$

For a given injected current I_i, the voltage V_c on the collector is proportional to the incident flux. The classical picture thus predicts a series of equidistant peaks in the collector voltage as a function of magnetic field.

In Fig. 7 (top) we show such a classical focusing spectrum, calculated for parameters corresponding to the experiment discussed below ($L = 3.0\mu m$, $k_F = 1.5 \times 10^8 m^{-1}$). The spectrum consists of equidistant focusing peaks of approximately equal magnitude superimposed on the Hall resistance (dashed line). The p-th peak is due to electrons injected perpendicularly to the boundary which have made p-1 specular reflections between injector and collector. Such a classical focusing spectrum is commonly observed in metals,[34] albeit with a decreasing height of subsequent peaks because of partially diffuse scattering at the metal surface. Note that the peaks occur in one field direction only; In

257

reverse fields the focal points are at the wrong side of the injector for detection, and the normal Hall resistance is obtained. The experimental result for a 2DEG is shown in the bottom half of Fig. 7 (trace a; trace b is discussed below). A series of five focusing peaks is evident at the expected positions. This observation by itself has two important implications:

- A point contact acts a monochromatic point source of ballistic electrons with a well-defined energy;

- The electrostatically defined 2D electron gas boundary is a good mirror with very little diffuse scattering.

On the experimental focusing peaks a fine structure is resolved at low temperatures (below 1 K). The fine structure is well reproducible but sample dependent. A nice demonstration of the reproducibility of the fine structure is obtained upon interchanging current and voltage leads, so that the injector becomes the collector and vice versa. The resulting focusing spectrum shown in Fig. 7 (trace b) is almost the precise mirror image of the original one (trace a) - although this particular device had a strong asymmetry in the widths of injector and collector. The symmetry in the focusing spectra is an example of the general reciprocity relation for four-terminal resistance measurements (which tells us that an interchange of current and voltage leads has no effect if one also reverses the sign of the magnetic field). If one applies the Landauer-Büttiker formalism to the electron focusing geometry,[6] one finds that the ratio of collector voltage V_c to injector current I_i is given by

$$\frac{V_c}{I_i} = \frac{2e^2}{h} \frac{T_{i \to c}}{G_i G_c}, \tag{8}$$

where $T_{i \to c}$ is the transmission probability from injector to collector, and G_i and G_c are the two-terminal conductances of the injector and collector point contact. Since[9] $T_{i \to c}(B) = T_{c \to i}(-B)$ and $G(B) = G(-B)$, this expression for the focusing spectrum is manifestly symmetric under interchange of injector and collector with reversal of the magnetic field.

The fine structure on the focusing peaks in Fig. 7 is the first indication that electron focusing in a 2DEG is qualitatively different from the corresponding experiment in metals. The main features of the experimental results can be explained in terms of quantum interference between the different skipping orbits from injector to collector, or alternatively in terms of interference of coherently excited edge channels, which are the quantum mechanical modes of this problem. Since the injector has a width below λ_F, it excites the modes coherently. The interference of modes at the collector is dominated by their rapidly varying phase factors $\exp(ik_nL)$. The wave number k_n corresponds classically to the separation of the center of the skipping orbit from the 2D electron gas boundary, according to $k_n = k_F \sin \alpha_n$, where α is the angle with the boundary normal under which the skipping orbit is reflected. The quantized values α_n may be obtained from the Bohr-Sommerfeld quantization rule. The resulting mode interference oscillations can become much

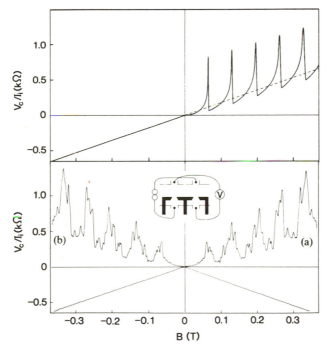

FIG. 7. Bottom: Experimental electron focusing spectrum (T = 50 mK, L = 3.0 μm) in the generalized Hall resistance configuration depicted in the inset. The two traces a and b are measured with interchanged current and voltage leads, and demonstrate the injector-collector reciprocity as well as the reproducibility of the fine structure. Top: Calculated classical focusing spectrum corresponding to the experimental trace a (50 nm wide point contacts were assumed). The dashed line is the extrapolation of the classical Hall resistance seen in reverse fields. (From van Houten et al.[6]).

larger than the classical focusing peaks. This has been shown in Refs. 29 and 7, where the transmission probability $T_{i \to c}$ was calculated in the WKB approximation with neglect of the finite width of the injector and detector. From Eq. (8) the focusing spectrum is then obtained in the form

$$\frac{V_c}{I_i} = \frac{h}{2e^2} \left| \frac{1}{N} \sum_{n=1}^{N} e^{ik_n L} \right|^2 ; \tag{9}$$

Here N is the number of edge channels, which equals the number of occupied Landau levels in the bulk of the 2D electron gas. No detailed agreement between the theoretical calculations and the experimental results may be expected, because of the uncertainties in shape of the electrostatic potential defining the point contacts and the 2D electron gas boundary. Nevertheless, it may be concluded that

- The experimental injector and collector point contacts resemble the idealized coherent point source/detector of the calculation;

- Scattering events other than specular scattering on the boundary can be largely ignored;

- The phase coherence is maintained over a distance of several microns to the collector.

It follows from Eq. (9) that if the interference of the modes is ignored the normal quantum Hall resistance $h/2Ne^2$ is obtained. This is *not* a general result, but depends specifically on the properties of injector and collector point contacts, as will be discussed in Sec. III.

3. ADIABATIC QUANTUM TRANSPORT

3.1. Suppression of Backscattering at a Quantum Point Contact

In the absence of a magnetic field, only a small fraction of the electrons injected by the current source into the 2DEG is transmitted through the point contact. The remaining electrons are scattered back into the source contact. This is the origin of the non-zero resistance of a ballistic point contact. In this sub-section we briefly discuss the effect of a relatively weak magnetic field on the point contact resistance.

Electron Optics in a Two-Dimensional Electron Gas

A *negative* magnetoresistance was found in a *four-terminal* measurement of the longitudinal point contact resistance R_L.[17] The voltage probes in this experiment are positioned on wide 2DEG regions, well away from the constriction (see the inset in Fig. 8). This allows the establishment of local equilibrium near the voltage probes, at least in weak magnetic fields, so that the measured four-terminal resistance does not depend on the properties of the probes. The experimental results are plotted in Fig. 8. The negative magnetoresistance is temperature independent (between 50 mK and 4 K), and is observed in weak magnetic fields once the narrow constriction is defined.

The mechanism causing the negative magnetoresistance is illustrated in Fig. 9. In a magnetic field the left- and right-moving electrons are spatially separated by the Lorentz force at opposite sides of the constriction. Backscattering thus requires scattering across the width of the constriction, which becomes increasingly improbable as l_{cycl} becomes smaller and smaller compared to the width. For this reason a magnetic field suppresses the constriction resistance in the ballistic regime.

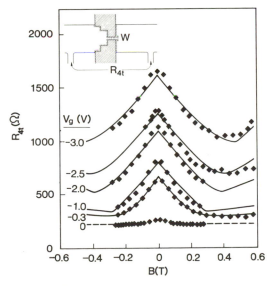

FIG. 8. Four-terminal longitudinal magnetoresistance R_L (labeled R_{4t} in the figure) of a constriction for a series of gate voltages. The negative magnetoresistance is temperature independent between 50 mK and 4 K. Solid lines are according to Eq. (10), with the constriction width as adjustable parameter. The inset shows schematically the device geometry, with the two voltage probes used to measure R_L (From van Houten et al.[23]).

261

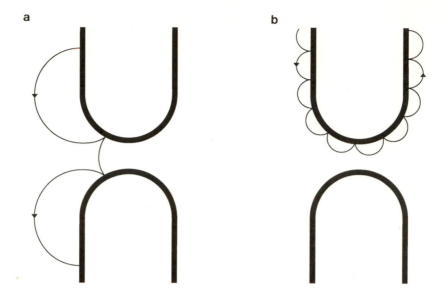

a   b

FIG. 9. Illustration of the reduction of backscattering by a magnetic field, which is responsible for the negative magnetoresistance of Fig. 8. Shown are trajectories approaching a constriction without a potential barrier, in a weak (a) and strong (b) magnetic field. (From Beenakker and van Houten[24]).

Quantum mechanically, the skipping orbits in Fig. 9 correspond to magnetic edge states. The theoretical result for the longitudinal resistance derived in Ref. 17 may be written in terms of the number of available edge states in the wide 2D electron gas regions, N_{wide} (equal to the number of bulk Landau levels, and proportional to $1/B$), and the smaller number of edge channels transmitted through the point contact, N_{min}, which is essentially constant in weak fields:

$$R_L = \frac{h}{2e^2} \left(\frac{1}{N_{min}} - \frac{1}{N_{wide}} \right). \qquad (10)$$

The negative magnetoresistance observed experimentally is well described by this result. In strong fields (corresponding to the semiclassical criterion that the cyclotron diameter $2l_{cycl} = 2mv_F/eB$ is less than the point contact width W_{min}) one has $N_{min} = N_{wide}$ so that R_L vanishes. [In reality, the point contact contains a potential barrier, in addition to a geometrical constriction. This causes a crossover to a positive magnetoresistance in strong fields, as seen in Fig. 8. This effect is also accounted for by Eq. (10).] The main implication of this study is that

- A magnetic field suppresses the geometrical backscattering at a point contact.

This ballistic effect is the weak field precursor of the adiabatic propagation of electrons in edge channels, discussed below.

3.2. Adiabatic Transport

A description of the quantum Hall effect[36] based on extended edge states and localized bulk states, as in Fig. 10, was first put forward by Halperin.[37] Generally, a local equilibrium was assumed at each edge. In the presence of a chemical potential difference $\delta\mu$ between the edges, each edge channel carries a current $(e/h)\delta\mu$ and thus contributes e^2/h to the Hall conductance. In this case of local equilibrium the two-terminal resistance R_{2t} of the Hall bar is the same as the four-terminal Hall resistance $R_H = R_{2t} = (h/e^2)N^{-1}$, see Fig. 10. The longitudinal resistance vanishes, $R_L = 0$. The distinction between a longitudinal and Hall resistance

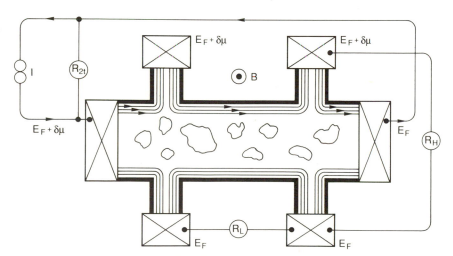

FIG. 10. Measurement configuration for the two-terminal resistance R_{2t}, the four-terminal Hall resistance R_H, and the longitudinal resistance R_L. The edge channels at the Fermi level are indicated, arrows point in the direction of motion of edge channels filled by the source contact at chemical potential $E_F + \delta\mu$. The current is equipartitioned among the edge channels at the upper edge, corresponding to the case of local equilibrium. (From Beenakker and van Houten[24]).

is topological: A four-terminal resistance measurement gives R_H if current and voltage contacts alternate along the boundary of the conductor, and R_L if that is not the case. There is no need to further characterize the contacts in the case of local equilibrium at the edge.

Both the quantum Hall effect[36] and the quantized conductance of a ballistic point contact[3,5] are described by one and the same relation $G = Ne^2/h$ between the conductance G and the number N of propagating modes at the Fermi level. A smooth transition from zero-field quantization to quantized Hall resistance follows from this relation, and has indeed been observed. The nature of the modes is very different, however, in weak and strong magnetic fields, leading to an entirely different sensitivity to scattering processes. Firstly, the zero-field conductance quantization is destroyed by a small amount of elastic scattering, while the quantum Hall effect is robust to scattering. This difference is a consequence of the *suppression of backscattering* by a magnetic field, which itself follows from the spatial separation at opposite edges of edge channels moving in opposite directions. Secondly, the spatial separation of edge channels at the *same* edge in the case of a smooth confining potential opens up the possibility of *adiabatic transport*, i.e. the full suppression of inter-edge channel scattering. In weak magnetic fields, adiabaticity is of importance within a point contact, but not on longer length scales. In a wide 2DEG region, scattering among the modes in weak fields establishes local equilibrium on a length scale given by the inelastic scattering length (which in a high-mobility GaAs-AlGaAs heterostructure is presumably not much longer than the elastic scattering length $l \sim 10\mu m$). The situation is strikingly different in a strong magnetic field, where the experimental observation of *selective* population and detection of edge channels (discussed below) has demonstrated the persistence of adiabaticity outside the point contact.

3.3. Quantum Point Contacts as Landau Level Selectors

In Sec. 2.3 we have seen how a quantum point contact can inject a *coherent* superposition of edge channels at the 2DEG boundary, in the coherent electron focusing experiment. In that section we restricted ourselves to weak magnetic fields. Here we will show how in the QHE regime the point contacts can be operated in a different way as *selective* injectors (and detectors) of edge channels. We recall that electron focusing can be measured as a generalized Hall resistance, in which case the structure due to mode interference is superimposed on the weak-field Hall resistance. As shown in Fig. 11, if the weak-field electron focusing

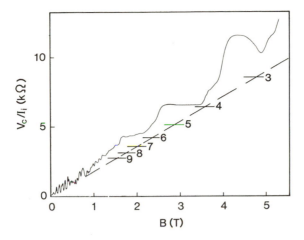

FIG. 11. Transition from weak-field electron focusing to high-field quantum Hall effect. Quantum Hall-plateau values corresponding to h/ie^2 are indicated for i = 3,4,...,9. The observed plateaux occur consistently at lower magnetic field values than expected from the normal 2D electron gas Hall resistance (indicated by the dashed line). (From van Houten et al.[6]).

experiments are extended to stronger magnetic fields, a transition is observed to the quantum Hall effect, provided the injecting and detecting point contacts are not too strongly pinched off. The oscillations characteristic of mode interference disappear in this field regime, suggesting that the coupling of the edge channels (which form the propagating modes from injector to collector) is suppressed, and adiabatic transport is realized. The experimental results of Fig. 11 show, however, quantized plateaux in the Hall resistance which do not occur at the magnetic field values expected (as indicated in Fig. 11). These deviations were first reported by Van Wees et al.,[40] and have come to be known as the *anamalous* quantum Hall effect.

To understand these deviations, it is no longer sufficient to model the point contacts by a point source/detector of infinitesimal width (as was done in Sec. 2.3), but a somewhat more detailed description of the electrostatic potential $V(\psi,y)$ defining the point contacts and the 2DEG boundary between them is required. Schematically, $V(x,y)$ is represented in Fig. 12. Fringing fields from the split gate create a potential barrier in the point contacts, so that V has a saddle form as shown. The heights of the barriers in the injector and collector can be determined from the two-terminal conductances of the individual point contacts. The point contact separation in the experiment is small (3 μm), so that one can assume fully adiabatic transport from injector to collector in strong magnetic fields. The electrons then move along equipotentials at the guiding center energy E_G, given by

$$E_G = E_F - \left(n - \frac{1}{2}\right)\hbar\omega_c. \tag{11}$$

Note that the edge channel with the smallest index n has the largest guiding center energy. In the absence of inter-edge channel scattering, edge channels can only be transmitted through a point contact if E_G exceeds the potential barrier height E_i (disregarding tunneling through the barrier). The injector thus injects $N_i \sim (E_F - E_i)/\hbar\omega_c$ edge channels into the 2DEG, while the collector is capable of detecting $N_c \sim (E_F - E_c)/\hbar\omega_c$ edge channels. Along the boundary of the 2DEG, however, a larger number of $N_{wide} \sim E_F/\hbar\omega_c$ edge channels, equal to the number of occupied bulk Landau levels in the 2DEG, are available for transport at the Fermi level. The selective population and detection of Landau levels leads to deviations from the normal Hall resistance.

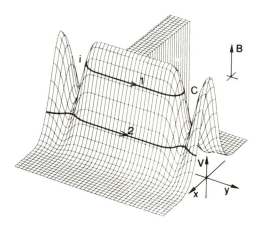

FIG. 12. Schematic potential landscape, showing the 2DEG boundary and the saddle-shaped injector and collector point contacts. In a strong magnetic field the edge channels are extended along equipotentials at the guiding center energy, as indicated here for edge channels with index n = 1,2 (the arrows point in the direction of motion). In this case a Hall conductance of $(2e^2/h)N$ with N = 1 would be measured by the point contacts - in spite of the presence of 2 occupied spin-degenerate Landau levels in the bulk 2DEG. (From Beenakker et al.[14]).

These considerations can be put on a theoretical basis by applying Eq. (8) to this problem. We consider the case that the barrier in one of the two point contacts is sufficiently higher than in the other, to ensure that electrons which are transmitted over the highest barrier will have a negligible probability of being reflected at the lowest barrier. Then $T_{i \to c}$ is dominated by the transmission probability over the highest barrier, which follows from the two-terminal conductance of that point contact according to $T_{i \to c} \sim (h/2e^2)\min(G_i, G_c)$. Substitution in Eq. (8) gives the remarkable result that the *Hall conductance* $G_H \equiv I_i/V_c$ measured in the electron focusing geometry can be expressed entirely in terms of the *contact conductances* G_i and G_c,

$$G_H \sim \max(G_i, G_c). \tag{12}$$

Eq. (12) tells us that quantized values of G_H occur not at $(2e^2/h)N_{wide}$, as one would expect from the N_{wide} populated Landau levels in the 2DEG - but at the smaller value of $(2e^2/h)\max(N_i, N_c)$. This explains the deviations shown in Fig. 11. A striking demonstration of the validity of Eq. (12) has been given by van Wees et al..[40] The implication of the experiments on the anomalous quantum Hall effect is that

- Point contacts can be used to *selectively* populate and detect Landau levels at the 2D electron gas boundary;

- Adiabatic transport (i.e. transport in the absence of inter-Landau level scattering) may persist over a distance of a few μm.

- Transmission probabilities in the adiabatic transport regime may be estimated from the simple picture of guiding center motion along equipotential lines at discrete values for the energy E_G.

These experiments support Büttiker's view[18] that the non-ideality of the coupling of reservoirs to the conductor affects the accuracy of the quantum Hall effect in the absence of local equilibrium. Extensions of this work have shown that the lack of local equilibrium may persist over macroscopic distances.[38,39] In addition, the Shubnikov-De Haas oscillations in the longitudinal resistance were found to be dramatically affected by the properties of the contacts,[40] establishing the failure of conventional theories in terms of a local resistivity tensor.

4. FROM ELECTRON WAVEGUIDE TO ELECTRON CAVITY

In the preceding sections, we have discussed ballistic and adiabatic transport in 2D and in 1D electron gases in terms of the independent motion of electrons at the Fermi level. We have called this discipline electron optics in the solid state. It is natural to ask whether the analogy between optics and electronic motion also applies to transport through a 0D structure, behaving like an *electron cavity* (also known as a quantum dot). The existence of zero-dimensional states requires a three-dimensional confinement, so that tunneling is the only possible transport mechanism at low temperatures.

The familiar problem of resonant tunneling through a 2D quantum well, pioneered by Chang, Esaki and Tsu,[41] is generally treated as the electronic analogue of optical transmission through a Fabry-Pérot resonator. This is quite adequate, at least in the regime of linear response of interest here, where the feed-back mechanism due to charge build-up in the well may be neglected.[42] Interestingly, resonant tunneling through a quantum *dot* rather than a well, is *not* adequately described in terms of a Fabry-Pérot type of treatment, even in the small-bias regime.

The reason is the increased significance of electron-electron interactions in small tunnel junctions. The requirement that an integer number of electrons is present in a metallic region between two tunnel barriers, in addition to the requirement that charge transfer proceeds by discrete amounts e, causes a *Coulomb-blockade of tunneling*,[43] studied in detail in the field of metallic tunnel junctions, where quantum size effects do not play a role. Such effects may also play an important role in semiconductor nanostructures, however.[44,45] The combined effects of level discreteness and single-electron charging have begun to be addressed only recently,[46,20,21] motivated in part by the availability of semiconductor nanostructures in which an appreciable 0D level-splitting may be realized experimentally (using techniques similar to those discussed in this paper). It is instructive to consider the following crude estimates, appropriate for a quantum dot of area A defined by means of split gates in a 2D electron gas.

The 0D level splitting in the dot is of order $\Delta E \sim 1/\rho A$, where $\rho = m/\pi\hbar^2$ is the 2D density of states. The electron-electron interaction is taken into account in the simplest approximation by assigning an effective capacitance C to the dot (C may include the effects of the electrostatic interactions between the dot, the leads, and the gate). The elementary charging energy, corresponding to a single excess electron in the dot, is then $e^2/2C$. Let us assume for simplicity that C is dominated by the dot-gate capacitance, which is of the order $\varepsilon\varepsilon_0 A/d$, where d denotes the distance between the dot and the gate. On inserting the appropriate numbers for GaAs($\varepsilon_0 = 13$, $m = 0.067m_e$), and taking a minimum realistic value for $d = 50$ nm, one finds that the charging energy $e^2/27$ may exceed the 0D level splitting ΔE by an order of magnitude. Note that this holds true for a wide range of dot areas (the ratio $(e^2/2C)/\Delta E = e^2 dm/\varepsilon_0 \varepsilon \pi \hbar^2$ does not depend on A).

As shown in Ref. 20, the combined effects of dot-charging and 0D level splitting may be accounted for by defining a renormalized level splitting

$$\Delta E^* = \Delta E + \frac{e^2}{2C}. \tag{13}$$

In a measurement of the conductance of a quantum dot as a function of gate voltage one will observe a sequence of oscillations, due to *Coulomb-regulated resonant tunneling* through the dot. These Coulomb-blockade oscillations may be observed at a much higher temperature (of order $\Delta E^*/k_B$) than expected on the basis of the bare level splitting. In the limit $e^2/2C \gg \Delta E$, a magnetic field has very little effect on the periodicity of the oscillations. Note that the spin-degeneracy is lifted by the Coulomb repulsion, so that the application of a magnetic field does not cause a splitting of the peaks. Indeed, the Coulomb blockade has been predicted to suppress quantum interference effects such as the Aharonov-Bohm effect in a quantum dot in the quantum Hall effect regime.[20]

These concepts have recently been applied[21] to the analysis of the remarkable periodicity of the conductance oscillations in a disordered quantum wire, first found experimentally in narrow Si inversion layer channels by Scott-Thomas et al.[47,48]

A necessary condition for the appearance of the Coulomb blockade is that the electrons are well-localized in the dot. This requires a junction resistance R exceeding h/e^2 (this condition[43] follows from the requirement that e^2/C should exceed the energy uncertainty h/RC associated with the leakage of charge out of the dot on a time scale RC). For tunneling through 0D levels this condition is satisfied,[50,51] so that the Coulomb blockade needs to be considered in interpreting the available experimental data.[52-57] For ballistic or adiabatic transport, on the other hand, the conductance exceeds $2e^2/h$, and a single-particle treatment should be adequate. Thus, one is led to conclude that

- Quantum ballistic and adiabatic transport in 2D or 1D electron gases may be viewed as electron optics at the Fermi level, but tunneling through 0D states in a quantum dot requires a consideration of electron-electron interaction effects.

ACKNOWLEDGEMENT

Most of the work described in this paper has been a joint effort between the Philips Research Laboratories at Eindhoven and at Redhill, and the Delft University of Technology. The results reproduced in Fig. 8 were obtained in collaboration with Cambridge University. It is a pleasure to acknowledge in particular the ongoing collaboration with Carlo Beenakker.

REFERENCES

1. Yu.V. Sharvin, Zh. Eksp. Teor. Fiz. **48**, 984 (1965) [Sov. Phys. JETP **21**, 655 (1965)]

2. A collection of early papers is given in Proc. 7th Int. Conf. Electronic Properties of Two-Dimensional Systems, Surf. Sci. **170** (1986).

3. B. J. van Wees, H. van Houten, C. W. J. Beenakker, J. G. Williamson, L. P. Kouwenhoven, D. van der Marel and C. T. Foxon, Phys. Rev. Lett. **60**, 848 (1988).

4. H. van Houten, B. J. van Wees, J. E. Mooij, C. W. J. Beenakker, J. G. Williamson and C. T. Foxon, Europhys. Lett. **5**, 721 (1988).

5. D. Wharam, T. J. Thornton, R. Newbury, M. Pepper, H. Ahmed, J. E. F. Frost, D. G. Hasko, D. C. Peacock, D. A. Ritchie and G. A. C. Jones, J. Phys. C **21**, L209 (1988).

6. H. van Houten, C. W. J. Beenakker, J. G. Williamson, M. E. I. Broekaart, P. H. M. van Loosdrecht, B. J. van Wees, J. E. Mooij, C. T. Foxon, and J. J. Harris, Phys. Rev. B **39**, 8556 (1989).

7. C. W. J. Beenakker, H. van Houten and B. J. van Wees, Festkörperprobleme/Advances in Solid State Physics **29**, 299 (1989).

8. R. Landauer, IBM J. Res. Dev. **1**, 223 (1957).

9. M. Büttiker, Phys. Rev. Lett. **57**, 1761 (1986).

10. C. W. J. Beenakker and H. van Houten, Phys. Rev. B **39**, 10445 (1989).

11. H. van Houten and C. W. J. Beenakker, in *Nanostructure Physics and Fabrication*, M. A. Reed and W. P. Kirk, eds., (Academic Press, New York, 1989).

12. L. W. Molenkamp, A. A. M. Staring, C. W. J. Beenakker, R. Eppenga, C. E. Timmering, J. G. Williamson, C. J. P. M. Harmans and C. T. Foxon, Phys. Rev. B **41**, 1274 (1990).

13. H. U. Baranger and A. D. Stone, Phys. Rev. Lett. **63**, 414 (1989).

14. C. W. J. Beenakker and H. van Houten, Phys. Rev. Lett. **63**, 1857 (1989).

15. J. Spector, H. L. Störmer, K. W. Baldwin, L. N. Pfeiffer, and K. W. West, Appl. Phys. Lett. **56**, 1290 (1990).

16. U. Sivan, M. Heiblum, C. P. Umbach and H. Shtrikman, Phys. Rev. B **41**, 7937 (1990).

17. H. van Houten, C. W. J. Beenakker, P. H. M. van Loosdrecht, T. J. Thornton, H. Ahmed, M. Pepper, C. T. Foxon and J. J. Harris, Phys. Rev. B **37**, 8534 (1988).

18. M. Büttiker, Phys. Rev. B **38**, 9375 (1988).

19. B. J. van Wees, E. M. M. Willems, C. J. P. M. Harmans, C. W. J. Beenakker, H. van Houten, J. G. Williamson, C. T. Foxon, and J. J. Harris, Phys. Rev. Lett. **62**, 1181 (1989).

20. C. W. J. Beenakker, H. van Houten and A. A. M. Staring, Phys. Rev. B, to be published.

21. A. A. M. Staring, H. van Houten, C. W. J. Beenakker and C. T. Foxon, in *High magnetic fields in semiconductor physics III*, G. Landwehr, ed. (Springer, Berlin, 1991).

22. H. van Houten, C. W. J. Beenakker and B. J. van Wees, in *Nanostructured Systems*, Ed. M. A. Reed, (A volume of *Semiconductors and Semimetals*, Willardson and Beer series eds.) (Academic Press, New York), to be published.

23. H. van Houten and C. W. J. Beenakker, in *Analogies in Optics and Microelectronics*, W. van Haeringen and D. Lenstra, eds. (Kluwer, Dordrecht, 1990).

24. C. W. J. Beenakker and H. van Houten, *Quantum Transport in Semiconductor Nanostructures,* Solid State Physics, **44**, 1 (1991).

25. H. van Houten, B. J. van Wees and C. W. J. Beenakker, in *Physics and Technology of Submicron Structures*, H. Heinrich, G. Bauer and F. Kuchar, eds., (Springer, Berlin, 1988).

26. H. van Houten, in Shinkinososhi [Future Electron Devices], **1**, 25 (1990).

27. Y. Imry, in: *Directions in Condensed Matter Physics*, Vol. 1, G. Grinstein and G. Mazenko, eds. (World Scientific, Singapore, 1986).

28. E. G. Haanappel and D. van der Marel, Phys. Rev. B **39**, 5484 (1989); D. van der Marel and E. G. Haanappel, Phys. Rev. B **39**, 7811 (1989).

29. L. I. Glazman, G. B. Lesovick, D. E. Khmel'nitskii, R. I. Shekhter, Pis'ma Zh. Eksp. Teor. Fiz. **48**, 218 (1988) [JETP Lett. **48**, 238 (1988)].

30. A. Yacoby and Y. Imry, Phys. Rev. B **42**, 5341 (1990).

31. D. A. Wharam, M. Pepper, R. Newbury, H. Ahmed, D. G. Hasko, D. C. Peacock, J. E. F. Frost, D. A. Ritchie and G. A. C. Jones, J. Phys. Condens. Matter **1**, 3369 (1989).

32. T. J. Thornton, M. L. Roukes, A. Scherer and B. P. van der Gaag, Phys. Rev. Lett. **63**, 2128 (1989).

33. C. W. J. Beenakker, H. van Houten and B. J. van Wees, Europhys. Lett. **7**, 359 (1988).

34. V. S. Tsoi, Pis'ma Zh. Eksp. Teor. Fiz. **19**, 114 (1974) [JETP Lett. **19**, 70 (1974)].

35. J. Spector, H. L. Stormer, K. W. Baldwin, L. N. Pfeiffer and K. W. West, Surf. Sci. **228**, 283 (1990).

36. K. von Klitzing, G. Dorda and M. Pepper, Phys. Rev. Lett. **45**, 494 (1980).

37. B. I. Halperin, Phys. Rev. B **25**, 2185 (1982).

38. B. W. Alphenaar, P. L. McEuen, R. G. Wheeler and R. N. Sacks, Phys. Rev. Lett. **64**, 677 (1990).

39. S. Komiyama, H. Hirai, S. Sasa and T. Fuji, Solid State Comm. **73**, 91 (1990).

40. B. J. van Wees, E. M. M. Willems, L. P. Kouwenhoven, C. J. P. M. Harmans, J. G. Williamson, C. T. Foxon and J. J. Harris, Phys. Rev. B **39**, 8066 (1989).

41. L. L. Chang, L. Esaki and R. Tsu, Appl. Phys. Lett. **24**, 593 (1974).

42. L. Eaves, in *Analogies in Optics and Microelectronics*, W. van Haeringen and D. Lenstra, eds. (Kluwer, Deventer, 1990).

43. K. K. Likharev, I.B.M. J. Res. Dev. **32**, 144 (1988), and references therein.

44. H. van Houten and C. W. J. Beenakker, Phys. Rev. Lett. **63**, 1893 (1989).

45. L. I. Glazman and R. I. Shekhter, J. Phys. Condens. Matter **1**, 5811 (1989).

46. D. V. Averin and A. N. Korotkov, Zh. Eksp. Teor. Fiz. **97**, 1661 (1990), [Sov. Phys. JETP **70**, 937 (1990)]; A. N. Korotkov, D. V. Averin and K. K. Likharev, Physica B **165 & 166**, 927 (1990).

47. J. H. F. Scott-Thomas, S. B. Field, M. A. Kastner, H. I. Smith, and D. A. Antoniadis, Phys. Rev. Lett. **62**, 583 (1989).

48. U. Meirav, M. A. Kastner, M. Heiblum and S. J. Wind, Phys. Rev. B **40**, 5871 (1989).

49. S. B. Field, M. A. Kastner, U. Meirav, J. H. F. Scott-Thomas, D. A. Antoniadis, H. I. Smith and S. J. Wind, Phys. Rev. B **42**, 3523 (1990).

50. V. Kalmeyer and R. B. Laughlin, Phys. Rev. B **35**, 9805 (1987).

51. W. Xue and P. A. Lee, Phys. Rev. B **38**, 3913 (1988).

52. C. G. Smith, M. Pepper, H. Ahmed, J. E. F. Frost, D. G. Hasko, R. Newbury, D. C. Peacock, D. A. Ritchie and G. A. C. Jones, Surf. Sci. **228**, 387 (1990).

53. B. J. van Wees, L. P. Kouwenhoven, C. J. P. M. Harmans, J. G. Williamson, C. E. T. Timmering, M. E. I. Broekaart, C. T. Foxon and J. J. Harris, Phys. Rev. Lett. **62**, 2523 (1989).

54. R. J. Brown, C. G. Smith, M. Petter, M. J. Kelly, R. Newbury, H. Ahmed, D. G. Hasko, J. E. F. Frost, D. C. Peacock, D. A. Ritchie and G. A. C. Jones, J. Phys. Condes. Matter **1**, 6291 (1989).

55. U. Meirav, M. A. Kastner and S. J. Wind, Phys. Rev. Lett. **65**, 771 (1990).

56. C. G. Smith, M. Pepper, H. Ahmed, J. E. F. Frost, D. G. Hasko, D. C. Peacock, D. A. Ritchie and G. A. C. Jones, J. Phys. C: Solid State Phys. **21**, L893 (1989).

57. P. A. Lee, Phys. Rev. Lett. **65**, 2206 (1990).

SOME RECENT DEVELOPMENTS IN THE PHYSICS OF RESONANT TUNNELING

L. Eaves

Department of Physics
University of Nottingham
Nottingham, NG7 2RD, U.K.

Abstract - The effect of a large magnetic field, B, applied parallel to the plane of the barriers on the resonant current-voltage characteristics of n-type and p-type double barrier structures is described. The effect of the magnetic field on the electron resonances in n-type structures can be understood using the effective mass approximation. The results for the p-type structures provide a novel magneto-tunneling spectroscopy technique for probing the complicated $\varepsilon(k_\parallel)$ dispersion curves of hole states in a quantum well. These experiments reveal directly the strong admixing of the light and heavy hole states and that some of the states correspond to negative hole effective mass for motion in the plane of the quantum well.

1. INTRODUCTION

Resonant tunneling of electrons in double barrier semiconductor heterostructures has remained a subject of great scientific and technological interest since its first observation by Chang, Esaki and Tsu in 1974.[1]

Review articles have appeared recently which discuss high frequency applications[2] and the fundamental physics[3] of this type of structure, including the "sequential versus coherent tunneling controversy," space charge buildup effects and the effect of quantizing magnetic fields. This article describes a new potential application of resonant tunneling in the presence of magnetic fields. The technique, "resonant tunneling magneto-spectroscopy," allows us to measure directly the anomalous and highly non-parabolic energy dispersion curves of hole states in quantum wells.

In bulk semiconductors, the dispersion curves of the light and heavy hole valence bands are parabolic at low hole energies. Their curvatures give the inverse masses of the light and heavy holes. However, when a quantizing magnetic field ($\underline{B} \parallel \underline{z}$) is applied to a bulk semiconductor, the quantized hole energies or Landau levels, as a function of k_z, are very complicated.[4] A confining quantum well potential in, for example, an AlAs/GaAs/AlAs heterostructure, also gives rise to complicated hole dispersion curves, as a function of the in-plane wavevector component, $k_\parallel$, even at zero magnetic field. In both cases, the complexity arises due to the admixing of light hole and heavy hole states by a term ($\underline{k}.\underline{J}$) in the valence band Hamiltonian which couples the hole momentum to its quasi-spin ($\underline{J}$). An excellent theoretical account of hole dispersion in quantum wells and superlattices is given by Bastard.[5] Recently, Wessel and Altarelli[6] have calculated hole states in resonant tunneling devices and have shown that the complex nature of the valence band strongly modifies the resonant tunneling current-voltage characteristics, $I(V)$. Mendez and co-workers[7] were, to our knowledge, the first to investigate p-type resonant tunneling devices. Their structures show strong resonant features in $I(V)$ and they were able to identify resonant tunneling into both light and heavy hole states of the quantum well with some evidence of mixing.

Despite the considerable interest in hole states in quantum wells, there have been remarkably few direct experimental investigations of the anomalous hole dispersion curves, since spectroscopic techniques tend to average over k-space. In this article, we will describe a novel experimental method of determining the hole dispersion curves in the quantum well of resonant tunneling devices using magnetotunneling spectroscopy in which a strong magnetic field is applied parallel to the plane of the tunnel barriers. Tunneling holes acquire a large $k_\parallel$ in the magnetic field through the action of the Lorentz force. The resulting shift of the voltage positions of the peaks in the current-voltage characteristics reveal directly the strong coupling between light and heavy hole states at finite $k_\parallel$ and show clearly that some of the states correspond to negative hole mass (hole energy *decreasing* with *increasing* $k_\parallel$).

We will preface our discussion of the holes by a description of the behavior of electrons in n-type resonant tunneling devices in the presence of a magnetic field applied parallel to the plane of the barriers.[8,9,10] Such measurements illustrate many of the basic features of resonant tunneling. In addition, these results show that the shift in voltage of the peak current of the electron resonance with magnetic field is quadratic and essentially reflects the parabolic nature of the $\varepsilon(k_\parallel)$ curves of the electrons. In contrast, we will see that the hole resonances shift non-quadratically, corresponding to the highly non-parabolic nature of the hole $\varepsilon(k\parallel)$ curves in the quantum well.

2. SAMPLE CHARACTERISTICS

Both the n- and p-type resonant tunneling devices used in this investigation were grown by molecular beam epitaxy. The structures incorporate an undoped spacer layer between each barrier and the heavily-doped contact layer adjacent to it. This feature appears to increase the peak-to-valley ratio in the I(V) characteristics, probably by reducing ionized impurity scattering. Three structures are considered in this article. The n-type resonant tunneling device (NU183) is shown schematically below. Two p-type structures were examined; the first (NU448), with a well thickness of 4.2 nm, is shown schematically below. The second (NU490) has a well thickness of 6.8 nm, but with all of the other layer thicknesses the same as for structure NU448. All three structures were processed into circular mesas. For the p-type devices, the lower electrical contact was to the 3.0 μm heavily doped p-layer immediately above the n+ substrate.

3. ELECTRON RESONANT TUNNELING IN A TRANSVERSE MAGNETIC FIELD

The layer structure of the asymmetric n-type resonant tunneling device (NU183) is given above. We consider the bias direction in which electrons are injected into the well through the thicker barrier (11.1 nm). They leave the well through the thinner (8.3 nm) barrier. Under these conditions there is little charge buildup in the well so that the resonance

NU183

0.5 μm GaAs, n = 2 x 10^{18} cm^{-3}, top contact
50 nm GaAs, n = 10^{17} cm^{-3}
50 nm GaAs, n = 10^{16} cm^{-3}
3.3 nm GaAs, undoped
11.1 nm Al$_{0.4}$Ga$_{0.6}$As, undoped (thick barrier)
5.8 nm GaAs, undoped (well)
8.3 nm Al$_{0.4}$Ga$_{0.6}$As, undoped (thin barrier)
3.3 nm GaAs, undoped
50 nm GaAs, n = 10^{16} cm^{-3}
50 nm GaAs, n = 10^{17} cm^{-3}
2 μm GaAs, n = 2 x 10^{18} cm^{-3}
n+ GaAs (001) substrate

NU448

0.6 μm GaAs, p = 2 x 10^{18} cm^{-3}, top contact
100 nm GaAs, p = 1 x 10^{18} cm^{-3}
100 nm GaAs, p = 5 x 10^{17} cm^{-3}
5.1 nm GaAs, undoped
5.1 nm AlAs, undoped (barrier)
4.2 nm GaAs, undoped (well)
5.1 nm AlAs, undoped (barrier)
100 nm GaAs, p = 5 x 10^{17} cm^{-3}
100 nm GaAs, p = 1 x 10^{18} cm^{-3}
3.0 μm GaAs, p = 10^{18} cm^{-3}, lower contact
n+ GaAs (001) substrate

is not distorted by space-charge effects.[11] Tunneling occurs from a two-dimensional electron gas (2DEG) which forms in the accumulation layer of the emitter contact when a bias is applied to the device. Figure 1(a) shows the current-voltage characteristics of a 200 μm diameter mesa at 4 K in magnetic fields B between 0 and 11.5 Tesla applied parallel to the plane of the barriers and perpendicular to the current flow $\underline{J}$. We define $\underline{B} \| \underline{z}$ and $\underline{J} \| \underline{x}$. The main effects are:

(a) There is a rapid decrease in the peak current with magnetic field; at 4 T the peak current is less than half the zero-field value. This trend is reversed at higher fields where there is a gradual increase.

(b) The peak position (V_p) moves to higher bias as the field increases. By 11 T, V_p has changed by more than 300 mV. The threshold voltage (V_{th}) above which the current increases rapidly to its resonant peak value also shifts to higher values of bias with increasing B, as does the cut-off voltage (V_v) beyond the resonant peak. Figure 2 plots the threshold, peak and cut-off voltages against B. The threshold and cut-off voltages were deduced by extrapolating the linear portions of I(V) on each of the resonant peaks to zero current.

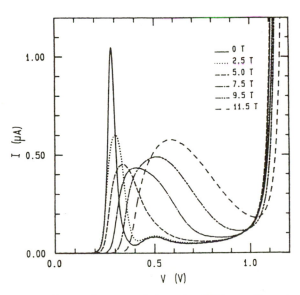

FIG. 1. Current-voltage characteristics of a 200 μm diameter mesa of NU183 at the first resonance in reverse bias at 4 K in magnetic fields $\underline{B} \perp \underline{J}$ up to 12 Tesla.

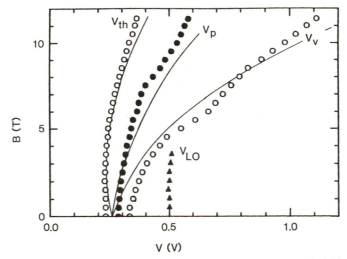

V (V)

FIG. 2. Plot of the threshold, peak and cut-off voltages for NU183 against magnetic field $\underline{B} \perp \underline{J}$, showing the broadening of the resonance and the diamagnetic shift in the peak position. The solid lines show the calculated field-dependence of the voltages discussed in the text. The position of the LO phonon peak (▲) does not change with field since this transition does not conserve transverse momentum.

(c) The resonance becomes very much broader. At B = 0, it is extremely sharp and has a full-width at half-maximum of less than 100 mV. This sharp resonance is characteristic of resonant tunneling from a two-dimensional state with little charge buildup in the well. At 11 T the resonance is more than 400 mV wide.

(d) The peak-to-valley current ratio falls dramatically with field, from ~26:1 at zero field to 1.8:1 at 11 T.

(e) At B = 0 the resonant peak is almost symmetrical. However, at higher fields the peak in the current is markedly closer to the threshold voltage than to the cut-off.

(f) The LO phonon replica peak at V ~ 500 mV[12] does not shift with magnetic field and disappears for B > 5 T. The position of the LO phonon peak is also plotted in Figure 2, marked ▲.

Resonant tunneling can only occur when there is an occupied state in the 2DEG of the emitter accumulation layer with the same values of total energy ε and components of canonical momentum k_y and k_z as an unoccupied state in the quantum well. Since both the accumulation layer potential and the quantum well are relatively narrow in this sample, the quantization is predominantly electrical and we can describe the effect of a transverse field on the energy levels using a perturbation approach. The results of this calculation, with the zero of energy taken at the center of the well, give the following values for the electron energy in the accumulation layer, ε_a, and well, ε_w:

$$\varepsilon_a = \varepsilon_0 + eE(b_1 + w/2) + \frac{3m^*\omega_c^2}{2a^2} + \frac{\hbar^2 k_z^2}{2m^*} + \frac{\hbar^2(k_y - k_{y0})^2}{2m^*}$$

and

$$\varepsilon_w = \varepsilon_1 + \frac{m^*\omega_c^2 w^2}{24} + \frac{\hbar^2 k_z^2}{2m^*} + \frac{\hbar^2 k_y^2}{2m^*},$$

where ε_1 is the quantum confinement energy in the well, ε_0 is the quantum confinement energy in the accumulation layer, as shown in Figure 3, E is the applied electric field, a is the variational parameter of the Fang-Howard wavefunction, b_1 and w are the thicknesses of the emitter barrier and quantum well respectively, and $k_{y0} = eB\Delta s/\hbar = eB(3/a + b_1 + w/2)/\hbar$. This corresponds classically to the change in the wavevector due to the action of the Lorentz force on the tunneling electron as it moves from its "quantum stand-off distance" $3/a$ in the emitter accumulation layer to the center of the well. Note that k_{y0} corresponds to the acquired in-plane momentum in the quantum well. Resonant tunneling occurs for $\varepsilon_a = \varepsilon_w$. If we assume that the effective masses are the same in the emitter and the well, the kinetic energies for motion in the z-direction are equal and so the conservation condition may be written as

$$\varepsilon_0 + eE(b_1 + w/2) + \Delta\varepsilon + \frac{\hbar^2(k_y - k_{y0})^2}{2m^*} = \varepsilon_1 + \frac{\hbar^2 k_y^2}{2m^*},$$

where $\Delta\varepsilon$ is the difference in the diamagnetic energy shifts, $m^*\omega_c{}^2(3/a^2 -w^2/12)/2$, which is small. At low temperatures, only states in the emitter with $|k_y - k_{y0}| < k_F$ are occupied where k_F is the Fermi wavevector of the 2DEG in the emitter accumulation layer ($\varepsilon_F = \hbar^2 k_F^2/2m^*$). Therefore, filled states can be represented in ε - k_y space as a section of a parabola centered at $k_y = k_{y0}$, between the points k_{y0} - k_F and k_{y0} + k_F.

The well states lie on a parabola centered at $k_y = 0$. The conservation conditions can be interpreted graphically by looking for intersections of these two curves, as illustrated in Figure 3.[8,13,14] At B = 0, $k_{y0} = 0$ so the two parabolae are centered about the same axis. At zero bias the emitter curve is below the energy level in the well and resonant tunneling cannot occur. As the bias is swept, the emitter curve moves up relative to the well state and when the voltage drop across the emitter region and the

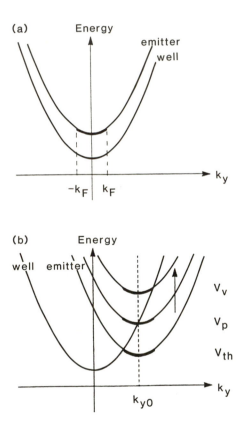

FIG. 3. Schematic diagram illustrating the conservation conditions which govern resonant tunneling of electrons in a transverse magnetic field. The occupied states in the emitter are represented by the portion of a parabola between $k_{y0} - k_F$ and $k_{y0} + k_F$. Resonant tunneling occurs for biases where this curve intersects the parabola representing the electron states in the quantum well. (a) B = 0: the parabolae have a common axis and only coincide for one value of bias, producing a sharply peaked resonance, (b) B > 0: the origin of the emitter parabola is displaced by $k_{y0} = eB(3/a + b_1 + w/2)/\hbar$ and the two parabolae intersect for a range of bias voltages between V_{th} where $k_y = k_{y0} - k_F$ and V_v where $k_y = k_{y0} + k_F$. This produces a much broader resonance.

282

first half of the quantum well is $eV_{e,th} = \varepsilon_1 + \varepsilon_F$, the device comes onto resonance. If we neglect the effects of space charge buildup in the well, this is the only applied bias for which resonant tunneling can occur, so at zero magnetic field the threshold, peak and cut-off voltages are coincident, $V_{e,p} = V_{e,th} = V_{e,v} = V_0$.

The application of a transverse magnetic field shifts the center of the emitter parabola to the right by k_{y0}. When $k_y = k_{y0} - k_F$, the threshold of the resonance is reached and electrons can tunnel into the well. As the bias is increased, the point of intersection moves to higher energy and k_y. There is a range of bias for which some of the states in the emitter are in resonance with states in the well. The cut-off occurs for the bias at which $k_y = k_{y0} + k_F$. If we neglect charge buildup in the quantum well and the variation in the transmission coefficient with k_y, the peak current occurs when there is the largest number of emitter states available to tunnel from, which is for $k_y = k_{y0}$. The corresponding voltage drops across the emitter are:

$$\text{Threshold } eV_{e,th} = \varepsilon_1' + \frac{\hbar^2(k_{y0} - k_F)^2}{2m^*}$$

$$= eV_0 + \frac{\hbar^2\left(k_{y0}^2 - 2k_{y0}k_F\right)}{2m^*} ; \tag{1}$$

$$\text{Peak } eV_{e,p} = \varepsilon_1' + \frac{\hbar^2\left(k_{y0}^2 + k_F^2\right)}{2m^*} = eV_0 + \frac{\hbar^2 k_{y0}^2}{2m^*} ; \tag{2}$$

$$\text{Cut} - \text{off } eV_{e,v} = \varepsilon_1' + \frac{\hbar^2(k_{y0} + k_F)^2}{2m^*}$$

$$= eV_0 + \frac{\hbar^2\left(k_{y0}^2 + 2k_{y0}k_F\right)}{2m^*} ; \tag{3}$$

$$\text{with } k_{y0} = eB(3/a + b_1 + w/2)/\hbar ; \tag{4}$$

where $\varepsilon_1' = \varepsilon_1 - \Delta\varepsilon$. This predicts a quadratic shift in the peak position and a linear increase in the width of the resonance with magnetic field. However, the Fermi wavevector k_F of the emitter 2DEG is a function of the applied bias since it depends on the number density in the emitter n_a, $k_F = (2\pi n_a)^{1/2}$. This can be obtained from the periodicity of the oscillations in the tunnel current or differential capacitance when the magnetic field is applied perpendicular to the interfaces ($\underline{B} \parallel \underline{J}$).[15] In addition, the voltage drop across the emitter region is not linearly related to the total applied bias. The distribution of the potential within the device varies due to the depletion of the collector contact; V_e can be approximately related to the total applied bias V by

$$V_e/V = (3/a + b_1 + w/2)/(3/a + \lambda_c + b_1 + b_2 + w + u),$$

where λ_c is the V-dependent collector screening length and u = 3.3 nm is the thickness of the undoped GaAs layer near the barriers. Since λ_c increases with bias whereas 3/a decreases, the proportion of the applied bias dropped across the emitter decreases as the voltage increases, so changing the dependence of V_p on B. Both of these factors tend to make the peak move more rapidly to higher bias than expected from equation (2). In particular, the increase in k_F with bias leads to the resonance being asymmetric about the peak, as observed, since the Fermi wavevector at cut-off is greater than that at threshold. Using this model, we can predict the variation with field of the total applied voltages (V) corresponding to $V_{e,th}$, $V_{e,p}$ and $V_{e,v}$. It demonstrates the usefulness of the effective mass-WKB approximation for treating the resonant tunneling of conduction electrons.[10]

To summarize this section the quadratic shift in voltage of the electron resonance with increasing transverse magnetic field arises from the second term in equation (2) above, in which k_{y0} corresponds to the in-plane momentum, $k_\parallel$, acquired from the action of the magnetic field. The quadratic shift reflects the parabolic nature of the electron dispersion relation $\varepsilon(k_\parallel)$ in the quantum well.

4. HOLE RESONANT TUNNELING IN A TRANSVERSE MAGNETIC FIELD

The valence band bending diagram of the structures is shown schematically in Figure 4 for an applied voltage (V), which gives rise to a hole accumulation layer adjacent to the emitter barrier. A two dimensional hole gas (2DHG) forms in the accumulation layer of the emitter with a quasi-bound hole state energy, ε_0. The quasi-bound hole states in the quantum well are given their conventional notation where LH and

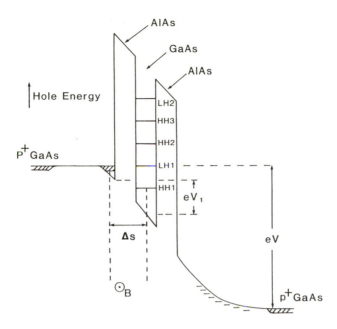

FIG. 4. Schematic diagram of p-type resonant tunneling device under an applied voltage V. Note that a two-dimensional hole gas forms in the accumulation layer adjacent to the emitter barrier. The applied magnetic field B causes the holes to acquire an additional in-plane momentum as shown schematically in the bottom part of the figure.

HH refer to light and heavy holes respectively. The states can be defined in this way only for $k_\parallel = 0$, since in this case there is no coupling between the light and heavy holes.

Figure 5 shows the I(V) characteristics obtained for structure NU448 at 4 K. Six distinct peaks are observed in I(V) over the measured voltage range up to 2.4 V. The figure shows the variation of the I(V) characteristics as a function of magnetic field applied parallel to the plane of the barriers. The devices were examined in Nottingham up to fields of 12 T, and in Grenoble up to fields of 26 T, using the CNRS-MPI hybrid magnet system. It can be seen that the voltage positions of the peaks and thresholds of the resonances shift with increasing magnetic field, but at different rates and in different directions. The voltage positions of the peaks in I(V) are plotted in Figure 6(a) for structure NU448 (4.2 nm well). Figure 7 shows a similar plot for structure NU490 (6.8 nm well). These curves bear a remarkable resemblance to the ε versus $k_\parallel$ dispersion curves of holes in quantum wells.

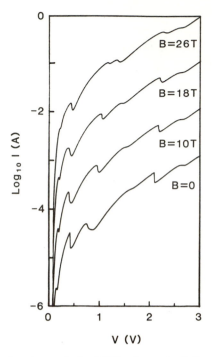

FIG. 5. Logarithmic plots of the I(V) characteristics of NU448 (mesa diameter 100 μm) at various in-plane magnetic field values, T = 4 K. The logarithmic vertical axis gives the current for B = 0. For clarity, the curves for 10, 18 and 26 T are displaced upwards by 1, 2 and 3 orders of magnitude respectively.

As can be seen by reference to Figure 3 and equations 2 and 4 above, the applied magnetic field allows us to scan the occupied hole states in the 2DHG of the emitter accumulation layer through the $\varepsilon(k_\parallel)$ dispersion curves of the quantum well. The V versus B plots in Figures 6(a) and 7 therefore map out the anomalous dispersion curves of the hole states in the quantum well.

From equation 4 we can estimate the $k_\parallel$ gained at a particular value of B. The corresponding $k_\parallel$ values are given on the upper axes of Figures 6(a) and 7 with the appropriate barrier and well widths for NU448. Using the Luttinger parameters for the valence band of GaAs,[16] we estimate a hole quantum stand-off distance of $3/a$ = 5 nm. It is worth stressing that equation 4 is the key to our analysis and corresponds to the value of in-plane momentum $\hbar k_\parallel$ which a tunneling hole acquires under the action of the vector potential (classically the Lorentz force) in its motion from emitter accumulation layer to quantum well.

Note that the energy of the tunneling hole in the quantum well is considerably less than the total applied voltage and is given by $\varepsilon_{hw} = eE(3/a + b_1 + w/2)$ where E is the electric field in the region of the emitter barrier (b_1 is the thickness of the tunnel barrier).

A particularly interesting feature of Figure 6(a) is the weak dispersion of the HH1 and LH1 curves, due to their strong admixing with increasing $k_\parallel$. In contrast, the HH2 and HH3 curves are strongly dispersed in opposite senses, approaching each other at low B (low $k_\parallel$) and starting to repel each other at B $\simeq$ 26 T ($k_\parallel \simeq 1.5\pi$ x 10^6 cm^{-1}). The curvature of the HH3 state corresponds to negative hole effective mass. Such a repulsive interaction is expected[5,17] at this value of $k_\parallel$ ($\simeq \pi/w$) for a well width of 4.2 nm. Note from Figure 5 that the strengths of

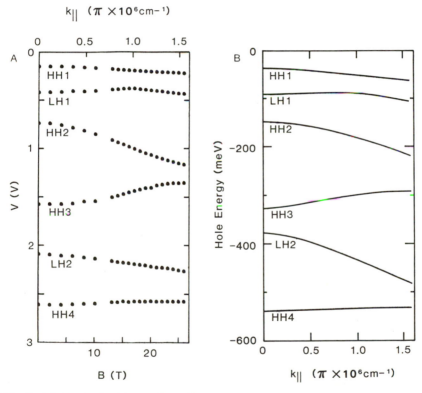

FIG. 6. (a) Variation in the voltage position of the peaks in the I(V) as a function of in-plane magnetic field for structure NU448 (4.2 nm well). The value of $k_\parallel$ is estimated using the procedure described in the text. (b) Calculated hole dispersion curves for an isolated AlAs/GaAs/AlAs quantum well of width 4.2 nm in zero electric and magnetic field.

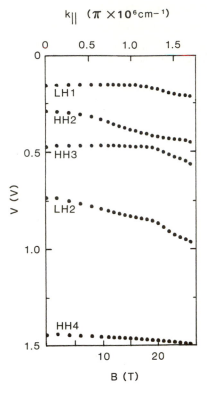

FIG. 7. Variation in the voltage position of the peaks in I(V) as a function of in-plane magnetic field for structure NU490 (6.8 nm well). The value of $k_\parallel$ is estimated using the procedure described in the text.

these two resonances in I(V) become comparable as the admixing increases. A similar repulsion between HH2 and HH3 is shown in Figure 7 for NU490. In this case, the repulsion occurs at lower values of B ($k_\parallel$) due to the wider well width (w = 6.8 nm for this structure). For NU490, the first HH1 resonance gives rise to too small a current to observe in the DC I(V) characteristics.

For a qualitative assessment of the data shown in Figure 6(a) we have derived the hole dispersion relations for an isolated AlAs/GaAs/AlAs quantum well of width 4.2 nm in zero electric and magnetic fields using a four component envelope function formalism.[18] This is shown in Figure 6(b). The calculated hole energies of the HH1, LH1, HH2 and HH3 states at $k_\parallel$ = 0 correspond closely to the relative voltage values of the first four resonance peaks in the I(V) curve at B = 0 for NU448. Furthermore, the theoretical results reproduce accurately the main features of the experimental curves, in particular the weak dispersion of the HH1, LH1 and HH4 resonances and the much stronger dispersion of the HH2 and HH3 resonances. The only major qualitative difference between Figures 6(a) and (b) is the much larger dispersion obtained from theory for the LH2 resonance.

This remarkable correspondence shows that the applied magnetic field allows us to observe directly the strongly non-parabolic dispersion curves of holes confined in a quantum well potential. As Wessel and Altarelli[6] have shown, the dispersion curves for an isolated hole quantum well do not differ appreciably from those for a well with finite barriers provided the barriers are of sufficient width, as is the case here.

For a more quantitative comparison with experiment a full quantum mechanical treatment of hole resonant tunneling in the presence of a magnetic field applied parallel to the barrier is required. This would need to consider the effect of the large electric field ($> 10^7$ V/m) on the hole states in the quantum well, and to describe the confined holes in the emitter accumulation layer. At fields B > 20 T, a non-perturbative approach is necessary since the magnetic length $(\hbar/eB)^{1/2}$ is comparable with the dimensions of electrical confinement. In addition, the finite spread of in-plane wave-vector ($-k_F < k_y < k_F$) in the emitter accumulation layer should be taken into account.

Finally, for a quantitative comparison between theory and the measured I(V) curves, it is necessary to model correctly the distribution of space charge throughout the device.

We have investigated the anisotropy of the resonances when the magnetic field is rotated in the plane of the barrier. Only very small differences are observed for B parallel to the [100] and [110] axes, indicating that anisotropy effects are relatively unimportant in the structures investigated here. In addition, we have not found any evidence for spin splitting of the hole state due to applied magnetic and electric fields.

Note that for sample NU448, the LH2 and HH4 resonances occur at high voltages, corresponding to resonant injection of holes near the top of the collector barrier.

This work on resonant magnetotunneling of holes is published in detail in ref. 19.

5. CONCLUSION

We have investigated resonant tunneling of electrons and holes in double barrier structures in the presence of a magnetic field applied parallel to the plane of the barriers. The magnetic field dependence of the electron resonances can be described using the effective mass approximation. Our measurements on the p-type structures have shown that the application of a magnetic field parallel to the plane of the barriers provides a new tunneling spectroscopy technique for directly investigating the complicated in-plane dispersion curves of holes in quantum well heterostructures.

ACKNOWLEDGEMENTS

I gratefully acknowledge the stimulating collaboration of my colleagues who have worked with me on the problem of resonant tunneling in a magnetic field - in Nottingham: Mr. R. K. Hayden, Dr. M. L. Leadbeater, Dr. D. K. Maude, Dr. E. C. Valadares, Dr. M. Henini and Dr. O. H. Hughes; in Grenoble - Prof. J. C. Portal and Mr. L. Cury. I also acknowledge invaluable discussions with Mr. R. Wessel, Professor M. Altarelli, Dr. F. W. Sheard and Dr. G. A. Toombs.

This work is supported by SERC, CNRS and the European Community.

REFERENCES

1. L. L. Chang, L. Esaki and R. Tsu, Appl. Phys. Lett. **24**, 593 (1974). See also R. Tsu and L. Esaki, Appl. Phys. Lett. **22**, 562 (1973). For a brief overview of the early history of this subject see L. Esaki, Proc. 17th Int. Conf. on the Physics of Semiconductors, San Francisco 1984, eds. J. D. Chadi and W. A. Harrison, published by Springer-Verlag, p. 473 (1985).

2. T. C. L. G. Sollner, E. R. Brown, W. D. Goodhue and H. Q. Le, "Microwave and Millimeter-Wave Resonant-Tunneling Devices," Chapter 6, in Physics of Quantum Electron Devices, ed. F. Capasso, p. 147 (1990).

3. E. E. Mendez, Proc. of NATO ASI on Physics and Applications of Quantum Wells and Superlattices, Erice, Sicily, 1987, eds. E. E. Mendez and K. von Klitzing (publ. by Plenum) vol. 170, p. 159 (1987); see also L. Eaves, F. W. Sheard and G. A. Toombs, "The

Investigation of Single and Double Barrier (Resonant Tunneling) Heterostructures Using High Magnetic Fields," Chapter 2, Physics of Quantum Electron Devices, ed. F. Capasso, p. 107 (1990); and L. Eaves, Analogies in Electronics and Optics ed. W. van Haeringen and D. Lenstra (publ. by Kluwer Academic), p. 227 (1990).

4. J. M. Luttinger and W. Kohn, Phys. Rev. **97**, 869 (1955). See also J. M. Luttinger, Phys. Rev. **102**, 1030 (1956).

5. G. Bastard, Wave Mechanics Applied to Semiconductor Heterostructures, Les Éditions de Physique (1988).

6. R. Wessel and M. Altarelli, Phys. Rev. **B39**, 12802 (1989).

7. E. E. Mendez, W. I. Wang, B. Ricco and L. Esaki, Appl. Phys. Lett. **47**, 415 (1985).

8. M. L. Leadbeater, L. Eaves, P. E. Simmonds, G. A. Toombs, F. W. Sheard, P. A. Claxton, G. Hill and M. A. Pate, Solid State Electronics **31**, 707 (1988).

9. E. S. Alves, M. L. Leadbeater, L. Eaves, M. Henini, O. H. Hughes, A. Celeste, J. C. Portal, G. Hill and M. A. Pate, Superlattices and Microstructures **5**, 527 (1989).

10. M. L. Leadbeater, E. S. Alves, L. Eaves, M. Henini, O. H. Hughes, A. Celeste, J. C. Portal, G. Hill and M. A. Pate, J. Phys. Condens. Matter **1**, 4865 (1989).

11. E. S. Alves, L. Eaves, M. Henini, O. H. Hughes, M. L. Leadbeater, F. W. Sheard, G. A. Toombs, G. Hill and M. A. Pate, Electronics Lett. **24**, 1190 (1988).

12. M. L. Leadbeater, E. S. Alves, L. Eaves, M. Henini, O. H. Hughes, A. Celeste, J. C. Portal, G. Hill and M. A. Pate, Phys. Rev. **B39**, 3438 (1989).

13. L. Eaves, K. W. H. Stevens and F. W. Sheard, The Physics and Fabrication of Microstructures and Microdevices, Springer Proceedings in Physics **13** (ed. M. J. Kelly and C. Weisbuch) (1986).

14. R. A. Davies, D. J. Newson, T. G. Powell, M. J. Kelly and H. W. Myron, Semicond. Sci. Technol. **2**, 61 (1987).

15. M. L. Leadbeater, E. S. Alves, F. W. Sheard, L. Eaves, M. Henini, O. H. Hughes and G. A. Toombs, J. Phys. Condens. Matter **1**, 10605 (1989).

16. Landolt-Börstein Numerical Data and Functional Relationships in Science and Technology, ed. V. Madelung (Springer, Berlin, 1982). Group III, Band 17.

17. R. Wessel, Diploma Thesis at Technische Universität Munchen, 1989.

18. L. C. Andreani, A. Pasquarello and F. Bassani, Phys. Rev. **B36**, 5887 (1987).

19. R. K. Hayden, D. K. Maude, L. Eaves, E. C. Valadares, M. Henini, F. W. Sheard, O. H. Hughes, J. C. Portal and L. Cury, Phys. Rev. Lett. **66**, 1749 (1991).

NONLINEAR OPTICS AND OPTOELECTRONICS IN QUASI TWO DIMENSIONAL SEMICONDUCTOR STRUCTURES

D. S. Chemla

Lawrence Berkeley Laboratory, University of California
Berkeley, California 94720

Abstract - In this article we review the most salient results obtained in the 80's in the fields of nonlinear optics and optoelectronics of semiconductor quantum well structures, with a particular emphasis on the properties which are specific to the low dimensionality of elementary excitations.

1. INTRODUCTION

Optical transitions near the band gap of semiconductors are dominated by excitonic effects. The coulomb correlation between photo-generated electron-hole pairs produce in the absorption spectrum a series of peaks below the gap and a strong enhancement of the optical transitions above the gap. These correspond respectively to the bound and scattering states of excitons. The exciton Bohr radius, a_0, and Rydberg, R_y, provide natural length and energy scales for the processes which are involved in the operation of many important optoelectronic devices such as lasers, detectors, modulators, photonic switches, etc. It is not often

Highlights in Condensed Matter Physics and Future Prospects
Edited by L. Esaki, Plenum Press, New York, 1991

recognized that excitonic effects are not only very large, they also extend over a wide range of energy, about fifty R_y, on each side of the gap.

In the most important direct gap III-V semiconductors: $a_0 \approx 50 \to 500\text{Å}$, corresponding to: $R_y \approx 15 \to 2\text{meV}$. Hence in ultrathin semiconductor layers, such as quantum wells (QW) and superlattices, which thickness L is comparable to a_0 quantum size effects modify significantly the excitonic structure as compared to that of the bulk parent materials. The confinement of electrons and holes in the low gap layers induces discrete levels for the motion perpendicular to the layers and strongly enhance their correlation producing several series of exciton resonances which are visible at and even above room temperature, as shown in Figure 1a. In our lecture we review nonlinear optics and optoelectronics in these quasi two dimensional (2D) excitons (Schmitt-Rink, Chemla, Miller 1989). We concentrate on the features which are the most significantly affected by 2D-confinement. In particular, large

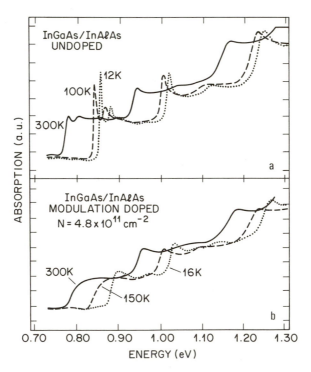

FIG. 1. Absorption spectra of an undoped (a) and an n-type (N = 4.8 x 10^{11}cm^{-2}) modulation-doped (b) InGaAs/InAlAs multiple quantum well structure (L = 100Å), as a function of temperature (from Livescu et al. 1988).

densities, $N \approx a_0^{-3}$, of excitons (X), electrons (e) and/or holes (h) can be introduced in QW-structures by photoexcitation, by current injection or by field induced displacement of carriers. These populations strongly modify excitonic resonances resulting in large changes in the optical absorption and refraction. In turn these effects can be used to investigate the physics of quasi 2D-excitons or can be exploited in device applications. Furthermore, because of the anisotropy of ultrathin layered structures the application of electric field produce electroabsorption with very distinct character depending on the direction of the electric field perpendicular or parallel to the layers. For parallel fields, the response is similar to that of the bulk, essentially a field ionization. For perpendicular field, however, the charges are squeezed against the QW walls but ionization is hindered. This translates by large shift of the absorption edge with distinct exciton features up to field magnitude much larger than the 3D ionization field. This effect is known as the Quantum Stark Effects (QCSE) and has found very important optoelectronic applications which we shall review briefly.

2. NONLINEAR OPTICAL PROPERTIES OF QUANTUM WELLS

The nature of the nonlinear optical response of semiconductors is different depending on whether virtual or real e-h populations are generated. Experimentally, virtual e-h pair effects are produced by excitation well below the absorption edge and the e-h plasma effects by excitation above it. At room temperature, and in III-V compounds, exciton effects are more difficult to select; not only resonant excitation must be used, but they must be observed in times short compared to the exciton ionization time.

In earlier experiments on low-temperature bulk GaAs, it was found that excitons produce much smaller nonlinear optical effects than an e-h plasma (Fehrenbach et al. 1982). Guided by the notion that the saturation of 3D excitons is due to screening, this was attributed to the fact at that screening in a dielectric exciton gas is much weaker than in a metallic e-h plasma. An exciton gas can be generated selectively by resonant excitation. At temperatures larger than the binding energy, they are unstable against thermal ionization. In the dilute limit, this is due to collisions with thermal LO phonons, and the ionization time can be estimated from the homogeneous absorption linewidth. In typical III-V QW's and at room temperature, it is found to be a fraction of a ps. Hence in a resonant excitation experiment one expects to see a rather small reduction of the excitons absorption as long as the e and h form bound states and then a stronger one as the excitons ionize into a e-h plasma. Experiments aimed at observing this effect lead to unexpected results. The effects of photocarriers are usually determined by measuring the change in sample transmission of a weak laser-probe beam, induced by a strong pump laser beam. In the small signal regime, the so-called differential transmission spectrum (DTS),

$$\frac{\Delta T}{T} = \frac{T(\text{pump on}) - T(\text{pump off})}{T(\text{pump off})} \tag{1}$$

reproduces the changes in the absorption of the sample, $\Delta T/T \sim -\Delta\alpha(\omega)$ x ℓ. Figure 2, shows room temperature DTS of a GaAs/AlGaAs multiple quantum well structure DTS of a GaAs/AlGaAs multiple quantum well structure (L = 96Å), measured with broad-band 50 femtoseconds probe pulses at 50 femtoseconds intervals before and after resonant generation of $n_z = 1$ heavy-hole excitons by a narrow-band 100 femtoseconds pump pulse (Knox et al. 1985). Initially, a strong bleaching of the $n_z = 1$

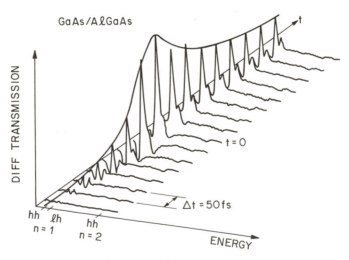

FIG. 2. Room temperature differential transmission spectra of a GaAs/AlGaAs multiple quantum well structure (L = 96Å), measured with broad-band 50 femtoseconds probe pulses at 50 femtoseconds intervals before and after excitation resonant with the $n_z = 1$ heavy-hole exciton by a narrow-band 100 femtoseconds pump pulse (from Knox et al. 1985).

heavy-hole exciton absorption lasting less than a ps occurs, then the absorption recovers partly and settles at the same value as it would under cw excitation. The dynamics of the $n_z = 1$ heavy-hole exciton bleaching can be described assuming the instantaneous generation of $n_z = 1$ heavy-hole excitons which, because of collisions with LO phonons, transform in ≈ 300 femtoseconds into free $n_z = 1$ e and h, which then live for long times. The transient excitons produce, before they ionize, a strong bleaching of the $n_z = 1$ heavy-hole exciton resonance about twice as large as that due to the free e and h at long times, while the changes at the $n_z = 2$ heavy-hole exciton resonance are again small. In addition, the lineshape of the differential spectra at short and long times is different (Chemla and Miller 1985).

This result is in seeming contradiction to commonly accepted ideas and deserves some discussion. This experiment presents two novel aspects, namely i) reduced dimensionality and ii) ultrashort optical pulse excitation at room temperature, both of which are important in the interpretation (Schmitt-Rink et al. 1985). First, it shows that the efficiency of free-carrier screening is reduced in 2D, which is consequently even more true in the case of excitons. It also indicates that the excitons do not behave as perfect bosons, they are sensitive to the underlying Fermi statistics of their constituents. The saturation of the exciton resonance is simply due to the consequences of the Pauli principle: Phase Space Filling (PSF or the fact that the area already occupied by an exciton cannot sustain more excitons) and Exchange (the modification of Coulomb interaction at very short distance due to exclusion). Between the absorption event and the first collision with a LO phonon, the resonantly excited excitons have not yet had time to thermalize and therefore, for this very short time of ≈ 300 femtoseconds, they are essentially at zero temperature. Then, because the LO phonon energy is much larger than the exciton binding energy, the first collision with a LO phonon ionizes the excitons and produces free e and h of rather large thermal energy. At long times, these thermally activated free e and h are essentially at room temperature and hence the experiment compares the effects of "cold" excitons (at short time) to those of a "warm" room temperature e-h plasma of the same density (at long times). The exclusion principle effects due to the "warm" e-h plasma are smaller than those due to the "cold" excitons, because they decrease as the ratio between the thermal wavelength to the Bohr radius decreases. This accounts then for the different bleaching of the $n_z = 1$ heavy-hole exciton resonance at short and long times.

In order to test the conjecture that in 2D the effects of screening are weak as compared to the effects of the exclusion principle, a direct experimental comparison of both effects at room temperature has been performed (Knox et al. 1986). The principle of these investigations is the following. Using femtosecond laser pulses it is possible to generate nonequilibrium carrier distributions in the continuum between the $n_z = 1$ and 2 exciton resonances and to observe their effects on the absorption spectrum before and during their relaxation. If the e and h excess energies are less than the LO phonon energy, exchange of energy with the lattice will take a rather long time and the carriers will first thermalize among themselves before reaching an equilibrium with the lattice. Immediately after excitation, the carriers do not occupy states out of which the $n_z = 1$ and 2 excitons are constructed and thus the Pauli principle effects are not effective, whereas screening sets in right away. As the carriers equilibrate among themselves, their distribution evolves toward a thermal one and they start to fill up the states at the bottom of the $n_z = 1$ subbands. Therefore, they produce PSF on the corresponding excitons. Only the high-energy tails of the thermalized distributions will extend up to the bottom of the $n_z = 2$ subbands, so that for these PSF remains basically ineffective. Therefore, the evolution of the absorption at the $n_z = 1$ excitons during thermalization will first show the effects of screening and then how PSF is turned on, whereas at the $n_z = 2$ heavy-hole exciton resonance there should be little (if any) change during thermalization. Then, on a much longer time scale, the carriers will equilibrate with the lattice and eventually recombine.

This process is shown in Figure 3, where room temperature DTS of a GaAs/AlGaAs multiple quantum well structure are presented. The spectra were measured with broadband 50 femtoseconds probe pulses at 50 femtoseconds intervals from 100 femtoseconds after arrival of a narrow band 100 femtoseconds pump pulse. The spectral distribution of the pump photons is shown at the bottom of the figure; the pump pulse produces approximately $N \approx 2 \times 10^{10} \text{cm}^{-2}$ e-h pairs. Immediately upon its arrival, one sees absorption changes at the $n_z = 1$ and 2 excitons as well as spectral hole burning in the continuum, corresponding to state filling by the nonequilibrium carrier distributions. The $n_z = 1$ excitons loose some oscillator strength and broaden slightly, whereas the $n_z = 2$ heavy-hole exciton experiences also a very small red shift due to the weak intersubband band gap renormalization (BGR). As the carriers thermalize, the spectral hole in the continuum changes shape and shifts to lower energies. Accompanying this relaxation, the bleaching of the n_z

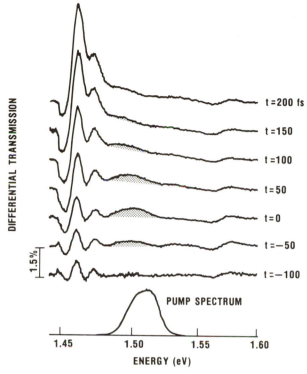

FIG. 3. Room temperature differential transmission spectra of a GaAs/AlGaAs multiple quantum well structure (L = 96Å), measured with broad-band 50 femtoseconds probe pulses at 50 femtoseconds intervals before and after excitation of a nonequilibrium e and h distributions by a narrow-band 100 femtoseconds pump pulse (from Knox et al. 1986).

= 1 resonances increases and finally settles after 200 femtoseconds at the same level as in the case of thermalized e-h plasma generation discussed previously. The changes at the $n_z = 2$ heavy-hole exciton resonance remain essentially the same, showing that the screening has not changed significantly during the thermalization of the carriers. The comparison of the $n_z = 1$ heavy-hole exciton bleaching at $\tau = 0$ and after $\tau = 200$ femtoseconds shows that the effects of the exclusion principle are several times larger than those of screening, confirming the reduced efficiency of screening in 2D. In addition, the data also contain a wealth of information about carrier relaxation in microstructures that is not directly relevant to our discussion (Knox et al. 1988).

The composite structure of exciton produces exciton-exciton interaction which has been observed by optical techniques. For example, the short-range hard core repulsion due to Pauli exclusion makes it more difficult to create excitons in the presence of other excitons and hence should produce a blue shift of the resonance at high density. In 3D exciton resonances hardly change their energy as the density of excitons increases because of an almost exact cancellation of the hard core blue shift by a red shift due to screening. And indeed in bulk GaAs only a very small shift (Schultheis et al. 1986) or no shift at all (Fehrenbach et al. 1982, 1985) have been reported. In 2D QW's, because of the quenching of screening, the blue shift due to Pauli exclusion is no longer balanced and thus is measurable (Schmitt-Rink et al. 1985). Measurements performed on QW's of various thicknesses have shown this blue shift when excitons are selectively generated or, at low temperatures, when they are formed after a few ps from photoexcited free e and h (Peyghambarian et al. 1984, Hulin et al. 1986, Masumoto et al. 1986). The shift is rather difficult to observe in thick layers, but is large enough to be observed in narrow QW's (<100Å), where the dimensionality approaches the pure 2D limit. A typical example is presented in Figure 4 which shows transmission spectra of a GaAs/AlGaAs multiple quantum well structure (L = 50Å) at T = 15k, measured with a broad-band 120 femtoseconds probe pulses before and after resonant $n_z = 1$ heavy-hole

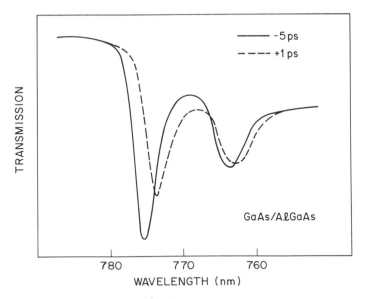

FIG. 4. Transmission of a GaAs/AlGaAs multiple quantum well structure (L = 50Å) at T = 15K, measured with broad-band 120 femtoseconds probe pulses before and after resonant nxubz = 1 heavy-hole excitation by a narrow-band 120 femtoseconds pump pulse (from Peyghambarian et al. 1984).

exciton excitation by a narrow-band 120 femtoseconds pump pulse (Peyghambarian et al. 1984).

An even more dramatic effect of exciton-exciton interaction on the nonlinear optical properties of QW have been observed recently in time-resolved Four-Wave-Mixing (FWM). FWM is a powerful and elegant method to study the dephasing process in the optical response. In this technique two pulsed laser beams with wavevectors k_1 and k_2 interfere in a sample to produce a diffracted beam in the direction $k_3 = 2 \times k_2 - k_1$. The magnitude of the diffracted signal in the direction k_3 is then recorded as a function of the time-delay, $T = t_2 - t_1$, between a pulse of beam-2 and a pulse of beam-1. The intensity of the signal exhibit usually a step like onset, reproducing that of the laser pulses, and an exponential tail. The results are interpreted in terms of the two level-model of Yajima and Taira (1979) to determine from the exponential decay the time, T_2, during which the system has not experienced incoherent scattering. It was found in high excitation femtoseconds FWM signal qualitatively disagrees with this behavior. In the case of GaAs/AlGaAs-QW, as seen in Figure 5, the signal indeed decay as e^{-2T/T_2} for positive time delay, $T > 0$, but for negative time delay, $T < 0$,

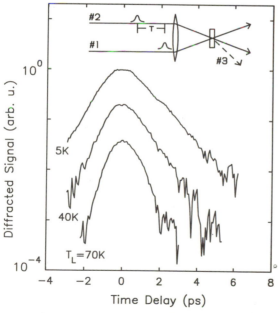

FIG. 5. Degenerate four wave mixing signal vs time delay for a L = 170Å GaAs/GaAlAs quantum well sample for three lattice temperatures. Inset: schematic of the experimental configuration (from Leo et al. 1990).

(a)

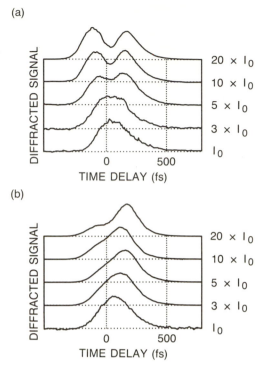

(b)

FIG. 6. Degenerate four wave mixing signal vs time delay for a L = 200Å InGaAs/InAlAs sample for increasing intensity $I_o \approx 3$ MW cm^{-2}), (a): 10meV, (b): 6meV detuning (from Wegener et al. 1990).

it raises as e^{-2T/T_2} ! (Leo et al. 1990). In the case of InGaAs/InAlAs-QW, Figure 6 shows the FWM signal for two detunings (10meV & 6meV), as the intensity of both beams is scaled up from $I_o = 3$MWcm^{-2} to 20xI_o. At the highest intensities the exciton density is $N_x \approx 10^{11}$cm^{-2}, comparable to the saturation density of the exciton. As the incident intensity is increased the strength of the signal for negative time delays increases. Then the single maximum time-profile at low intensities gradually evolves into a lineshape that exhibits two distinct maxima in time for the 10meV detuning case, and a strong asymmetry for the 6meV detuning case (Wegener et al. 1990).

In crystals, because of translation symmetry, the energy bands are quasi-continua of e and h levels corresponding to the Bloch wavevector k. Because the momenta of optical photons is negligible, optical transitions are vertical i.e. the light connects e and h levels with the same wavevector k. The coupling to the light field, E, is measured by the Rabi frequency $\mu_k E$ where μ_k is the interband matrix element. In semiconductors, however, the Coulomb interaction, $V_{k,k'}$, which is responsible for excitonic effects, couples the various levels in k-space. Hence in the presence of a strong laser field the optically connected e-h levels do not experience the applied field, rather they see the self-consistent "local field" which is the sum of the applied field and of the "molecular" field, due to all the other e-h pair amplitudes at different wavevector k' (Schmitt-Rink and Chemla 1986, Schmitt-Rink et al. 1988). The coupling with the local field at k is thus a function of the pair amplitude $\Gamma_{k'}$ at all the other wavevectors k'. It is measured by the renormalized Rabi frequency, $\Delta_k = \mu_k E + \sum_{k'} V_{k,k'} \Psi_{k'}$,. For ultrashort laser pulses, $E \sim \delta(t)$. The first term of Δ_k varies also as $\delta(t)$, hence in the absence of the Coulomb force the light can interact with the grating only when the later is formed i.e. at positive time delay $T > 0$. The second term of Δ_k, on the contrary, exhibits a step like raise and an exponential decay $\sim \exp(-t/T_2)$. The corresponding polarization, $P = \sum_k \mu_k^* \Psi_k$ can and does interact with the polarization grating for $T > 0$ as well as for $T < 0$. The negative time delay signal is thus due to the Coulomb mediated interaction between the levels in k-space. The dip seen at very high density originate from the saturation of the excitonic transition when the maximum of the two pulses overlap in the sample. The theoretical description of the general case of finite pulse duration and high density can be solved non-perturbatively only by numerical methods (Wegener et al. 1990). Figure 7 shows the theoretical results for the diffracted signal, calculated for Gaussian pump and probe pulses of 110 femtoseconds duration tuned 3 Ry below the exciton resonance. The

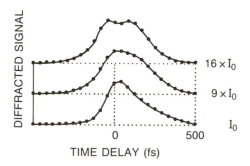

FIG. 7. Calculated Degenerate four wave mixing signal vs time delay for three different intensities. The picture is result of a numerical solution to all order of the coherently interacting Calculated absorption spectra for pure 2D excitons for 10 Ry detuning and increasing pump intensities (from Ell et al. 1988).

material parameters are those of bulk GaAs and a dephasing time of $T_2 = 0.4Ry^{-1}$ was assumed. Results for three different pump intensities are presented, corresponding to peak Rabi frequencies of 0.2, 0.6 and 0.8 Ry. They agree qualitatively with the experimental data. At low intensity (I_0), the time-integrated signal decays as expected with a time constant $T_2/2$ for positive time delay. For negative time delay, however, it exhibits a rising wing with a $T_2/4$ time constant. As the excitation intensity increases, the negative time delay signal becomes more pronounced (9 × I_0), and a temporal profile with two distinct maxima emerges (16 × I_0) as seen in the case of InGaAs/InAlAs-QWs. The numerical solutions confirm that the unusual FWM time profile is due to the Coulomb interaction which mediates a nonlinear coupling of populations and pair amplitudes.

To observe coherent light-matter interactions in semiconductors, such as the renormalization of optical transitions in an intense laser field (Optical or AC Stark effect) it is necessary to use ultrashort laser pulses and nonresonant excitation of virtual excitons below the fundamental absorption edge, where the dephasing rates are exponentially small. Because of the extended nature and strong interactions of the elementary excitations in semiconductors, the excitonic AC Stark effect (Mysyrowicz et al. 1986, Von Lehmen et al. 1986) is more complex than the AC Stark effect seen in atomic systems (Liao and Bjorkholm 1975, We et al. 1977). The theory reveals profound similarities between coherent virtual excitons and Bose condensed gases or superconductors (Schmitt-Rink and Chemla 1986, Schmitt-Rink et al. 1988, R. Zimmermann 1988, Zimmerman and Hartmann 1988). Following the early work this effect has attracted much attention (for a recent review see Chemla et al. 1989). For significantly nonresonant excitation i.e., when the detuning and the Rabi frequency are larger than the dephasing / collision rates it is possible to neglect collisions altogether and derive a closed kinetic equation for the density matrix based on the Hartree-Fock (HF) approximation (Schmitt-Rink and Chemla 1986, Schmitt-Rink et al. 1988). The HF theory treats on equal footing the effects of the laser field and the Coulomb interaction resulting, for a simple two-band semiconductor, in the renormalization of the e and h energies, $\varepsilon_k \rightarrow \varepsilon_k - \sum_k V_{k,k'} n_{k'}$, and the Rabi frequency, $\mu_k E \rightarrow \Delta_k$. Again, both changes express the fact that in the presence of Coulomb interactions an e-h pair with a given k does not only experience the external field but also a significant internal one associated with e-h pairs created at k'. For a monochromatic pump field the full kinetic equations have been solved

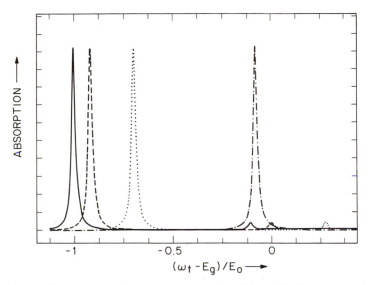

FIG. 8. Absorption spectra in arbitrary units: The full lines give the unperturbed exciton absorption in 2D, while the dashed and dotted lines correspond to increasing pump intensities (from Ell et al. 1988, 1989).

numerically (Ell et al. 1988, 1989, Schaefer 1988 and Schaefer et al. 1988). Some of the results of Ell at al. are shown in Figure 8. The full lines give the unperturbed exciton absorption in 2D, while the dashed and dotted lines correspond to increasing pump intensities. Strong AC Stark shifts of the 1s exciton are found, not unlike those of a two-level atom. However, in striking contrast to two-level atoms, the 1s exciton oscillator strength is nearly constant. The origin of this strange behavior is that AC Stark shift of the band gap is always larger than that of the 1s exciton, because of the larger spatial extent of scattering states. This increased AC Stark shift of the band gap corresponds to an effective increase of the exciton binding energy and thus of the oscillator strength, which cancels out the decrease due to PSF. The pure shift of exciton was actually observed experimentally by comparing the DTS to the derivative of the linear absorption (Knox et al. 1989). In the limit of pure AC Stark shift, the two lineshapes should be identical as shown in Figure 9b, where the low excitation DTS (heavy solid line) of a L = 74Å GaAs-QW sample excited $\approx$ 50 meV below the exciton at low intensity (Rabi frequency < Ry) is compared to the derivative (light solid line) of the linear absorption Figure 9a. Within the experimental accuracy the two profiles are almost identical, in agreement with the theory.

305

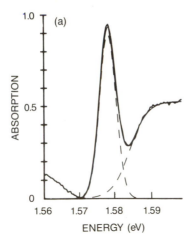
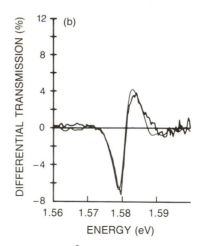

FIG. 9. (a) Linear absorption spectrum of a L = 74Å GaAs/GaAlAs quantum well sample in the vicinity of the $n_z = 1$ exciton resonance. (b) Differential transmission spectrum of the sample pumped at low intensity 50meV below below the resonance (heavy solid line) compared to the derivative of the linear absorption spectrum (light solid line) (from Knox et al. 1989).

When the pump intensity is very large (Rabi frequency $\gg$ Ry) both the DTS profile and dynamics are strongly modified (Knox et al. 1989). This is shown in Figure 10, where the full time course of the DTS is presented. At negative delays i.e., when the probe precedes the pump, an oscillating DTS is seen, in agreement with theory and other measurements on molecules (Brito Cruz et al. 1986) and semiconductors (Joffre et al. 1988, Fluegel et al. 1987 and 1988). Around T = 0 a rapid initial transient with a duration close to that of the pump-probe cross correlation is observed, both at the heavy-hole and light-hole excitons. The profile of this initial transient corresponds to a combination of both shift and bleaching. The DTS, however, does not recover after the pump pulse is over, but stabilizes to a long-lived component (many ps), the magnitude of which is only about two times smaller than that of the maximum at T = 0. The lineshape of the DTS evolves as well, as shown in the inset of Figure 10. Curve (a) shows the DTS around the 1s heavy-hole exciton resonance at delay T = 264 femtoseconds. The AC Stark effect does not contribute to the signal at this delay, since the pump pulse is only 100 fs long. The DTS profile is remarkably symmetric, in contrast to the DTS profile at T = 0. Curve (b) shows the integrated DTS around the 1s heavy-hole exciton resonance at T = 264 femtoseconds. The DTS integrated over the exciton line is nearly zero, which corre-

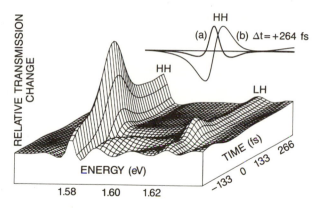

FIG. 10. Time evolution of the differential transmission spectrum of the same sample as in Fig. 9, pumped 80meV below the resonance at high intensity ($I_p \approx 3$GWcm^{-2}). Inset (a) Differential transmission profile near the resonance at large time delay (T > 264 femtoseconds). (b) Integral of the differential transmission spectrum demonstrating that the area is conserved (from Knox et al. 1989).

sponds to a pure broadening without any loss of oscillator strength. This line shape is reminiscent of that produced by dynamical screening / collisional in a e-h plasma (Chemla et al. 1984). Since the sample is excited just below the gap E_g, it is thus possible to generate an electron-hole plasma by nonresonant two-photon absorption (TPA) at energies close to $2E_g$. Initially, such a plasma does not occupy states at the band edge, because the relaxation of carriers by emission of phonons takes many ps. Nevertheless, it may be able to produce some form of nonequilibrium collisional broadening. To demonstrate the presence and origin of the plasma, the photocurrent in a QW p-i-n diode was studied during an AC Stark effect experiment. A photocurrent with a quadratic dependence on the pump intensity was observed, which unambiguously proves the presence of real carriers in the sample with the right density dependence for TPA generation (Knox et al. 1989). The effects of the nonequilibrium electron-hole plasma on the excitonic absorption change qualitatively with time. Initially, as mentioned above, the carriers excited by TPA have excess energy $2\hbar\omega_p - E_g \sim 1.5$ev, far larger than the confinement potential in the QW (~ 200meV). Therefore, right after their generation, they are free to move in 3D and able to screen the 2D excitons effectively. This prevents the full recovery of the excitonic absorption. At early times, they do not occupy states in the QW, and it takes a few ps for them to relax down to the band edge. They then become confined and their screening of the 2D excitons qualitatively changes. Eventually they occupy the states out of which the 2D excitons are constructed and bleach the exciton line in such a way that the integrated DTS will no longer be zero. Finally, they recombine in a few ns, and the excitonic absorption fully recover. A theoretical description of this very complex dynamics is not available at the present time.

Because of the pronounced differences in optical properties of quasi-2D and 3D semiconductors, there have been numerous attempts to fabricate 1D or 0D systems. Up to this day, however, only very little progress has been reported despite some recent results (Heitmann 1990). The main difficulties reside in the increased importance of surface related defects and of the unavoidable size fluctuations when processing or growing such systems. One elegant way to investigate the physic of the 2D → 0D transition in uniform material with excellent optical and electronic quality, is to apply a strong magnetic field, H, perpendicular to the plane of a QW-structure. The magnetic field imposes on the carriers a quadratic potential $V_{mag} = (eHr)^2/8mc^2$. It confines their motion within areas characterized by the cyclotron radius, $l_c = (c/eH)^{1/2}$. In the absence of Coulomb interaction the e-h eigenstates are the discrete, equally spaced, Landau levels. In actual semiconductors, the Coulomb potential, $V_{coul} = -e^2/\varepsilon_o r$ is active as well and the e-h eigenstates are Magneto-excitons (M-X). The relative strength of the two forces is measured by the dimensionless parameter $\lambda = (\alpha_o/l_c)^2$ (Akimoto and Hasegawa 1967, Shinada and Tanaka 1970, McDonald and Richie 1986, Yang and Sham 1987). It can be shown then for $\lambda \to \infty$ the 2D M-X behaves exactly like non-interacting point Bosons (Lerner and Lozovik 1981, Paquet et al. 1985) which optical properties are those of perfect isolated two-level systems (Stafford et al. 1990). Hence, in principle, the M-X in QW-structures can be tuned continuously from quasi-2D to 0D by varying $\lambda = 0 \to \infty$.

The evolution of the absorption spectrum of a 70Å GaAs QW-structure at low temperature, 4K, and for $\lambda = 0 \to 3$ (H = 0 Tesla → 12 Tesla) is shown in Figure 11 (Stark et al. 1990a). The highest field corresponds to $l_c \approx 70$Å i.e., a regime of strong magnetic confinement. The optically active M-X originate from the X-bound states. The ground state (1s) only experience a small diamagnetic shift as seen on the figure for the 1s heavy-hole and light-hole (Tarucha et al. 1984). The higher bound states ((n > 1) disperse strongly and at high field, exhibit the usual linear dispersion of the Landau levels. At $\lambda = 0$ the heavy-hole (n > 1)s-excited states are located under the light-hole 1s-ground state. In

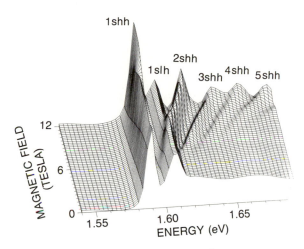

FIG. 11. Linear absorption spectra of a L = 70Å GaAs/AlGaAs quantum well sample versus magnetic field H = 0 → 12Tesla (from Stark et al. 1990a).

this sample, the (n > 1)s heavy-hole states move energetically above the 1s light-hole state for λ > 1.5 (H = 6Tesla), as seen on the figure. The (n >1)s light-hole states are too weak to be resolved. The nonlinear responses of M-X and X are compared in Figure 12 (Stark et al. 1990a). At H= 0 the resonant generation of 1s-X (Figure 12 b) produces a blue shift and a loss of oscillator strength of the resonance as already discussed. At H = 12Tesla (Figure 12 c) only the loss of oscillator strength remains observable, the hard core blue shift has completely disappeared, confirming that the inter M-X interactions have been quenched by the magnetic confinement (Lerner and Lozovik 1981, Stafford et al. 1990, Bauer 1990). The H = 12Tesla spectrum is much richer, since the higher M-X are now visible. They experience significant collisional broadening and small red shifts, due to their interaction with the 1s M-X. At zero time delay, excitation resonant, with the 2s M-X (Figure 12d) evidence a red shift of the 1s and (n>2)s M-X. The signal at the 2s M-X is almost symmetric, indicating a broadening and loss of oscillator strength at constant area. The femtoseconds-dynamic of the absorption spectrum in this situation is presented in Figure 13, where 13 DTS for time delay T = -300 femtoseconds → + 300 femtoseconds are shown. Although the signal at the 2s and (n >2)s M-X rise simultaneously following the integral of the laser pulse, a clear delay in the rise of the signal at the 1s M-X is observed. This corresponds to the kinetics of relaxation of the excited M-X toward the lowest level. One important aspect of the magnetic field is that it can be used to control the carrier relaxation by changing the

309

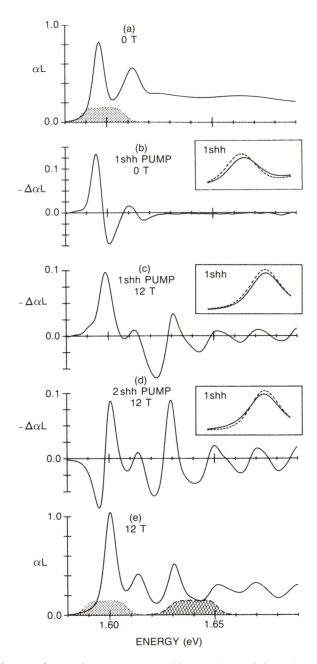

FIG. 12. Linear absorption spectra at H = 0Tesla (a) and H = 12Tesla (e). Differential transmission spectra for excitation resonant with the $n_z = 1$ 1s-heavy hole excitons at H = 0Tesla (b) and H = 12Tesla (c) and for excitation resonant with the $n_z = 1$ 2s-heavy hole excitons at H = 12Tesla (d) (from Stark et al. 1990a).

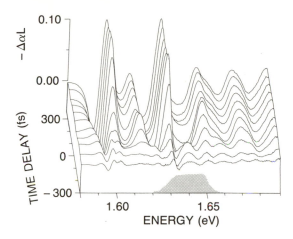

FIG. 13. Differential transmission spectra for excitation resonant with the $n_z = 1$ 2s-heavy hole excitons at $H = 12$Tesla versus time delay (from Stark et al. 1990a).

separation between the states and hence, the phase space available in the emission of phonons. The investigations of M-X nonlinear optics are just starting and are still under progress (Stark et al. 1990b).

3. DIFFERENTIAL ABSORPTION SPECTROSCOPY OF ELECTRONIC DEVICES

Semiconductor heterostructures exhibit remarkable new electronic transport phenomena, such as the fractional quantum Hall effect (Tsui et al. 1982, Tsui 1990) and Resonant Tunneling (Chang et al. 1974). They are also extremely attractive for device applications. Since the presence of carriers strongly modify their optical properties, it is possible to use optical techniques, such as differential absorption spectroscopy, to obtain direct information on electronic transport. In this section we discuss investigations of charge and voltage distribution and dynamics within Field Effect Transistor (Bar Joseph et al. 1987, Chemla et al. 1987, 1988) and the determination of charge distribution and density in Double Barrier Diodes (DBD), with a particular emphasis on the coherent or sequential nature of the transport (Bar Joseph et al. 1990, Woodward et al. 1991).

In modulation-doped heterostructures (Dingle et al. 1978, Gossard and Pinczuk 1985), the carriers are spatially separated from the dopants, which results in extremely high carrier mobilities (up to several 10^6 cm^2/Vs). The optical properties have been widely studies using such techniques as luminescence, luminescence excitation (Pinczuk et al. 1984, Miller and Kleinman 1985, Sooryakumar 1985, 1987, Meynardier et al. 1986, Skolnick et al. 1987, Delalande et al. 1986, 1987, Delalande 1987, Livescu et al. 1988), and direct absorption measurements (Lee et al. 1987, Chemla et al. 1987, 1988, Bar-Joseph et al. 1987, Livescu et al 1988, Huang et al. 1988). We will concentrate on the absorption measurements which present a double advantage for understanding the physics of transport: i) the linear and nonlinear absorption cross sections can be determined experimentally accurately, thus providing the measurements with absolute calibration, ii) absorption can be performed at extremely low intensity, thus being essentially non-invasive. Figure 1b shows absorption spectra of an n-type ($N \approx 5x10^{11}$ cm^{-2}) modulation-doped InGaAs/InAlAs multiple quantum well structure (L = 100Å), as a function of temperature (Livescu et al. 1988). The modulation-doped sample has nominally the same structure as the undoped sample, which spectra were shown in Figure 1a, and yet its absorption reveals striking differences. Although the n_z = 2 and 3 transitions can be clearly identified also, the sharp excitonic n_z = 1 peaks have disappeared even from the low-temperature spectrum, being replaced by a single broad feature, blue shifted by a significant amount (about 40 meV). This blue shift is maintained over the whole temperature range. As the temperature is raised, the absorption edge peak broadens and disappears rapidly. At 150K, only a "knee" is left, which barely changes its shape up to room temperature. The n_z = 2 and 3 higher energy peaks are observed at energies close to the corresponding peaks in the undoped sample, but they exhibit a somewhat larger width even at low temperature. Luminescence experiments show a large Stokes shift between emission and absorption spectra, which decreases with increasing temperature or decreasing doping. Although detailed explanation of these results require a theory outside the scope of this review (see for example Schmitt-Rink, Chemla and Miller 1989), the main features can be explained as follows. During growth, the dopants are restricted to the barrier material. In order to maintain a constant Fermi energy throughout the sample, they ionize with the free carriers migrating to the QW's where they occupy the lower energy states and hence, inhibit the n_z = 1 optical transition up to the Fermi energy. The broad peak seen at the onset of absorption is the so called Fermi-edge singularity originally introduced by Mahan (1967). This effect is the PSF due to the gas of electron. In one-electron theory, it gives rise to the well-known Burstein (1954) blue shift of the absorption edge. The charge transfer creates significant electric fields which shift the $n_z = 2$ and 3 resonances.

In a Modulation Doped Field Effect Transistor (Mod-FET), which conducting channel is a QW, the application of a voltage onto the gate deplete the QW underneath it and switches off the device. This changes the QW absorption spectrum from one similar to that shown in Figure 1b (device on) to one similar to that of Figure 1a (device off). Monitoring the absorption of the device during switching thus allows to probe directly and in situ the charge build-up. Figure 14 shows the low temperature DTS measured on a ($L = 100\text{Å}$) InGaAs/InAlAs Mod-FET through the transparent InP-substrate (Chemla et al. 1987, 1988, Bar Joseph et al. 1987b) for a series of gate-to-source voltages. The electron concentration in these DTS varies from $N = 0$ to the maximum value $N_{max} \approx 8 \times 10^{11}$ cm^{-2}. At the $n_2 = 1$ transitions, the DTS is positive and increases with N until the heavy-hole and light-hole exciton resonances appear clearly. This behavior corresponds to the progressive quenching of the absorption as the electrons fill up the lowest conduction band. At

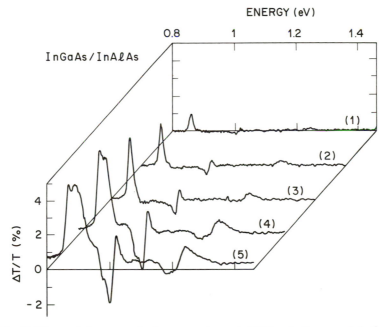

FIG. 14. Differential transmission spectra at 10K of the modulation-doped QW-conducting channel of a InGaAs/InAlAs Field Effect Transistor. The spectra are measured under the gate and correspond to increasing gate-to-source voltage (from Chemla et al. 1988).

the $n_z = 2$ and 3 transitions, the signal is smaller, with a differential lineshape showing positive and negative parts indicative of red shift, due to the electrostatic field, and collisional broadening, due to the electrons in the lowest subband. Clear shoulders also develop as the bias voltage increases; they are due to transitions which are forbidden in the empty square well and become allowed in the full biased well. The analysis of the DTS gives thus access to the charge density and the value of the electric field under the gate. It is found that the band gap renormalization varies as $N^{1/3}$, as also found in luminescence and luminescence excitation experiments (Delalande et al. 1986, 1987, Delalande 1987). The dynamics of the Mod-FET switching up to $\approx$ 20GHz were investigated by sampling techniques (Wiesenfeld et al. 1989). One of the most interesting results of these experiments is shown in Figure 15. Taking advantage of the electro-optic properties of semiconductors, it is possible to measure the voltage under electrodes (Valdmanis and Mourou 1986, Kolner and Bloom 1986). Figure 15 shows i) the electro-optic signal under the gate, ii) the signal due to the charge built up under the gate, iii) the electro-optic signal under the drain, when the Mod-FET is switched by a 50ps electrical pulse. The charge density reaches its maximum 11ps

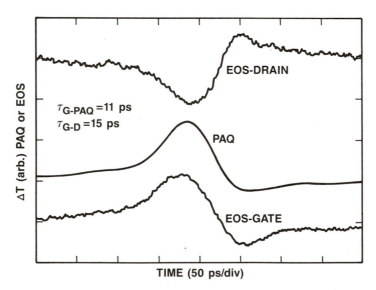

FIG. 15. Measured propagation delays of a 50ps electrical pulse switching a field effect transistor. Lower curve, electro optic signal measuring the voltage under the gate, center curve, phase-space absorption quenching signal due to the charge accumulation under the gate, upper curve electro optic signal measuring the voltage under the drain (from Wiesenfeld et al. 1989).

after the applied voltage on the gate, and surprisingly it takes another 4ps for the voltage to appear on the drain. This represents the first in situ measurement of the switching time of the Mod-FET (15ps) and its resolution in a charging time (11ps) and drain RC-time (4ps).

The Double Barrier Diode (Chang et al. 1974) is another quantum device which has received much attention, both because of its potential application to high speed electronics (Brown et al. 1989) and because it allows to investigate the phenomenon of quantum tunneling (Eaves 1990, for a recent review see for example Capasso 1990). The current-voltage characteristics of DBD's exhibit strong negative differential resistance (NDR) which have been explained by two models. In one approach, the electronic transport is described as a coherent tunneling through the whole DBD and the NDR is explained as due to interferences similar to thos involved in optics in Fabry-Perot interferometers (Ricco and Azbel 1984). In the other model, the NDR is explained as a sequential tunneling from a 3D density of states in the emitter electrode to the 2D density of states in the QW (Luryi 1985). In this case, the overall transport is incoherent and scattering events can interrupt the phase of the electrons as they transit through the DBD. To help answering these questions, it is important to determine the charge density and distribution inside the device as it is brought through the NDR. Numerous attempts have been made based on luminescence techniques (see for example Young 1988), but the accuracy of these methods is too poor to give reliable results (Frensley 1989). Here again DTS, by providing absolute calibration, is a powerful means of investigation of the charge distribution (Bar Joseph et al. 1989, 1991, Woodward et al. 1991).

An example of a DTS seen through a DBD, as the voltage is switched from $V = 0$ to the voltage V_{max} corresponding to the peak current just before the NDR, is shown in Figure 16c. The sample consists of a 45Å InGaAs-QW between two 70Å InAlAs-barriers and two $N = 5 \times 10^{17} \text{cm}^{-3}$ electrodes each 3000Å wide. Three distinct spectral regions are easily identified. From low to high energy, one distinguishes a broad negative hump starting at 0.8eV and centered around 0.83eV, a positive one located around 0.9eV and two narrow peaks near 0.97eV. To have an unambiguous assignment of these features, Figure 16a and b present respectively the absorption spectrum of a 1 μm thick InGaAs sample and of a 50-period multiple QW sample with 45Å InGaAs-QW and 70Å InAlAs-barriers nominally identical to that of the DBD. The high energy peaks clearly correspond to the heavy-hole and light-hole

315

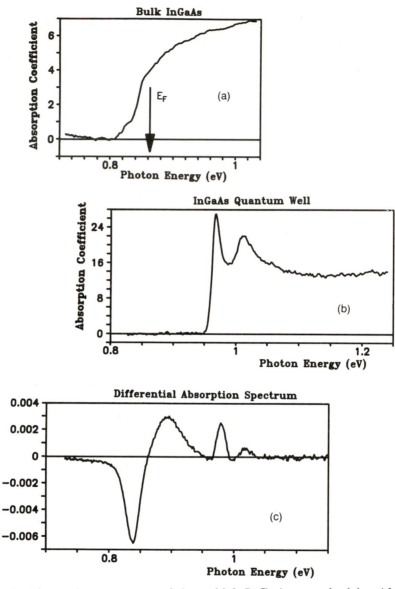

FIG. 16. Absorption spectrum of 1μm thick InGaAs sample (a). Absorption spectrum of a InGaAs/InAlAs multiple quantum well sample (L = 45Å Barrier thickness = 70Å) (b). Differential transmission spectrum of a resonant tunneling diode with same quantum well and barrier structure as the multiple QW-sample (c).

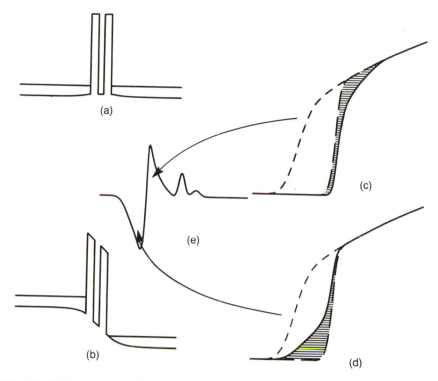

FIG. 17. Schematic explanation of the origin of the differential transmission spectrum due to the accumulation and the depletion layers (from Bar Joseph et al. 1989).

QW transition. The onset of the first negative feature corresponds to the gap of the bulk InGaAs and the cross over to the positive hump corresponds to the Fermi energy, $E_{fs} \approx 55\text{meV}$, shown by the arrow above this gap in Figure 16a. The two broad features originate from the depletion region of the collector (negative hump) and from the accumulation region of the emitter (positive hump). To understand this, consider Figure 17, the conduction band profile of the DBD is depicted in (a) for $V = 0$ and in (b) for $V = V_{max}$. The separation between the Fermi level and the bottom of the conduction band in the accumulation region, is larger under bias than with no applied voltage, and conversely it is smaller in the depletion region. Hence, under bias the accumulation region becomes more transparent just above E_{fs} (c) and the depletion region becomes more opaque just below E_{fs} and eventually down to the gap (d). These changes translate into the two humps with the cross over at E_{fs}. Let us note that besides providing an in situ measurement of the

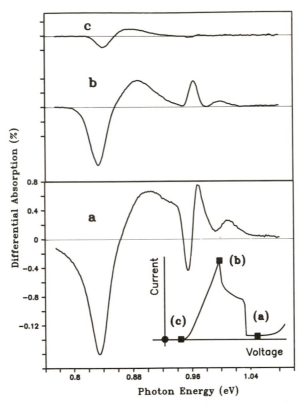

FIG. 18. Differential transmission spectrum of the resonant tunneling diode for the bias voltage shown on the J(V) characteristic.

actual charge density in the electrodes, the DTS profile directly relates to the conduction band profile in these uneasily accessible regions. More important for the discussion of the origin of the device operation is the QW-DTS. Figure 18 shows three DTS obtained by switching the voltage between V = 0 and the onset of the current (c), the peak of the current (b) and in the NDR valley (a) as indicated on the J(V) curve in the inset. The QW-DTS (c) is the signature of a small red shift indicating a weak field across the QW without charge accumulation. The QW-DTS (b) corresponds to the bleaching of the excitons and thus to a large charge density in the QW. Since above the exciton the DTS is essentially zero, there are *no electrons at energies above* the states out of which the excitons are constructed. Finally, the QW-DTS (a) which is somehow superimposed to the signal from the accumulation region indicates a strong field across the QW without charge inside. These DTS illustrate the three phases of the operation of the device and demonstrate that a charge accumulates in the QW when current flows through the DBD.

Using nonlinear optical techniques mentioned in the previous section, it is possible to calibrate the DTS to determine accurately the charge density under various bias conditions and in a variety of structures (Bar Joseph et al. 1989, 1991, Woodward et al. 1991). It is important to note, however, that direct insight on the scattering can be obtained by simple inspection of the DTS's. The states out of which the excitons are built are within one binding energy, E_b, of the band extrema. Hence, the bleaching of the exciton resonances is due to electrons within this energy above the bottom of the conduction band i.e., $E_b \approx 9.5 \text{meV}$ in the case of 45Å InGaAs QW. At the peak, current electrons with excess energy equal or larger than the Fermi energy $E_{formula} = 55 \text{meV}$, are injected in the QW and yet, as already noted, the DTS signal is null above the excitons. This means that electrons injected at high energy in the QW relax down to the bottom of the first conduction subband within the well before exiting in the collector. This proves unambiguously that inelastic intrasubband scatterings occur inside the QW and hence, that the phase of the electrons is interrupted many times during the transport. This important result has been confirmed by two independent sets of experiments. First, it has been shown that the charge density in the QW and the current through the DBD are essentially independent of temperature from 10K to 300K, leading to an almost constant transit time $\tau = \text{Ne}/J \approx 65 \text{ps}$ through the QW (Bar Joseph et al. 1990). NDR is observed despite the fact that at this temperature and for this material exciton-LO phonon collisions occur on average every 200 femtoseconds (Wegener et al. 1989)! Furthermore, it was recently shown (Woodward et al. 1991) in DBD with $n_z = 2$ subband, there is no charge in this $n_z = 2$ subband but significant charge in the $n_z = 1$ lowest level. Hence, NDR can occur also with intersubband scattering!

4. OPTOELECTRONIC APPLICATIONS

The special properties of QW structures have been exploited in numerous electronics and photonic devices, some of which are already commercially available. Electronic devices include high speed field effect transistors and hetero-bipolar transistors, resonant tunneling diodes (Brown et al. 1989, Capasso 1990), and other not electron resonant tunneling devices (Luryi 1987). Photonic devices include high performance waveguide (Dutta 1987, Yariv 1989) and surface emitting (Jewell et al. 1990) semiconductor lasers, infrared photodetectors and modulators (Levine 1989), as well as high performance near band gap PIN-photodetectors, modulators and self electro-optic devices (SEED).

Because the same growth technology can be used to make all these devices, they can be in principle integrated and indeed several examples of novel multifunction integrations have been reported (Koren et al. 1989). A comprehensive review of all optoelectronic applications of heterostructures is far beyond the scope of this paper, we will limit ourselves to a brief summary of the near band gap modulators and SEED's.

The most developed modulators exploit the special electroabsorption properties of QW (Wood 1988). It is well known that electric fields affect the linear optical properties near the fundamental absorption edge of semiconductors (Franz 1958, Keldysh 1958). The mechanisms governing QW electroabsorption are very different depending on whether the electric field is parallel or perpendicular to the layers. For fields parallel to the layers the behavior is qualitatively similar to the case of 3D electroabsorption i.e., Franz-Keldysh-like oscillations above the band edge, a weak absorption tail below the edge, and broadening of excitonic resonances due to field ionization. Experiments on QW's (Miller et al. 1985, Knox et al. 1986b) are found in quantitative agreement with theory (Lederman and Dow 1976). For field perpendicular to the layers the QW electroabsorption exhibit a qualitatively different behavior. It is found, contrary to the 3D and parallel field cases,

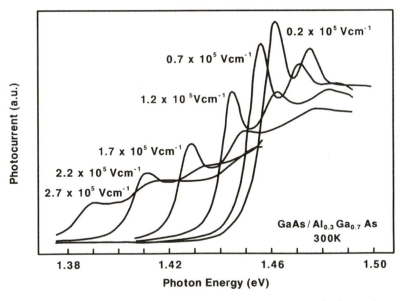

FIG. 19. Quantum Confined Stark Effect measured by photocurrent spectroscopy, showing the absorption of a L = 95Å GaAs/AlGaAs modulator for electric field F = 0 → 2.7 x 10^5 V/cm.

that the excitons do not broaden significantly with field, and that the peaks experience strong red shifts which can be larger than the binding energy by one to two orders of magnitude, Figure 19. For this reason, this effect has received the special name of QCSE (Miller et al. 1984a, 1985a, Brum and Bastard 1985). The key physical reason for the persistence of the excitonic resonance in the perpendicular field QW case is that the walls of the quantum well prevent the field ionization of excitons. As field is applied, the electron and hole (in the lowest confined states) are pulled to opposite sides of the well, but they are restrained by the walls from going any further. Thus, the exciton is not "ripped apart" by the field; it is very strongly polarized by the field with the electron and hole orbits being partly separated spatially, but at distance $\approx a_o$. Consequently, it is still a well-defined correlated e-h system with a well-resolved optical absorption resonance without significant additional lifetime-broadening.

There are several qualitatively new aspects of QCSE modulators. Firstly, the changes of absorption that can be achieved with moderate applied voltage, ($\Delta\alpha > 10^4 \text{ cm}^{-1}$), are so large that large contrast can be obtained even in a single pass through ≈ 1 μm of QW material. This results in low operating energy ($\approx 10^2$ femto Joules for $10 \times 10 \times 1$ μm^3 modulator) that becomes comparable to that of electronic devices. Secondly, the physical mechanisms (resonant tunneling and thermoionic emission) suggest that very fast (< 1 ps) recovery times are possible. And, indeed, devices operating at 20 GHz have been reported (Kotaka et al. 1989). Thirdly, the QW-modulators are simultaneously excellent photodetectors with essentially unity quantum efficiency. As discussed below, this possibility of dual operation can be exploited in novel devices that are neither completely electronic nor photonic. Finally, the last important aspect is, once again, the flexibility of the technology for multiple function integration. In the case of waveguide geometry, a number of practical devices using QCSE have been reported (Wood 1988). And even the first examples of integrated subsystems have already been demonstrated, including continuously tunable QW Distributed Bragg Reflector (DFB) lasers (Koch et al. 1989) and QW integrated heterodyne receivers (Koch et al. 1989, 1990, Takeuchi et al. 1989). An example of the complexity of the structures which can be fabricated nowadays is shown in Figure 20. This balanced heterodyne receiver is comprised of two QW-PIN detectors, one QW-DFB laser acting as a local oscillator, a QW-phase modulator and a QW-amplifier, and all the buried passive waveguides necessary to connect them.

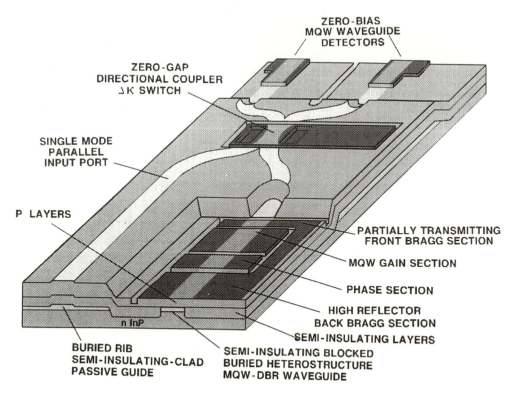

FIG. 20. Example of a lightwave communication Photonic Integrated Circuit (balanced heterodyne receiver) involving several quantum well optoelectronic devices (from Koch et al. 1989).

The other possible modulator geometry is that where the optical beams propagate perpendicular to the QW, it is very interesting for applications where large arrays of devices are required. To exploit the strong QCSE electroabsorption in this geometry, QW modulators have been grown by Molecular Beam Epitaxy onto semiconductor multilayer dielectric mirrors to make surface reflection modulators (Boyd et al. 1987). This presents the advantage of doubling the thickness of the active QW-material. By doping the mirrors, the electric connectivity can be preserved. Further regrowth of dielectric mirrors onto the QW-modulator has been used to form a high finesse Fabry-Perot resonator with very high contrast due to the increased number of passes (Yan et al. 1989, Whitehead et al. 1989).

Although the QCSe has received most of the attention in terms of practical applications (Miller 1987), there are, however, other interesting electroabsorption effects in more complex QW-structures. The simplest of these structures is the pair of coupled wells where the applied field spatially separate the electron and the hole, which become localized in different wells. The wavefunction overlap is reduced and the oscillator strength is quenched (Islam et al. 1987). An important recent development of electroabsorption in coupled systems is the Wannier-Stark localization in superlattices (Bleuse et al. 1988, Mendez et al. 1988). In superlattice, the wavefunctions are extended, the eigen states form minibands somehow intermediate between 2D and 3D (Bastard 1990), and the absorption edge is not very abrupt. The application of an electric field such that the potential difference over one period becomes comparable to the width of the miniband, localizes the particles within individual wells. The structure thus recovers the sharp absorption edge of quasi-2D structures and large change of absorptions are seen near the gap (Bastard and Ferreira 1989).

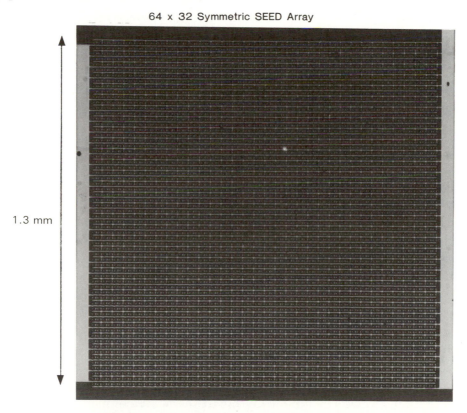

64 x 32 Symmetric SEED Array

1.3 mm

FIG. 21. Array of 2048 = 64 x 32 symmetric-SEED's. Each optical logic gate consists of a couple 10 x 10 μm devices (from Miller 1990).

A very interesting aspect of QW-modulators, which was already mentioned, is that they can act as photodetectors as well. The photocurrent traversing the QW-modulator gives a mean to measure the internal field and a signal for feedback by an external circuit. This is the principle of the SEED's (for a comprehensive review see Miller 1990). Again by integration, the dimension of the feedback loop can be reduced to a few μm keeping the motion of charges local. Various feedback implementations have been proposed and demonstrated to make optical bistable elements. The SEED's are complete optical logic devices with time-sequential-gain, low switching energy, and sub-nanosecond response time. They can perform many logic operations, including OR, AND, NOR and NAND. Large arrays of 2048 = 64 x 32 devices are now commercially available. An example of such arrays is shown in Figure 21. The logic elements are reflection type (10 μm x 10 μm)-mesas with excellent response uniformity. Their switching energy ranges from 1 pJ to 2.5 pJ for contrast ratio, respectively 3/1 and 5/1, and supply voltage of 6V and 15V (Lentine et al. 1990). Proof of principle of simple system operations with arrays of SEED's have been performed including ring-counter (Lentine et al.), shift-register (McCormick et al. 1989a), and more than 200 SEED's on one array have been operated simultaneously with a single light source (McCormick et al 1989b).

5. CONCLUSION

One of the tasks assigned to us by the organizers of this NATO Science Forum is to draw some conclusions from the achievements of condensed matter physics in the eighties and to try to project directions for future research. This is, of course, a very dangerous task; history has taught us that important discoveries usually come as surprises. Condensed matter physics has always been associated in a symbiotic manner with technology. Discoveries leading to new technologies and new technologies providing the Scientists new tools for their research. Thus, it is natural to think that the most interesting directions of research are those which will have the biggest impact on technology. Among the many technologies that we need to develop for the next century, the only one about which I have a little experience, are the Information Technologies i.e., those involved in acquiring, processing, transporting and storing information. Thus, I will limit myself to them. A complete explanation of the reasons behind my choice of important directions of research would require a very lengthy discussion and will overlap with the report of the panel. Therefore, I have chosen to just give here a list of subjects that, in my opinion, request some attention. I have distinguished three areas where breakthroughs are needed and I have indicated some specific and outstanding problems to solve.

i) GROWTH AND FABRICATIONS

a. Controlled and reliable method for electronic confinement in all dimensions (without introduction of defects).

b. Controlled and reliable method for dielectric confinement in all dimensions (without introduction of defects).

ii) NEW MATERIALS

a. New semiconductor systems.

b. New hybrid systems (semiconductor / superconductor / insulator / metal).

c. Engineering of self-organizing molecular and supra-molecular systems.

iii) NEW FUNCTIONS

a. Generic method for local ($< 1\ \mu$m) and high density integration of optics and electronics.

b. New concepts for ultra high speed (> 1 THz) optoelectronics.

c. New concepts for high speed, high contrast, low driving voltage optoelectronic parallel processing (to try solving the "interconnects" bottleneck).

ACKNOWLEDGEMENTS

Most of the work presented in this review was performed at AT&T Bell Laboratories in the eighties where I have benefited from the collaboration and support of many colleagues and friends.

I would like to take the opportunity of this article to express my deep appreciation to: Israel Bar Joseph, Tao Chang, Al Cho, Art

Gossard, Wayne Knox, Karl Leo, Gaby Livescu, David Miller, Aron Pinczuk, Stefan Schmitt-Rink, Jag Shah, Jason Stark, Ann Von Lehmen, Martin Wegener, Joe Weiner and Jane Zucker.

REFERENCES

1. O. Akimoto, H. Hasegawa, J. Phys. Soc. Jpn. **22**, 181 (1967).

2. G. E. W. Bauer, Phys. Rev. Lett. **64**, 60 (1990).

3. I. Bar-Joseph, J. M. Kuo, C. Klingshirn, G. Livescu, T. Y. Chang, D. A. B. Miller and D. S. Chemla, Phys. Rev. Lett. **59**, 1357 (1987).

4. I. Bar-Joseph, T. K. Woodward, D. S. Chemla, D. L. Sivco, A. Y. Cho, Phys. Rev. **B41**, 3264 (1989).

5. I. Bar-Joseph, Y. Gedalyahu, A. Yacoby, T. K. Woodward, D. S. Chemla, D. L. Sivco, A. Y. Cho, to be published (1991).

6. G. Bastard, these proceedings (1991).

7. G. Bastard, R. Ferreira, <u>Spectroscopy of Semiconductor Microstructures</u>, ed. G. Fasol, A. Fasolino, P. Lugli, NATO ASI Series **B-206**, 333 (1989).

8. J. Bleuse, G. Bastard, P. Voisin, Phys. Rev. Lett. **60**, 220 (1988).

9. G. D. Boyd, D. A. B. Miller, D. S. Chemla, S. L. McCall, A. C. Gossard, J. H. English, Appl. Phys. Lett. **50**, 1119 (1987).

10. E. Burstein, Phys. Rev. **93**, 632 (1954).

11. J. A. Brum and G. Bastard, Phys. Rev. **B31**, 3893 (1985).

12. C. H. Brito Cruz, R. L. Fork, W. H. Knox, C. V. Shank, Chem. Phys. Lett. **132**, 341 (1986).

13. E. R. Brown, T. C. L. G. Sollner, C. D. Parker, W. D. Goodhue, C. L. Chen, Appl. Phys. Lett. **55**, 1777 (1989).

14. F. Capasso, ed. Physics of Quantum Electron Devices, Springer Series in Electronic and Photonics, **28** (1990).

15. L. L. Chang, L. Esaki, R. Tsu, Appl. Phys. Lett. **24** 593 (1974).

16. D. S. Chemla, I. Bar-Joseph, C. Klingshirn, D. A. B. Miller, J. M. Kuo and T. Y Chang, Appl. Phys. Lett. **50**, 585 (1987).

17. D. S. Chemla, I. Bar-Joseph, J. M. Kuo, T. Y. Chang, C. Klingshirn, G. Livescu, D. A. B. Miller, IEEE J. Quant. Electron. **24**, 1664 (1988).

18. D. S. Chemla, D. A. B. Miller, P. W. Smith, A. C. Gossard and W. Wiegmann, IEEE J. Quant. Electron. **20**, 265 (1984).

19. D. S. Chemla, W. H. Knox, D. A. B. Miller, S. Schmitt-Rink, J. B. Stark, R. Zimmermann, J. Lum. **44**, 233 (1989).

20. C. Delalande, Physica Scripta T **19**, 129 (1987).

21. C. Delalande, G. Bastard, J. Orgonasi, J. A. Brum, H. W. Liu, M. Voos, G. Weimann and W. Schlapp, Phys. Rev. Lett. **59**, 2690, (1987).

22. C. Delalande, J. Orgonasi, M. H. Meynardier, J. A. Brum and G. Bastard, Solid State Commun. **59**, 613 (1986).

23. R. Dingle, H. L. Stoermer, A. C. Gossard and W. Wiegmann, Appl. Phys. Lett. **33**, 665 (1978).

24. N. K. Dutta, Heterojunction Band Gap Discontinuities Capasso and Margaritondo, North-Holland, Amsterdam, (1987).

25. L. Eaves, these proceedings, (1990).

26. C. Ell, J. F. Mueller, K. El Sayed, H. Haug, Phys. Rev. Lett. 62, 304 (1989).

27. C. Ell, J. F. Mueller, K. El Sayed, L. Banyai, H. Haug, Phys. Stat. Sol. **B150**, 393 (1988).

28. G. W. Fehrenbach, W. Schaefer and R. G. Ulbrich, J. Lumin. **30**, 154 (1985).

29. B. Fluegel, N. Peyghambarian, G. Olbright, M. Lindberg, S. W. Koch, M. Joffre, D. Hulin, A. Migus, A. Antonetti, Phys. Rev. Lett. **59**, 2588 (1987).

30. B. Fluegel, J. P. Sololoff, F. Jarka, S. W. Koch, M. Lindberg, N. Peyghambarian, M. Joffre, D. Hulin, A. Migus, A. Antonetti, C. Ell, L. Banyai, H. Haug, Phys. Stat. Sol. **B150**, 357 (1988).

31. W. Franz, Z. Naturforschg **13a**, 484-9 (1958).

32. W. R. Frensley, M. A. Reed, J. H. Luscombe, Phys. Rev. Lett. **62**, 1207 (1989).

33. A. C. Gossard and A. Pinczuk, <u>Synthetic Modulated Structures</u>, edited by L. L. Chang and B. C. Giessen (New York: Academic), p. 215 (1985).

34. D. Heitmann, these proceedings, (1990).

35. D. Huang, H. Y. Chu, Y. C. Chang, R. Houdre, H. Morkoc, Phys. Rev. **B38**, 1246 (1988).

36. D. Hulin, A. Mysyrowicz, A. Antonetti, A. Migus, W. T. Masselink, H. Morkoc, H. M. Gibbs and N. Payghambarian, Phys. Rev. **B33**, 4389 (1986).

37. M. N. Islam, R. L. Hillman, D. A.B. Miller, D. S. Chemla, A. C. Gossard, J. H. English, Appl. Phys. Lett. **50**, 1098 (1987).

38. J. L. Jewell, Y. H. Lee, A. Scherer, S. L. McCall, N. A. Olsson, J. P. Harbison, L. T. Florez, Optical Engineering **29**, 210 (1990).

39. M. Joffre, D. Hulin, A. Migus, A. Antonetti, C. Benoit a la Guilleaume, N. Peyghambarian, M. Lindberg, S. W. Koch, Opt. Lett. **13**, 276, (1988).

40. L. V. Keldysh, Zh. Eksp. Teor. Fiz. 34, 1138 (1958): Sov. Phys. JETP **34**, 788 (1958).

41. T. L. Koch, U. Koren, R. P. Gnall, F. S. Choa, F. Hernandez-Gil, C. A. Burrus, M. G. Young, M. Oron, B. I. Miller, Electron. Lett. **25**, 1621 (1989).

42. T. L. Koch, U. Koren, R. P. Gnall, C. A. Burrus, B. I. Miller, Electron. Lett. **24**, 1431 (1989).

43. B. H. Kolner, D. M. Bloom, IEEE J. Quant. Electron. **QE–22**, 69 (1986).

44. U. Koren, T. L. Koch, B. I. Miller, Tech. Dig. IGWO '89 Houston, TX, 68 (1989).

45. I. Kotaka, K. Wakita, O. Mitomi, H. Asai, Y. Kawamura, IEEE Phot. Tech. Lett. **1**, 100 (1989).

46. W. H. Knox, R. L. Fork, M. C. Downer, D. A. B. Miller, D. S. Chemla, C. V. Shank, A. C. Gossard and W. Wiegmann, Phys. Rev. Lett. **5**, 1306 (1985).

47. W. H. Knox, C. Hirlimann, D. A. B. Miller, J. Shah, D. S. Chemla,

C. V. Shank, A. C. Gossard and W. Wiegmann, Phys. Rev. Lett. **56**, 1191 (1986).

48. W. H. Knox, D. A. B. Miller, T. C. Damen, D. S. Chemla, C. V. Shank and A. C. Gossard, Appl. Phys. Lett. **48**, 864 (1986b).

49. W. H. Knox, D. S. Chemla, G. Livescu, J. E. Cunningham, J. E. Henry, Phys. Rev. Lett. **61**, 1290 (1988).

50. W. H. Knox, D. S. Chemla, D. A. B. Miller, J. B. Stark, S. Schmitt-Rink, Phys. Rev. Lett. **62**, 1189 (1989).

51. F. L. Lederman and J. D. Dow, Phys. Rev. **B13**, 1633 (1976).

52. J. S. Lee, Y. Iwasa, N. Miura, Semicond. Sci. Technol., **2**, 675 (1987).

53. A. L. Lentine, H. S. Hinton, D. A. B. Miller, J. E. Henry, J. E. Cunningham, L. M. F. Chirovsky, IEEE J. Quant. Electron. **25**, 1928 (1989).

54. K. Leo, M. Wegener, J. Shah, D. S. Chemla, E. O. Gobel, T. C. Damen, S. Schmitt-Rink, W. Schaefer, Phys. Rev. Lett. **65**, 1340 (1990).

55. I. V. Lerner, Y. E. Lozovik, Zh. Eksp. Toer. Fiz, 80, 1488 (1981): Sov. Phys. JETP **53**, 763 (1981).

56. B. F. Levine, G. Hasnian, C. G. Bethea, N. Chand, Appl. Phys. Lett. **54**, 2704 (1989).

57. P. F. Liao, J. Bjorkholm, Phys. Rev. Lett. **34**, 1 (1975).

58. G. Livescu, D. A. B. Miller, D. S. Chemla, M. Ramaswamy, T. Y. Chang, N. Sauer, A. C. Gossard and J. H. English, IEEE J. Quant. Electron. **24**, 1677 (1988).

59. S. Luryi, Appl. Phys. Lett. **47**, 490 (1985).

60. S. Luryi, in Heterojunction Band Gap Discontinuities, Capasso and Margaritondo, North-Holland, Amsterdam.

61. F. B. McCornick, A. L. Lentine, L. M. F. Chirovsky, L. A. D'Asaro, in OSA Proceeding on Photonic Switching, **192** (1989).

62. F. B. McCornick, A. L. Lentine, R. L. Morrison, S. L. Walker, in Proceeding of the Annual Meeting of the OSA, MII4, (1989).

63. A. H. McDonald, D. S. Richie, Phys. Rev. **B33**, 8336 (1986).

64. G. D. Mahan, Phys. Rev. **153**, 882 (1967a).

65. Y. Masumoto, S. Tarucha and H. Okamoto, J. Phys. Soc. Jpn. **55**, 57 (1986).

66. E. E. Mendez, F. Agullo-Rueda, J. M. Hong, Phys. Rev. Lett. **60**, 2426 (1988).

67. M. H. Meynadier, R. E. Nahory, J. M. Worlock, M. C. Tamargo and J. L. de Miguel, Phys. Rev. Lett. **60**, 1338 (1988).

68. D. A. B. Miller, Opt. Eng. **26**, 368 (1987).

69. D. A. B. Miller, Opt. Quant. Electron. **22**, S61 (1990).

70. D. A. B. Miller, D. S. Chemla, T. C. Damen, A. C. Gossard, W. Wiegmann, T. H. Wood and C. A. Burrus, Appl. Phys. Lett, **45**, 13 (1984b).

71. D. A. B. Miller, D. S. Chemla, T. C. Damen, A. C. Gossard, W. Wiegmann, T. H. Wood and C. A. Burrus, Phys. Rev. **B32**, 1043 (1985a).

72. D. A. B. Miller, D. S. Chemla, T. C. Damen, A. C. Gossard, W. Wiegmann, T. H. Wood and C. A. Burrus, Phys. Rev. **B32**, 1043 (1985).

73. D. A. B. Miller, J. S. Weiner and D. S. Chemla, IEEE J. Quantum Electron. **QE–22**, 1816 (1986b).

74. R. C. Miller and D. A. Kleinman, J. Lumin. **30**, 520 (1985).

75. A. Mysyrowicz, D. Hulin, A. Antonetti, A. Migus, W. T. Masselink, H. Morkoc, Phys. Rev. Lett. **56**, 2748 (1986).

76. D. Paquet, T. M. Rice, K. Ueda, Phys. Rev. **B32**, 5208 (1985).

77. N. Peyghambarian, H. M. Gibbs, J. L. Jewell, A. Antonetti, A. Migus, D. Hulin and A. Mysyrowicz, Phys. Rev. **53**, 2433 (1984).

78. A. Pinczuk, J. Shah, R. C. Miller, A. C. Gossard and W. Wiegmann, Solid State Commun. **50**, 735 (1984).

79. R. Ricco, M. Y. Azbel, Phys. Rev. **B29**, 1970 (1984).

80. W. Schaefer, K. H. Schuldt, R. Binder, Phys. Stat. Sol. **B150**, 407 (1988).

81. W. Schaefer, Adv. Solid State Phys. **28**, 63 (1988).

82. S. Schmitt-Rink, D. S. Chemla, D. A. B. Miller, Adv. Phys. **38**, 89 (1989).

83. S. Schmitt-Rink, D. S. Chemla, D. A. B. Miller, Phys. Rev. **B32**, 6601 (1985).

84. S. Schmitt-Rink, D. S. Chemla, Phys. Rev. Lett. **57**, 2752 (1986).

85. S. Schmitt-Rink, D. S. Chemla, H. Haug, Phys. Rev. **B**, 941 (1988).

86. L. Schultheis, J. Kuhl, A. Honold, T. W. Tu, Phys. Rev. Lett. **57**, 1797 (1986), Phys. Rev. Lett. **57**, 1635 (1986).

87. M. Sinada, K. Tanaka, J. Phys. Soc. Jpn. **29**, 1258 (1970).

88. M. S. Skolnick, J. M. Rorison, K. J. Nash, D. J. Mowbray, P. R. Tapster, S. J. Bass and A. D. Pitt, Phys. Rev. Lett. **58**, 2130 (1987).

89. R. Sooryakumar, A. Pinczuk, A. C. Gossard, D. S. Chemla and L. J. Sham, Phys. Rev. Lett. **58**, 1150 (1987).

90. R. Sooryakumar, D. S. Chemla, A. Pinczuk, A. C. Gossard, W. Wiegmann and L. J. Sham, Solid State Commun. **54**, 859 (1985).

91. C. Stafford, S. Schmitt-Rink, W. Schaefer, Phys. Rev. **B41**, 10000, (1990).

92. J. B. Stark, W. H. Knox, D. S. Chemla, W. Schaefer, S. Schmitt-Rink, C. Staford, submitted to Phys. Rev. Lett., (1990a).

93. J. B. Stark, W. H. Knox, D. S. Chemla, S. Schmitt-Rink, to be published, (1990b).

94. S. Tarucha, H. Okamoto, Y. Iwasa, N. Miura, Sol. Stat. Comm. **52**, 815 (1984).

95. D. C. Tsui, H. L. Stoermer and A. C. Gossard, Phys. Rev. Lett. **48**, 1559 (1982).

96. D. C. Tsui, these proceedings, (1990).

97. J. A. Valdmanis, G. A. Mourou, IEEE J. Quant. Electron. **QE-22**, 79 (1986).

98. A. von Lehmen, D. S. Chemla, J. E. Zucker, J. P. Heritage, Opt. Lett. **11**, 609 (1986).

99. M. Wegener, I. Bar Joseph, G. Sucha, M. N. Islam, N. Sauer, T. Y. Chang, D. S. Chemla, Phys. Rev. **B39**, 12794 (1989).

100. M. Wegener, D. S. Chemla, S. Schmitt-Rink, W. Schaefer, Phys. Rev. **A42**, 5675 (1990).

101. M. Whitehead, G. Parry, Electron. Lett. **25**, 566.

102. J. M. Wiessenfeld, M. S. Heutmaker, I. Bar-Joseph, D. S. Chemla, J. M. Kuo, T. Y. Chang, C. A. Burrus, J. C. Perino, Appl. Phys. Lett. **55**, 1109 (1989).

103. T. H. Wood, J. Lightwave Tech. **6**, 743 (1988).

104. T. K. Woodward, D. S. Chemla, I. Bar-Joseph, H. U. Baranger, D. L. Sivco, A. Y. Cho, to be published in Phys. Rev. B (1991).

105. F. Y. Wu, S. Ezekiel, M. Ducloy, B. R. Mollow, Phys. Rev. Lett. **38**, 1077 (1977).

106. T. Yajima, J. Taira, J. Phys. Soc. Jpn. **47**, 1620 (1979).

107. R. H. Yan, R. J. Simes, L. A. Coldren, IEEE J. Quant. Electron. **25**, 2273 (1989).

108. S. R. E. Yang, L. J. Sham, Phys. Rev. Lett. **58**, 2598 (1987).

109. A. Yariv, IEEE Circuits and Devices Magazine **5**, 25 (1989).

110. J. F. Young, B. M. Wood, G. C. Aers, R. S. L. Devine, H. C. Liu, D. Landheer, B. Buchanan, A. J. Springthrope, P. Madeville, Phys. Rev. Lett. **60**, 2085 (1988).

111. R. Zimmermann, Phys. Stat. Sol. **B146**, 545 (1988) and Proc. 19th Int. Conf. Phys. Semicond., Warsaw, (1988).

112. R. Zimmermann, R. Hartmann, Phys. Stat. Sol. **B150**, 365 (1988).

CORRELATED TUNNEL EVENTS IN ARRAYS OF ULTRASMALL JUNCTIONS

T. Claeson and P. Delsing

Physics Department
Chalmers University of Technology
Göteborg, Sweden

1. INTRODUCTION

Modern microelectronics technology enables the fabrication of ultrasmall tunnel junctions. A large number of identical elements can be made. We have, thus, possibilities of studying arrays of very small tunnel junction and, in particular, correlated tunnel events in these, the subject of this report.

Tunneling in normal size junctions is usually a stochastic process. The current in the leads is due to a continuous charge transport, the (surface) charge on the tunnel junction, which can be considered as a leaky capacitor, C, is built up continuously until a tunneling event of a discrete charge takes place. These events are random, they are due to thermal fluctuations, as the capacitive energy associated with the tunneling of an electron, $e^2/2C$, is much smaller than the thermal energy, $k_B T$. In a $10\mu m \times 10\mu m$ junction, C is typically about 3pF, $e^2/2C$ corresponds to a temperature of about $3 \bullet 10^{-4}$K, and the tunneling is completely dominated by thermal fluctuations even at helium temperatures. Decreasing the junction area four to five orders of magnitude and

Highlights in Condensed Matter Physics and Future Prospects
Edited by L. Esaki, Plenum Press, New York, 1991

arriving at the present art of fabricating ultrasmall junctions, we will find that new tunnel effects will occur.[1] The voltage fluctuation due to a tunneled charge is now of the order of a millivolt and the change of the capacitive energy corresponds to a few Kelvin.

A Coulomb blockade of tunneling will prevent electrons from tunneling when the bias voltage is small, $eV>e^2/2C$ if the capacitive energy is larger than (i) thermal and (ii) quantum fluctuations:

(i) $\qquad e^2/2C >k_BT$,

(ii) $\qquad e^2/2C >h/\tau$, where τ is typically the discharging RC-time.

The latter condition can be rewritten as

$$R > R_Q = h/4e^2(\approx 6.5k\Omega)$$

where R is the smallest of the tunnel resistance and the shunt resistance.

The I-V curve of an ultrasmall tunnel junction is characterized not only by a blockade range but also by an offset voltage, $2V_{off}=e/C$, between the asymptotes at large positive and negative bias.

Furthermore, the tunneling events are no longer stochastic but may be periodic, i.e. there is a time correlation between single electron tunneling (SET) events.[1-6] The capacitor is charged during a time determined by the (constant) current, $\Delta t=e/I$, an electron tunnels and the junction capacitor is discharged, it is charged again, etc. The relaxation oscillator has an average frequency of $f=I/e$. The linewidth of the oscillation is determined by, e.g., the temperature, the environment and the bias value (such that the oscillation is broadened and becomes ill defined at high bias). A condition, besides those of low temperature and high tunneling resistance, is that the oscillation period is larger than either the charging (discharging) RC-time or the relaxation/tunneling time. In general, the oscillation becomes less pronounced with higher frequency, and is no longer well defined above the RC cut-off frequency, $f>1/RC$. A calculated I-V curve and oscillation waveforms for different bias points are shown in Fig. 1 for a single unshunted tunnel junction. Note that the time scale is normalized to the oscillation period and that the oscillation decays at large bias.

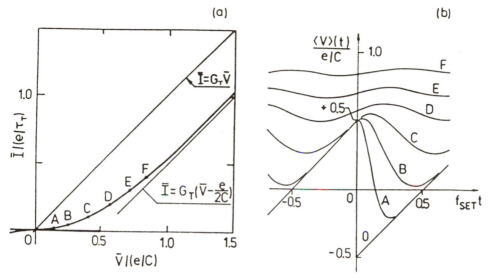

FIG. 1. SET effects in an unshunted tunnel junction with non-superconducting electrodes in the low temperature limit: (a) dc I-V curve, (b) oscillation waveforms at several biases as marked in (a). From Ref. 1.

The environment of a small tunnel junction is of great importance. A shunt capacitance C_s (e.g., caused by stray capacitances from the leads to the ground) may easily become larger than the junction capacitance. As it is coupled in parallel, it will add to the total capacitance and destroy the single electron tunneling effects. Likewise, the leads will act as transmission lines with an impedance, Z_e, seen by the junction that is of the order of the free space impedance of 377Ω, effectively shunting the high impedance of the small junction. Actually, the dc I-V curve (Coulomb blockade, etc.) is determined by excitations at high frequency, i.e., the high frequency impedance of the surrounding circuitry,[7-14] typically determined by the frequency corresponding to the bias. Structure will be obtained at a bias corresponding to the excitation frequency. We can compare with, e.g., the phonon structure seen in a tunneling curve due to finite lifetimes of excitations in superconductors.

A single junction has to be protected from the low impedance shunting environment. One way to do so is to place it closely connected to high resistance strips. Another way, which will be discussed in this article, is to use one or several other junctions coupled in series. A pair of junctions, for example, formed by an intermediate particle inside an oxide junction or located at an STM tip, is the most studied configuration up till now.

1.1. Correlation in Space for Double Junctions

The voltages over two junctions in series distributes according to the impedances, mainly as the capacitances in high resistance junctions at low bias and as the resistances at high bias.[15] No current flows due to the Coulomb blockade at low bias. As the threshold voltage over one of the junctions is exceeded, an electron tunnels. The middle electrode, has a net capacitance of $C_\Sigma = C_1 + C_2$, C_1 and C_2 being the capacitances of the two junctions. It is charged (with a large charging energy) and the threshold voltage of the second junction is exceeded. The electron that had entered through the first junction rapidly leaves the electrode through the second junction. There is a "space" correlation between the tunneling events.

The I-V curve of the two unequal junctions displays a zero current part below the threshold, the current rises rapidly as the threshold is exceeded, it flattens out again and there is formed a periodic pattern, a so called "Coulomb staircase." If the junctions are strongly asymmetric such that the resistance of one is much larger than the other (e.g.,

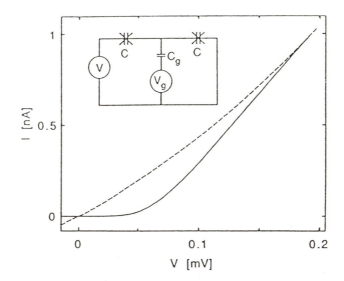

FIG. 2. Modulation of the current through two junctions by a gate voltage. The two curves display maximum (full curve, no gate voltage) and minimum (dashed, with gate current) Coulomb gaps. This presumably corresponds to an integer and a half-integer additional charge on the middle electrode, resp. The traces were recorded at T=10 mK. R=58 kΩ $C_g \approx 0.03C$. Superconductivity of the A1 electrodes has been suppressed by a magnetic field of 2 T. From Ref. 18.

336

$R_1 \gg R_2$), then the period of the staircase is determined by the capacitance of the junction with the larger resistance (e/C_1). Sharp staircase steps are obtained for grossly unequal resistances. At large biases, the asymptotes are offset, $V_{off} = e/C_\Sigma$.

If an additional, continuous charge Q_o is injected from the outside onto the middle electrode, the staircase is shifted[15-17]. It is varying between two extremes, the period being $Q_o = e$. A charge Q_o of $e/2$ (or $n \cdot e/2$) has shifted the periodic pattern by a phase of π. The original staircase is recovered when $Q_o = e$ (and $n \cdot e$). The surface charge on a capacitance is a continuous variable. It depends upon the distribution of charges in the electrodes and varies continuously with the continuously varying current in the leads. Examples of experimentally determined[18] current-voltage curves with $Q_o = ne$ and $(n+1/2)e$ are given in Fig. 2. The periodic variation of the voltage across two junctions with voltage applied to a gate coupled to the middle electrode is shown in Fig. 3. The latter curves are taken as a function of temperature. Note that the periodic variation is smeared considerably by increased temperature.

The Coulomb-blockade threshold is hence controlled by the charge on the middle electrode. If the two junctions are not well shielded from the outside, if they are unsymmetric and influenced differently by the

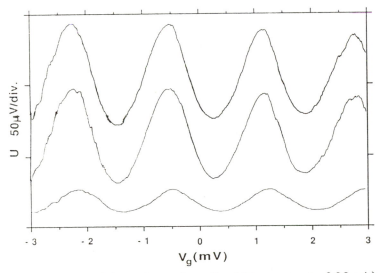

FIG 3. Modulation of the voltage (at a fixed bias current of 30 pA) of two 100 nm × 100 nm series coupled junctions as a function of gate voltage. T=40, 100, and 250 mK counted from the top. $R_1 + R_2 = 310$ kΩ. $2V_{off} = 425$ μV. $C_g = 10^{-16}$F is obtained from the period of modulation ($\Delta V_g = e/C_g$). The superconductivity of the electrodes is suppressed by a magnetic field.

current, or if mechanical vibrations shake the charge distributions, the tunnel current can oscillate between the two extremes in the I-V characteristic.

The additional degree of freedom represented by the voltage of the middle electrode means that the double junction can be biased by a constant voltage supply (i.e., have a low shunt impedance) and still display Coulomb charging and other SET effects. This is not true for a single junction which does not permit a constant voltage source but demands a constant current source for SET effects to occur.

The first observations of the Coulomb charging effects were made already in the end of the 60's in tunnel junctions containing a large number of small particles.[19-21] There, tunneling between the two electrodes often occured via an intermediate particle that was charged up. Since then, experiments have been performed using one dominant grain[16], planar double junctions,[17] and, in a number of cases, an STM[22,23] where tunneling occurs, for example, between the tip and a sample through an intermediate droplet, a particle localized on the tip, etc. Recently, Coulomb staircases have been observed in series coupled, one-dimensional electron gas channels formed in a GaAs/AlGaAs heterojunction.[24] Hence the Coulomb charging effects and the spatial correlation of tunneling events have been demonstrated in a number of different two-junction systems.

1.2. Charge and Flux: Conjugated Variables

In a Josephson junction, with coupling energy $E_J (=(\Phi_o/2\pi)I_J$, $\Phi_o =h/2e)$ larger than the charging energy E_c $(=Q^2/2c$, $Q=2e)$ and $R_J << R_Q$, the phase $(4\pi e/h)\int V dt$ is a well defined entity (Correspondingly, we can say that the flux, $\int V dt$, or fluxoid, in a closed loop containing the junction has a well defined value). The charge, on the other hand is indeterminate with a fluctuating number of Cooper pairs. The normal Josephson effects dominate. In the opposite limit, $E_J << E_c$, $R >> R_Q$, the charge is the well defined entity and the charging effects dominate. Charge and flux are conjugate variables obeying the Heisenberg uncertainty relation. There are a number of variables, effects and applications in the single electron tunneling that have corresponding entities in Josephson tunneling. For example, charge in SET junctions corresponds to flux in Josephson junctions, and in particular, the electron charge, e, to the quantized flux, $\phi_o = h/2e$; current and voltage are corresponding variables; capacitance and Josephson inductance; resistance and conductance; series and parallel coupled junctions; time correlation of charge tunneling events gives rise to voltage (SET) oscillations while time correlated flux transfer gives current (Josephson) oscillations; sensitive electrometers based on series coupled SET junctions and sensitive magnetometers (SQUID's) based on parallel Josephson junctions, etc.

1.3. Bloch Oscillations

Correlated tunneling of Cooper pairs,[1] that are oscillations of the Bloch type, can take place in small Josephson junctions where the Josephson coupling energy is of the same order as the charging energy and the quasiparticle resistance is large (of the order or R_Q). These oscillations have not been observed yet and we will not discuss them further in this text.

1.4. Orthodox Theory of SET

A relatively simple, so called "orthodox," theory predicts the properties that have been described in this introduction.[1] It assumes an equilibrium situation between each tunneling event and it is in remarkably good accord with observed effects. It neglects a number of factors including finite dimensions of the comprised elements and energy quantizations in the electrodes. It should be revised when the classical (RC) and quantum (h/E_c) time scales of recharging in sufficiently small junctions become as short as the tunneling "traversal" time and the reciprocal plasma frequency that governs the relaxation of surface charge after tunneling. In this article we will only discuss results obtained from the "orthodox" theory.

2. ARRAYS OF JUNCTIONS - "CHARGE SOLITONS"

Consider several (N) tunnel junctions coupled in series. If one of the electrodes of a junction of the array is charged, the surfaces of that electrode will be charged (the additional charge will come up to the surface in order to maintain electro-neutrality in the interior). The neighboring electrodes will then be polarized, etc. The polarization chain would be infinite if charge was not lost to the surroundings by stray capacitances. With real structures, a charge pattern of finite length will be formed. It can be considered as a topological single-electron soliton.[25] Within a simple model that only considers electrostatic coupling of the electrodes to tunneling capacitances, C, and stray capacitances, C_o, the soliton field decays exponentially and the extension of the soliton has been calculated to be of the order of $2M \approx 2(C/C_o)^{1/2}$ junctions. It is typically 5 - 10 junctions long for the arrays that we will consider here. The potential distribution, i.e. the charge on the stray capacitances, around the center of a soliton is shown in Fig. 4.

In order to inject a soliton into the array, a certain threshold voltage, $V_t(\approx e/(CC_o)^{1/2}$ if $M<<N)$, has to be applied. Below that voltage, the junctions remain in a Coulomb-blockade state and no current passes along the array. After the soliton has entered the array edge, it will drift along the array of junctions. Solitons of similar charge

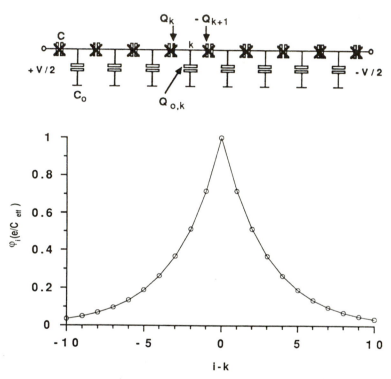

FIG. 4. A charge, e, on one junction, that is a member of a long array of series coupled junctions, affects the charge distributions on surrounding junctions. A charge soliton is formed. Its charge equals the supplied charge, e, and its extension is given by the junction and stray capacitances. It typically is of the order of a few junctions for common array parameters. The upper part of the figure shows an array of junctions, each of which has a capacitance of C and with stray capacitance C_o to ground from each intermediate electrode. The charge distribution on electrode k is shown. The lower part of the figure shows the potential distribution on the electrodes around the charged electrode k of a long homogeneous array with $M=3$.

repel each other. A new soliton will enter the array first when the first one has traveled some distance from the edge. A quasi-periodic one-dimensional lattice of solitons will slide along the array giving rise to correlation of tunneling events in both space and time. At large N and M, narrow band SET oscillations with frequency $f=I/e$ will occur.[25]

The offset voltage at large bias is $V_{off}=(N-M)e/2C$ when the number of junctions, N, is much larger than the extension, M, of the soliton.

The corresponding entity in a long Josephson junction (which can be considered as an array of small junctions in parallel) is a fluxon (soliton). Solitons on long Josephson junctions have been extensively studied. A sliding lattice of fluxons gives rise to radiation and a dc voltage $V=hf_J/2e$ across the junction according to the Josephson relation. We will return later to the analogy in discussing the locking of oscillations to microwave radiation and applications in the form of current and voltage standards.

3. SAMPLE FABRICATION

A suspended bridge technique developed by Dolan[26-28] can be used to fabricate ultrasmall junctions. It is based upon two resist layers exposed by an e-beam. The mask is formed in the upper layer which rests on the lower, undercut layer. The substrate is tilted at two different angles during the evaporation of the bottom and top electrodes in order to give the desired overlap, see Fig. 5. A tunneling barrier is formed by oxidizing the bottom electrode before the top one is deposited. Using aluminum, this results in individual junction resistances of the order of 5-1000kΩ per junction, as for the experiments[29-32] to be described, with an electrode line width and overlap of the order of 80 nm. The capacitance of such a junction is typically $(2-3) \cdot 10^{-16}F$ giving a charging energy corresponding to 3-4K. The stray capacitance to the ground from each electrode is small - of the order of $10^{-17}F$. The inductance of the intermediate strip is also small.

The number of junctions in an array can be varied over a large range. To avoid boundary effects, the length should be larger than 10-20 junctions. The total voltage over all the junctions of the array can be measured easily. The voltage across a single junction of the array can be determined by leads consisting of arrays of similar, high resistance tunnel junctions.[13] These are more easily fabricated in the same process than conventional high resistance strips next to the single junction. Charge

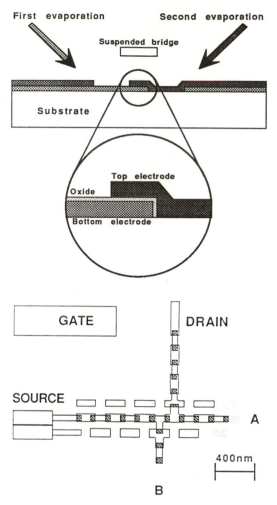

FIG. 5. The method of hanging resist bridges, in a two-layer resist technology, and evaporation from different angles is used to form overlap tunnel junctions as sketched above. The distances and angles are not to scale. The lower, undercut resist layer, upon which the bridge rests, is not shown. A possible configuration of an array of tunnel junctions is shown in the lower right hand corner of the figure. The array extends from the source to the drain. It is fabricated by evaporation from two angles through two sets of mask holes (the additional sets of metal patches do not disturb the measurements). Additional arrays connected to points A and B give a possibility to measure the voltage of a single junction with low lead capacitance.

can be coupled into intermediate electrodes capacitively via a gate or resistively. In the latter cases we can, again, use an array of junctions. A typical configuration is shown in Fig. 5.

The SET effects have to be measured at low temperature, prefereably in a dilution refrigerator reaching T<0.1K, unless the junctions are extremely small. To avoid the pick-up of noise, the measuring circuit should be symmetrically arranged and the leads well filtered. Microwaves may be fed into the junctions in crude ways.

4. COULOMB BLOCKADE IN 1-D ARRAYS

The current-voltage curve for a nineteen-junction array[31] is shown in Fig. 6. As T is lowered, the dc I-V curve becomes nonlinear, with a drastic suppression of the conductance below a threshold value, $V_t=2.0$ mV. There is no detectable current at low voltage bias (I<50 fA), note the expanded current scale. The resistance at zero bias is larger than 100 GΩ, i.e. at least 4 orders of magnitude higher than at large bias. V_t is definitely less than NV_Δ, where $V_\Delta=2\Delta/e\approx0.35$mV is the superconducting gap voltage. At large bias $V_{off}\approx7$mV. The Coulomb blockade is very pronounced at 50 mK but is also discernible at 4.2K.

The I-V curves can be compared with calculated curves using the "orthodox" theory[1] with charge solitons moving along the array.[25] Such plots are given in Fig. 6b. The calculations assumed a uniform array with identical junctions and reasonable parameters extracted from the experiment,[31] namely an asymptotic resistance of 210 kΩ per junction, $C = 2.4 \cdot 10^{-16}$F per junction, $C_o=1.2 \cdot 10^{-17}$F per electrode, a sub-gap resistance per junction of 1.4MΩ (if there were no charging effects in the junction) and $2\Delta(0)=0.38$meV. The calculated $V_t=2.4$meV is close to the experimental value. The discrepancy between the theoretical and experimental curves at voltages just above the threshold "supervoltage," where the experimental curve is not as steep as the theoretical one may be due to noise smearing, background charge, and an assumption of a constant tunnel resistance below the gap in the calculations.

The curves were only slightly affected by the superconductivity of the Al electrodes occurring at $T_c\approx1.2$K. The current rise at the voltage corresponding to the superconducting energy gap is rapidly smeared at increased tunneling resistance. In a sample with a resistance per junction of 820 kΩ it was not seen at all. The superconducting gap structure can be suppressed by a magnetic field. No Josephson current was observed

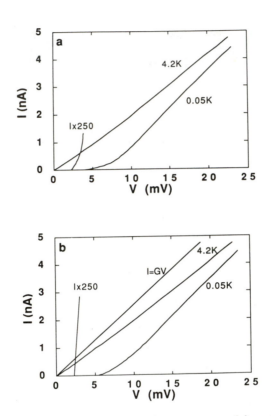

FIG. 6. dc I-V curves from an array of nineteen tunnel junctions for two temperatures. (a) Experimental curves for 70nmx90nm A1/I/A1 junctions. A
blow-up of the current scale for the 50 mK curve shows that the current is practically zero at low bias below $V_t \approx 2mV$. (b) Calculations using the theory of
Ref. 25. $C_o=1.2 \cdot 10^{-17}$ F, $C = 2.4 \cdot 10^{-16}$ F, R=210 kΩ per junction. The
superconducting gap has also been taken into account. From Ref. 31.

in any of fourteen arrays with junction resistances ranging from 33 to 1100KΩ (per junction).[32]

The large scale I-V characteristic was unaffected by an increase of the temperature up to about 0.5K. Above that temperature it was gradually smeared. An offset voltage is seen even at 4.2K, see Fig. 6.

The I-V curve was also smeared by applied microwaves. Additional fine structure could be distinguished in a few of the samples at low temperature. This will be treated next.

5. TIME CORRELATED TUNNELING IN ARRAYS

The power of the high frequency SET oscillations in high resistance junctions is weak and will be difficult to detect directly. The large impedance mismatch between the junctions and the environment makes the task even more difficult. However, we can apply ideas from corresponding Josephson effects. The first experimental verification of the ac Josephson effect was the observation of so called Shapiro current steps in the I-V curve of a superconducting tunnel junction irradiated by microwaves. The internally generated Josephson oscillation is mixed with the external radiation, or harmonics of it, and whenever the resulting difference frequency is zero, there will be an additional current contribution. The steps occur at $V_n = hf_J/2e = nhf_{ext}/2e$.

Likewise, we expect voltage steps to appear in the I-V curve of an ultrasmall, high resistance tunnel junction that is irradiated by microwaves. These steps should occur at currents of $I_n = nef_{ext}$ (note current, not voltage!). To amplify the structure, we can observe the differential resistance, dV/dI, as a function of bias current when high frequency radiation is applied. We expect peaks in the dynamic resistance at $I_n = nef_{ext}$.

Fig. 7a shows the tunneling resistance as a function of bias current for the same nineteen-junction array[31] as in Fig. 6a. Pronounced peaks appear at bias currents corresponding to $I = nef_{ext}$ when the array is exposed to $f_{ext} = 0.75$ GHz microwaves. The peak magnitude depends upon the microwave power (in a Bessel function dependence, compare with Shapiro steps in Josephson junctions). The peak location, however, remains fixed in bias current. The voltage location, on the other hand, depends on microwave power, temperature, magnetic field, etc. The peak structure is most easily seen at low temperature; 50 mK in Fig. 7, but it can be distinguished for temperatures up towards 1K.

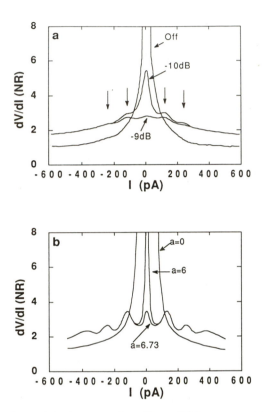

FIG. 7. The dynamic resistance as a function of bias current for the same array as in Fig. 6 at 50 mK for three values of the microwave power (f_{ext}=0.75GHz). It is normalized to the array resistance (NR). (a) experimental curves; (b) calculated ones. The difference in amplitude, a, between the two pumped curves in (b) corresponds to 1 dB in power, i.e. a similar difference as in (a). Arrows show nominal positions of the resistance peaks as expected from I=nef$_{ext}$. From Ref. 31.

Very similar, but even more pronounced, resistance peaks show up in calculated curves, see Fig. 7b. The calculations used the charge soliton model[25,31] and the same parameters as those in the theoretical Coulomb blockade curves of Fig. 6b. Curves are given for two microwave powers and no power. The difference between the microwave amplitudes has been chosen the same as for the experimental traces. The theoretical and experimental sets of curves have, overall, features that are similar in shape and location. The peaks in the experimental curves are broadened somewhat. This may be due to a number of factors: a finite ac-modulation of about 5 pA r.m.s. to measure the dynamic resistance; quantum fluctuations, that are estimated to be about 0.15K; pick-up of noise (most pronounced peaks were measured at nights when competing activities are low); as well as inhomogeneities in junction parameters and background charge.

Fourteen arrays of 15 to 53 Al junctions with resistances of 50-210 kΩ per junction were studied in ref. 32. Three of the arrays displayed well distinguishable peaks. Other arrays did not show the microwave induced structure. This may be caused either by too few and too low resistance junctions in a few of the arrays, or, in other cases, non-uniform arrays with internal charging effects in particularly weak junctions. The charge solitons will be pinned by junctions with a smaller capacitance. A random scattering of R and C parameters of more than 30% is expected to smear the microwave induced structure considerably. Internal charging effects give rise to Coulomb staircases and it may be difficult to distinguish weak radiation induced peaks in that structure. It can be noted that the arrays that did display resistance peaks were fabricated at times when the filament of the e-beam writing instrument was fresh and gave finer linewidths.

The magnitude of a resistance peak depends on the applied microwave frequency. It grows weaker as the frequency is increased. We could detect microwave induced peaks in bands ranging from 0.75 to 5 GHz, determined by the inductive microwave coupling between a three turn coil and the array. The peak structure decreased and became very small at $f_{ext}=5$GHz which is approaching the RC cut-off frequency of the array.

Stronger photon induced structure may be seen in previously unpublished data. The microwave coupling was improved by using a capacitive coupling instead of the inductive one. This gives a wider frequency coupling range.

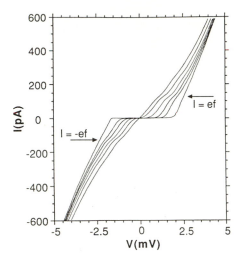

FIG. 8. I-V characteristics of an array consisting of 25 Al junctions irradiated by microwaves at a frequency of 0.8 GHz. The relative microwave powers for the curves are -∞, -26, -24, -22, -20 and -18 dBm (referred to the microwave generator). Superconductivity of the Al electrodes was quenched by a magnetic field. T=50 mK. R=60 kΩ per junction.

Note that steps are seen even directly in the I-V curve of Fig. 8. Microwaves of 0.8 GHz and at several power levels were applied to the 25-junction array. The superconductivity of the aluminum electrodes was quenched by a magnetic field. The absence of superconductivity seems to strengthen the SET oscillations; the steps could not be seen directly in the I-V curve when there was no magnetic field applied.

Up to fourth order peaks could be distinguished in the differential resistance curves as shown in Fig. 9. The higher order peaks are due to phase locking of harmonics of the external radiation. There is even a structure (or a weak peak) at a bias current of about 70 pA. This may be a subharmonic peak. The resistance curve of Fig. 9 was not symmetric. The peaks at negative current bias were much less pronounced than those for positive bias. However, it was possible to affect both the symmetry and the amplitudes of the peaks by applying a voltage to a gate electrode that was situated about 10 μm away from the array. By tuning

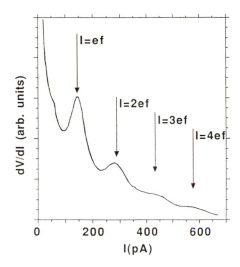

FIG. 9. Differential resistance as a function of current for a 25-junction array with microwaves applied at a frequency f=0.9 GHz. The peaks fall close to their expected values, I=nef. Note that there is even a sign of a subharmonic peak at about 70 pA. Superconductivity was quenched by a magnetic field. T=50 mK.

the gate voltage, it was possible to maximize the amplitude of a given peak. This indicates that there is a strong influence of the background charge on the SET process.

We can plot the current values of the peaks against applied frequency. This is done in Fig. 10 for four peaks in the three arrays with inductive coupling to the microwaves. Note that the points fall close to lines given by I=nef$_{ext}$ with n=±1 and ±2. This was also found for the capacitively coupled 25-junction array over the whole frequency range (0.1-1.5 GHz) that was tested. The good agreement supports the notion of time correlated tunneling events and SET oscillations. Even if we have not observed free running oscillations, we can claim that the period of the tunneling is locked to the external signal frequency.

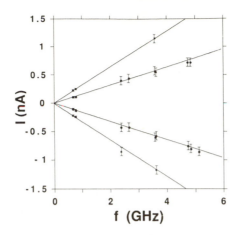

FIG. 10. Frequency dependence of the dc current positions of the $n=\pm1$ and $n=\pm2$ resistance peaks for three arrays with $N=12$, 15 and 19. The lines have slopes $\pm e$ and $\pm2e$. The data points fall close to these lines in good accord with the theory of time correlated tunneling. From Ref. 31.

6. SPACE CORRELATED TUNNELING IN ARRAYS

A tunneling event in one junction charges next electrode and is expected to be shortly followed by an event in next junction, and so on. The charge soliton will be transported along the array. Like the case of two series coupled junctions, which we discussed in the introduction, the tunneling and the Coulomb blockade will be strongly influenced by the charge distribution on the intermediate electrodes. The I-V curve will display the Coulomb blockade behavior with no current for $V<V_t$ when the injected charge $Q_o=0$ or ne. It is shifted maximally when $Q_o=(n+1/2)e$ for which there is no blockade at zero bias.

The response of the voltage U across a two-junction array, at a constant bias current, as the gate voltage is varied was shown in Fig. 3. The gate is coupled capacitively to the array. The period corresponds to a change of the (determining) intermediate electrode charge of e. Going to three junctions there will be additional splits in the U-V_g relation corresponding to different periods if the junctions are not identical. Similar relationships can be traced for multi-junction arrays as we will discuss later under applications.

6.1. Turnstiles and Charge Pumps

SET oscillations were induced by an rf field applied perpendicular (via a capacitive gate) to an N=4 array of junctions in a recent experiment[33] by the Delft and Saclay groups. The device called a "turnstile" is depicted in Fig. 11. When a positive voltage is induced in the middle electrode, one electron tunnels from the bottom to the middle electrode in Fig. 11a. During next half-cycle, when the middle electrode voltage is negative, the electron continues upward. The flow of electrons is locked to the "turnstile" frequency. The dc I-V curve of the device is shown in

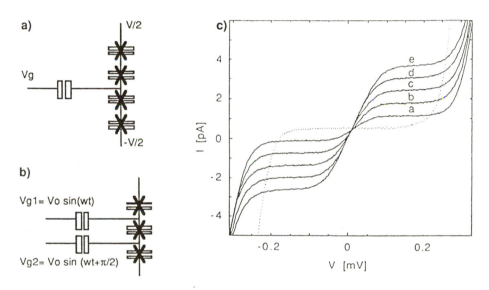

FIG. 11. (a) Schematic picture of the so called turnstile for transfer of single electrons through a linear array of 4 small tunnel junctions controlled by an rf-voltage applied to a gate (Geerligs et al., Ref. 33). (b) The charge "pump" (Pothier et al., Ref. 34) employs two gates with phase shifted rf-signals. (c) I-V curves of a turnstile: dotted curve is for the case with no rf-signal to the gate, curves a)-e) with frequencies 4-20 MHz in 4 MHz steps. Pronounced current steps are seen at I=ef. They are flat for f<10 MHz. C=0.5 fF. Cg=0.3 fF. R=340 kΩ. The superconductivity of the electrodes was suppressed. T=10 mK. From Ref. 33.

Fig. 11c for several values of the rf frequency, $f = n \cdot 4\text{MHz}$. Well pronounced voltage steps are seen at $I=ef$. The deviation from this relation was less than $1:10^3$.

The Saclay group has further developed the device[34] to a "charge pump." They noticed that is is possible to transfer electrons through the array without applying a dc voltage across it. Two coherent, $90°$ out-of-phase rf signals are applied as shown in Fig. 11b. During the first quarter period, an electron is drawn from one end of the array to the first gate electrode. It continues to the second gate electrode during the second quarter period and is later transferred to the other end of the array. Correspondingly one can achieve zero current steps at $V_n = nhf/2e$ in Josephson junctions.

7. OTHER TYPES OF JUNCTIONS

We have hitherto considered metal/oxide/metal tunnel junctions. However, we can also envisage other types of junctions for the SET effects.

Guinea and Garcia[35] suggested a type of "turnstile" based on a vibrating STM and an intermediate small droplet. When the STM tip is at its closest approach to the droplet one electron can tunnel resonantly and charge the droplet. The latter is then discharged during the relatively long time that the tip is withdrawn. The current is limited by the mechanical cut-off frequency of the vibrating tip.

A higher cut-off frequency should be possible using an optically modulated medium, e.g., changing the dielectric constant in a junction or the photoconductivity of a weak link by short laser pulses with a high repetition rate. Difficulties in coupling the optical signal into the high impedance device might be overcome using antenna or grating designs.

A double junction behavior has recently been observed[24] in a GaAs heterojunction 2-D electron gas where constructions were formed by a split gate with protrusions. Hence a confined channel with two weak points was formed. One can foresee many further developments of 1-D electron gas structures in semiconductors or semimetals. Point contacts with resistances higher than R_Q are needed, they do not need to be of the conventional tunnel junction type.

8. POSSIBLE APPLICATIONS OF SINGLE ELECTRON TUNNELING

Several types of applications of SET can be envisaged.[1,36] Here we will give a couple of examples that are based on the time and space correlations. We will not consider digital devices, like "single-electron-logic" ones, but refer to the literature.[37,38]

8.1. Dc Current Standard

The unity of volt is now mainained with the aid of the Josephson ac effect. Current steps are induced with a voltage separation, $\Delta V = (h/2e)f_{ext}$, by applied microwaves, f_{ext}. Frequency can be measured with extremely high accuracy. Long arrays of Josephson junctions and high microwave frequencies enable voltages of about 10V without comparators. It is presently possible to determine the voltage unit with an accuracy close to 10^{-8} V using an accepted, long term agreed value of $h/2e$.

Likewise, it may be possible to use microwave induced SET voltage steps to maintain a current standard. Sharp steps at not too low currents, much sharper than those observed up till now, are needed to give well defined current values. A detailed analysis of the precision that may be achieved with the current standard has not been carried out, but in principle this precision should be of the same order as the one for the Josephson standard.

A problem is the range of currents that can be maintained by the SET standard, particularly at the high current end. The induced step structure becomes less well defined as the frequency is increased - and the charging time, or other limiting time scales, are approached. If we assume that microwave frequencies up to about 10 GHz can be used, we would be able to get of the order of 10^{-9} A for a single 1-D array. The phase locking of Bloch oscillations of superconducting pairs may give an extension of the current scale of the order of two magnitudes. Coupling a large number of 1-D arrays in parallel may give another 2-3 orders. Finally, the current can be increased further by using superconducting comparators[39] without a loss of precision. Presently, it seems difficult reaching 1 A. On the other hand, the prospects of reaching 1 V in Josephson junctions, without using limiting comparators, were also dim during the first years of experiments. And now we can obtain 10 V.

An extension of the range to small currents may be feasible using the turnstile (or pump) device. Sharp steps and a precision of the order of 10^{-4} has already been achieved during the first experiments in the low MHz range. Parasitic macroscopic quantum tunneling of charge, i.e. a passage of charge through many junctions at the same time, has to be suppressed.[40,41] Small currents from a stable source should be possible to calibrate by injecting them into the middle electrode of an array and counting oscillations of the I-V curve in time. It should be possible to determine currents in the range 10^{-21} to 10^{-15} A in that way.

Basic electro-dynamic relations can be tested by forming the so called quantum metrology triangle. It combines the SET (or Bloch) oscillations, that give a relation between current and frequency, with the Josephson voltage standard (frequency to voltage) and the quantum Hall resistance, giving a quantized voltage/current (resistance) relation. Preferably, these should be contained on the same chip.

8.2. Sensitive Electrometer

We have seen that the I-V curve of two, or more, series coupled junctions is shifted by changes in the charge on a middle electrode. By keeping the current constant and observing the change in the source-to-drain voltage as a capacitively coupled gate voltage is varied, we can estimate the response. An example is shown in Fig. 12 for a thirteen-junction array.[29] A periodic response is registered. The period corresponds to a change in charge of one electron on an intermediate electrode. The charge sensitivity at the steepest part of the response curve corresponds to a few times 10^{-4} $\bullet e/\sqrt{Hz}$ at a lock-in amplifier frequency of 10 Hz, where the 1/f noise cannot be neglected. Theoretically, estimates of the charge sensitivity in the white noise limit have given $10^{-5}e/\sqrt{Hz}$ as the best value.[36,42]

Obviously, the sensitivity is extremely high. Commercially available electrometers are about 5 to 6 orders of magnitude less sensitive. The difficulty is to couple charge to the middle electrode. The object to be measured will have a much larger capacitance than the one coupling to the middle electrode. A large portion of the sensitivity will be lost.

That problem has been solved for the Josephson counterpart - the dc-SQUID that consists of two junctions connected in parallel and a flux transformer coupling magnetic flux to the loop. The SQUID magnetometer has a flux sensitivity much better than the quantized flux, $\phi_o = h/2e \approx 2 \bullet 10^{-15} Vs$.

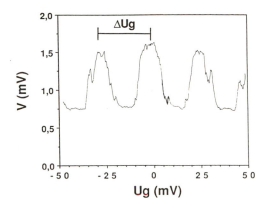

FIG. 12. The response to a gate voltage is illustrated. A 13-junction array was biased with a fixed current (≈ 5 pA) at T=1.3 K and the voltage over the array is plotted as a function of applied gate voltage. A periodic response is seen - the period corresponds to a change of one electron charge on an intermediate electrode. By measuring the noise level at 10 Hz, biased at the steepest part of the response curve, we deduce a sensitivity for change in charge of $2 \cdot 10^{-4} e/\sqrt{Hz}$. The transistor "gain," i.e. the relation between changes in the output and gate voltages, is of the order of 0.2 for this curve. From Ref. 29.

8.3. Transistors - Capacitively and Resistively Coupled

The response of the voltage from source to drain to changes in gate voltage at a constant bias current as shown in Fig. 12 can also be used in a three-terminal, transistor type device. The gate can be capacitively coupled to the array, as in Fig. 12, or resistively coupled, where charge is injected into the middle electrode via a large resistor.[42]

The gain, i.e., the ratio between output voltage at constant current and input gate voltage, is 0.2 at the optimal high slope region of Fig. 12. This is about as high as expected for a capacitively coupled transistor with the array parameters in question. Higher voltage gain may be more easily obtained for resistively coupled transistors.

We have tested a resistively coupled SET transistor where the gate lead contains an array of high resistance junctions.[43] A complication in the interpretation of the results is that no charge is transported through the gate array until a Coulomb blockade threshold voltage is exceeded.

T. Claeson and P. Delsing

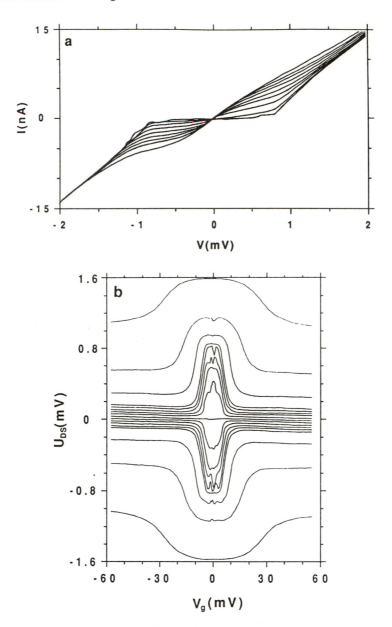

FIG. 13. A resistively coupled transistor where charge is injected via a gate resistor consisting of 13 A1/I/A1 tunnel junctions fabricated in the same process as the source (S) and drain (D) junctions. (a) Current vs source-to-drain voltage for gate voltage increasing in 2.5 mV steps. (b) Source-to-drain voltage as a function of gate voltage for different bias currents. T=50 mK. R_D+R_S=0.11 MΩ. R_g=3.0 MΩ. A maximum voltage gain of 0.15 and power gain of 4.1 were estimated for this device.

The response of a resistively coupled transistor is shown in Fig. 13. I-V curves are shown for several gate voltages in Fig. 13a. Choosing a constant current, it is possible to map the gate voltage dependence. This is done in the lower part of the figure which shows the drain-to-source voltage as a function of gate voltage for several bias currents. A voltage gain of 0.85 has been obtained for a resistively coupled device.

A theoretical analysis based on the "orthodox" theory of correlated tunneling was performed by Kazacha and Likharev.[43] They could qualitatively reproduce the features of the array coupled transistor, see Fig. 14. Junction parameters were determined from experiments on similar junctions.[31]

The current gain of a SET transistor can be considerably larger than unity as the gate impedance is much larger than the resistance between source and drain. A capacitively coupled transistor can have an infinite current gain. Power gains of the order of 4 were registered for the resistively coupled transistors.[30,43] A transistor with gain can be used, e.g., as a logic element. However, other devices that more effectively use the unique ability of shuffling around single electrons in arrays of ultrasmall junctions seem preferable.[37,38] Again, we can compare with fluxon based Josephson logic circuits.

9. CONCLUSION

Single electron tunneling effects have been conclusively proven during the last couple of years. It has been possible to fabricate ultrasmall tunnel junctions in a controlled way. In particular, it has been possible to decouple them from the parasitic capacitance of the measurement leads. The charging energy of the junction capacitance can be made larger then thermal and quantum fluctuation energies.

In this survey, we have shown that tunnel events in series coupled junctions can be correlated in space and in time. The sensitivity of the I-V curve to changes in the charge distribution on intermediate electrodes can be used to control the tunneling of single electrons by sub-electron charges. Hypersensitive Coulomb meters and SET transistors are devices that could utilize these effects if coupling problems are solved.

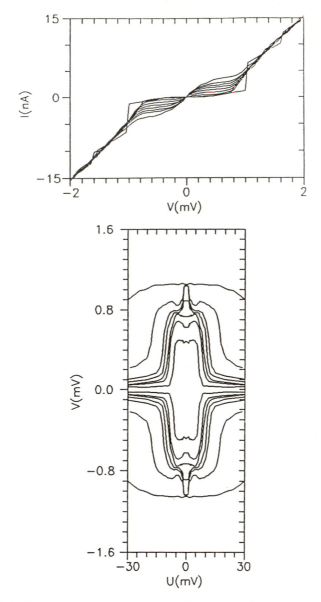

FIG. 14. Calculated curves corresponding to the resistively coupled SET-transistor of Fig. 13. The "orthodox" theory of single-electron tunneling was used to calculate the I-V curves for gate voltage increments of 2.5 mV (top part). The lower diagram shows the source-to-drain voltage as a function of gate voltage for several bias currents. $C=0.3$ fF and $C_o=0.02$ fF were used. The maximum voltage gain is estimated to 0.25. From Ref. 43.

A free running, well defined SET oscillation has been predicted but has not yet unequivocally been proven experimentally. However, phase locking of tunnel events to an applied microwave signal gives induced steps in the I-V curve that follow the fundamental relation $I=nef_{ext}$. All observations are fully in line with theoretical concepts (the "orthodox" theory of SET) that include free running SET oscillations. This may give rise to quantum current and capacitance standards in the future if sharp enough structures and high enough currents can be realized. The "turnstile" and "charge pump" are interesting devices that may give high precision at the low current end. In those experiments, the charge transfer is triggered by the applied signal. There is no free running oscillation in that case. This seems to be different from the case of phase locked oscillations.

Most of the effects and applications that have been discussed here have their counterparts in Josephson tunneling. There exists a deep, but incomplete, analogy between SET and Josephson effects.

Many problems remain to be solved. Complications due to size quantization and finite time scales for tunneling and relaxation should be taken into account. Even smaller and better controlled junctions that are well protected from the environment need to be made. New electronic devices that utilize unique features of the SET concept can be envisaged. The observation of possible Bloch oscillations of Cooper pairs in superconducting junctions is another challenge. The transition between single electron and pair tunneling, as the ratio between the Coulomb charging and Josephson energies is varied, remains to be mapped out. The field is expected to be very lively in the 90's.

ACKNOWLEDGEMENT

We thank the Moscow State University group for their helpful cooperation, and in particular, L. Kuzmin and K. K. Likharev, who are co-authors of many of the papers referenced here.

REFERENCES

1. For reviews, see K.K. Likharev, "Correlated Discrete Transfer of Single Electrons in Ultrasmall Tunnel Junctions," IBM J. Res. Dev. **32**, 144 (1988); D. V. Averin and K. K. Likharev, "Single-Electronics: A Correlated Transfer of Single Electrons and Cooper Pairs in Systems of Small Tunnel Junctions," in Quantum Effects in

Small Disordered Systems, B. L. Altshuler, P. A. Lee and R. A. Webb, eds., Elsevier, to be published; K. K. Likharev, Dynamics of Josephson Junctions and Circuits, Gordon and Breach Sci. Publ., N.Y., 1986; G. Schön and A. D. Zaikin, "Quantum Coherent Effects, Phase Transitions, and the Dissipative Dynamics of Ultra Small Tunnel Junctions," Physics Reports, to be published.

2. D. V. Averin and K. K. Likharev, "Possible Coherent Oscillations at Single-Electron Tunneling," in SQUID'85, H. Lübbig and H.-D. Hahlbohm, Eds., W. de Gruyter, Berlin 1985, p. 197.

3. D. V. Averin and K. K. Likharev, "Coulomb Blockade of the Single-Electron Tunneling, and Coherent Oscillations in Small Tunnel Junctions," J. Low. Temp. Phys. **62**, 345 (1986); Zh. Eksp. Teor. Fiz. **90**, 733 (1986) [Sov. Phys.-JETP **83**, 427 (1986)].

4. E. Ben-Jacob and Y. Gefen, "New Quantum Oscillations in Current Driven Small Junction," Phys. Lett. **108A**, 289 (1985).

5. R. Brown and E. Simanek, "Transition to Ohmic Conduction in Ultrasmall Tunnel Junctions," Phys. Rev. B **34**, 2957 (1986).

6. F. Guinea and G. Schön, "Coherent Charge Oscillations in Tunnel Junctions," Europhys. Lett. **1**, 585 (1986).

7. Yu. V. Nazarov, "Anomalous Current-Voltage Characteristics of Tunnel Junctions," Zh. Eksp. Teor. Fiz. **95**, 975 (1989) [Sov. Phys. JETP **68**, 561 (1989)]; D. V. Averin and Yu. V. Nazarov, "Coulomb Fingerprints on the I-V Curves of the Normal Tunnel Junctions," to be published.

8. M. H. Devoret, D. Estève, H. Grabert, G. -L. Ingold, H. Pothier and C. Urbina, "Effect of the Electromagnetic Environment on the Coulomb Blockade in Ultrasmall Tunnel Junctions," Phys. Rev. Lett. **64**, 1824 (1990).

9. S. M. Girvin, L. I. Glazman, M. Jonson, D. R. Penn and M. D. Stiles, "Quantum Fluctuations and the Single-Junction Coulomb Blockade," Phys. Rev. Lett. **64**, 3183 (1990); K. Flensberg, S. M. Girvin and M. Jonson, "Quantum Fluctuations and Charging Effects in Small Tunnel Junctions," to be published.

10. M. Ueda and S. Kurihara, "Quantum Theory of Ultrasmall Tunnel Junctions," Proc. LT-19 Satellite Workshop on Macroscopic

Quantum Phenomena (ed. T. D. Clark), World Sci., Singapore, 1990, to be published.

11. G. -L. Ingold, "Influence of the Electrodynamical Environment on Tunnel Junctions," NATO Adv. Res. Workshop: Rate Processes in Dissipative Systems: 50 Years After Kramers, to be published.

12. A. N. Cleland, J. M. Schmidt and J. Clarke, "Charge Fluctuations in Small-Capacitance Junctions," Phys. Rev. Lett. 64, 1565 (1990).

13. P. Delsing, K. K. Likharev, L. S. Kuzmin and T. Claeson "Effect of High-Frequency Electrodynamic Environment on the Single-Electron Tunneling in Ultrasmall Junctions," Phys. Rev. Lett. 63, 1180 (1989).

14. T. Claeson, P. Delsing, L. S. Kuzmin and K. K. Likharev, "Two Fundamental Results from Low-Temperature Experiments with One-Dimensional Arrays of Ultrasmall Tunnel Junctions," Proc. 3rd Int. Symp. Foundations of Quantum Mechanics (Eds. S. Kobayashi, H. Ezawa, Y. Murayama and S. Nomura, eds.), The Phys. Soc. Japan, Tokyo, 1990, p. 255.

15. K. Mullen, E. Ben-Jacob, R. C. Jaklevic and Z. Schuss, "I-V Characteristics of Coupled Ultrasmall-Capacitance Normal Tunnel Junctions," Phys. Rev. B37, 98 (1988).

16. L. S. Kuzmin and K. K. Likharev, "Direct Observation of the Discrete Correlated Single Electron Tunneling," Pis'ma v. Zh. Eksp. Teor. Fiz. 45, 389 (1987) [JETP Lett. 45, 495 (1987)].

17. T. A. Fulton and G. J. Dolan, "Observation of Single Electron Charging Effects in Small Tunnel Junctions," Phys. Rev. Lett. 59, 109 (1987).

18. L. J. Geerligs, V. F. Anderegg, J. Romijn and J. E. Mooij, "Single Cooper-Pair Tunneling in Small-Capacitance Junctions," Phys. Rev. Lett. 65, 377 (1990).

19. C. A. Neugebauer and M. B. Webb, "Electrical Conduction Mechanism in Ultrathin Evaporated Metal Films," J. Appl. Phys. 33, 74 (1962).

20. H. R. Zeller and I. Giaever, "Tunneling, Zero-Bias Anomalies and Small Superconductors," Phys. Rev. 181, 789 (1969).

21. J. Lambe and R. C. Jaklevic, "Charge Quantization Studies Using a Tunnel Capacitor," Phys. Rev. Lett. **22**, 1371 (1969).

22. P. J. M. van Bentum, R. T. M. Smokers and H. van Kempen, "Incremental Charging of Single Small Particles," Phys. Rev. Lett. **60**, 2543 (1988).

23. R. Wilkins, E. Ben-Jacob and R. C. Jaklevic, "Scanning-Tunneling-Microscope Observations of Coulomb Blockade and Oxide Polarization in Small Metal Droplets," Phys. Rev. Lett. **63**, 801 (1989).

24. U. Meirav, M. A. Kastner and S. J. Wind, "Single Electron Charging and Periodic Conductance Resonances in GaAs Nanostructures," Phys. Rev. Lett. **65**, 771 (1990); L.P. Kouwenhoven, B. J. van Wees, R. Kraayeveld and K. J. P. M. Harmans, "Electronic Transport in an Artificial One Dimensional Crystal," to be published.

25. N. S. Bakhvalov, G. S. Kazacha, K. K. Likharev and S. I. Serdyukova, "Single-Electron Solitons in One-Dimensional Tunnel Structures," Zh. Eksp. Teor. Fiz. **95**, 1010 (1989) [Sov. Phys. JETP **68**, 581 (1989)]; K. K. Likharev, N. S. Bakhavalov, G. S. Kazacha and S. I. Serdukova, "Single-Electron Tunnel Junction Array: An Electrostatic Analog of the Josephson Transmission Line," IEEE Trans. Magn. **MAG–25**, 1436 (1989).

26. G. J. Dolan, "Offset Masks for Lift-Off Photoprocessing," Appl. Phys. Lett. **31**, 337 (1977).

27. A variant of the technology, but with a supported fiber above the substrate, was used by J. Niemeyer, P.T.B. Mitt. **84**, 251 (1974).

28. G. J. Dolan and J. H. Dunsmuir, "Very Small ($\geq$ 20 nm) Lithographic Wires, Dots, Rings and Tunnel Junctions," Physica B **152**, 7 (1988).

29. L. S. Kuzmin, P. Delsing, T. Claeson and K. K. Likharev, "Single-Electron Charging Effects in One-Dimensional Arrays of Ultra-Small Tunnel Junctions," Phys. Rev. Lett. **62**, 2539 (1989).

30. P. Delsing, "Single Electron Tunneling in Ultrasmall Tunnel Junctions," thesis, Chalmers Univ. Techn., Göteborg, 1990.

31. P. Delsing, K. K. Likharev, L. S. Kuzmin and T. Claeson, "Time-Correlated Single-Electron Tunneling in One-Dimensional Arrays of Ultrasmall Tunnel Junctions," Phys. Rev. Lett. **63**, 1861 (1989).

32. P. Delsing, T. Claeson, K. K. Likharev and L. S. Kuzmin, "Observation of Single-Electron-Tunneling Oscillations," Phys. Rev. B.

33. L. J. Geerlings, V. F. Anderegg, P. A. M. Holweg, J. E. Mooij, H. Pothier, D. Estève, C. Urbina, and M. H. Devoret, "Frequency-Locked Turnstile Device for Single Electrons" Phys. Rev. Lett. **64**, 2691 (1990).

34. H. Pothier, P. Lafarge, P. F. Orfila, C. Urbina, D. Esteve and M. H. Devoret, "Single Electron Pump Fabricated with Ultrasmall Normal Tunnel Junctions," to be published; L. S. Kuzmin and K. K. Likharev, "Direct Observation of the Discrete Correlated Single Electron Tunneling," Pis'ma v. Zh. Eksp. Teor. Fiz. **45**, 389 (1987) [JETP Lett. **45**, 495 (1987)]

35. F. Guinea and N. Garcia, "Scanning Tunneling Microscopy, Resonant Tunneling and Counting Electrons: A Quantum Standard of Current," Phys. Rev. Lett. **65**, 281 (1990).

36. K. K. Likharev, "Single-Electronics: Correlated Transfer of Single Electrons in Ultrasmall Junctions, Arrays and Systems," to be published.

37. K. K. Likharev, "A Possibility to Design Analog and Digital Integrated Circuits Based on Discrete Single-Electron Tunneling," Mikro-elektronika **16**, 195 (1987) [Sov. Micro-electron.].

38. K. K. Likharev and V. K. Semenov, "Possible Logic Circuits Based on the Correlated Single-Electron Tunneling in Ultrasmall Junctions," Extended Abstracts, International Superconducting Electronics Conference, Tokyo, 1987, p. 182.

39. R. F. Dziuba and D. B. Sullivan, "Cryogenic Direct Current Comparators and their Applications," IEEE Trans. Magn. **MAG-11**, 716 (1975).

40. D. V. Averin and A. A. Odintsov, "Macroscopic Quantum Tunneling of the Electric Charge in Small Tunnel Junctions," Phys. Lett. A **140**, 251 (1989); L. J. Geerligs, D. V. Averin and J. E.

Mooij, "Observation of Macroscopic Quantum Tunneling through the Coulomb Energy Barrier," Phys. Rev. Lett., to be published.

41. D. V. Averin, "Quantum Tunneling of the Electric Charge," NATO Adv. Res. Workshop <u>Rate Processes in Dissipative Systems: 50 Years After Kramers</u>, to be published.

42. K. K. Likharev, "Single-Electron Transistors: Electrostatic Analogs of the DC SQUIDS," IEEE Trans. Magn. **MAG–23**, 1142 (1987).

43. P. Delsing, T. Claeson, G. S. Kazacha, L. S. Kuzmin and K. K. Likharev, "1-D Array Implementation of the Resistively-Coupled Single Electron Transistor," IEEE Trans. Magn., to be published.

HIGH Tc SUPERCONDUCTORS: A CONSERVATIVE VIEW

J. Friedel

Laboratoire de Physique des Solides
Université Paris XI
91405 Orsay Cedex
France

Abstract - Classical BCS superconductive couplings leading to small T_c's rely on phonon mediated weak coupling of delocalized electrons in three dimensional structures. Attempts to produce or explain higher T_c have considered the breakdown of one or several of these hypotheses. The possible role of quasilowdimensionality for delocalized electrons will be stressed, and the potential superiority of fine mesh three dimensional networks of chain conductors argued from that point of view.

In the case of the CuO_2 planes, the possible roles of covalency and antiferromagnetism will be discussed.

1. INTRODUCTION

This introduction to a complex field will try and set the problem of high T_c superconductivity in a historical perspective.

Since Bardeen, Cooper and Schrieffer's work, superconductivity is attributed to a coupling of delocalized electronic carriers through a polarization field.

Highlights in Condensed Matter Physics and Future Prospects
Edited by L. Esaki, Plenum Press, New York, 1991

If the electrons are fermions with a *density of states* n(E) which varies *smoothly* near the Fermi level, as occurs necessarily in reasonably isotropic 3d structures, and if the *relative coupling strength* $\lambda = V_\eta(E_M) \cong V/E_M$ is *weak*, one obtains

$$T_c \cong T_o \exp(-1/\lambda) \tag{1}$$

where $k_B T_o$ is the Debye Temperature T_D, of order 100^K. Thus T_c much larger than 10^K cannot be explained that way.

This difficulty was realized quite early on. Various ways of circumventing it have been tried, testing the three fundamental assumptions leading to (1). We replace the problem of oxide superconductors in this perspective before discussing in more details the possible interplay of covalency and magnetism in the description of correlations and transport properties along the CuO_2 planes.[1]

2. THE PROBLEM OF HIGH T_C'S

2.1 Strong Couplings

The first idea that comes to mind is to consider stronger (phonon) couplings λ. If equation (1) were still valid, T_c could reach very sizeable values, of order T_D.

This however is not right, as seen already in McMillan's formula, which gives the first correction for increasing λ's. This replaces λ by $\lambda/(1 + \lambda)$, thus divides T_c by a factor $e \cong 2.5$. But for really large values of λ, one goes into the a boson like *polaron* regime, where coupling must be made in real space. With increasing λ's, these polarons take increasing masses and localization, thus decreasing coupling strength; thus in this limit T_c. decreases towards zero for increasing λ. As a result, T_c goes through a maximum for an intermediary value of λ, where it is far below T_D.

The possibility of superconductivity by creation of bipolarons was proposed early on, in competition with the BCS scheme; it has been developed systematically and led Müller to explore the field of peroxides, i.e. ionic solids with large polarisabilities and sizeable electronic conductivities. However, at least in its simplest form where emphasis is on couplings by atomic displacements, this does not seem very promising: it would lead to very modest T_c's and, for the oxides, is difficult to reconcile with the fermion properties of the electronic carriers as evidenced by NMR Knight shift and photoemission spectroscopy in the normal state above T_c.

2.2. Electron Couplings

In the weak coupling limit, where equation (1) holds, $k_B T_o$ is now an electronic cut off energy, which can be much larger than $k_B T_D$. One can therefore expect a large T_c, without going to very large coupling strengths, where the same difficulties can be expected as for ordinary polarons.

Although this idea was first propounded without success for the d electrons of transitional metals, two cases are now known where such couplings probably predominate.

In Steglich's *heavy fermions* compounds, the small f shells of lanthanide or actinide elements exchange slowly electrons with the d conduction band. This broadens the f states into narrow virtual f states, building at low temperature a narrow and incompletely filled band, with large values of $n(E_M$ and large effective masses. The smallness of the f orbitals leads to very strong ff correlations which both increase the effective mass and provide a possible superconductive coupling with specific symmetry of the gap. Indeed some of these compounds show such superconductivity. But the T_c's observed remain very small, probably because of the large masses of the carriers and of the poor stability of the f band with respect to thermal magnetic decoupling of the f shells.

In Jerome's *organic superconductors* of the $(TMTSF)_2X$ family, electrical conductivity is made possible along stacks of flat TMTSF unsaturated molecules by electron transfer to neighboring X (negative) ions. Because of the quasi-one-dimensional nature of these stacks, most of these compounds show a low temperature Fröhlich instability, leading to antiferromagnetism because of a sizeable ratio of electron interactions over Fermi energy, even if these two terms have smaller values than in metals. If pressure or doping are applied, a superconductive phase can be observed at low temperature, in direct contact with antiferromagnetism. A study of the (large) fluctuations above the triple point between T_N and T_c agrees with the idea that in such compounds superconductivity arises through a coupling via some kind of antiferromagnetic fluctuations. Here again however, T_c remains of order 1^K, because T_N is of that order. And Little's initial idea that in organic compounds T_c could be large due to strong electronic polarization of the molecules by moving carriers does not seem to hold.

In *superconductive oxides*, electron interactions are also strong enough to induce antiferromagnetism, at least in a number of compounds. This has led most people to think that electron interactions are responsible for the large T_c's observed. This might well be true and we shall discuss below what model of electron correlations is most likely to occur for oxides. The inspection of phase diagrams shows however a strong difference with organic superconductors, where T_c has its maximum value at the triple point with T_N; in oxides in the contrary, T_c is strongly depressed near the region of antiferromagnetism. This suggests to me that in oxides antiferromagnetism and superconductivity are more likely competitive instabilities, and not helping each other as in the organic compounds. If one wants to develop a model of superconductive coupling through antiferromagnetic fluctuations, it must refer to very specific fluctuations, with an optimal coherence length of a few interatomic distances. It seems strange to me that such a very specific condition would apply in all known cases of oxide superconductivity.

2.3. Strong (Quasi 1 or 2D) Anisotropy

Equation (1) assumes $n(E)$ to vary smoothly near the Fermi level E_M. It can then be taken as its average value $n(E_M)$ outside the integral that relates T_c to λ

This is not necessarily the case for stacking of chains or planes. Indeed, in 1 or 2 dimensional compounds, $n(E)$ has infinite van Hove anomalies which subsist as very substantial humps in quasi low dimensional situations where electrons can jump slowly between chains or planes or when electron scattering is included. If the Fermi level is near enough to such a hump, T_c can take values much larger than in (1). Furthermore T_c depends much less on the cutoff energy T_0. Thus large values of T_c can be expected for suitable dopings in the weak coupling regime; little dependance of T_c is expected on isotope masses even if coupling is through phonons.

The effect described is maximum for (quasi 1D) *chain compounds*, where the optimum superconductive coupling varies as a power law of λ. But 1d thermal fluctuations strongly reduce T_c from such a meanfield estimate. Such fluctuations are observed in the organic $(TMTSM)_2X$ superconductors, as well as in chain compounds with antiferromagnetic

(SDW) or lattice modulation (CDW) instabilities. This is therefore *not* the best geometry to explore.

In (quasi 2D) *plane compounds*, the van Hove anomaly effect is still very sizeable. Thermal fluctuations do not depress T_c appreciably. Organic superconductors such as $(BEDTTTF)_3Cu(SCN)_3$ with $T_c \cong$ 10^K, seem to fall within this category: their stacks of unsaturated molecules are arranged in parallel planes, where the chains are nearer to each other within each plane than between planes; and the corresponding quasi-cylindrical surfaces of constant energy show a van Hove anomaly near to the Fermi level. The CuO_2 planes of the high T_c oxides also show, in band structures that neglect magnetic complications, one or several strong van Hove anomalies, which could easily explain T_c of order 100^K or above for suitable dopings using very standard strengths of phonon mediated interactions. However, if the interplane coupling is small and if a magnetic field is applied normal to the planes, the lattice of vortex lines which is formed between H_{c1} and H_{c2} seem to "melt" already below T_c, owing to the strong anisotropy of the vortex line tension; and this seems to limit the practical usefulness of the best of such compounds well below T_c.

If one follows this line of thought, it becomes clear that the best solution would correspond to a chain structure, if one could avoid low d fluctuations. This is possible if the parallel *chains* are *coupled in three dimensional arrays*, with a mesh less than the coherence length ξ of superconductivity along the chains. For high T_c superconductors, where ξ is inversely proportional to $k_B T_c$ thus small, this means in practice a mesh of *molecular dimension*, and this periodic coupling must furthermore not spoil the quasi 1d van Hove anomalies of the independent chains. This set of somewhat contradictory conditions might be approximately fulfilled in the A15 compounds, with their interpenetrating chains of transitional metals, and explain their relatively large T_c's. This might also apply to the cubic oxides $Ba(PbBi)O_3$ and $(Ba,K)BiO_3$, with sets of parallel chains which use different orbitals on the atoms where they cross. It would surely be worthwhile to explore much more systematically the effect of doping on the Fermi surface and T_c in such compounds.

3. CORRELATIONS AND COVALENCY IN COPPER OXIDE PLANES

More detailed analyses of transport in the CuO_2 planes of the oxide superconductors rely on a tight binding description of the electronic structure.

At the most elementary level, this involves the Cu 3d and O 2p orbitals that build the partially occupied band responsible for transport. We call E_d, E_p their energies (for given ionicities), t the dp transfer energy, U_d and U_p the Coulomb on site repulsions. For clarity,

E_d is the energy of the corresponding d electron in Cu^{++};

$2 E_d + U_d + U_d$ that of the two d electrons in Cu^+;

E_p is the energy of the corresponding p electron in O^-;

$2 E_p + U_p$ that of the two p electrons in O^{--}.

Such a simple description is sufficient to cover a variety of models and stress what I believe is an important point concerning the metallic ones. I am of course aware that it misses some finer points deemed by some to be essential.

Then, if $\varepsilon_d = - E_d - U_d$ and $\varepsilon_p = - E_p - U_p$ are the (positive) energies to create a positive hole in the full Cu 3_d^{10} and O 2_p^6 shells respectively, the energy to create Cu^{++} from Cu^+ is ε_d while that to create Cu^{+++} from Cu^{++} is $\varepsilon_d - U_d$; similarly, the energy to create O^- from O^{--} is ε_p, while that to create O^o from O^- is $\varepsilon_p - U_p$. This language in terms of hole creation energies ε_d, ε_p simplifies the discussion.

3.1. Ionic Models

In strongly ionic compounds, the condition

$$\Delta = \varepsilon_p - \varepsilon_d = E_d - E_p + U_p >> t \tag{2}$$

prevails. The electronic holes essentially concentrate on the Cu sites. They can jump from copper site to copper site with a frequency given by

$t_{eff} \cong t^2/\Delta$.

In undoped planes, with one hole per copper site, Verwey-Mott insulation occurs, with a Hubbard gap U_d and an exchange coupling J related to T; for doped planes, the excess holes or electrons are mobile.

Such a scheme has indeed been used in this context. The main problems are as follows:

- Magnetic state for undoped planes. It seems to be antiferromagnetic, with probably large quantum fluctuations, in agreement with experiments on $La_2 Cu O_4$ and $Y Ba_2 Cu_3 O_6$. A number of other solutions have been proposed, notably Anderson's resonating valence bond (RVB) states, made of coherent combinations of singlet states formed by exchanges between two sites. More recently, flux states have been considered, where circulation of charges along the bonds of closed "plaquettes" replaces the simple one bond exchanges of the RVB state.

- Effect of doping. At very small dopings, it is probable that the most stable state is that of a magnetic polaron where the excess carrier polarizes locally the undoped plane into a ferromagnetic region. There are then many unresolved questions: there is no evidence for the superparamagnetic moments thus predicted; if one can neglect the potential of the dopant atoms, one would expect a phase separation where all excess carriers go into a single large ferromagnetic region, except if their long range Coulomb repulsions are strong enough to keep them apart, a somewhat unlikely event. At finite dopings, other possibilities have been explored: a periodic modulation of the antiferromagnetic structure, or a stabilization of the RVB or flux phases mentioned above for the undoped phase.

- Ensuing superconductive couplings. There is obviously, in the doped and conductive regime, a wealth of possible electron-electron couplings, which need not be detailed because this picture is anyway too extreme.

3.2. Charge Transfer Models

On doping CuO_2 planes with holes, one creates increasing and large amounts of p holes on the oxygen ions. This has induced most people to shift to a ionocovalent model with:

$$U_d, U_p \gg \Delta \gg t.$$

In undoped CuO_2 planes, the holes are still localized essentially on the Cu, in a Mott insulator model with a charge transfer gap Δ. They are coupled by an exchange interaction J' related to an effective transfer integral:

$$T' \cong t^2/U_d.$$

Much the same analysis of the ensuing antiferromagnetism holds as in the ionic model.

With hole doping, excess holes are created in the O 2p states, leading to hole conduction, as qualitatively observed. Because Δ is relatively small, the holes created on the O ions have sizeable covalent dp character which makes them easily excited from the Cu sites. A satellite line in X rays absorption spectra is related to the creation of a second hole in a Cu ion, thus an energy shift $U_d - \Delta$.

The most elaborate picture of this kind interprets spectroscopic data with a cluster model which minimizes the role of t. It gives [2]

$$\Delta \cong 3.5 \text{ eV}; \ U_d \cong 8.8 \text{ eV}; \ U_p \cong 6.0 \text{ eV}; \ t \cong 1.3 \text{ eV}.$$

In this picture, the doping holes in the O2p states build a weakly correlated gas of low density, which resonates with highly correlated magnetic d states. The situation is reminiscent of the heavy fermions case and could lead to similar superconductive couplings especially if one takes into account direct transfer integrals between O2p orbitals on neighboring Cu O bonds.

All charge transfer models however predict a fundamental asymmetry between holes doping and electrons doping which is very difficult to reconcile with the great similarity of the phase diagrams of say $La_{2-x} Sr_x Cu O_4$ and $Nd_{2-x} Ce_x Cu O_4$.

3.3. Delocalized Models

The preceding models, if conceptually clear, do not necessarily apply to the $Cu O_2$ planes of oxide superconductors. The reason is that the effect of t is far from small, even according to the promotors of the charge transfer models. I strongly believe this action suppresses any Hubbard or charge transfer gap in these oxides and makes them essentially metallic.

Let us first consider the simple but somewhat extreme case

$$\Delta = 0.$$

The p and d orbitals offer essentially equivalent sites for the holes. Whatever the value of U, *no* Mott localization can occur, either in undoped planes with one hole for one d and two p orbitals, or in any reasonably doped planes. There is no gap at the Fermi level *in the paramagnetic phase* and the compounds are metallic.[1]

A similar delocalized state without pseudogap is expected to remain as long as[2]

$$\Delta < w$$

where w is the width of the partly occupied 3d - 2p band. In the *covalent* condition, w itself is not related to T', but directly to t; it is especially large because both the upper and lower halves of the partly occupied 3d-2p band must be considered, as they are only separated by a gap smaller than their width.[1] This all out band width is roughly of order

$$w \cong 5\,t$$

if the U's are small, and this order of magnitude estimate must be conserved even for $U \gg w$ because, in this range of small Δ's, the concentration of holes per (Cu or O) site is small. Indeed a recent study of the paramagnetic phase using a mean field CPA like approximation[3] gives the metallic condition for large U's as

$$\Delta < 6.7t(1 + 6.7t/4U). \tag{3}$$

Values of the parameters cited above as deduced from the charge transfer model fulfill clearly such conditions.

This result also agrees with the covalent character found in band structures, where electron correlations are neglected. The Hartree Fock one electron on site energies are then related to E_d and E_p (thus to ε_d and ε_p), but not equal to them. The (small) differences are related to the exact ionicity of the CuO_2 planes. More precisely, in the covalent limit where

$$E_d{}^{HF} - E_p{}^{HF} \ll w,$$

the copper ions should have on the average $1/2$ positive hole, thus an ionicity near to $Cu^{+3/2}$, while each oxygen ion should have $1/4$ positive hole, thus an ionicity $O^{-7/4}$. Then

$$E_d{}^{HF} = E_d + 3/4\,U_d$$

$$E_p{}^{HF} = E_p + 7/8\,U_p$$

Thus

$$E_d{}^{HF} - E_p{}^{HF} = E_d - E_p + 3/4\,U_d - 7/8\,U_p = \Delta - 1/4\,U_d + 1/8\,U_p.$$

With the values reported above, this would give

$$E_d{}^{HF} - E_p{}^{HF} = \Delta - 1.45^{ev} \cong 2^{ev}.$$

The difference $E_d{}^{HF} - E_p{}^{HF}$ appearing in band structure is indeed of the order of a very few eV's, implying again Δ less than w.

If, as I believe, condition (3) is fulfilled, a number of consequences follow.

In undoped CuO_2 planes, insulation should be due at low temperatures to the long range antiferromagnetism, which opens a gap at the Fermi level. Above T_N, insulation can come from localization of carriers by the observed magnetic disorder. Indeed the fairly long range antiferromagnetic fluctuations observed in this range should scatter coherently the carriers: a magnetic pseudogap should replace the antiferromagnetic one present below T_N; and scattering by magnetic disorder should localize the states inside this pseudogap near the Fermi level.

In conclusion, there is "Slater Anderson" insulation, and not "Verwey Mott" insulation, because of the strong covalency. In the antiferromagnetic state, one then expects holes to exist in the O 2p orbitals, as indeed seems to show at the proper energy in spectoscopic data.[2] Thus also the average number of holes on each copper is less than unity, leading to a reduced magnetic moment even without quantum fluctuations.

For small Δ's, as discussed here, one expects a fundamental symmetry in the effects on phase diagrams of doping by holes and by electrons, as seems the case.[4] Doping introduces supplementary carriers which, at low temperatures, can either be localized in impurity states in the antiferromagnetic gap or, at larger dopings, lead to a shifting of wave length and long range disordering of antiferromagnetism. Above T_N, which is expected to decrease fast with doping as observed, the fairly long range antiferromagnetic fluctuations also observed should still produce a magnetic pseudogap, strong enough to localize in that range all but the doping carriers. When doping becomes large enough to kill the long range magnetic fluctuations or their effect on Fermi electrons, one expects a disappearance of this pseudogap and all carriers to take part in conduction. This would explain qualitatively the Hall effect which, in $La_{2-x}Sr_xCuO_4$, corresponds to an increasing number of holes up to $x \cong 0.25$, then switches to electron conduction;[5] similar (but opposite) behavior with doping are reported[4] in electron doped compounds such as $Nd_{2-x}Ce_xCuO_4$; they can be similarly explained.

A more elaborate analysis of transport and magnetism in doped planes would require a quantitative study of the effects of the pseudogap. This has *not* been done so far, even in the limit of weak U_d and U_p. In particular, neither the Coherent Potential Approximation inherent in all simple treatments of the Hubbard models nor the simple paramagnon approach treat the strong and coherent magnetic scattering responsible for the pseudogap in a satisfactory way.

Renormalization techniques used in organic conductors could be useful, but should be extended to quasi 2d dimensions. Also as in organic conductors, umklapp electron electron scattering due to $U \neq 0$ could produce a gap function of $U_d - U_p + (U_d + U_p)(\Delta/w)$ at the Fermi level of the undoped paramagnetic phase.[6] But in the present range $\Delta \ll w$ and $U_d - U_p \ll U_d$, this should be a small correction to the antiferromagnetic (pseudo) gap due to U. It must be stressed finally that the values of U_d, U_p deduced from spectoscopic data are probably too large, because the charge transfer model used underestimates the effect of the band width. As in transitional metals, one can expect the value of U_d deduced from the position of the charge satellite to decrease significantly for the band width w considered here.

As stressed above, the aspect of phase diagrams suggests that superconductively is by phonon coupling in these oxides. From that point of view, one can understand that a magnetic pseudogap can perturb strongly the effect of a van Hove anomaly: a maximum of T_c could take place when antiferromagnetic correlations become short range while the Fermi level is still near enough to a van Hove anomaly for this to be active in superconductivity. It might be however that superconductivity is through some kind of coupling through antiferromagnetic fluctuations. From that point of view, models such as Schrieffer's spin bag or the models developed for organic superconductors might possibly apply.

4. CONCLUSION

We have stressed that the possible interest of quasi low dimensionality should induce more work on the cubic oxides or similar compounds with 3 intersecting families of conducting chains.

In the normal (planar) oxides, we have stressed that electron-phonon couplings could play the leading role in what we believe is a covalent and essentially metallic situation. But even then the exact interplay of magnetism and transport remains to be understood in details.

Oxide superconductors thus provide one more example of the rich variety of phenomena made possible by the interplay of covalency and correlation.

REFERENCES

1. References up to 1989 for this paper can be found in J. Friedel, J. Phys. Cond. Matter **1**, 7757 (1989); a short experimental review is

found in the MRS Bulletin XV no. 6 (June, 1990).

2. H. Eskes and G. A. Sawatzky, Phys. Rev. B, to be published.

3. M. Cyrot and B. Mayou, Phys. Rev. **B41**, 4033 (1990) and private communication.

4. M. B. Maple, MRS Bulletin XV, 6 p. 60 (1990).

5. H. Tagaki, Y. Tokura and S. Uchida in Mechanisms of High Temperature Superconductivity, ed. H. Kamimura and A. Oshiyama (Springer Heidelberg 1989) Springer Series in Materials Science, p. 238.

6. V. J. Emery, R. Bruinsma and S. Barisic, Phys. Rev. Lett. **48**, 1039 (1982).

HIGH–ENERGY SPECTROSCOPY STUDIES OF HIGH Tc SUPERCONDUCTORS

J. Fink,[1] N. Nücker,[1] H. Romberg,[1] M. Alexander,[1]
P. Adelmann,[1] R. Claessen,[2] G. Mante,[2]
T. Buslaps,[2] S. Harm,[2] R. Manzke[2] and M. Skibowski[2]

[1] KFZ. Karlsruhe, Institut für Nukleare
Festkörperphysik, Karlsruhe, Germany
[2] Institut für Experimentalphysik, Universität Kiel
Kiel, Germany

ABSTRACT - Investigations of the electronic structure of high-T_c super-conductors and related compounds are reviewed. In particular we report on electron-energy-loss and direct and inverse photoemission studies. Information on the character, the symmetry and the dynamics of charge carriers is provided.

1. INTRODUCTION

Four years after the discovery of cuprate superconductors by Bednorz and Müller[1] the mechanism for high-T_c superconductivity is still completely unclear. One reason for this is that even the normal-state properties of these highly correlated systems are not understood, although a large amount of experimental and theoretical work has been dedicated to their investigation. At present there is a lively debate, whether these systems behave as a normal Fermi liquid, a marginal Fermi liquid[2] or a Luttinger liquid.[3] High-energy spectroscopy studies have revealed many normal state properties of high-T_c superconductors and their undoped parent compounds. In this contribution we review those studies. For convenience, we mainly concentrate on our own investigations.

The undoped parent compounds, e.g., La_2CuO_4 or Nd_2CuO_4, are antiferromagnetic (AFM) insulators due to strong correlation effects on the Cu sites. As shown in Fig. 1, antiferromagnetism is destroyed upon p- or n-type doping in La_2CuO_4 or Nd_2CuO_4, respectively. Close to or near the concentration where the magnetic order disappears, a metal-insulator transition occurs and a high-T_c superconducting phase appears which disappears at even higher dopant concentrations. The phase diagram is in some way misleading since it suggests that antiferromagnetism and superconductivity are well separated phases. However, it is well known from neutron scattering and NMR experiments[5,6] that the AFM ranges given in Fig. 1 shows only the range of 3-dimensional order of the CuO_2 planes and that there exist spin fluctu-

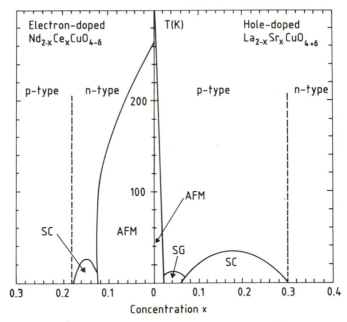

FIG. 1. Phase diagram of $La_{2-x}Sr_xCuO_{4+y}$ and $Nd_{2-x}Ce_xCuO_{4-y}$. AFM: antiferromagnetic phase; SG: spin-glass phase; SC: superconducting phase. (after Ref. 4).

ations with low spin fluctuation frequencies ($E_{sf} \sim 10$ meV) and coherence lengths considerably larger than the lattice constants ($\xi_{sf} \sim 15$ Å). This may indicate that correlation effects are also important in the superconducting range. To understand the normal state in the superconducting regime it is probably crucial to study the evolution of the electronic structure when going from the undoped system to the highly doped compound.

The important layers in the cuprate superconductors are the CuO_2 layers in which a strong covalent bonding between Cu $3d_{x^2-y^2}$ orbitals and O $2p_{x,y}$ orbitals occurs. Thus, an antibonding CuO $dp\sigma$ band is formed which in a single-particle picture (and according to LDA band-structure calculations) should be half filled. Due to strong correlations on the Cu sites a charge-transfer insulator[7] is formed. As shown in Fig. 2 the valence band is composed mainly of O $2p$ states and some admixture of Cu $3d$ states. The conduction band, also called upper Hubbard band, and the lower Hubbard band (below the valence band, not shown) are predominantly formed of Cu $3d$ states with some admixture of O $2p$ states. There are various models for the change of the electronic structure upon p- or n- type doping. They are illustrated in Fig. 2. In a rigid-band-like model for the charge-transfer insulator, holes in the valence band with mainly O $2p$ character and electrons in the conduction band with mainly Cu $3d$ character will be formed upon p- or n-type doping,

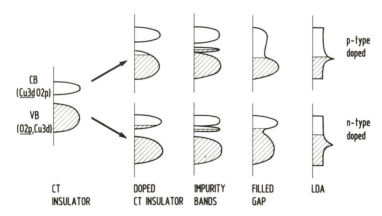

FIG. 2. Illustration of the electronic structure of cuprate superconductors and their parent compounds. Left panel: charge transfer insulator. Middle panel: Various models for p- and n-type doped systems. Right panel: Schematic density of states of the antibonding CuO $dp\sigma$ band according to LDA band-structure calculations.

respectively. Other models predict the formation of partially filled impurity bands, due to the potential of the dopant atoms[8] or due to a many-body effect when removing or adding charges of the CuO_2 layers.[9] Further models predict the filling of the gap with the Fermi level located at the bottom or at the top of the gap for p- and n-type doping, respectively.[10] At present, the metal-insulator transition and the electronic structure of cuprates at intermediate dopant levels is rather unclear. It is generally assumed that at very high dopant concentrations a normal metal is formed which probably can be described by LDA band structure calculations. The electronic structure of these systems are illustrated on the right panel in Fig. 2.

2. RESULTS AND DISCUSSION

2.1 Evolution of the Electronic Structure as a Function of Dopant Concentration

Core excitations from the O 1s level provide information on the local density of unoccupied states with 2p character at the O sites. The measurements can be performed by high-energy electron-energy-loss spectroscopy in transmission (EELS) or by x-ray absorption spectroscopy (XAS) using synchotron radiation. In the field of high-T_c superconductivity those measurements were first performed by EELS on the system $La_{2-x}Sr_xCuO_4$,[11] by XAS on $YBa_2Cu_3O_7$,[12] and on both systems by EELS[13.] In Fig. 3 we show recent systematic EELS measurements of the O 1s absorption edges in $La_{2-x}Sr_xCuO_{4+\delta}$.[14] In the lowest spectrum for the undoped compound with x = 0 and $\delta \sim 0$ the peak at 530.2 eV is ascribed to transitions into O 2p states admixed to the conduction band (upper Hubbard band) (see Fig. 2). The wave function of this band can be written in the form $| \Psi > = \alpha | d^9 > + \beta | d^{10} \underline{L} >$ where $\underline{L}$ denotes holes on the ligands and β^2 determines the admixture of O 2p states. The intensity of this peak is therefore a measure of covalency of the CuO bond in the CuO_2 layers. The spectrum for x = 0 and $\delta \sim 0.015$ shows some spectral weight at lower energies with a threshold at about 528.5 eV. This spectral weight is explained by unoccupied O 2p states in the gap formed upon p-type doping due to excess O. Upon doping with Sr (x > 0) there is a continuous increase of spectral weight at the dashed line shown in Fig. 3. This line corresponds for the metallic systems (x > 0.05) to the Fermi level since the binding energy of the O 1s level, as determined by XPS measurements,[11] is close to 528.5 eV. The Fermi level appears 1.5 eV below the threshold of the

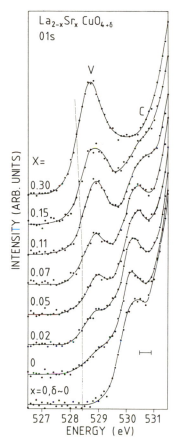

FIG. 3. O 1s absorption edges of $La_{2-x}Sr_xCuO_{4+\sigma}$ for $0 \leq x \leq 0.3$ For x=0, spectra for $\delta \sim 0$ and $\delta \sim 0.015$ are shown.

conduction band which is roughly the value of the charge-transfer gap (1.8-2 eV).[15-16] This indicates that for p-type doping the Fermi level is formed at the bottom of the charge-transfer gap. The continuous increase of unoccupied O 2p density of states suggests the formation of hole states on O sites, the number being proportional to the dopant concentration. No discontinuity is observed at the dopant concentration x = 0.06 where the metal-insulator transition occurs. For x < 0.06, probably the states at the Fermi level are localized, e.g., by the potential of the disordered Sr atoms. Finally, it is remarkable that the peak C related to the upper Hubbard band is still present (but slightly reduced in the superconducting range at x = 0.11 and x = 0.15). At x = 0.3, where superconductivity disappears, peak C has almost disappeared. The data indicate that correlation effects are still important in the superconducting range in agreement with the above mentioned neutron scattering and NMR results. The disappearance of peak C can be interpreted in terms of the formation of a normal metal at higher dopant concentrations (x > 0.3). Probably, there is no more a splitting of the CuO dpσ band into three bands in this range. There are other effects which can lead to a reduction of peak C. Recent cluster calculations[17] on the O 1s absorption edges of the CuO_2 plane resulted in a reduction of peak C by a hybridization of the conduction band C with the valence band in the final state. However, this effect leads to a disappearance of peak C close to x = 0.7. Also sum rule effects may lead to a reduction of peak C but probably this effect is even smaller than the hybridization effect. Therefore, the <u>strong</u> reduction of peak C probably indicates a screening out of correlation effects at higher dopant concentrations. The cluster calculations in Ref. 17 also indicate that the calculated spectra are almost not influenced by the core hole potential for reasonable values of the core hole potential. Recently, similar measurements have been performed on $La_{2-x}Sr_xCuO_4$ (x $\leq$ 0.15) by XAS using a bulk sensitive fluorescence yield detection mode and results have been obtained similar to the above mentioned EELS results.[18]

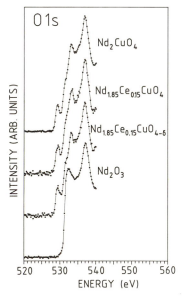

FIG. 4. O 1s absorption edges of Nd_2CuO_4, $Nd_{1.85}Ce_{0.15}CuO_4$, superconducting $Nd_{1.85}Ce_{0.15}CuO_{4-\delta}$ and Nd_2O_3.

For the n-type doped system $Nd_{2-x}Ce_xCuO_{4-\delta}$ similar O 1s absorption spectra were measured by EELS[19] (see Fig. 4). Upon n-type doping with Ce and upon a further treatment in a reducing atmosphere to obtain a superconducting material with $\delta > 0$ there is almost no change in the pre-peak structure. This indicates that the conduction band states are almost not changed. The results also exclude the existence of "impurity" bands as shown in Fig. 2. The phase diagram of Fig. 1 indicates that at the end of the superconducting range there is a transition from n- to p-type conductivity for $Nd_{2-x}Ce_xCuO_4$. Therefore, those "impurity" bands would be expected to be half filled at x ~ 0.15. A partially filled "impurity" band would cause a further pre-peak at lower energies which is not observed. Therefore, the Fermi level is probably close to the top of the charge-transfer gap. The results for the p- and n-type doped systems favor a model in Fig. 2 which is labeled "filled band." Probably O 2p (Cu 3d) states from the valence (conduction) band are pushed into the gap upon p(n)-type doping and the character of the charge carriers is

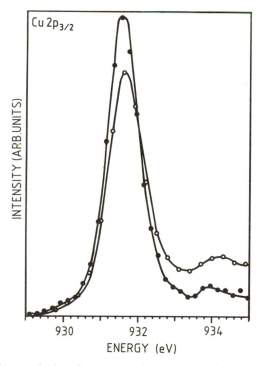

FIG. 5. Comparison of the Cu $2p_{3/2}$ absorption edges normalized to the Nd $3d_{5/2}$ absorption line of Nd_2CuO_4 (closed circles) and $Nd_{1.85}Th_{0.15}CuO_{4-\delta}$ (open circles).

predominantly O 2p (Cu 3d) like. This would imply that upon n-type doping the number of unoccupied Cu 3d states will be reduced. In fact this is observed in the Cu 2p absorption edges[19] which probe unoccupied Cu 3d states. The spectra (see Fig. 5) show a 14% reduction of the absorption line, i.e., unoccupied Cu 3d states when going from the undoped Nd_2CuO_4 to the n-type doped system $Nd_{1.85}Th_{0.15}CuO_{4-\delta}$.

2.2 Symmetry of States Close to the Fermi Energy

Orientation dependent EELS measurements of core excitations in single crystals provide information on the symmetry of unoccupied states in high-T_c superconductors.[20] In Fig. 6a we show O 1s absorption edges of a $Bi_2Sr_2CaCu_2O_8$ single crystals for momentum transfer **q** parallel and perpendicular to the CuO_2 (a,b) planes. For **q** $\parallel$ **a,b**: and for **q** $\parallel$ **c** O $2p_{x,y}$ and O $2p_z$ are probed, respectively. The spectra clearly demonstrate that there are only O $2p_{x,y}$ states at the Fermi level and no O $2p_z$ states. There was a considerable discussion on the symmetry of holes on O sites. The data shown in Fig. 6a and similar results obtained by XAS[21,22] rule

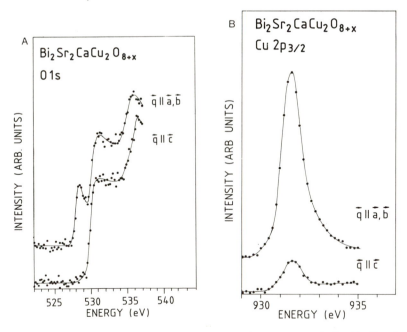

FIG. 6. Orientation dependent absorption edges of a $Bi_2Sr_2CaCu_2O_8$ single crystal (a) O 1s absorption edges (b) Cu 2p absorption edges.

out all models for high-T_c superconductivity based on out-of-plane π-holes in the CuO_2 planes[23] and $2p_z$ holes on the apex O atoms[24,25]. On the other hand, these measurements cannot differentiate between O $2p\sigma$ holes (as expected from LDA band-structure calculations) or in-plane π-holes (as derived from cluster calculations.[26] In Fig. 7 we show similar measurements on $YBa_2Cu_3O_7$ single crystals.[20] In this case, also O $2p_z$ states close to E_F are detected, having the lowest threshold energy. Band structure calculations[27,28] predict the smallest O 1s binding energy for O(4), the atom between the Cu atoms of the chains and the planes. Therefore, these O $2p_z$ states are assigned to holes in O(4) 2p orbitals having a σ bond with the Cu $3d_{y^2-z^2}$ orbital of the chain. The O $2p_{x,y}$ states detected for $\mathbf{q} \parallel \mathbf{a,b}$ are then due to empty O $2p_{x,y}$ orbitals in the plane and due to O $2p_y$ orbitals of the O(1) atoms in the chain. The orientation dependent measurements on $YBa_2Cu_3O_7$, together with the calculations of the chemical shift, therefore, favor σ holes on O in agreement with recent NMR measurements.[29,30] In Fig. 6b orientation dependent Cu $2p_{3/2}$ absorption edges of a $Bi_2Sr_2CaCu_2O_8$ single crystal are shown. They probe the symmetry of unoccupied Cu 3d states, i.e., the 3d states in the upper Hubbard band. As expected, the spectra

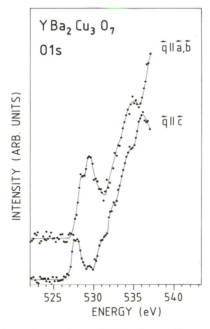

FIG. 7. Orientation dependent O 1s absorption edges of a single crysatal of $YBa_2Cu_3O_7$.

reveal that the unoccupied Cu states have predominantly $3d_{x^2-y^2}$ character. The spectral weight in the $\mathbf{q} \| \mathbf{c}$ spectrum has been ascribed to a 10% admixture of probably $3d_{3z^2-r^2}$ character. Another explanation has been given by Bianconi et al.[31] assigning this spectral weight to a partially unoccupied $3d_{3z^2-r^2}$ band. This explanation seems to be supported by a 400 meV shift between the $\mathbf{q} \| \mathbf{c}$ and the $\mathbf{q} \| \mathbf{a,b}$ line detected in their XAS measurements (see also Ref. 32). However, in our EELS measurements we have never detected such an energy shift within error bars ($\Delta E = 50$ meV).

2.3 "Bands" Close to the Fermi Level

Angular resolved photoemission spectroscopy (ARUPS) is probably a crucial experiment for the determination of the electronic structure of cuprate superconductors. In ARUPS the imaginary part of the single-particle Green's function is measured which is called the spectral function $A(\omega k)$. In a Fermi liquid, the spectral function is dominated by a well-defined single particle excitation which is due to quasiparticles being long-lived at low binding energies. This part has for given k a total weight Z_k which is called the quasi-particle weight. It can be described by

$$A(\omega, k) = \frac{1}{\pi} \frac{\mathrm{Im}\Sigma(\omega, k)}{[\omega - \Sigma_k - \mathrm{Re}\Sigma(\omega, k)]^2 + [\mathrm{Im}\Sigma(\omega, k)]^2} \tag{1}$$

where $\Sigma(\omega,k)$ is the self energy correction. Its imaginary part causes a broadening and its real part a shift of the single particle excitation at ε_k. In a Fermi liquid, when moving to the Fermi level, the quasiparticle width disappears as the second power of ε_k. In addition, in the spectral function $A(\omega,k)$ there is a part proportional to $1-Z_k$, which is caused, e.g., by the creation of multiple particle-hole pairs and which is called the incoherent part.

In Fig. 8 we show typical ARUPS spectra on a single crystal of $Bi_2Sr_2CaCu_2O_8$ measured with an energy resolution of 60 meV.[33] Near the Γ point (zero emission angle) there is a broad feature with a maximum at $E_B \sim 0.2$ eV. With increasing emission angle (higher k values) the width of this feature is reduced and the maximum moves to the Fermi level. At $\vartheta = 8°$ (along the Γ-X direction and at $\vartheta \sim 20°$ (along the Γ-M direction) there is a crossing of this "band" through the Fermi level.

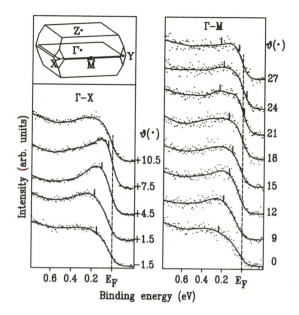

FIG. 8. Angle-resolved photoemission spectra of $Bi_2Sr_2CaCu_2O_8$ along the ΓX (left panel) and ΓM (right panel) direction taken with 18 eV photon energy at 300 K. The solid lines result from a least squares fit. The position of the peaks near E_F are indicated by ticks. The insert shows the Brillouin zone.

In Fig. 9 we show typical angular-resolved inverse photoemission spectra (ARIPES) along ΓX of a $Bi_2Sr_2CaCu_2O_8$ crystal taken with a resolution of 0.6 eV.[34] Although structures could not be resolved as good as in the ARUPS data, there is a clear indication that at incidence angles ϑ between $30°$ and $40°$ there is an increase of intensity at E_F. It is probably caused by the unoccupied part of the "band" crossing the Fermi-level. The wave vectors k_F where the occupied and the unoccupied "bands" cross the Fermi level agree within error bars.

The results from photoemission and inverse photoemission measurements and similar data from other groups[35,36] have been interpreted in different ways. Assuming that the "bands" are due to bands of single particle energies renormalized by correlation effects these "bands" can be compared with band-structure calculations. In Fig. 10 we show the ARUPS and the ARIPES data together with LDA band structure calculations on $Bi_2Sr_2CaCu_2O_8$.[37] In Fig. 11 the calculated Fermi surface is compared to experimental values of ARUPS and ARIPES. It is evident

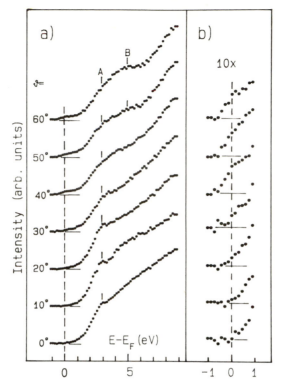

FIG. 9. Angle-resolved inverse photoemission spectra of $Bi_2Sr_2CaCu_2O_8$ along the ΓX direction (left panel). Right panel: Spectra enlarged close to the Fermi level.

that the experimental data close to E_F are well reproduced by LDA band structure calculations. Deviations at higher binding energies may be explained by a shift of the single-particle energies due to strong correlation effects, expressed by a non-zero real part of the self energy correction $\Sigma(\omega, k)$. This can be expressed also by an enhanced effective mass $m^*/m \sim 4$ of the charge carriers. Similar enhanced effective masses (although much less pronounced) have been reported for other systems such as alkali metals[38] or Ni metal,[39] too. In this interpretation, the imaginary part of the self energy correction $\Sigma(\omega, k)$ leads to a strong broadening at higher binding energies, i.e. a reduction of the lifetime of the quasi particles. It is still unclear, whether the strong "background" centered around 0.5 eV is due to the incoherent part of the spectral function or whether it is caused by secondary electrons or by primary electrons that have suffered a small energy-loss and a small change in momentum

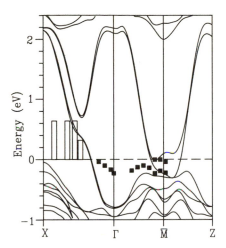

FIG. 10. Band structure of $Bi_2Sr_2CaCu_2O_8$ according to Ref. 37. The solid squares denote occupied states as derived from angular resolved photoemission spectroscopy. The empty bars represent unoccupied states as derived from angular resolved inverse photoemission spectroscopy.

at or near the surface. Olson et al.[35] have assumed that the background is extrinsic and caused by the latter process and not caused by an intrinsic one, i.e. by the incoherent part of the spectral function. With this assumption they received a better fit of their data for a broadening proportional to $|E-E_F|$ than for a width proportional to $(E-E_F)$.[2] The latter is expected for a normal Fermi liquid while for a marginal Fermi liquid[2] a linear energy dependence of the width is predicted. In this second explanation of the photoemission data on single crystalline

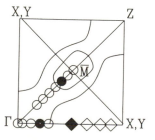

FIG. 11. Basal plane of the Brillouin zone including the Fermi surface as calculated in Ref. 37. Circles and diamonds denote points in k-space measured by angular-resolved photoemission and angular-resolved inverse photoemission spectroscopy, respectively. Solid circles denote the observation of a crossing of "bands" through the Fermi level.

$Bi_2Sr_2CaCu_2O_8$ the spectral weight at the Fermi level is no more explained by an excitation of quasi particles, since Z_k goes to zero when E_k approaches the Fermi level. The peak at E_F is then entirely incoherent and the normal Fermi liquid picture fails.

A third explanation of the photoemission experiments is given in terms of a Luttinger liquid,[3] in which charge and spin degrees of freedom are separated and the spectral weight is explained by an incoherent excitation of holons and spinons. Experiments with improved energy <u>and</u> momentum resolution are needed to decide between the three different explanations.

It is interesting to note that at present no comparable measurements on the lifetime broadening exist to our knowledge in normal metals. The reason for this is that in three-dimensional systems the peak widths are determined by the inverse lifetime of both photoelectrons and holes. In most cases the lifetime broadening due to photoelectrons is dominant. In two-dimensional systems, only the hole lifetime contributes to the broadening.[40] Therefore, comparable measurements in simple two-dimensional metals are highly demanded.

2.4 The Superconducting Gap

In order to obtain information on the **k**-dependent changes of the electronic structure between normal and superconducting states, ARUPS measurements have been performed on single crystals of $Bi_2Sr_2CaCu_2O_8$ for various temperatures.[41] Similar angle-integrated UPS measurements have been performed first by Imer et al.[42] In Fig. 12 we show typical ARUPS measurements with a $K_{\parallel}$ vector along the ΓX direction where, according to band-structure calculations and according to the experimental data described above, a Cu-O band intersects the Fermi level. While at 125 K an edge, i.e. the turning point of the emission onset, is situated right at E_F, below T_c at 77 K and at 63 K the edge is slightly shifted by $\Delta = 30 \pm 4$ meV to higher binding energies. In addition, spectral weight is piled up at an energy labeled with S. These experiments have shown for the first time an opening of a "BCS-like" gap at k_F. The size of the gap, i.e. $2\Delta/k_BT_c \sim 8$, which is about twice as large as the weak coupling BCS value, indicates that the high-T_c cuprates are strong coupling superconductors. Recently, Olson et al.[43] have performed similar measurements with an improved resolution ($\Delta E_{1/2} = 30$ meV) and the appearance of a well pronounced "BCS-like" singularity and the opening of a gap with slightly smaller size has been observed.

The measurements would support a BCS mechanism for the high-T_c cuprates. On the other hand, opponents against a BCS mechanism have realized that the sum rule is not fulfilled.[44] There are considerably less states removed from the gap region than appear in the singularity. These additional states must come from lower binding energies or from wavevectors different from k_F.

In ARUPS measurements along the ΓM direction where, according to band-structure calculations, a BiO band should cross the Fermi level, a gap of almost the same size has been detected as along the ΓX direction.[45] Recent ARUPS measurements on single crystals on which a small amount of Au has been deposited indicate that the singularity along

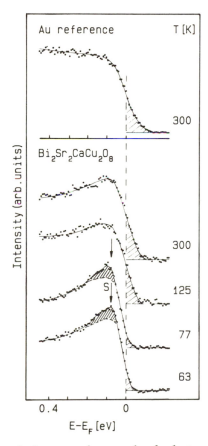

FIG. 12. High resolution angular resolved photoemission spectra on a single crystal of $Bi_2Sr_2CaCu_2O_8$ above and below T_c. A wavevector $k_{\parallel}$ was chosen close to a value where the Cu-O "band" crosses the Fermi level. For comparison a spectrum of evaporated Au taken under identical conditions is shown.

the ΓX direction (CuO bands) remains below T_c while that along the ΓM direction (BiO bands) disappears upon Au evaporation on the surface. Since these crystals have been cleaved between the two BiO planes[34] it was concluded from these measurements that the first layer, namely the BiO layer is already superconducting.[36]

2.5 Surface Electronic States in $YBa_2Cu_3O_7$

It is a puzzling problem, why in $YBa_2Cu_3O_7$ no superconducting gap could be detected by ARUPS. Early ARUPS measurements could

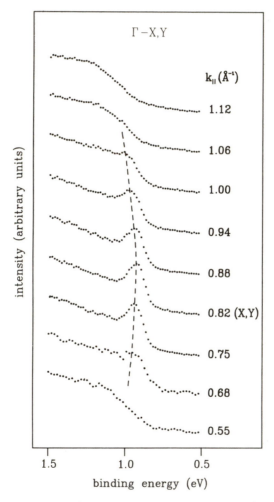

FIG. 13. Angle-Resolved photoemission spectra of a single crystal of $YBa_2Cu_3O_{6.8}$ taken with 21.22 eV photon energy along the Γ-X,Y direction in the first and in the second Brillouin zone at T<20 K.

not even observe a Fermi edge in these materials. Arko et al.[46] have shown that the surface is not stable in vacuum at room temperature but that the surface remains quite stable in vacuum if the sample is kept at very low temperatures. Under these conditions a Fermi edge was observed for certain **k** values but no superconducting gap has been realized.

During a systematic ARUPS study of the Fermi surface and of occupied dispersives bands close to E_F we have observed an intrinsic surface-derived electronic band near the X(Y)-point of the Brillouin zone at a binding energy of 0.9 eV[47]. In Fig. 13 we show typical energy distribution curves for this state between 0.5 and 1.5 eV taken along the Γ-X,Y direction at emission angles corresponding to wavevectors $k_\parallel$ around the Brillouin zone boundary. The extra-ordinary low value for the width of this feature ($\Delta E_{1/2} \sim 100$ meV) is reminiscent of the widths found for surface states in metals, e.g., Cu metal. Further evidence for the surface nature of the 0.9 eV peak is given by its sensitivity to adsorbates displayed in Fig. 14. With increasing contamination due to

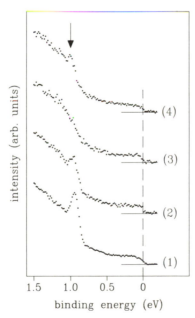

FIG. 14. Angle resolved photoemission spectra of the surface-derived peak in $YBa_2Cu_3O_{6.8}$ near the X,Y-point recorded 1.5, 6 and 13 hours after cleavage (spectra 1,2 and 3 respectively). Spectrum 4 shows the effect of flashing the sample to 50 K for 5 minutes causing a partial evaporation of adsorbates and the reappearance of the 0.9 eV peak.

the residual molecules in the vacuum chamber the 0.9 eV peak disappears and reappears after flashing the sample up to 50 K, i.e. partially removing the adsorbate. Finally, no dispersion for wavevectors perpendicular to the surface have been observed which is also typical of surface derived states.

The appearance of a surface-derived state indicates a redistribution of charges on the surface. This can be explained by the fact that any cleavage in these compounds inevitably requires the breaking of strong ionic or covalent bonds between layers. Since in the bulk there is a charge transfer from the CuO_2 layers to the one-dimensional CuO_3 ribbons, the formation of a surface on any layer should lead to a rearrangement of the electronic charge and therefore to a change of the electronic structure at the surface compared to that in the bulk. The situation is different from $Bi_2Sr_2CaCu_2O_8$ because there the cleavage occurs between two adjacent Bi-O layers which are coupled only weakly by a van der Waals-like bonding with almost no valence charge between them. There is a further difference between $Bi_2Sr_2CaCu_2O_8$ and $YBa_2Cu_3O_7$, however, the two surfaces formed upon cleavage must be always different. It is remarkable that in 8 surfaces investigated so far, always the same surface-derived electronic states have been derived. The reason for this is not clear at present. Among the possible crystal terminations the Ba-O(4) surface with a Cu(1)-O(1) sublayer is the least polar one and therefore this surface termination has been suggested. The missing CuO_2-Y-CuO_2 layer complex with a charge of about -1 per formula unit above the BaO layer may lead to a reduction of holes in the O(4) atoms and therefore to an almost neutral BaO surface layer. This picture is in agreement with the calculations of Calandra et al.[48] which predict a very strong O(4)-derived peak in the surface density of states at 0.8 eV binding energy.

The observation of a surface derived peak in $YBa_2Cu_3O_7$ indicating a rearrangement of electronic charges at the surface probably explains why in this compound no superconducting gap could be observed. On the other hand, Fermi edges and "bands" have been detected on surfaces where also the 0.9 eV peak was realized. Hence, a mapping of volume-derived bands by ARUPS remains probably a meaningful experiment.[49,50]

REFERENCES

1. J. G. Bednorz and K. A. Müller, Z. Phys. **B64**, 189 (1986).

2. C. M. Varma, P. B. Littlewood, S. Schmitt-Rink, E. Abrahams and A. E. Ruckenstein, Phys. Rev. Lett. **63**, 1996 (1989).

3. P. W. Anderson, Phys. Rev. Lett. **64** 1839 (1990).

4. M. B. Maple, MRS Bulletin, June 1990, p. 60.

5. R. J. Birgenau, Y. Endoh, K. Kakurai, Y. Hidaka, T. Murakami, M. A. Kastner, T. R. Thurston, G. Shirane and K. Yamada, Phys. Rev. **B39**, 2868 (1989).

6. A. J. Millis, H. Monien and D. Pines, Phys. Rev. **B42** 167 (1990).

7. A. Fujimori and F. Minami, Phys. Rev. **B30**, 957 (1984); J. Zaanen, G. A. Sawatzky and J. W. Allen, Phys. Rev. Lett. **55**, 418 (1985); S. Hüfner, Z. Phys. **B61**, 135 (1985).

8. A. Fujimori, Y. Tokura, H. Eisaki, H. Takagi, S. Uchida and M. Sato, Phys. Rev. **B40**, 7303 (1989).

9. S. Ishihara, H. Matsumoto and M. Tachiki, preprint.

10. P. C. Pattnaik and D. M. Newns, Phys. Rev. **B41**, 880 (1990); C.A.R. Sá de Melo and S. Doniach, Phys. Rev. **B41**, 6633 (1990).

11. N. Nücker, J. Fink, B. Renker, D. Ewert, C. Politis, P. J. W. Weijs and J. C. Fuggle, Z. Phys. **B67**, 9 (1987).

12. J. A. Yarmoff, D. R. Clarke, W. Drube, U. O. Karlsson, A. Taleb-Ibrahimi and F. J. Himpsel, Phys. Rev. **B36**, 3967 (1987).

13. N. Nücker, J. Fink, J. C. Fuggle, P. J. Durham and W. M. Temmerman, Phys. Rev. **B37**, 5158 (1988).

14. H. Romberg, M. Alexander, N. Nücker, P. Adelmann and J. Fink, Phys. Rev. **B42**, 8768 (1990).

15. S. Tajima, H. Ishii, T. Nakahashi, T. Takagi, S. Uchida, M. Seki, S. Suga, Y. Hidaka, M. Suzuki, T. Murakami, K. Oka and H. Unoki, J. Opt. Soc. Am. **B6**, 475 (1989).

16. J. Orenstein, G. A. Thomas, D. H. Rapkine, C. G. Bethea, B. F. Levine, B. Batlogg, R. J. Cava, D. W. Johnson, Jr. and E. A. Rietman, Phys. Rev. **B36**, 8892 (1987).

17. E. Eskes and G. A. Sawatzky, Phys. Rev. **B43**, 119 (1991).

18. C. T. Chen, F. Sette, Y. Ma, M. S. Hybertsen, E. B. Stechel, W. M. C. Foulkes, M. Schluter, S.-W. Cheong, A. S. Cooper, L. W. Rupp, Jr., B. Batlogg, Y. L. Soo, Z. H. Ming, A. Krol and Y. H. Kao, Phys. Rev. Lett. **66**, 104 (1991).

19. M. Alexander, H. Romberg, N. Nücker, P. Adelmann, J. Fink, J. T. Markert, M. P. Maple, S. Uchida, H. Takagi, Y. Tokura, A. C. W. P. James and D. W. Murphy, Phys. Rev. B, **43**, 333 (1991).

20. N. Nücker, H. Romberg, X. X. Xi, J. Fink, B. Gegenheimer, and Z. X. Zhao, Phys. Rev. **B39**, 6619 (1989).

21. F. J. Himpsel, G. V. Chandrashekhar, A. B. McLean and M. W. Shafer, Phys. Rev. **B38**, 11946 (1988).

22. P. Kuiper, M. Grioni, G. A. Sawatzky, D. B. Mitzi, A. Kapitulnik, A. Santaniello, P. de Padova and P. Thiry, Physica C **157**, 260 (1989).

23. K. H. Johnson, M. E. McHenry, C. Counterman, A. Collins, M. M. Donovan, R. C. O'Handley and G. Kalonji, Physica C **153–155**, 1165 (1988).

24. H. Kamimura, Jpn. J. Appl. Phys. **26**, 6627 (1987).

25. A. Fujimori, Phys. Rev. **B39**, 793 (1989).

26. Y. Guo, J.-M. Langlois and W. A. Goddard III, Science **239**, 896 (1988).

27. W. E. Pickett, Rev. Mod. Phys. **61**, 433 (1989).

28. J. Zaanen, M. Alouani and O. Jepsen, Phys. Rev. **B40**, 837 (1989).

29. H. Alloul, T. Ohno and P. Mendels, Phys. Rev. Lett. **63**, 1700 (1989).

30. M. Takigawa, P. C. Hammel, R. H. Heffner, Z. Fisk, K. C. Ott and J. D. Thompson, Phys. Rev. Lett. **63**, 1865 (1989).

31. A. Bianconi, M. de Santis, A. di Cicco, A. M. Flank and P. Lagarde, Physica **B158**, 443 (1989); M. de Santis, A. di Cicco, P. Castrucci, A. Bianconi, A. M. Flank, P. Lagarde, H. Katayama-Yoshida and C. Politis, Physica **B158**, 480 (1989).

32. M. Abbate, M. Sacchi, J. J. Wnuk, L. M. M. Schreuers, Y. S. Wang, R. Lot and J. C. Fuggle, preprint.

33. G. Mante, R. Claessen, T. Buslaps, S. Harm, R. Manzke, M. Skibowski and J. Fink, Z. Phys. **B80**, 181 (1990).

34. R. Claessen, R. Manzke, H. Carstensen, B. Burandt, T. Buslaps, M. Skibowski and J. Fink, Phys. Rev. **B39**, 7316 (1989).

35. C. G. Olson, R. Liu, D. W. Lynch, R. S. List, A. J. Arko, B. W. Veal, Y. C. Chang, P. Z. Jiang and A. P. Paulikas, Phys. Rev. **B42**, 381 (1990).

36. B. O. Wells, Z. -X. Shen, D. S. Dessau, W. E. Spicer, C. G. Olson, D. B. Mitzi, A. Kapitulnik, R. S. List and A. J. Arko, preprint.

37. H. Krakauer and W. E. Pickett, Phys. Rev. Lett. **60**, 1665 (1988).

38. E. Jensen and E. W. Plummer, Phys. Rev. Lett. **55**, 1912 (1985).

39. A. Liebsch, Phys. Rev. Lett. **43**, 1431 (1979); D. E. Eastman, F. J. Himpsel and J. A. Knapp, Phys. Rev. Lett. **40**, 1514 (1978).

40. T. C. Chiang, J. A. Knapp, M. Aono and D. E. Eastman, Phys. Rev. **B21**, 3515 (1980); B. J. Slagsvold, J. K. Grepstad and P. O. Gartland, Phys. Scr. **T4**, 65 (1983).

41. R. Manzke, T. Buslaps, R. Claessen and J. Fink, Europhys. Lett. **9**, 477 (1989).

42. J. -M. Imer, F. Patthey, B. Dardel, W. -D. Schneider, Y. Baer, Y. Petroff and A. Zettl, Phys. Rev. Lett. **62**, 336 (1989).

43. C. G. Olson, R. Liu, A. -B. Yang, D. W. Lynch, A. J. Arko, R. S. List, B. W. Veal, Y. C. Chang, P. Z. Jiang and A. P. Paulikas, Science **245**, 731 (1989).

44. P. W. Anderson and Y. Ren, in High Temperature Superconductivity, edited by K. S. Bedell, D. Coffey, D. E. Meltzer, D. Pines and J. R. Schrieffer (Addison-Wesley, Redwood City, 1990) p. 3; P. W. Anderson, Phys. Rev. **B42**, 2624 (1990).

45. C. G. Olson, R. Liu, D. W. Lynch, R. S. List, A. J. Arko, B. W. Veal, Y. C. Chang, P. Z. Jiang and A. P. Paulikas, Phys. Rev. **B42**, 381 (1990).

46. A. J. Arko, R. S. List, Z. Fisk, S. -W. Cheong, J. D. Thompson, J. A. O'Rourke, C. G. Olson, A. -B. Yang, T. W. Pi, J. E. Schriber and N. D. Shinn, J. Magn. Magn. Mater. Lett. **75**, L1 (1988).

47. R. Claessen, G. Mante, A. Huss, R. Manzke, M. Skibowski, Th. Wolf and J. Fink, preprint.

48. C. Calandra, F. Manghi, T. Minerva and G. Goldoni, Europhys. Lett. **8**, 791 (1989).

49. J. C. Campuzano, G. Jennings, M. Faiz, L. Beaulaigue, B. W. Veal, J. Z. Liu, A. P. Paulikas, K. Vandervoort, H. Claus, R. S. List, A. J. Arko and R. J. Bartlett, Phys. Rev. Lett. **64**, 2308 (1990).

50. G. Mante, R. Claessen, A. Huss, R. Manzke, M. Skibowski, M. Knupfer, Th. Wolf and J. Fink, to be published.

THE ROLE OF THE SHORT AND ANISOTROPIC COHERENCE LENGTH

G. Deutscher

School of Physics and Astronomy
Raymond and Beverly Faculty of Exact Sciences
Tel Aviv University, Ramat Aviv, Israel

Abstract - On the phenomenological level, the most striking character-istic of the high Tc oxides is their short and anisotropic coherence length. As deduced for instance from critical field measurements, the in plane coherence length is about equal to 3 to 4 lattice constants, while the out of plane coherence length is rather of the order of the distance between neighboring atomic planes. We discuss the consequences of this unique situation. The small coherence *volume* results in large fluctuation effects and a correspondingly small pinning energy for the vortices in the Shubnikov phase. The short coherence *length* renders the supercon-ducting order parameter very sensitive to defects on the lattice or atomic scale, with for instance new boundary conditions at surfaces and inter-faces. The large *anisotropy* allows an easy decoupling of the strongly superconducting CuO_2 planes. But the short coherence length is also the basis of a number of unique applications of the high Tc oxides.

Highlights in Condensed Matter Physics and Future Prospects
Edited by L. Esaki, Plenum Press, New York, 1991

399

1. INTRODUCTION

The discovery of Bednorz and Muller[1] is presenting us with two major challenges. The most fundamental one is of course to establish the pairing mechanism responsible for the high critical temperature, which includes an understanding of the normal state properties of the oxides. The second one is technological: will high Tc oxides play an important role in the power and electronic industries in the next century? As we all know, high hopes have been raised by the discovery and much is at stake in the answer to this question.

Fortunately, progress in the technology of the oxides does not require a preliminary understanding of the microscopic mechanism responsible for the high Tc, although such understanding would clearly be highly beneficial particularly for the discovery of other and possibly higher Tc materials. But what is absolutely essential for the development of the technology is a good understanding of the *phenomenology* of the new superconductors. Early transport and magnetic measurements showed quickly that this phenomenology is quite different from that of the low Tc superconductors. These measurements indicated critical current densities that were low and highly sensitive to an applied field in polycrystalline samples, and a reversible magnetic behavior in a broad range of magnetic fields below the upper critical field H_{c2}. Some of these features, like the reversible magnetic behavior, persist in single crystals, showing that they are of a fundamental character and not just due to poor ceramic quality.

These findings raised considerable concern about the possibility of making useful high Tc wires for high current-high field applications. We had realized at an early stage, with Alex Muller, that the short coherence length is at the origin of the above difficulties[2]. We now understand them in more detail, in terms of the small *coherence volume*, of the *short coherence length*, and of its *large anisotropy*. The values of these parameters in the oxides are quite different from what they are in the low Tc superconductors; we shall show that they govern their phenomenology.

The small coherence volume and the large anisotropy limit the high temperature performances. Nevertheless, high field-high current applications are possible at moderate temperatures in the less anisotropic oxides such as YBaCuO, and at low temperatures in the more anisotropic oxides such as BiSrCaCuO as recently demonstrated[3].

Another important challenge is the fabrication of weak links in the form of controlled junctions, preferably Superconductor / Insulator / Superconductor (S/I/S) junctions. Understanding of the behavior of junctions in the oxides has gained greatly from the study of controlled grain boundary junctions by the IBM group[4]. Again, this behavior is quite different from that of low Tc junctions, and we shall show that it is also due to the short coherence length.

2. COHERENCE VOLUME, CRITICAL REGION AND PINNING ENERGIES

In the low Tc superconductors, the condensation energy per coherence volume is much larger than the thermal energy $k_B T_c$, except in a very small range of temperatures near T_c. This is the fundamental characteristic of the conventional superconductors that determines their phenomenology: thermodynamic fluctuations of the order parameter can safely be neglected, so that the mean field approximation is excellent to describe the superconducting phase transition; vortices are effectively pinned by defects with a volume of the order of the coherence volume ξ^3, thus their motion is well described in terms of flux creep.

The situation is quite different in the oxides, where the condensation energy per coherence volume is rather of the order of $k_B T_c$: one then expects a large critical region[5] and weak pinning at temperatures of the order of Tc[6], both effects being closely related to each other as we shall see.

We define $u = [U(O)/k_B T_c]$, where $U(O)$ is the condensation energy per coherence volume at zero temperature. In an isotropic BCS superconductor:

$$U(O) = \frac{1}{2} N(O)\Delta^2(O)\xi^3(O) \tag{1}$$

where $N(O)$ is the normal state density of states, $\Delta(O)$ and $\xi(O)$ are respectively the zero temperature gap and coherence length. For a free electron gas model:

$$u \cong \left(n\, \xi^3(O) \right).\left[\frac{\Delta(O)}{E_F} \right] \tag{2}$$

where n is the density of electrons. For the pure low Tc Type II super-conductor Nb, (n $\xi^3(O)$) is of the order of 1.10^6, [$\Delta(O)/E_F$] of the order of 1.10^{-3}, hence u $\approx 1.10^3$.

For the high Tc oxides, it is preferable to calculate u directly from experimental data obtained in the superconducting state, rather than to use a free electron gas model. The condensation energy per unit volume ΔF can be calculated from the measured jump of the heat capacity at Tc, ΔC, through the relation:

$$\Delta F(O) = 8\pi^2.\Delta C.T_c \tag{3}$$

This classical relation is obtained by deriving twice the free energy to obtain the heat capacity, and by assuming that in the oxides the thermodynamical critical field follows the same temperature dependence as in the conventional superconductors[7]. For YBCO, the jump in the heat capacity is now well established, $\Delta C = 75$ mj/k^{2}mole[8]. The coherence volume can be determined from the coherence lengths obtained from measurements of the upper critical fields. These have been now well determined for YBCO by magnetization measurements, $\xi_{ab} = 15\text{Å}$ and $\xi_c = 3\text{Å}$[9]. Taking into account the experimental margins of uncertainty we then estimate for YBCO $1 < u < 10$. No comparably detailed data exist yet for the more anisotropic Tl and Bi oxides, but it is likely that the main difference with YBCO is a shorter ξ_c, which would give $u \leq 1$.

2.1. Critical Region

The parameter u determines both the width of the critical region and the rate of flux creep at temperatures of the order of Tc. The width of the critical region is the range of temperatures where the condensation energy per coherence volume is of the order of $K_B T_c$ or smaller (Ginzburg criterion). For a three dimensional superconductor this leads to:

$$u \simeq \varepsilon_c^{-1/2} \tag{4}$$

where ε_c is the width of the critical region expressed in units of the reduced temperature scale [(Tc-T)/Tc]. This relation gives us a means to obtain directly u from an experimental determination of ε_c. The very detailed heat capacity measurements of Salamon et al.[10] on a single crystal of YBCO show definitely large fluctuation effects within a couple of degrees from Tc (Fig. 1), fully confirming our previous estimate for u that corresponds to a measurable but small critical region. Penetration depth measurements in this oxide are also consistent with a mean field description in most of the temperature range[11].

In BiPbSrCaCuO, the range of strong fluctuations is maybe 10 times larger[12], but a meaningful comparison between the two materials must take into account a possible difference between their respective dimensionalities. We shall come back to this point later when we discuss strong anisotropy effects.

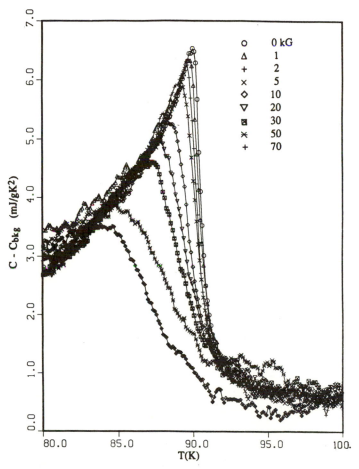

FIG. 1. Heat capacity of an YBCO untwinned crystal under various fields applied parallel to the c axis. Notice the λ like anomaly in zero field and its suppression by an applied field. This is interpreted as a 3D to 2D crossover (see text). Data from Salamon et al., Ref. 10.

2.2. Pinning

As stated above, the parameter u also determines the ability of crystallographic defects to pin vortices. Following Anderson[13], we consider a typical pinning center as being a normal region of size ξ. In this region the order parameter is zero whether the vortex passes through it or not, and therefore the condensation energy per coherence volume U is also the energy barrier that must be overcome to free a vortex line from this center. The rate at which a vortex line will free itself from this potential well is determined by the Boltzman factor. The necessary condition for efficient pinning at temperatures of the order of Tc is thus u >> 1. In other words, *vortices cannot be pinned inside the critical region.* In a superconductor with a large critical region, vortices can only be pinned at low temperatures T << Tc.

We thus expect the width of the critical region to be well correlated with the rate of flux creep and with the range of temperature and field where the magnetic behavior is reversible. This is well borne out by the existing data: the range of reversible magnetic behavior and the rate of flux creep are much smaller in YBCO than in BiSrCaCuO[14].

Detailed measurements of the critical behavior is thus both of great fundamental and practical interest. It allows a determination of the potential of a new superconductor for high field-high current-high temperature application, without engaging in a detailed study of the pinning efficiency of various defects.

So far, we have taken into account the anisotropy only in as much as it reduces the coherence volume, thereby increasing the width of the critical region and reducing the pinning energy, i.e. our approach has been three dimensional. Several experimental features cannot be understood in this framework:

1) the 3D approach leads to the prediction that the field beyond which the magnetic behavior becomes reversible, H_{IRR} must vary as a power law of the temperature $H_{IRR} \propto (Tc-T)^n$, with $n=3/2$[15]. This is well verified in YBCO, but not in BiSrCaCuO for which it has been found that $H_{IRR} \propto \exp-(T/T_o)$[14] (Fig. 2).

2) the width of the resistive transition is greatly enhanced when a magnetic field is applied parallel to the c axis, irrespective of the direction of the current[16]. Hence this broadening cannot be due to flux flow or creep.

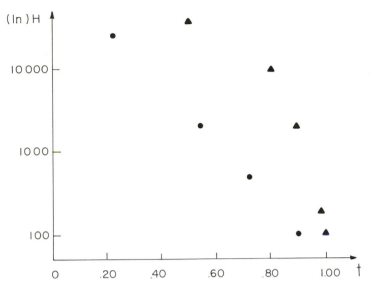

FIG. 2. Temperature dependence of the irreversibility field H_{IRR} in YBCO (triangles, after Ref. 15) and in BSCCO (circles, after Ref. 14).

We shall argue below that points 1) and 2) can be understood in terms of extreme anisotropy effects. But we first discuss the influence of the short coherence length on boundary properties.

3. COHERENCE LENGTH AND BOUNDARY EFFECTS

It is now well established that large angle grain boundaries (or at least most of them) behave as Josephson junctions in the high Tc oxides, contrary to the case of conventional low Tc superconductors where they do not reduce the critical current. This junction behavior is observed even in clean boundaries[17], and must therefore have a fundamental origin.

3.1. Critical Current of Junctions

The origin of extrinsic critical currents was assigned early on to the short coherence length[2]. It results in a depression of the order parameter at surfaces and interfaces, for instance at grain boundaries. Following the formalism first introduced by de Gennes[7], this depression can be described in terms of an *extrapolation length*, equal to the inverse of the logarithmic derivative of the order parameter, in the superconducting bank, at the interface (Fig. 3).

405

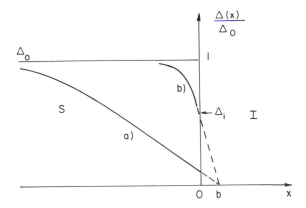

FIG. 3. Boundary condition for the superconducting order parameter. The boundary depression of the order parameter is stronger at high temperatures (a), because $\xi(T)$ diverges at Tc, while the extrapolation length b is temperature independent.

A fundamental characteristic of this length is that it is *temperature independent*. The depression is a function of the parameter $[b/\xi_s(T)]$. In the limit where this parameter is small, the depression is given by:

$$\frac{\Delta_i}{\Delta_o} = \frac{b}{\xi(T)} \tag{5}$$

where Δ_i and Δ_o are respectively the value of the order parameter at the boundary and in the bulk. When the material that constitutes the junction is an insulator, a semiconductor or a very poor conductor, and the superconductor is a conventional low Tc, large coherence length one, b $>> \xi(0)$. Then the limit $[b/\xi(T)]<<1$ never applies in practice, expect extremely close to Tc. But in the case of the oxides, it has been shown[2] that Eq. 5 applies over a significant range of termperatures near Tc. Accordingly, the critical current of the junction, proportional to $(\Delta_i)^2$, must vary as $(Tc-T)^2$, rather as $(Tc-T)$ as is the case for a conventional Ambegaokar-Baratoff junction[18].

This prediction has now been fully confirmed by the detailed measurements of the IBM group on Grain Boundaries Junctions (GBJs)[4] near Tc. The full temperature dependence of the critical current is also well described by the model[19].

3.2. IcR Product

Another characteristic property of GBJ's is that the product of their critical current by their normal state resistance (IcR product) scales with their critical current density, while IcR is a constant for Ambegaokar-Baratoff junctions. For a wide variety of large angle GBJ's it is observed that IcR α $J_c^{1/2}$ or equivalently that IcR α R^{-1}.[20]

Again, this can be understood on the basis of the short coherence length, as realized (actually during this meeting (!)), see Ref. 21. For the range of junctions studied, and at low temperatures, 0.1 meV < IcR < 10 meV, i.e. IcR is always significantly smaller than the bulk pair potential (presumably equal to at least 20 meV). Since $Ic(0)R = \Delta_i(0)$, Eq. 5 applies. According to electron microscopy observations, the junction is only a few Å thick and it is reasonable to assume that it is thinner than the decay length of the order parameter, ξ_n. In this limit, b α (ξ_n^2/d_n).

The V(I) characteristics can be well fitted to the resistively shunted junction model. The values of R thus obtained indicate that the junction has a high effective resistivity ρ_n of the order of 1 Ωcm. The meaning of this value is probably that the junction is made of an insulating material with a localization length somewhat larger than the thickness d_n of the junction. Then both ρ_n and ξ_n are determined by the value of the (finite) coefficient of diffusion $D(d_n)$ on scale d_n: ρ_n α ξ_n^{-2} α D^{-1}. Thus b α $(\rho_n d_n)^{-1}$ = RA, where A is the cross section of the junction. The scaling law IcR α R^{-1} then follows from Eq. 5.

3.3. Polycrystalline Samples

The special boundary condition for the order parameter due to the short coherence length is thus well established through the laws I_c α $(Tc-T)^2$ and I_cR α R^{-1} observed in GBJs. There are indications that twin boundaries may also limit the critical current with an IcR product of the order of 5 to 10 meV.[22]

In a non textured sample, there will be a broad distribution of R values, and accordingly of Ic values. As is well known, the macroscopic resistance of a random resistor network is determined by that of the "typical" resistor, i.e. that of the last resistor necessary to establish an infinite cluster starting from the lowest R values. This typical resistor will also determine the macroscopic critical current density j_c. Because of the scaling law Ic α R^{-2} we expect $j_c\alpha$ $(\rho_n)^{-2}$ in polycrystalline samples.

The low critical current in polycrystalline samples is thus reasonably well understood. We want nevertheless to stress that the "transparency" of the grain boundaries as expressed by their normal state resistance R may vary considerably from one oxide to another. For instance, boundaries might be more transparent in BiSrCaCuO than in YBaCuO.[3]

4. STRONG ANISOTROPY EFFECTS

The extreme anisotropy of the upper critical field in BiSrCaCuO leads to estimates of the out of plane coherence length smaller than 1Å. This is a clear sign that this compound is basically 2 dimensional. The coupling between the superconducting stacks of CuO_2 planes is very weak, and one can expect 3D coherence to be easily destroyed. Description of the critical behavior and pinning in terms of a coherence *volume* becomes then inappropriate; the quasi 2D character must be taken into account explicitly. Since in all the known layered high Tc oxides ξ_c is smaller than the lattice spacing along the c axis, this quasi 2D character must to some extent be present in all of them.

Loss of 3D coherence when a magnetic field is applied parallel to the c axis has been clearly observed in single crystals of LaSrCuO[15]. In these experiments, the broadening of the resistive transition was observed to occur even when the current was applied parallel to the mag-

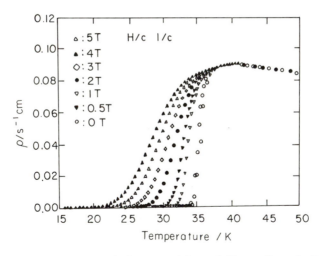

FIG. 4. Brodening of the resistive transition of $(La_{0.94}Sr_{0.06})_2CuO_4$ crystal, with both the magnetic field and the current applied parallel to the c axis. (After Ref. 16a).

netic field, i.e. in a geometry where the vortices are not subjected to a Lorentz force (Fig. 4).

A similar interpretation can be given to the heat capacity experiments of Salamon et al.[9], which show that under a field applied parallel to the c axis, the transition is *broadened rather than shifted* (Fig. 1). In particular, the divergence of the heat capacity at Tc, characteristic of a 3D λ transition, is suppressed by the applied field. The broad transition observed in strong fields, without the λ overshoot, is characteristic of a reduced dimensionality for which there is no true phase transition. The observed width is consistent with the 2D result $\varepsilon_c \approx u^{-1}$.

4.1. Field Induced 3D to 2D Crossover

The experimental data thus strongly suggests that a 3D to 2D crossover takes place when a magnetic field is applied parallel to the c axis.

It has been proposed that this crossover is due to thermal fluctuations that result in the displacement of vortex cores between successive CuO planes, or stacks of planes[23]. Instead of being straight and parallel to the applied field, the vortex cores can follow a "zig-zag" path, with added stretches parallel to the ab planes (Fig. 5). The phases of the order parameter on opposite sides of these stretches differ by π, over a

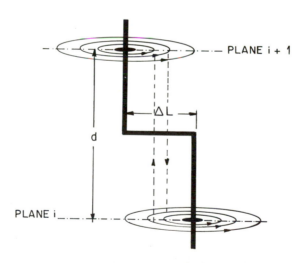

FIG. 5. Thermal fluctuations can result in added vortex stretches intercalated between successive (ab) planes. These stretches lead to local phase decoupling between the planes over an area ΔS (see text). Resistance *along the c axis* is restored when the vortex density is of the order of $(\Delta S)^{-1}$.

certain area ΔS that we shall shortly estimate. When the vortex density (B/Φ_o), where B is the induction and Φ_o the flux quantum, is of the order of $(\Delta S)^{-1}$, 3D phase coherence is lost.

Decay of the phase difference between successive planes is governed by a Sine Gordon equation, with a characteristic length scale r_g:

$$r_g = \xi_{ab}(O)(g)^{-1/2} \tag{6}$$

where $g = (\xi_c (O)/d)^2$, with d the distance between CuO planes[24]. In BiSrCaCuO, $r_g \approx 500\text{Å}$ The thermal average of the length ΔL of the added vortex stretches is given by[25]:

$$<\Delta L> \approx \lambda_{ab} \frac{\lambda_c(4\pi)^2 k_B T}{(\phi_o)^2} \tag{7}$$

We first note that $<\Delta L>$ diverges near Tc and goes to zero when T goes to zero. In order of increasing temperatures, we thus meet the following situations:
i) $<\Delta L> < \xi_{ab}$. No decoupling effect.
ii) $\xi_{ab} < <\Delta L> < r_g$. Then $\Delta S = (r_g)^2$, and the decoupling field is given by:

$$H_d = \frac{\phi_o}{(r_g)^2} \tag{8}$$

iii) $r_g < <\Delta L>$. Then $\Delta S = <\Delta L>$. r_g, and

$$H_d = \frac{\phi_o}{<\Delta L> .r_g} \tag{9}$$

Nearer to Tc, other expressions are obtained, with a stronger temperature dependence.[23] The above relations are of course only rough estimates. They do not take into account the series of stretches that occur along a vortex line, nor the interaction between vortices.

From Eq. 7, $<\Delta L>$ follows a power law near Tc, $<\Delta L> \alpha \lambda_{ab}\lambda_c$, and therefore so does H_d. Using mean field coefficients for $\lambda(T)$, we obtain $H_d \alpha (Tc-T)$ from Eq. 9, valid no too close to Tc $(\Delta L<\lambda_{ab})$. This is in agreement with the observed broadening of the resistive transition when the magnetic field and the current are parallel to the c axis[16] (Fig. 4).

The 3D to 2D crossover should also be accompanied by a change in the pinning regime, which remains to be studied. Assuming that pinning is less efficient in the 2D than in the 3D regime, it is tempting to identify H_d and H_{IRR}. The low temperature exponential behavior of the later would imply (see Eq. 8) that the anisotropy coefficient g is strongly temperature dependent in the very anisotropic oxides. This might be the case if the BiO planes were actually metallic, with an order parameter induced by the proximity effect. The metallic character of the BiO planes, suggested sometime ago to explain precisely the anamalous temperature dependence of H_{IRR}[14], seems to have been confirmed by recent photoemission experiments showing an induced gap on the BiO plane.[25] The matter remains nevertheless a subject of controversy at the time of writing, in view of contradicting STM results.[26]

5. CONCLUSIONS

The recent experimental results that we have reviewed here confirm the dominant role played by the short coherence length in the phenomenology of the oxides. Related theoretical developments in areas such as the properties of the vortex state are currently being pursued actively. A microscopic understanding of boundary effects is also required. Another promising area for experimental and theoretical work is the proximity effect. Up to now, most experiments have been performed using normal metals but there are reasons to believe that it might be more interesting to study the properties of superconducting-semiconducting interfaces.

We have emphasized many of the adverse effects of the short coherence length. But it can also be the basis for a number of unique applications of the oxides. An obvious one is to make use of the extremely high $H_{c2}(0)$ for the generation of very high magnetic fields, maybe of the order of 100T. This is possible at low temperatures where 3D coherence is firmly established, as shown by the quoted exponential increase of H_{IRR} and recent experiments on Bi wires. Another one is the possibility of manufacturing three terminal devices using ultra thin films, where superconductivity can persist precisely because of the short ξ (see Fisher in these proceedings). We have discussed at length grain boundary junctions. They are also unique to the short ξ oxides and form the basis of highly successful high Tc SQUIDS. I have no doubt that further unique applications will follow.

ACKNOWLEDGEMENTS

I am deeply indebted to Alex Mueller for many in depth discussions on short coherence length effects in the cuprates. This work was supported in part by the GIF and by the Oren Family Chair of Experimental Physics.

REFERENCES

1. J. G. Bednorz and K. A. Muller, Z. Phys. **B64**, 189 (1986).

2. G. Deutscher and K. A. Muller, Phys. Rev. Lett. **59**, 1745 (1987).

3. Y. Yamada et al., Jpn. J. Appl. Phys. **29**, L450 (1990).

4. R. Gross et al., Phys. Rev. Lett. **64**, 228 (1990).

5. G. Deutscher, in Novel Superconductivity, S. A. Wolf and V. Kresin eds. (Plenum, New York 1987), p. 23.

6. G. Deutscher, in the Proceedings of the International Conference on Transport Properties of Superconductors, May 1990, Rio de Janeiro (Ed. R. Nicolsky).

7. P. G. de Gennes, Superconductivity in Metals and Alloys, (W. A. Benjamin, New York 1966).

8. A. Junod et al., Physica **C162–164**, 482 (1989).

9. U. Welp et al., Physica **C161**, 1 (1989).

10. M. B. Salamon et al., Physica **A168**, 283 (1990).

11. L. Krusin Elbaum et al., Phys. Rev. Lett. **61**, 217 (1988).

12. K. Okazaki et al., Phys. Rev. **B41**, 4296 (1990).

13. P. W. Anderson and Y. B. Kim, Rev. Mod. Phys. **36**, 39 (1964).

14. P. de Rango et al., J. de Physique **50**, 2857 (1989).

15. Y. Yeshurun and A. Malozemoff, Phys. Rev. Lett. **60**, 2202 (1988).

16. This has been observed in a) LaSrCuO by K. Kitazawa et al., Jpn. J. Appl. Phys. **28**, L555 (1989); b) in YBCO by K. C. Woo et al.,

Phys. Rev. Lett., **63** 1877 (1989); c) in BiSrCaCuO by K. Kadowaki et al., Physica **C161**, 313 (1989).

17. D. Dimos et al., Phys. Rev. Lett. **61**, 219 (1988).

18. V. Ambegaoar and A. Baratoff, Phys. Rev. Lett. **10**, 4861 (1963).

19. G. Deutscher, IBM J. Res. Develop., **33**, 293 (1989).

20. R. Gross et al., Phys. Rev. B in press; S. E. Russek et al., Appl. Phys. Lett. **57**, 1155 (1990).

21. G. Deutscher and P. Chaudari, to be published.

22. S. Anlage, private communication.

23. G. Deutscher and A. Kapitulnik, Physica **A168**, 338 (1990).

24. S. Doniach, Los Alamos report (1989).

25. B. O. Wells et al., to be published.

26. T. Hasegawa and K. Kitazawa, Jpn. J. Appl. Phys. **29**, L434 (1990).

ARTIFICIALLY GROWN SUPERLATTICES OF CUPRATES

O. Fischer[1], J.-M. Triscone[1], O. Brunner[1],
L. Antognazza[1], M. Affronte[2] and L. Mieville[1]

[1]Dept. de Physique de la Matière Condensée
Genève 4, Switzerland

[2]Dept. de Physique, I.M.O. Ecole Polytechnique
Fédérale, 1015 Lausanne, Switzerland

Abstract - The high temperature superconductors are layered materials whose properties are related to the stacking sequence of the various metal oxide layers. This paper describes recent efforts to modify artificially this stacking sequence and thus to produce new superlattice structures. In particular, we have grown $YBa_2Cu_3O_7$/ $PrBa_2Cu_3O_7$ superlattices with modulation wavelengths as small as 24Å We review here recent experiments on this system and we demonstrate how these superlattices can be used to study the role of anisotropy in the thermally activated flux creep behavior of the high temperature superconductors.

Since the discovery of the high temperature superconductors by Bednorz and Muller[1] and with the subsequent discoveries of $YBa_2Cu_3O_7$ (YBCO)[2], $Bi_2Sr_2Ca_nCu_{1+n}O_{6+2n}$ (BiSCCO)[3] and $Tl_2Ba_2Ca_nCu_{1+n}O_{6+2n}$ (TlBCCO)[4], it has been found that there are several families of high temperature superconductors which all have in common a laminar crystal structure with very anisotropic physical properties. Their structure can be visualized as a stacking of atomic metal or metal-oxide planes with the crystallographic c direction perpendicular to these planes.[5] The CuO_2 planes have been found to be mainly responsible for the superconducting properties. However, the other planes separating the conducting CuO_2 planes are also of importance and determine in particular the anisotropy of the compounds. Generally speaking, the various high temperature superconductors differ in the stacking sequence and the choice of the planes.

Systematic studies of these materials require investigations on materials where the properties can be varied in a controlled manner. In view of the layered nature of these materials it is an exciting challenge to produce artificial superlattices from these compounds, both to be able to obtain model systems and genuine new materials. Since 1989, it has been known that certain superlattices of such oxides can be made.[6,7,8,9,10]

In this paper we report on the preparation and on the properties of such artificial materials. Our films were fabricated using thin film multilayer deposition techniques. In particular, ultrathin layers of $YBa_2Cu_3O_7$ (YBCO) and of $REBa_2Cu_3O_7$ (REBCO, RE = Dy, Pr) were deposited alternately. Superlattices with individual layers as thin as one unit cell have been produced, i.e. every second Y-plane has been replaced by a Dy or a Pr plane. In the case of the DyBCO/YBCO superlattices we find superconducting properties which are very close to those of YBCO films.[6] This is expected for an ideal superlattice since YBCO and DyBCO have very similar superconducting properties. This result thus demonstrates that the artificial build up of the structure does not in itself produce any defects or imperfections that are detrimental to superconductivity. In the case of PrBCO/YBCO superlattices we observe dramatic changes in the superconducting properties related to the fact that PrBCO is not a metallic conductor.[7]

For a complete description of the preparation of our superlattices we refer to refs[6,11]. Here we shall only present briefly some of the main points. We use a dc magnetron sputtering technique with REBCO targets. The deposition temperature is situated between 700°C and 800°C and highly oriented c-axis multilayers have been prepared both on MgO and $SrTiO_3$ substrates. The individual YBCO or REBCO layer thicknesses were always chosen to be a multiple of the c-axis lattice parameters (c = 11.62Å for YBCO). In the following we shall refer to the multilayers by the nominal thickness which we assume for simplicity to be a multiple of 12Å In all experiments we used the exact lattice parameters. X-ray investigations have shown that the films are oriented both in the c-direction and in the a,b plane[11,12] In Fig. 1 we show one of the early results demonstrating with the sharp satellite peaks that a coherent modulation is obtained in these superlattices. An analysis of the satellite peaks in various superlattices revealed that there is very little mixing between the YBCO and the PrBCO layers with possible some "mechanical" interdiffusion between the two RE layers at the interface.

We have recently carried out a detailed TEM study of some of our films[13,14>]. An important result of these investigations is that the individual YBCO and PrBCO layers are clearly separated and that the interface between the YBCO and PrBCO is atomically sharp and no evidence

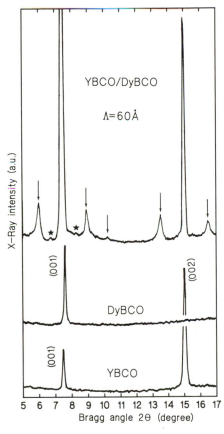

FIG. 1. X-ray diffractograms for an YBCO/DyBCO multilayer with a modulation wavelength of 60Å and for DyBCO and YBCO thin films. The arrows indicate the multilayer satellite peaks. [6].

417

for interdiffusion can be seen. However, the individual YBCO and PrBCO planes are slightly wavy with an amplitude of a few unit cells. This probably reflects slight corrugations in the substrate surface and may explain the slight "mechanical" interdiffusion observed in the x-ray data. In mixed a-axis and c-axis films deposited at lower temperatures we observed similar sharp interfaces in the a-axis grains showing that the interdiffusion is small even along the a,b plane[14]. Another important result of this study is that although defects in the substrate produce a perturbed area in the first 100Å we find that as the film grow thicker the 123 phase grows continuously over the imperfect area and thus the quality of the layering is not strongly dependent on the substrate quality. Other groups have recently also reported TEM studies on such oxide multilayers and found good layering[8,10]

The YBCO/PrBCO superlattices offer unique possibilities to systematically investigate certain properties and their relation to the anisotropy of the superconductor. Since PrBCO is known to be an insulator it is expected that inserting thin PrBCO layers between the YBCO layers will partly decouple the latter and thus make the material more anisotropic. When the PrBCO layers are thick we expect that the individual YBCO layers become independent and thus we have a means to investigate ultrathin YBCO layers with thicknesses down to a single unit cell dimension. In the latter case we have a single sheet with a double CuO_2 plane.

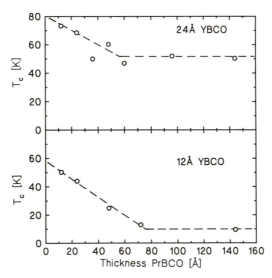

FIG. 2. T_c versus d_{PrBCO} for $d_{YBCO} = 12Å$ and $24Å$

We have studied the critical temperature of several such series of superlattices. Figure 2 shows T_c as a function of the thickness of the individual PrBCO layers (d_{PrBCO}) for two values of the thickness of the YBCO layers (d_{YBCO} = 12Å and 24Å)[7,11,15.]. We observe a continuous decrease of T_c up to d_{PrBCO}= 60 - 70Å and then a saturation for higher PrBCO thickness. We interpret this saturation as a result of a complete decoupling of the individual layers. It demonstrates that one single CuO_2 double layer (one unit cell) is superconducting but with a reduced T_c of about 10 K (this includes possibly some of the mechanical interdiffusion effects discussed above). For two double CuO_2 layers (d_{YBCO} = 24Å) T_c saturates around 50 K. We do not yet understand the origin of the slow decrease of T_c with increasing d_{PrBCO}. However, the data shows that the superconducting layers somehow couple across the relatively thick PrBCO layers up to a thickness of about 60 to 70Å. Ariosa and Beck[16] have proposed a model to account for these data. They assume that an individual CuO_2 plane is granular and that phase fluctuations reduce T_c. Electrostatic coupling between the layers reduce the fluctuations and restores the bulk T_c. Further investigations are under way to test these ideas as well as very recent theoretical work by Schneider et al.[17] Similar results to ours have now been reported.[9,10] These groups report somewhat higher saturation values for T_c, an effect partly related to their definition of T_c. It is also interesting to study how the critical temperature of the individual, decoupled YBCO layers is restored when d_{YBCO} is increased. In Fig. 3 we show T_c as a function of d_{YBCO} for a serie with d_{PrBCO} = 144Å. We observe a gradual increase of T_c and only for d_{YBCO} of the order of 200Å do we observe a critical temperature equal to the one of YBCO.

In order to check whether the layering influences the number of carriers and to test whether this possibly could explain the decrease in T_c, we have investigated the resistivity and the Hall effect of our multilayers[18]. Figure 4 shows the Hall number (= $1/eR_H$, where $R_{<H>}$ is the Hall coefficient.) as a function of temperature for four YBCO/PrBCO multilayers: 12Å/12Å, 24Å/24Å, 36Å/36Å and 96Å/96Å The data have been normalized to the total YBCO thickness anticipating that the Hall effect is mainly determined by these YBCO layers. For comparison we show also the result for a YBCO film and a YBCO crystal. It is striking to notice that the four multilayers give essentially the same result. The result of the YBCO film is close to the result for the YBCO crystal and give a somewhat higher Hall number. A

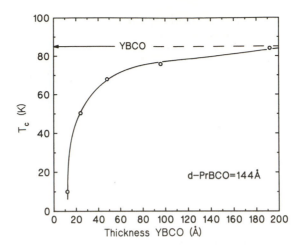

FIG. 3. T_c versus d_{YBCO} for $d_{PrBCO} = 144$Å

possible origin of this may be that all four multilayers grow presumably coherently and are therefore strained, which may possibly affect the Hall coefficient. In any case these results demonstrate that interdiffusion effects and/or interface effects are small, even in the 12Å/12Å sample, and we conclude that the carrier density does not change significantly

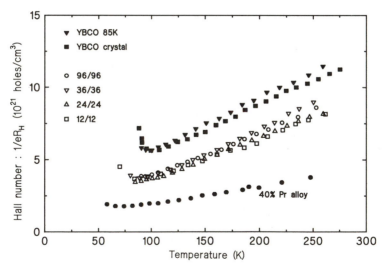

FIG. 4. The Hall number as a function of temperature for four multilayers: 12Å/12Å, 24Å/24Å, 36Å/36Å and 96Å/96Å as well as for an YBCO film an YBCO crystal and an $Y_{.6}Pr_{.4}BCO$ film.

when we reduce the layer thicknesses from 96Å to 12Å. The other striking result is that contrary to the $Y_{1-x}Pr_xBa_2Cu_3O_7$ alloys (we show in Fig. 4 the x = 0.4 case) the multilayers display a clear $1/T$ dependence of R_H as for clean YBCO crystals. This also shows that the additional scattering effects in the multilayer have no effect on R_H contrary to the case of the resistivity of these same multilayers which show a less pronounced T dependence as the modulation wavelength decreases. From these results it is clear that the T_c variation cannot be explained in terms of a change in the charge density, and that another explanation, probably related to the reduced dimensionality as in Ref. 16,17 must be sought.

One of the striking properties of the oxide superconductors is the unusual properties of the mixed state, where thermal fluctuation effects on the magnetic structure lead to a complex and unusual behavior. This leads to the observation of an "irreversibility line" in the H-T phase diagram separating a reversible and a hysteretic region[19,20] and to a relatively fast relaxation of the magnetization in the critical state.[21] Another remarkable manifestation of this behavior is the broadening of the resistive transition in a magnetic field as shown for a 24Å/96Å superlattice in Fig. 5a. It is generally supposed that in the "visible part" of the transition i.e. for $\rho > 10^{-2}\rho_n$, the broadening results from a complex flux flow behavior whereas the tail of the transition ($\rho < 10^{-2}\rho_n$) show a thermally activated behavior[22]:

$$\rho(T,H) = \rho_o \exp(-U/k_BT) \tag{1}$$

Here ρ_o is a resistivity of the order of the normal state resistivity and U the temperature and field dependent activation energy. This is thought to correspond to a thermally activated flux flow (TAFF)[23] and the reason for the relatively small values of U is believed to be related to the anisotropy of these materials. With our multilayers we can systematically study these questions. In previous papers we have reported the resistive transitions of various multilayers in magnetic fields parallel or perpendicular to the a,b plane.[12,24] Let us first consider the case of a magnetic field parallel to the a,b plane. A striking effect is that when d_{YBCO} is small (< 24Å) and d_{PrBCO} large (> 24Å) a parallel magnetic field of 9 Tesla has nearly no effect on the resistive transition. This is illustrated in Fig. 5b where the single transition cover those in fields of 0T, 1T, 3T, 6T and 9T. Note the difference when the field is perpendicular to the a,b plane, shown in Fig. 5a. This insensitivity extends to the tail of

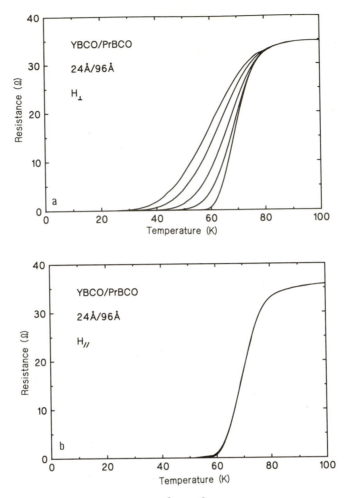

FIG. 5. Resistive transitions for a 24Å/96Å multilayer in magnetic fields of 0, 1, 3, 6, and 9 tesla: a) field parallel to the c-axis, b) field parallel to the a,b plane.

the transition. This is illustrated in Fig. 6 where we plot $\log\rho$ versus $1/T$. Since the individual YBCO layers in this sample contain only two double CuO_2 planes we believe that in this case the vortex picture breaks down and we have probably only weak screening currents in the layers and thus also no activated flux flow. Only when d_{YBCO} becomes very large (> 200Å) does one recover the bulk behavior of YBCO. Since this thickness is much larger that the c-axis coherence length these results confirm that the influence of the magnetic field on the resistive transition is related to the vortex motion rather than to the upper critical field H_{c2}.

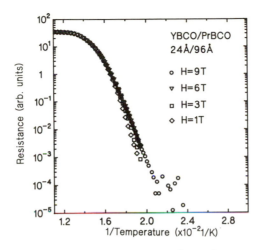

FIG. 6. Resistivity ρ(T,H) versus 1/T for a 24Å/96Å superlattice for different fields (1, 3, 6 and 9 tesla) parallel to the a,b plane.

We now turn to the case where the magnetic field is perpendicular to the a,b plane, and we start with a general expression for the activation energy U:

$$U(T,H) = \frac{\rho\mu_o H_c^2(T)V_c}{2} \qquad (2)$$

Here V_c is the correlation volume involved in the flux movement and p represents the corresponding fraction of the condensation energy density. We write $V_c = R_c^2 L_c$ where R_c and L_c are the correlation lengths perpendicular and parallel to the magnetic field. L_c represents in a certain sense the stiffness of the flux line lattice and depends on the tilt modulus C_{44}.[26] An isotropic material is expected to have a large L_C, whereas an anisotropic material with the high mass direction parallel to the field should have a small L_C. In a recent paper Palstra et al.[26] propose that the large activation energies found for YBCO compared to other high temperature superconductors like $Bi_2Sr_2CaCu_2O_8$ are essentially due to the fact that the larger anisotropy in the latter compounds make L_c much smaller and thus U much smaller. Their results implies that L_c for YBCO should be several hundred Å.

We have used our multilayers to test these ideas[27] by measuring the resistive transitions in magnetic fields perpendicular to the a,b plane.

Following the ideas of Palstra et al. we would expect that in multilayers with large d_{PrBCO} L_c should be limited to d_{YBCO} and for smaller d_{PrBCO} we would expect an intermediate L_c. Figure 7 show a plot of $\log(\rho(T))$ versus $1/T$ for various magnetic fields for a YBCO film and for two multilayers: 24Å/24Å and 96Å/96Å. One immediately sees that the

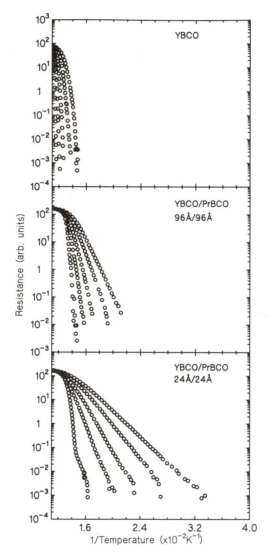

FIG. 7. Resistivity $\rho(T,H)$ versus $1/T$ for an YBCO film, a 96Å/96Å superlattice and a 24Å/24Å superlattice. The various curves are, counting from left, for magnetic fields parallel to the c- axis of 0, 1, 3, 6 and 9 tesla, respectively.

activation energies for YBCO are about 5 times larger than for the 96Å/96Å multilayer. Assuming that 96Å PrBCO is enough to decouple the YBCO layers and that L_c is several hundred Å, this is exactly what one would expect. Consider now the 24Å/24Å sample. Compared to the other superlattice we have not changed the amount of supercon-ducting material, but just divided it more finely. U again decreases but not by the expected factor of four. The reason for this is that 24Å PrBCO is not enough to decouple the YBCO layers. In fact, keeping $d_{YBCO} = 24$Å and increasing the PrBCO thickness to 48Å or to 96Å decouples the layers completely and U decreases a factor of four com-pared with the 96Å/96Å sample. We have extracted the activation ener-gies $\overline{U}$ from these Arrhenius plots. In principle, these values should be corrected for the temperature dependence of U. However, we find that the temperature dependence is close to $U(t)=U_o(1-t)^q$ with $q=1-1.25$ and that $\overline{U} \cong U_o$. We refer the reader to Ref. 27 for a discussion of this point. In Fig. 8 we show the obtained U values as a function of magnetic field. We note that the field dependence is very closely the same in all samples. We therefore plot U normalized to the value for YBCO, U_{YBCO}, for the samples 96Å/96Å, 24Å/48Å and 24Å/96Å. As can be seen in Fig. 9 the resulting values are largely field independent and there is the expected factor of four between the values for $d_{YBCO} = 96$ Å and

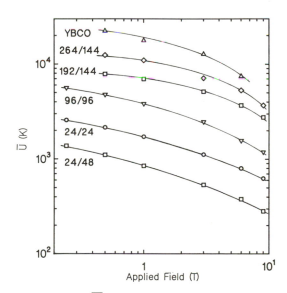

FIG. 8. Activation energies $\overline{U}$ versus magnetic field for an YBCO film, and five superlattices: 264Å/144Å, 192Å/144Å, 96Å/96Å, 24Å/24Å, 24Å/48Å

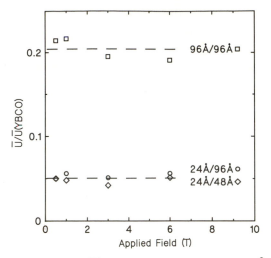

FIG. 9. Activation energies $\overline{U}$ of three superlattices: 24Å/96Å, 24Å/48Å and 96Å/96Å, normalized to the values obtained for YBCO, versus applied magnetic field.

$d_{YBCO} = 24\text{Å}$. The proportionality between U and d_{YBCO} is even better illustrated in Fig. 10 where we plot the reduced values $\overline{U}/\overline{U}_{YBCO}$ as a function of d_{YBCO} for three different field values. From this plot we see that U=const. d_{YBCO}. We furthermore find that L_c for the YBCO film is about 450Å and that this value is largely field independent.

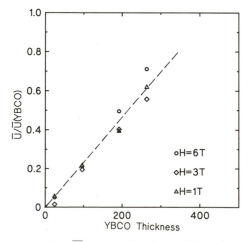

FIG. 10. Activation energies $\overline{U}$ normalized to the values for YBCO, versus thickness of the individual YBCO layers for magnetic fields of 1, 3 and 6 tesla.

From Fig. 8 we see that the magnetic field dependence of $\overline{U}$ is not a power law. In fact, a logarithmic plot show that $U/d_{YBCO} = -\alpha\ln B + \beta$. This is illustrated in Fig. 11 where we have normalized U so that all samples fall roughly on one curve: $U = \overline{U}(L_{cYBCO}/d_{YBCO})$. The curve shown is a fit with $\alpha=13K/\text{Å}$ and $\beta=45K/\text{Å}$. The exact explanation for this behavior remains to be fully understood. However, it is interesting to note that a logarithmic dependence of U on B can be obtained if we assume that the activation process corresponds to the formation of a dis-location pair in the vortex lattice[27]. Following Feigel'man et al.[28] we note that in the collective pinning model the pinning leads to a finite range of the order a_o^2/ξ of the interaction between the dislocations in a pair. Here a_o is the flux line lattice spacing and ξ the coherence length. For a very thin film in the 2D limit we can then write[28]: $U = (d\Phi^2/16\pi^2\mu_o\lambda^2) \ln(a_o/\xi)$. where Φ_O is the flux quantum, d the layer thickness ($= d_{YBCO>}$ in our case) and λ the penetration depth. Assuming $\lambda=1400\text{Å}$ and $\xi = 20\text{Å}$ we find $U[K]/d[\text{Å}] = -3.73\ln B + 23.7$. Taking into account the complexity of the dynamics of the vortex structure, the correspondence with the experimental results is remark-able. Both the field and the thickness dependence correspond and the expected temperature dependence is close to what is observed. Further-more the numerical constants are within roughly a factor two of the observed values. This suggests that this type of process may be impor-

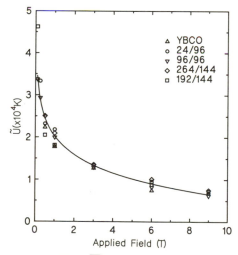

FIG. 11. Activation energies $\overline{U}(L_{cYBCO}/d_{YBCO})$ versus magnetic field for an YBCO film, and four different superlattices: 24Å/96Å, 96Å/ 96Å, 264Å/144Å, 192Å/144Å. The line represents the expression $\overline{U}[K]/d[\text{Å}] = -13\ln B + 45$.

tant in the observed flux flow behavior. We note finally that a loga-rithmic behavior has also been predicted by Inui et al.[29].

In conclusion, it is now established that certain high temperature superconductors can be used to produce artificial superlattices with sharp interfaces on the level of one unit cell. This opens up a whole field of future investigations, both on the fundamental properties of these mate-rials and on the tayloring of these materials to specific properties, for instance in the field of new devices. The work reported here has concen-trated on the growth characteristics and on the study of certain basic properties of the high temperature superconductors. In particular, we have studied the superconducting properties of ultrathin YBCO layers and we have been able to determine the flux line lattice correlation length L_c for YBCO and to demonstrate the role played by the anisotropy for the flux flow behavior. We hope to have illustrated how such artificial structures can be used and that the investigation of these and other superlattices of the cuprates has a large potential for future research.

ACKNOWLEDGEMENTS

It is a pleasure to acknowledge discussions with M. Karkut, M. Decroux, A. D. Kent, Ph. Niedermann and R. Griessen. We are grateful for technical assistance by T. Boichat and A. Stettler.

REFERENCES

1. J. G. Bednorz and K. A. Muller, Z Phys. **B 64**, (1986).

2. M. K. Wu, J. R. Ashburn, C. J. Torng, P. H. Hor, R. L. Meng, L. Gao, Z. J. Huang, Y. Q. Wang and C. W. Chu, Phys. Rev. Lett. **58**, 908 (1987).

3. H. Maeda, Y. Tanaka, M. Fukutomi and T. Asano, Japan J. Appl. Phys. **27**, L209 (1988).

4. Z. Z. Zeng and A. M. Hermann, Nature **332**, 55 (1988).

5. See for example K. Yvon and M. Francois, Z. Phys. B 76, **413** (1989).

6. J.-M. Triscone, M. G. Karkut, L. Antognazza, O. Brunner and O. Fischer, Phys. Rev. Lett. **63**, 1016 (1989).

7. J.-M. Triscone, O. Fischer, O. Brunner, L. Antognazza, A. D. Kent and M. G. Karkut, Phys. Rev. Lett. **64**, 804 (1990).

8. A. Gupta, R. Gross, E. Olsson, A. Segmuller, G. Koren and C. C. Tsuei, Phys. Rev. Lett. **64**, 3191 (1990).

9. Q. Li, X. X. Xi, X. D. Wu, A. Inam, S. Vadlamannati, W. L. McLean, T. Venkatesan, R. Ramesh, D. M. Hwang, J. A. Martinez and L. Nazar, Phys. Rev. Lett. **64**, 3086 (1990).

10. D. H. Lowndes, D. P. Norton and J. D. Budai, Phys. Rev. Lett. **65**, 1160 (1990).

11. J. M. Triscone, O. Fischer, L. Antognazza, O. Brunner, A. D. Kent, L. Mieville and M. G. Karkut. To appear in Science and Technology of Thin Film Superconductors 2, edited by R. D. McDonnel and S. A. Wolf.

12. O. Fischer, J. M. Triscone, L. Antognazza, O. Brunner, A. D. Kent, L. Mieville and M. G. Karkut. To appear in the Proceedings of the E-MRS Spring Meeting, Strasbourg, 1990.

13. O. Eibl, H. E. Hoenig, J. M. Triscone, O. Fischer, L. Antognazza and O. Brunner. Physica C, **172**, 365 (1990).

14. O. Eibl, H. E. Hoenig, J. M. Triscone and O. Fischer. Physica C, **172**, 373 (1990).

15. L. Antognazza, J. M. Triscone, O. Brunner, M. G. Karkut and O. Fischer, Physica **B 165–166**, 471 (1990).

16. D. Ariosa and H. Beck, Phys. Rev. B **43**, 344 (1991).

17. T. Schneider, Z. Gedik, and S. Ciraci, Europhys. Lett. **14**, 261 (1991).

18. M. Affronte, J. M. Triscone, O. Brunner, L. Antognazza, L. Mieville, M. Decroux and O. Fischer, Phys. Rev. **B43**, 11484 (1991).

19. K. A. Muller, M. Takashige and J. G. Bednorz, Phys. Rev. Lett. **58**, 1143 (1987).

20. Y. Yeshurun and A. P. Malozemoff, Phys. Rev. Lett. **60**, 2202 (1988).

21. A. C. Mota, G. Juri, P. Visani and A. Pollini, Physica C **162–164**, 1152 (1989).

22. T. T. M. Palstra, B. Batlogg, L. F. Schneemeyer and J. V. Waszczak, Phys. Rev. Lett. **61**, 1662 (1988).

23. P. H. Kes, J. Aarts, J. van den Berg, C. J. van der Beek, J. A. Mydosh, Supercond. Sci. Technol. **1**, 242 (1989).

24. O. Brunner, J.-M. Triscone, L. Antognazza, M. G. Karkut and O. Fischer, Physica **B 165–166**, 469 (1990).

25. P. H. Kes and J. van den Berg in Studies of High Temperature Superconductors, ed A. Narlikar, (NOVA Science Publishers, New York, 1990), Vol. 5, p. 83.

26. T. T. M. Palstra, B. Batlogg, R. B. van Dover, L. F. Schneemeyer and J. V. Waszczak, Phys. Rev. **B41**, 6621 (1990).

27. O. Brunner, L. Antognazza, J.-M. Triscone, L. Mieville and O. Fischer, to be published.

28. M. V. Feigel'man, V. B. Geshkenbein, A. I. Larkin, Physica **C 167**, 177 (1990).

29. M. Inui, P. B. Littlewood and S. N. Coppersmith, Phys. Rev. Lett. **63**, 2421 (1989).

EXCITATIONS AND THEIR INTERACTIONS IN HIGH Tc MATERIALS

J. R. Schrieffer

Department of Physics
University of California
Santa Barbara, CA 93106

Abstract - The nature of electronic excitations in high Tc materials and in their undoped parent compounds is discussed. In the weak doping limit, hole-like excitations are found to correspond to antiferromagnetic polarons or bags whose structure varies smoothly from the weak coupling (U < 4t) SDW case to the strong coupling (U > 4t) Mott-Hubbard-Heisenberg case. These excitations exhibit an attractive pairing interaction in the 2D planes. For weak and intermediate doping x, Fermi liquid behavior breaks down and a new approach is required. Several possible scenarios are discussed, including the question of whether pairing theory is valid for this system, regardless of the nature of the effective pairing potential. Experiments currently underway are essential in resolving whether exotic condensates, such as those involving anyons, flux phases, etc., occur in these materials, or if pairing theory can correctly describe the condensate if one generalizes the approach to include pairing due to strong spin and charge fluctuations between hole or electron excitations which have anomalous dressing.

Highlights in Condensed Matter Physics and Future Prospects
Edited by L. Esaki, Plenum Press, New York, 1991

1. INTRODUCTION

Superconductivity is one of the most widespread phenomena in nature. The critical temperature T_c for this state is observed to span thirteen orders of magnitude, from 10^{-3}°K for ^{3}He to 10^{10}°K for atomic nuclei and neutron stars, as shown in Figure 1. If one includes the quark condensation which occurred approximately one second following the big bang, the temperature range expands to sixteen orders of magnitude. In each case, an attractive effective potential V between fermions causes a "pairing" condensation in which the pair amplitude $[c^+_{k\uparrow}c^+_{-k\downarrow}]$ takes on a nonzero-value. The order parameter $\Delta_k V[c^+_{k\uparrow}c^+_{-k\downarrow}]$ has a symmetry which corresponds to pair orbital angular momentum ℓ and spin S as shown. The nature of the pairing potential is also indicated.

The remarkable discovery by Bednorz and Mueller[1] of superconductivity in cuprates above 30°K has been extended to 125°K and may well go higher. While it is unlikely that T_c's in this range arise

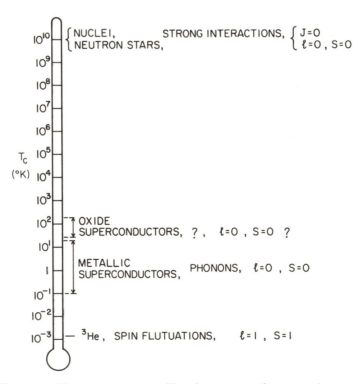

FIG. 1. The transition temperature T_c of superconductors, the nature of the pairing interaction and the quantum numbers of the condensate.

from the electron phonon occuring involved in conventional metallic superconductors, electronic mechanisms involving spin and/or charge degrees of freedom likely play a central role in raising Tc in these novel materials.

A crucial question is whether a pairing condensation in fact occurs in the cuprate superconductors, or alternatively, if a totally new phenomenon is involved which is in no way related to pairing. As we discuss below, progress is being made along both lines, i.e. extending the pairing theory to strongly interacting fermions (novel) as well as quite new concepts such as anyons, chiral phases, etc. (exotic). Fortunately, experimental studies have rapidly increased in scope and reliability, placing strong constraints on any valid theoretical framework. Other lectures in this forum address such experimental results in greater detail.

2. EXPERIMENTAL FACTS

We list only a few key properties[2] which set the cuprates aside from conventional superconductors. Many of these have to do with the normal state properties.

(1). A universal schematic phase diagram of the cuprates is shown in Figure 2, in which the hole concentrations $n_h(x)$ is a monotonically

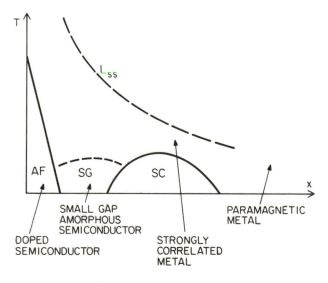

FIG. 2. Schematic phase diagram for high temperature superconducting materials, where x is doping concentration. The antiferromagnetic (AF), spin glass (SG) and superconducting (SC) phases are indicated. L_{ss} is the spin spin correlation length.

increasing function of the doping concentration x, as in the layered perovskite materials $La_{2-x}(Ba,Sr)_x CuO_4$ (so-called "214") and $Y Ba_2Cu_3O_{6+x}$ ("123"). The x=0 parent compounds are commensurate layered antiferromagnets, with sublattice magnetization of order $0.6\mu_B$. Doping sharply decreases the Néel temperature T_N, presumably due to each added hole suppressing antiferromagnetism over an area containing 5 - 10 neighboring sites, with $T_N \rightarrow 0$ for $n_h \sim 0.03$. We note that in the 123 compound, Cu chains act as a reservoir for added holes and only for $x \geq 0.4$ do holes substantially enter the CuO_2 planes, which are central to superconductivity. Optical studies indicate a gap in the electronic spectrum in the low n_h regime which broadens as one dopes past the superconducting regime. As indicated in Figure 2, the spin-spin correlation length L_{ss} decreases with increasing x,[3] showing finite range antiferromagnetic spin correlations exist in the superconductor. Broadly speaking, doping leads to a transition from antiferromagnetic insulating, to amorphous semiconductor, to strongly correlated metal to paramagnetic metal behavior, raising the much discussed question of the metal insulator transition. While there is general agreement on the physics of the insulating and paramagnetic metal limits, current theoretical arguments center on the nature of the strongly correlated metal.

(2). Some anomolous properties (underlined in the table) of the normal phase are

	Doped Cuprates	Conventional Metal
a. electrical resistivity ρ	$\underline{AT} + B$	$AT^3 + BT^2 + C$
b. NMR $T_1{}^{-1}$	$AT + \underline{B}$	AT
c. tunneling dI/dV	$\underline{A|V|} + B$	$AV + B$

d. Photoemission quasi particle-like spectra are observed, however the level width $\Gamma(E)$ appears to vary as $\underline{AE}$ versus BE^2 for conventional metals. Experimental uncertainties remain regarding the background subtraction and the reality of large energy tails of the difference spectra between the normal and superconducting phases.

e. For $0 < \chi < 0.3$, the Hall effect R_H varies as $+1/\chi$ vs. $-1/1-\chi$ as for conventional metals showing hole-like behavior.

(3). Despite these anomolies, the superconducting properties are in many respects analogous to conventional superconductors i.e. the pairing theory; e.g.

a. The Ginzburg-Landau theory, as extended to layered materials gives a good account of macroscopic properties of the high Tc superconductors.

b. An energy gap 2Δ is observed, which appears to be broadened somewhat and has a magnitude $2\Delta/kT_c$ which is larger than the weak coupling pairing theory value of 3.52, or order 5-6.

c. While the evidence is not definitive, experiments such as the muon spin rotation determination of the penetration depth $\lambda(T)$, photoemission and the temperature dependence of the surface resistance $R_s(T)$ argue for nodeless Δ_k on the fermi surface, i.e. "$\ell = 0$" and S=0 pairing. The flux quantum is observed to be hc/2e, consistent with pairing, although certain other approaches produce this result at well.

d. The superconducting coherence length appears to be very small $\sim 12\text{Å}$ along the ab plane and the interplanar spacing in the c direction. Remarkably, fluctuation effects are rather small in spite of the fact that the number of pairs per coherence volume is not large compared to unity.

e. The Hebel-Slichter peak in the NMR $T^{-1}{}_1$ below Tc is not observed, however this may be due to strong damping, as in the broadening of the energy gap.

3. THEORIES: NOVEL VERSUS EXOTIC

The present theories of high T_c can be divided into two limiting categories. In the first category a novel mechanism, other than phonons, couples fermions and leads to a pairing condensation much as in the original pairing theory, but the theory is generalized to account for the fact that a fermi liquid description of the normal state excitations does not hold. Mechanisms which have been advanced include spin and charge fluctuations, spin polarons and pseudo-gaps. The second category, termed exotic, invokes totally new approaches based on concepts absent in the pairing approach, such as normal phase excitations exhibiting fractional statistics, charge and spin separation, etc., with ground states violating macroscopic symmetries such as parity, P, and/or time reversal T. At present, both types of approaches are being developed with no definitive experimental resolution in hand as to which, if either, approach is fundamentally correct. In fact, many key experiments can be accounted for by both schemes, while others have not (or can not) be explained by one or another approach.

In addition to these limiting schemes, there are intermediate theories, such as the "marginal fermi liquid" approach[4] discussed below. In this scheme, a specific form of the dynamic susceptibility is assumed, from which the self energy of a hole is calculated. It is found that at low temperature the quasi-particle weight factor Z vanishes as one approaches the fermi surface, so that fermi liquid behavior breaks down. Some of the anomalous properties of the normal phase of the cuprates can be accounted for in this way. Whether a fully consistent explanation of all major experiments can be constructed in such phenomenological models remains for the future.

A fundamental issue which sets the language of the theories is whether the valence electrons are best thought of as being itinerant, as in band theory, or localized in real space as in the Mott-Hubbard state. The simplest framework which describes both regimes is the Hubbard model,[5] defined by the Hamiltonian.

$$H = - \sum_{(ij)s} t\, c_{js}^{+} c_{is} + U \sum_{i} n_{i\uparrow} n_{i\downarrow}, \tag{1}$$

where c_{is} destroys an electron in a Wannier orbital in cell i with z component of the spin s, and $n_{is} = c_{is}^{+} c_{is}$. The sum involving the hopping t is over nearest neighbor pairs, (ij). U is the on-site effective Coulomb repulsion. H is readily extended to a three band model, corresponding to the Cu - $d_{x^2-y^2}$ and the two oxygen p_σ orbitals per unit cell. Also, two body interactions between different sites can be added to make the model more realistic. It is likely that in the end, all of the low energy physics is contained in a one hand model, with t and U effective couplings which can be derived by integrating out the two other bands.

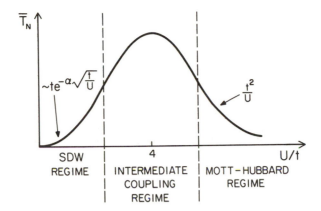

FIG. 3. The phase diagram for the undoped (one electron per cell) Hubbard model as a function of coupling strength U/t where $\overline{T}_N$ is the mean field Néel temperature.

As mentioned above, an important issue is whether the cuprates are in the weak ($U/4t < 1$) or strong ($U/4t > 1$) coupling regime? A phase diagram for mean field antiferromagnetic order is sketched in Figure 3, showing that the Néel temperature peaks for $U/4t \sim 1$. There is growing evidence that the physical regime of interest in the cuprates is $U/4t \sim 1$ so that the correct physics may be a "slush" phase, describable in either the electronic liquid or solid language. As we will see, many physical properties evolve smoothly between these limits. The same is likely to be true in both undoped and doped systems, unless a phase transition to an exotic (P,T) phase occurs.

4. WEAK COUPLING

4.1. Antiferromagnetic Phase

For simplicity we consider a one band model which is half filled (one electron per cell) for the undoped parent compound, $x=0$. This compound is observed to be a commensurate antiferromagnet with a magnetic wave vector $Q = (\pi/a, \pi/a)$ in the a - b plane. For $U/4t \ll 1$, this phase would correspond to a spin density wave (SDW) with the Bloch states $\psi_{ks\ell}$, being large on even sites and small on odd sites for $s=+1$, and the reverse for $s=-1$. The amplitude S of the SDW can be estimated by a mean field calculation with Gaussian fluctuation corrections.[6] One finds the ordered moment $| < S_{iz} > |$ varies with U/t as

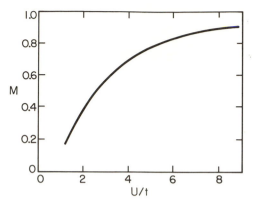

FIG. 4. The sublattice magnetization M of the undoped antiferromagnet as a function of U/t within the mean field approximation.

shown in Figure 4. While the mean field calculations predicts that $|\langle S_{iz} \rangle|$ approaches 1/2 for U $\ll$ t, Gaussian fluctuations, corresponding to spin flips arising from $S_{i+}S_{j-}$ type exchange terms, reduce this value to approximately 0.6 x 1/2, consistent with numerical studies on the 2-d Heisenberg model. From Figure 4 it is clear that for U/4t ~ 1, the ordered moment is a sizeable fraction of its large U limit. Experimentally, the ordered moment is observed to be approximately 0.6 x 1/2 in the undoped compound and varies slowly with doping, resembling local moment behavior. The second property of the antiferromagnet we mention is the spin wave velocity, v_s, given by the slope $d\omega/dq$ of the transverse spin wave frequency near the SDW wavevector Q. Figure 5

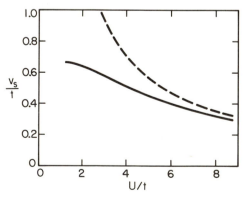

FIG. 5. The RPA spin wave velocity (solid line) and the corresponding quantity for the Heisenberg model plotted as a function of U/t.

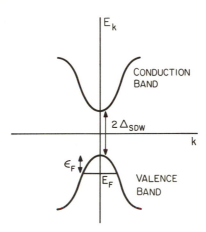

FIG. 6. The energy band of electrons in the SDW state. At finite doping ε_F is the Fermi energy of the holes.

shows how v_s evolves from the fermi velocity for $U/t \ll 1$ to the two site exchange integral $J \sim 4t^2/U$ for $U/t \gg 1$. Again completely smooth behavior is found.

Finally we note that because of exchange Bragg scattering the CDW induces an energy gap 2Δ in the electronic spectrum, as illustrated in Figure 6. Within the mean field approximation $2\Delta_{SDW}$ is proportional to the ordered moment and smoothly evolves from the SDW gap $2\Delta_{CDW} \sim t\exp\left(-\alpha\sqrt{t/U} \right)$ to the Mott-Hubbard limit $2\Delta_{MH} = U$.

While the 2-d antiferromagnet has long range order at $T = 0$, it has finite range order at all $T \neq 0$. However, weak 3-d exchange coupling raises the Néel temperature to of order the 2-d mean field T_N and we refer to 2-d analysis in this sense. The same is true for superconductivity, where 3-d effects will be included later to suppress 2-d phase fluctuations.

4.2. One Hole State

Consider adding one hole to the $x=0$ antiferromagnet.[6] Since the largest lattice magnetization occurs for precisely one electron per site on average, it is clear that the hole will locally reduce the sublattice magnetization in its vicinity. The situation is better understood by defining a staggered order parameter, M_n.

$$M_n \equiv < S_{nz} > \cos Q \cdot R_n, \qquad (2)$$

The cosine factor removes the rapid sign oscillation of the spin density with M_n being $+1$ for up spins living on the even sites and the down spins on the odd sites, while $M_n = -1$ corresponds to the time reversed state. The antiferromagnetic gap parameter is given within the mean field approximation by

$$\Delta_n = UM_n, \tag{3}$$

While Δ_n is uniform for the undoped system, it is reduced in magnetitude in the vicinity of an added hole, as shown in Figure 7. The spatial extent of the depressed value of Δ can be determined in the adiabatic approximation by minimizing the total energy of the system

$$E = E_{ex} + E_{hK} + E_{hex}, \tag{4}$$

where E_{ex} is the exchange energy of the undoped system, E_{hk} is the uncertainty principle kinetic energy of the added hole and $E_{x'}$ is the coupling of the hole to the exchange potential. Minimizing E with regard to the hole wave function ψ_h and Δ_n leads to the optimal shape of the internal structure of this excitation, which we term a "bag." The minimization condition can be phrased as the pressure due to antiferromagnetic spin fluctuations acting inward on the hole is exactly balanced by the uncertainty principle pressure of the hole trying to expand outward to lower its kinetic energy, similar to the bubble formed by an electron in liquid ^{4}He. One finds that for reasonable parameters, the size of the bag

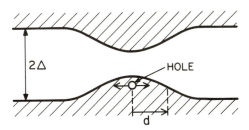

FIG. 7. The energy gap depression or bag surrounding an added hole. The hole is self consistently trapped in the comoving bag, the size of which, d, decreases to the scale of the lattice spacing for $U/t \gg 1$. While this spin bag is analogous to a spin polaron which occurs when an itinerant hole in band 1 couples to localized spins in band 2, in the cuprates the hole and spins are in the same band. This leads to the coupling between the hole and the spins being of order the "self exchange" U rather than the far smaller interband exchange interaction J.

is approximately 3 to 4 lattice spacings in the plane and the effective mass is of order 3 to 5 times the band mass. We note that the spin bag has a minimum energy at the Fermi surface.

It is interesting to note that in the large U limit the corresponding hole excitation has a minimum energy also at the Fermi momentum, namely at $k_F = (\pi/2 \, a, \pi/2 \, a)$ as shown by Shraiman and Siggia[7], and other authors.[8,9] Thus, we see a remarkable continuity in passing from the weak to strong coupling limits, namely that the minimum energy hole excitation has precisely the same momentum despite the fact that in the large U limit the Fermi surface *per se* no longer exists. Rather, in this limit, the hole's hopping matrix element is modified by a phase factor arising from the overlap of the many electron spin states of the background system before and after the hole hops. This phrase can be expressed in terms of a guage field which shifts the zero of energy from k = 0 to the Fermi surface of the weak coupling problem. Hence, the Fermi statistics of the weak coupling problem goes over to a guage field effect in the strong coupling problem.

Further, we note that the longitudinal spin fluctuations discussed above are supplemented by transverse spin fluctuations leading to a dipolar twist of the spin density wave field surrounding the hole. This twist has been shown by Frenkel and Hänke[10] to be of the same form as that deduced by Shraiman and Siggia, providing another link between the weak and strong coupling results. In fact, the numerical studies as well as these analytic studies strongly suggest that there is smooth evolution of all physical properties between the weak and strong coupling limits with one a sharp phase transition corresponding to the onset of Mott-Hubbard localization. Rather, localization sets in gradually in the antiferromagnetic phase by up spins preferring to sit on the even sites and odd spins on the odd sites with the degree of spin separation growing gradually from a very small amplitude to strength unity in the large U limit. It is with this in mind that we seek to understand the strongly correlated metals from both the small and large U limits with the expectation that in the physically relevant intermediate coupling regime the physics will be, in fact, the same regardless from which limit one begins.

A quantitative treatment of the weak coupling spin bag has been given by Wen, Zhang and the author (SWZ)[11] who started from the Bloch states in the presence of the spin density wave and calculated the hole self energy within the one loop approximation, using the correct spin wave excitations in the broken symmetry phase. Transverse spin waves

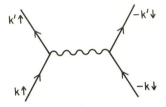

FIG. 8. The pairing interaction $V_{kk'}$ between SDW quasi particle states (k ↑, -k ↓), due to the exchange of one spin wave.

give rise to the dipolar spin twist about the wave vector k of the spin bag as shown by Frenkel and Hänke.

The interaction between two holes was also calculated by SWZ. As shown in Figure 8 the scattering potential calculated within the one spin wave approximation starting with the SDW quasi particles leads to a potential very different that the corresponding one loop potential starting with free electron Bloch states coupled to spin fluctuations. While the potential arising from the exchange of one spin fluctuation in the paramagnetic phase is repulsive for all momentum transfer as shown by Scalapino, et al.[12] and sketched in Figure 9a, a dramatic change takes place when one includes the effects of exchange Bragg scattering so that the Bloch states now are mixtures of momentum k and k+Q, where Q is the nesting wave vector (π/a, π/a). In this case one finds a deep basin of attraction for momentum transfer less than reciprocal of the bag size. This is precisely what one would expect from the Born approximation for the scattering between two particles which interact via an attractive effective potential whose range is the bag size. The fact that V is repulsive for large momentum transfer reflects the short wave exchange repulsion present in the weakly correlated paramagnetic phase, as in Figure 9a.

Another way of thinking of the attraction is in terms of the bipolaron problem where the bags can be thought of as spin polarons. Qualitatively, while two separated bags must each pay the exchange energy to reduce the order parameter in their vicinity, when they are closely spaced the polarons share a common bag thereby lowering their total energy. This is very similar to the so-called mattress effect which underlies the phonon attraction between electrons in conventional low temperature superconductors. In analogy, we call this attraction the spin mattress effect in the spin fluctuation problem.

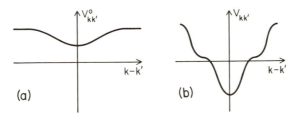

FIG. 9. (a) The one spin wave pairing interaction $V^0_{kk'}$ between bare holes, without the SDW potential effects. Note $V^0_{kk'}$ is repulsive for all $k - k'$. (b) The pairing interaction $V_{kk'}$ between quasi particles (bags) in the presence of the SDW potential. Note that $V_{kk'}$ is attractive for momentum transfer less than $1/d$, where d is the bag size.

While this attractive effective potential occurring in the antiferromagnetic phase is suggestive, one must explore whether such attractive effective interactions also occur between excitations in the strongly correlated paramagnetic phase which is appropriate for the superconducting material.

4.3. Paramagnetic Phase

We turn to the nature of the excitations in the paramagnetic phase and their interactions. A. Kampf and the author (KS) considered the problem of starting with the limit in which the hole concentration is sufficiently large to lead to a weakly correlated Fermi liquid type phase. As the hole concentration is decreased, the correlations begin to build up and lead to large amplitude spin fluctuations which are precursors to the metal insulator transition and ultimately lead to the antiferromagnetic phase. The central questions we will discuss are 1) do spin bags exist in the strongly correlated paramagnetic regime, and 2) if so, do these bags attract? The answer to both questions is affirmative, showing a strong similarity between the excitations in the small and large doping regimes.

As in the SWZ calculations KS calculate the hole self energy diagramatically. However, in this case, to obtain the relevant physics one must go to the two loop level rather than the one loop level discussed above. In essence the static SDW potential becomes dynamic in the paramagnetic phase, leading to a second loop which accounts for the fluctuating quasi-Bragg scattering.

Another way to phrase the problem is that one loop establishes a pseudo-gap and the second loop locally suppresses this pseudo-gap, just as in the SDW phase. Thus, the first issue is to investigate how the SDW gap goes over to a pseudo gap in the correlated paramagnetic phase. This can be seen within the weak coupling approach through the one loop hole self energy shown in Figure 10. KS assume a spin susceptibility $\chi(q,\omega)$ which is fit to experimental neutron scattering data, with χ taken of the form

$$\text{Im}\chi(q, \omega) = \sum_{Q_0} \frac{\Gamma/\pi}{(q - Q_0) + \Gamma^2} f(\omega), \tag{5}$$

where Q_0 is the nesting wave vector and f is a frequency distribution which is taken to be linear in ω with a cutoff frequency ω_0. The width Γ is of order $1/L_{ss}$ and tends to zero as the material is undoped to produce long range antiferromagnetism. When Γ is of order the reciprocal of the lattice spacing, spin fluctuations are rapidly damped out in space.

It is interesting to note that the one loop self energy already gives rise to a pseudo gap which goes over to the SDW gap as Γ approaches zero. The origin of the pseudo gap can be traced to the two time orderings of the Feynman diagram shown in Figure 10. For the time ordering Figure 10a the intermediate state contains to an electron above the Fermi surface, while in Figure 10b the immediate state contains a hole as well as the incoming and outgoing fermion lines. It is diagram

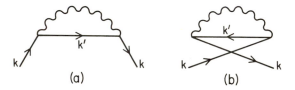

(a)   (b)

FIG. 10. The time ordered hole self energy Σ. The forward diagram (a) describes a hole scattering in an intermediate hole state k' below the Fermi surface, corresponding to a conventional polaron. In (b) the intermediate state corresponds to an electron in k' above the Fermi surface. The polaron type contribution Σ_a leads to a mass enhancement $m^*/m > 1$ and a build up of states at the Fermi surface while Σ_b leads to a reduction of m^*/m and a pseudo gap in the density of states near the Fermi energy. Physically, Σ_b arises from the Pauli principle exclusion of spin fluctuation which otherwise lower the ground state energy in the absence of the added hole. As the spin-spin correlation length L_{ss} increases, Σ_b dominates Σ_a and a pseudogap is formed, which leads to bag formation even though no long range spin order exists.

10b which becomes strongly enhanced with decreasing x as antiferromagnetic spin fluctuations grow in amplitude. The essential point is that the slope of Σ with respect to ω is negative for 10a as is the standard result for Fermi liquid theory. However, when the exchange interactions become strong, 10b becomes dominant and the slope of Σ with respect to ω becomes <u>positive</u> except very near the chemical potential. It is this positivity which leads to the pseudo gap, in that states of positive energy are raised and states of lower energy are lowered, making the effective mass become much smaller and reducing the density of states near the Fermi surface.

In Figure 11, the one particle spectral function $A(k,\omega)$ is plotted for several values of the parameter Γ, illustrating how the spectrum evolves from the Landau picture for Γ of order 0.3 to a structure with two main peaks and a small weight quasi particle peak surviving for Γ of order 0.1. Thus, as L_{ss} increases, the Landau picture of the normal phase breaks down and the spectral function for a given k is replaced by a sum of two peaks corresponding to the upper and lower spin density wave bands suitable broadened due to finite range spin order, plus a weak remnant quasi particle peak of strength Z. While the Fermi liquid picture may formally apply for very long wave length properties, it is clear that such quasi particles are useless as the basis for high temperature

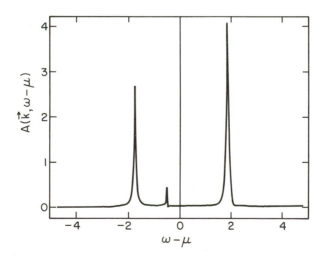

FIG. 11. One particle spectral function $A(k, \omega - \mu)$ plotted as a function of energy for k near the Fermi surface, with the parameter $\Gamma = 0.3$. The strength Z of the Landau quasi particle peak is very weak.

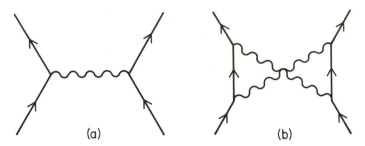

FIG. 12. The pairing potential between (a) bare particles due to exchange of one spin wave and (b) between bags (a two spin wave process).

superconductivity, since short wave length response must be considered in forming the pairing state in view of the fact that the superconducting coherence length is on the scale of 10-15Å.

Turning to the pairing interactions in the paramagnetic state, the potential due to the exchange of one fluctuation as shown in Figure 12a is repulsive, as mentioned above. However, if one goes to the next order and considers the exchange of two crossed line spin fluctuations, one obtains an attractive potential just as in the SWZ calculation for the antiferromagnetic state. The pairing potential looks very similar to that shown in Figure 9b, again showing the smoothness of physical properties as one goes from small to large x.

While much remains to be done to complete a quantitative theory of superconductivity starting from the paramagnetic phase, one can gain a qualitative understanding of why superconductivity peaks as a function of x, as shown in Figure 12. As spin fluctuations decrease in amplitude with increasing x, the pairing potential V also drops off in strength. However, as x decreases the bags become localized in the glassy phase thereby destroying the density of states as the Fermi surface, an essential property for superconductivity. Hence, the product $N(\mu)V$ is expected to exhibit a peak as a function of x due to the fall off of the density of states for small x and the fall off of V for large x. If as it has been suggested that phase separation occurs, this same argument would explain for T_c in the equilibrium phase. An interesting feature of these calculations is that they predict the chemical potential remains in the x=0 gap for substantial doping x.[14] This results from holes being doped into the quasi particle peak (fully relaxed bag), with one state being brought into the gap for each doped hole, as observed by Allen et al.[15]

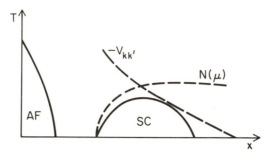

FIG. 13. The density of states $N(\mu)$ and pairing interaction $V_{kk'}$ as a function of x. $N(\mu)$ $V_{kk'}$ peaks at a value of x between the doping at which the spin glass melts into a conducting liquid and the value of x beyond which the spin fluctuations are weak.

4.4. Charge Density Waves and Phonons

The above discussion, which is based on a modulation of a pseudo gap by a carrier, would also appear to be relevant to charge density wave fluctuation systems, such as the bismuthates. Of course, in this case, the phonons are admixed with the charged density wave fluctuations since they carry the same quantum numbers. Whether the charged density wave component dominates the phononic contribution in such systems is unclear at present. Since the isotope effect in these materials is intermediate between that of low temperature superconductors and the high T_c materials, it may well be that superconductivity in the bismuthates is substantially influenced by charged density wave fluctuations.

In addition, however, there may be other electronic excitations which carry spin zero and play an important role in the high Tc systems, as has been suggested by a number of authors. Nevertheless, it is our belief that the spin fluctuation mechanism predominates in the superconductors which show a very small isotope effect with transition temperatures above 50°K.

5. EXOTIC SCHEMES

The first theoretical suggestion to account for high Tc superconductivity was advanced by P. W. Anderson[16] soon after Bednorz and Mueller's discovery. On the basis of earlier work concerning frustrated antiferromagnetics, Anderson suggested that a qualitatively new ground state of a frustrated spin system might occur in which pairs of spins are coupled to total spin zero with a strong resonance between mates of different spin pairs. Anderson proposed that for such a quantum spin liquid excitations (spinons) carrying the quantum number spin half and charge zero must exist. The result was foreshadowed by work of Pomaranchuk and Landau in the early 1940's. As proposed by Kivelson, Rokhsar and Sethna,[17] the spinon excitations are complemented by holon excitations having charge +e and spin zero. Anderson argued that the addition of a hole would bifurcate into a holon and a spinon, so that a hole gas decomposes into interacting holon and spinon gases. Since spinons carry no charge, superfluid current within this scheme must be carried by the holon gas. It is found, however, that holons repel each other in the plane so that Anderson proposed that a pair of holons and a pair of spinons collectively tunnel to a neighboring plane and back again, producing an effective attraction which leads to superconductivity. This mechanism is quite distinct from those discussed above in which the attraction already exists within the 2-d planar system and is converted to three dimensional long range order largely by the hopping of superfluid pairs between the planes locking the overall phase into a coherent superconducting state.

In a related approach, Laughlin,[18] as well as Wen, Wilczek and Zee[19] and others have proposed that high temperature superconductivity arises from the condensation of particles termed "anyons," having fractional statistics. As opposed to bosons or fermions whose wave function is either invariant or changes sign on the interchange of two such objects, anyons acquire a phase other than 0 or π on interchange. While the fundamental statistics of the underlying particles is that of fermions, namely, the statistics of electrons, in the spin liquid phase the excitations move in the background of localized magnetic moments whose state vector is altered as the hole hop from site to site as mentioned above. Since the net amplitude for a hole hop involves a factor which is the overlap of the spin wave function before and after the hope has occurred, this overlap leads to a phase in the hopping matrix element. Ultimately this spin overlap phase becomes reflected in the statistical phase corresponding to the anyon statistics.

It is straightforward to show that in these two dimensional problems the anyon statistical phase can be eliminated by attaching a flux tube Φ to each particle where Φ is proportional to the change in anyon phase. Thus if one has a so-called semion, i.e. half way between a boson and fermion, one can transform such semions to the fermion basis if one includes the phase arising from one particle moving in the statistical gauge potential of another particle. Thus, one can transmute the statistics in favor of an interaction between the particles which corresponds to the vector potential generated by the point flux tubes tied to each particle. From this point of view anyon superconductivity would be a condensation by the conventional BCS mechanism but with a velocity dependent pairing potential generated by the flux-tube interactions.

An alternate point of view is that one refers the semions to the boson basis. In this case the interaction is a repulsive interaction between bosons and leads to bosonic superfluidity similar to that occurring in superfluid ^{4}He.

Ultimately such an anyonic state can exist only if the ground state violates time reversal symmetry. Recent experiments[20,21], in particular the possible existence of circular dichroism and/or optical rotation in high Tc materials, searching for this property in the cuprates have been contradictory, Since in some of these experiments lattice strains, magnetic impurities, etc., can lead to such effects, it is of the utmost importance that one establish whether the observed optical rotation is due to parity violation as in sugar, or actually involves a broken time reversal symmetry. No doubt careful materials preparation and more extensive physical measurements will resolve this question. There is another concern, however, that an anyon gas of charged particles exhibits orbital magnetic moments which generate internal magnetic fields. While these fields have been estimated to be in the vicinity of 100 gauss, muon spin rotation experiments have set an upper limit of such fields at the muon site of approximately 0.8 gauss, calling in question the existence of such an anyonic system. Finally, since the statistics are propagated by spin waves of velocity v_s which is slow compared to the velocity of the holon and/or spinons, retarded statistics must be included.

6. MARGINAL FERMI LIQUID

The anomalous properties of the normal phase discussed in Section II have been partially accounted for by Varma, et al.[4] who assume that the dynamic spin susceptibility is given by the sum of a regular part χ_0 plus an anomalous part χ_1, where χ_1 is modelled by

$$\text{IM } \chi_1(q, \omega) = \begin{cases} c\omega/T & |\omega| < T \\ c & |\omega| > T. \end{cases} \tag{6}$$

While Fermi liquid theory predicts that Im_χ is proportional to ω for small ω, the coefficient is proportional to the reciprocal of the Fermi temperature rather that the reciprocal of the temperature. This difference leads, for example, to the electrical resistivity varying as T rather than at T^2 at lower temperature. Similar changes occur in other physical properties as seen experimentally.

An interesting aspect of the theory is that the quasi particle residue

$$Z \equiv 1/\left(1 - \frac{\delta\Sigma}{\delta\omega} \right) \tag{7}$$

vanishes at the Fermi surface, in contrast to the well known Fermi liquid result where Z at the Fermi surface approaches a constant smaller than unity. The small value of Z is characteristic of a number of approaches and reflects the strongly correlated properties of the electronic systems.

Recently the consistency of this approach has been discussed by Bedell and Zimanyi,[22] who have carried out a renormalization group calculation indicating that higher order diagrams lead to an instability of the theory when the theory is viewed in the most straightforward manner. Other properties, such as the electron tunneling density of states and the oxygen spin resonance relaxation rate as predicted by the model also appear to be in conflict with experiment. Nevertheless a number of features of the experimental situation are well accounted for by the theory. Perhaps further refinements of the theory will bring the approach in closer conformity with experiment.

7. CONCLUSIONS AND THE FUTURE

The cuprates are rewriting the text books of the 90's for condensed matter physics. The traditional concepts of the Fermi liquid, weak coupling, fermion-boson interactions, and weak renormalizations are being replaced by strongly interacting fermions, large amplitude boson fluctuations, evolution from metallic to pseudo gap to spin glass and insulating antiferromagnetic behavior as a function of doping and temperature. One of the most important questions is whether the strongly correlated metal regime exhibits some qualitatively new order parameter leading, for example, to parity and time reversal symmetry breaking and anyons? Alternatively, the strongly correlated metal may correspond to a strongly fluctuating version of the paramagnetic metal phase but having localized magnetic moments and bag-like excitations with large damping rates.

Many chapters of this story have been completed but the climax to the mystery is yet to be revealed. One way or another, the next several years are likely to witness dramatic changes in the way we think of interacting systems.

ACKNOWLEDGEMENTS

This work was partially supported by NSF Grant DMR89-18307, EPRI Grant EPRIRP-8009-18, and DOE Grant DOE85ER45197.

REFERENCES

1. G. Bednorz and K. A. Müller, Z. Phys. B, **64**, 189 (1986).

2. B. Batlogg, The Los Alamos Symposium, High Temperature Superconductivity Proceedings, Eds. K. Bedell, et al., p. 37 (1989).

3. G. Shirane, et al., Phys. Rev. Lett. **59**, 1613 (1987); J. Tranquada, ibid. **60**, 156 (1988).

4. C. M. Varma, et al., Phys. Rev. Lett. **63**, 1996 (1989).

5. J. Hubbard, Proc. R. Soc. London, Ser. A **276**, 283 (1963).

6. J. R. Schrieffer, X.-G. Wen and S.-C. Zhang, Phys. Rev. Lett. **60**, 944 (1988).

7. B. I. Shraiman and E. Siggia, Phys. Rev. **B40**, 9162 (1989).

8. For a review of computational methods and results for the 2-d Hubbard model, see D. J. Scalapino, The Los Alamos Symposium, High Temperature Superconductivity Proceedings, Eds. K. Bedell, et al. p. 314 (1989).

9. E. Dagotto, et al., Phys. Rev. **40**, 8945, 10977 (1989).

10. D. Frenkel and W. Hänke, to be published.

11. J. R. Schrieffer, X.-G. Wen and S.-C. Zhang, Phys. Rev. **B39**, 11663 (1989).

12. D. J. Scalapino, et al., Phys. Rev. **B34**, 8190 (1986); ibid. **35**, 6694 (1984).

13. A. Kampf and J. R. Schrieffer, Phys. Rev. **B41**, 6399 (1990); **B42**, 7967 (1990).

14. A. Kampf and J. R. Schrieffer, to be published.

15. J. W. Allen, et al., Phys. Rev. Lett. **64**, 595 (1990).

16. P. W. Anderson, Science **235**, 1196 (1987).

17. S. A. Kivelson, D. S. Rokhsar and J. P. Sethna, Phys. Rev. **B35**, 8865 (1987).

18. R. B. Laughlin, Phys. Rev. Lett. **60**, 2677 (1980).

19. X. -G. Wen, F. Wilczek and A. Zee, Phys. Rev. **B39**, 11413 (1989).

20. K. B. Lyons, Et al., Phys. Rev. **B64**, 2949 (1990).

21. A. Kapitulnik, et al., to be published.

22. K. Bedell and G. Zimanyi, to be published.

SOME RELEVANT PROPERTIES OF CUPRATE SUPERCONDUCTORS

K. Alex Müller

IBM Research Division
Zurich Research Laboratory
8803 Rüschlikon, Switzerland
and
University of Zurich
Physics Department, 8001 Zurich

Abstract - The stacking of CuO_2 layers in compounds of interest is related to their transition temperatures. Furthermore, the importance of out-of-CuO_2-plane apex oxygens is discussed, based on results from both statical and dynamical experiments and theoretical simulations.

1. STRUCTURE-RELATED PROPERTIES

Structural aspects in the high-T_c superconductors are quite important and will be reviewed first. The cuprates discovered early in 1986 form a new class of superconductors.[1] They all have layered structures with one or several CuO_2 layers sandwiched between so-called block layers X or Y, with X = Y or X ≠ Y, as shown in Fig. 1. The latter are oxides or consist of ions (such as Y^{3+}) of substantial variety, ranging from LaO with sodium-chloride structure over Cu-O single or double chains to Bi_2O_2, Tl_2O_2 bilayers, TlO single layers and much more complex ones discovered recently.[2] Compounds with up to four CuO_2 layers per unit cells have been synthesized. The progression of T_c as a

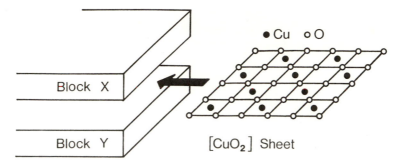

FIG. 1. Schematic structure of cuprate superconductors: $[CuO_2]$ sheet sandwiched by block layers X and Y. After Ref. 2.

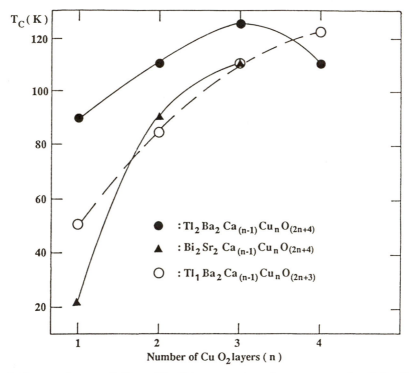

FIG. 2. Dependence of the critical temperature for superconductivity T_c as a function of the number of CuO_2 layers per unit cell (n) for various Tl and Bi compounds. The values reported are the maxima among the ensemble of samples measured. From Ref. 3.

function of the layer number n is shown in Fig. 2 for the TlO, Tl_2O_2 and Bi_2O_2 block layers.[3] The highest T_c of 125 K is seen to occur in $Tl_2Ca_2Ba_2Cu_3O_{10}$.

Electronically, the block layers accept or donate electrons as is also known in intercalated graphites. However, in contrast to the graphites, the block layers in most of the compounds do not conduct. In a way, the cuprates are the microscopic analogs to the artificial semiconductor heterostructures with their quantum wells. However, nature gives us the cuprates for free. In all high-T_c materials except one, electrons are accepted by the block layers, and the CuO_2 layers are hole conductors.[2] For one layer ($n=1$), 100% of the holes are located in that layer. For $n=2$, the two CuO_2 layers share 50% of the holes each. For $n>2$, the majority of the carriers are in the layer next to the acceptor block layers. In the $n=3$ compound, for example, 45% of the charge is sited in the

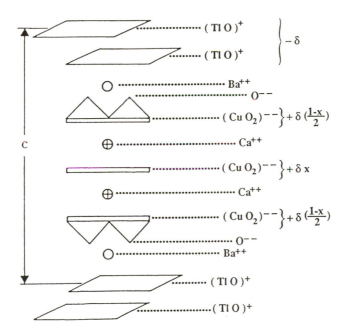

FIG. 3. Schematic picture of the layer structure of the stage-3 compond $Tl_2Ba_2Ca_2Cu_3O_{10}$ that corresponds to the presently highest value of $T_c = 125$ K. The nominal charges would correspond to half filling of the Cu-O bands analogous to the undoped La and Y compounds. The TlO bands actually cross the Fermi level and acquire an extra charge density $- \delta$. This induces a density δ of holes to be shared (non-homogeneously) among the three CuO_2 layers. Since holes are induced in the Cu-O bands in the stoichiometric compound, such a system may be considered as self-doped. From Ref. 3.

two outer layers and only 5-10% in the central one,[3] see Fig. 3. This recently recognized property, i.e., the charge disparity in inequivalent CuO_2 layers, seems to be relevant in reaching the highest T_c's. It results from the competition between charge delocalization in the CuO_2 layers and the Coulomb attraction of the charge in the block and those in the CuO_2 layers.

In the two theoretical presentations of this forum, the microscopic electronic two-dimensional character of the CuO_2 layers has been discussed. Here, we especially recommend that the reader look again at the delocalized models of Friedel.[4] In those models, the charge-transfer or Hubbard gaps are closed, as shown by Fink et al.[5] in the lecture on high-energy spectroscopic studies (see Fig. 2 "Filled Gap" thereof). In their lecture, experiments on the location of the charges on the atoms are reviewed that can be understood in terms of such delocalized models. There are indications of hole charges on apex oxygens in the $YBa_2Cu_3O_{7-\delta}$ compound. And it was in this very compound that back in 1988 a correlation between the distance of pyramidal apex oxygen to the CuO_2 plane and the transition temperature T_c was detected. This is a function of the oxygen stoichiometry δ.[6] Since then, this relation has been confirmed in many experiments, and thus we shall discuss the static microscopic findings related to it first.

2. PYRAMIDAL APEX OXYGEN

The oxygen stoichiometry dependence of the atomic distances and displacements has recently been reinvestigated by Cava et al.,[7] using high-resolution neutron scattering. In Fig. 4 their data are reproduced as evaluated in the bond valence sum, a quantity mainly determined by the apex oxygen to CuO_2 planar distance. It is seen that its progression maps that of T_c almost exactly. From this, one concludes that the apex-oxygen position is indeed relevant. This conclusion is further supported by the results on charge distributions described at the beginning of this paper. The hole charge is mainly located in the CuO_2 layer *next* to the block sheets, where also the apex oxygens are sited (see Fig. 3). In fact, this is the case for all cuprates with T_c's above the nitrogen boiling point. On the other hand, the compounds with no oxygen-copper pyramids have lower T_c's as do the electron-doped ones, whose T_c's are not higher than 22 K.[2] From this, it follows that the CuO_2 metalic-like layers show superconductivity but that the oxygen on out-of-plane octahedral or, better, pyramidal sites appear as boosters for the high-T_c compounds.

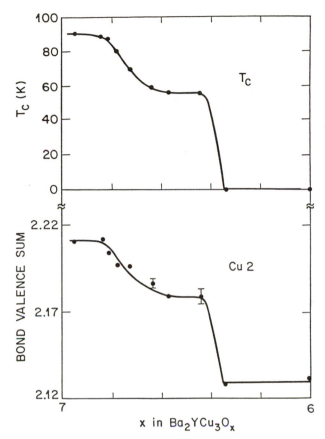

FIG. 4. Comparison of T_c and band valence sum around the plane copper as a function of oxygen stoichiometry. From Ref. 7.

Hydrostatic pressure experiments corroborate the above conclusions. Recently, upon applying hydrostatic pressure p to the "double-chain" $YBa_2Cu_4O_8$ compound with otherwise similar structure as $YBa_2Cu_3O_7$, a large enhancement of $T_c dT_c/dp=0.55$ K/kbar was found.[8] Conjointly with this finding, an accurate neutron structural analysis proved a substantial shift of the apex oxygens towards the CuO_2 planes by $\delta_z=(0.07\pm0.01)\text{Å}$ at 10 kbar. This was taken as clear evidence of charge-transfer changes as proposed earlier by Bishop et al.[9] The $T_c\simeq60K$ superconductor $YBa_2Cu_3O_{6.6}$ shows an even larger dT_c/dp of about 9 K/GPa as compared to that of $YBa_2Cu_3O_7$ of only 1 to 1.2,[10] i.e., five times smaller than for $YBa_2Cu_4O_8$. In the latter, the double chains appear to be more rigid. Before discussing recent theoretical results, we note than an absence of the pressure effect on T_c in the electron-doped $Nd_{1.85}Ce_{0.15}CuO_4$ has recently been reported and, in another paper, even a decrease has been observed for Sm and Eu compounds.[11] As is well known, this oxide has no apex oxygens. Therefore these results support our own view. From a compilation of $dlnT_c/dp$ graphs of the various compounds, Ohta et al.[12] obtained strong evidence that the relative position of the apex oxygens essentially governs T_c. Indeed, they calculated the Madelung potential ΔV_A of the O(4) p_z to the CuO_2 planar p_x and p_y orbitals to scale the optimum T_c's of *all known hole-doped superconductors*, see Fig. 5.

Substantial support of the apex hypothesis stems from very recent upper-edge X-ray absorption spectroscopy (XAS) in $YBa_2Cu_3O_7$: these

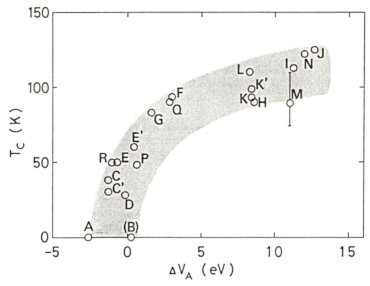

FIG. 5. The T_c versus ΔV_A correlation for hole superconductors, from Ref. 12a. The compounds collected are

A: $La_2SrCu_2O_{6.2}$, B: $La_2CaCu_2O_6$,
C: $La_{1.85}Sr_{0.15}CuO_4$,
C': $La_{1.84}Ba_{0.16}CuO_4$,
D: $Nd_{2-x-y}Sr_xCe_yCuO_{4-\delta}$,
E: $Y_{0.8}Ca_{0.2}Ba_2Cu_3O_{6.11}$,
E': $YBa_2Cu_3O_{6.5}$, F: $YBa_2Cu_3O_{6.8}$,
G: $Pb_2Sr_2Y_{0.5}Ca_{0.5}Cu_3O_8$,
H: $Tl_2Ba_2CuO_{6-\delta}$, I: $Tl_2Ba_2CaCu_2O_8$,
J: $Tl_2Ba_2Ca_2Cu_3O_{10}$,

K: $Bi_2Sr_2Ca_{0.9}Y_{0.1}Cu_2O_{8.24}$,
K': $Bi_{1.6}Pb_{0.4}Sr_2CaCu_2O_y$,
L: $Bi_2Sr_2Ca_2Cu_3O_{10+\delta}$,
M: $Tl_{0.5}Pb_{0.5}Sr_2CaCu_2O_7$,
N: $Tl_{0.5}Pb_{0.5}Sr_2Ca_2Cu_3O_9$,
P: $(Ba_{1-x}Nd_x)_2(Nd_{1-y}Ce_y)_2Cu_3O_{8+z}$,
Q: $Y_{0.9}Ca_{0.1}Ba_2Cu_4O_8$, and
R: $PbBaSrY_{1-x}Ca_xCu_3O_7$

data indicate a double-well potential on the oxygen in question along the c-axes at higher and low temperatures, which becomes anharmonic single well near T_c, i.e. in the fluctuation region.[13] The XAS data are in excellent agreement with the information from infrared excited Raman scattering of Taliani's group.[14] The Fourier transform spectrum is dominated by multiple-order scattering from the 505 cm^{-1} Cu(1)-O(4) totally axial-symmetric stretching mode. The considerable intensities of the overtones observed are taken as clear evidence of the strong coupling with local electronic excitations. The perpendicular polarization data prove the existence of a charge-transfer transition *perpendicular* to the CuO_2 *ab* plane.

The counterpart of the above A_g Raman active mode is the asymmetric IR active 155 cm^{-1} mode investigated by Genzel et al.[15] It involves an almost rigid motion of an entire O(4)-Cu(1) O(1)-O(4) cluster between CuO_2 sheets against Ba ions.[16] Its oscillatory strength is considerably larger than predicted by pure lattice-dynamical calculations.[15] This property was ascribed to chain-plane charge fluctuations, in line with the earlier proposal;[9] thus to some extent other degrees of freedom, here the Ba against CuO_3 motion, may be anharmonic as well. However, the oxygen anharmonic local potential together with electronic charge transfer appears to be the primary one in a recent theory[16] on the IR and Raman modes for the O_6 to O_7 stoichiometry with the observed correct O(4) static distortion.[6] Kakihana et al.[17] independently arrived at the same conclusion in their Raman study on substituting Co into $YBa_2Cu_3O_{7-\delta}$ to change the chain-plane charge transfer. A determination of the phonon density of states (PDOS) has shown that the highest modes of 70 to 89 meV present in insulating $YBa_2Cu_3O_6$ soften in the superconducting $YBa_2Cu_3O_{6.9}$.[18] This softening is accompanied systematically by a drastic enhancement of the PDOS in the 40 to 60 meV region. Moreover, inelastic neutron scattering in single crystals has revealed a "ghost" mode in the superconductor at an energy of 50 to 60 meV, occurring between the middle and the edge of the Brillouin zone along the $<\xi 00>$ direction.[19] The term "ghost" has been used owing to a quasi-split feature of this symmetric branch. The disappearance of the 80 meV peak in going from O_6 to O_7 has been ascribed to the softening of Cu(2)-O(2/3) and Cu(1)-O(4) band stretching modes.

With these views in mind, it has recently been attempted to parameterize the measured oxygen isotope shift exponent α_O by setting $\alpha_O = r\alpha_a$, where α_a is defined as the apex shift exponent and assumed to be due to anharmonic coupling of the pyramidal apex oxygen to the in-plane CuO_2 charge transfer modes. Here r is the ratio of apex to total oxygen ion content per unit cell, which is conjectured to be relevant. Table 1 gives the measured oxygen isotope shifts for ^{18}O substitution of ^{16}O, δT_c, the isotope shift exponents α_O, and the calculated apex shift exponent α_a from Ref. 20. One sees that a certain parameterization of the superconductors containing pyramids is possible, but this may be fortuitous.

Table 1. Measured oxygen isotope shifts for ^{18}O substitution of ^{16}O, δT_c, calculated oxygen isotope shift exponent α_O and the apex oxygen shift exponent α_a. Adapted from Ref. 20.

Compound	$T_c[K]$	$\delta T_c [K]$	α_O	α_a
La_2CuO_4	38		0.13-0.16	0.26-0.32
$YBa_2Cu_3O_7$	91	0.46(6)	0.040(6)	0.14(4)
$YBa_2Cu_3O_7$	91	0.20(5)	0.018(5)	0.06
$Bi_2Sr_2CaCu_2O_8$	75	0.32	0.034	0.14
$Bi_2Sr_2Ca_2Cu_3O_{10}$	110	0.34	0.023	0.12

3. THEORETICAL SIMULATIONS

The early emphasis[9] on an anharmonic double-well potential at the apex oxygen along c as being important for the pairing mechanism in high-T_c superconductors stimulated Morgenstern and collaborators[21] to undertake a large-scale quantum Monte Carlo simulation. As model Hamiltonian $\mathcal{H}$ they chose

$$\mathcal{H} = \mathcal{H}_H + \mathcal{H}_{BMT}, \tag{1}$$

where $\mathcal{H}_H = U\sum_i n_{i\uparrow}n_{i\downarrow} - t \sum_{<ij>\sigma} A_{i\sigma}^+ A_{j\sigma}$ is the well-known Hubbard Hamiltonian with on-site energy U and the kinetic energy term with hopping to nearest sites. The operators have their usual meanings; i and j denote sites, σ spin.

$$\mathcal{H}_{BMT} = g \sum_i n_i S_i^z - \Omega \sum_i S_i^x$$

is the double-well Hamiltonian with $S = 1/2$ spin coordinates of half quantum number, originally used by Brout, Müller and Thomas (BMT) for order-disorder ferroelectrics;[22] g is the depth of the potential and Ω the tunneling frequency between the two wells. The sum i goes over all the apex oxygens present. Thus, if there are half as many in the cell, the shift in T_c will be halved upon replacing all ^{16}O by ^{18}O, the constants g and Ω in $\mathcal{H}_{BMT}$ changing only little.

Using Eq. (1) the Cooper pair correlations

$$\chi = \frac{1}{N} \sum_{ij} < \Delta_i^+ \Delta_j >$$

were investigated with

$$\Delta_i^+ = c_{i\uparrow}^+ c_{i\downarrow}^+,$$

varying U and g as a function of both Ω and the 2D system size. With $\mathcal{H}_{BMT} = 0$, χ remained constant when the size of the system was increased, indicating that no tendency towards pairing occurs for the one-band Hubbard Hamiltonian alone. However, for $\mathcal{H}_{BMT} \neq 0$, enhanced s-type superconductivity correlations could be found in a resonant regime where $g \simeq \Omega = 1$, see Fig. 6. This resonant coupling to the

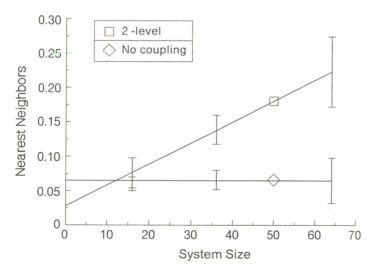

FIG. 6. Two level system. Extended s-wave susceptibility shows increase with system size. No increase for pure Hubbard model (Morgenstern et al., Ref. 21a).

electrons manifests itself in a reduction of the hopping element in $\mathcal{H}_H$. However, for too large values of the Hubbard parameter U again no tendency towards pairing could be deduced.

The above model therefore needs extension in terms of band models. This naturally follows from the chemical aspects and the correlated electronic structure of these materials especially dealt with by Eschrig.[23] In the case of the 123 structure, he found a metallic oxygen in-plane hole conductivity and *at the same time* a collectivization of copper *d*-electrons. The partial delocalization of the *d*-electron, i.e. a reduction of on-site *d*-correlation, results from charge transfer and is reminiscent of heavy-fermion behavior. Eschrig also concludes that "in all high T_c oxocuprates there is an oxygen ion outside of the $[CuO_2]^{-2}$ plane, doping it and at the same time feeling strong lattice anharmonicity."

4. FINAL REMARKS

The CuO_2-in-plane 2D electronic and magnetic properties of the cuprate superconductors, to which two theoretical and a part of the experimental contributions at this forum have been devoted, clearly are quite fundamental. However, per se they only yield transition temperatures in the 20 K range. Therefore, complementary to these 2D interactions, the out-of-plane ones, especially those in which apex oxygens are involved, appear to be important for boosting the T_c's to values three to five times higher than those which the mere in-plane interactions yield. The results summarized in this report speak quite a clear language.

REFERENCES

1. J. G. Bednorz and K. A. Müller, Z. Phys. B **64**, 189 (1986); J. Bednorz, M. Takashige and K. A. Müller, Europhys. Lett. **3**, 379 (1987).

2. Y. Tokura and T. Arima, Jpn. J. Appl. Phys. **29**, 2388 (1990).

3. M. Di Stasio, K. A. Müller and L. Pietronero, Phys. Rev. Lett. **64**, 2827 (1990).

4. J. Friedel, these proceedings.

5. J. Fink, N. Nücker, H. Romberg, M. Alexander, P. Adelmann, R.

Claessen, G. Mante, T. Buslaps, S. Harm, R. Manzke and M. Skibowski, these proceedings.

6. R. J. Cava, B. Batlogg, S. A. Sunshine, T. Siegrist, R. M. Fleming, K. Rabe, L. F. Schneemeyer, D. W. Murphy, R. B. van Dover, P. K. Gallagher, S. H. Glarum, S. Nakahara, R. C. Farrow, J. J. Krajewski, S. M. Zahurak, J. V. Waszczak, J. H. Marshall, P. Marsh, L. W. Rupp, Jr., W. F. Peck and E. A. Rietman, Physica C **153-155**, 560 (1988); R. J. Cava, B. Batlogg, K. M. Rabe, E. A. Rietman, P. K. Gallagher and L. W. Rupp, Jr., ibid **156** 523 (1988). In the latter paper, direct evidence of charge transfer from chains to planes in $Ba_2YCu_3O_7$ due to O(4) motion is presented.

7. R. J. Cava, A. W. Hewat, E. A. Hewat, B. Batlogg, M. Marezio, K. M. Rabe, J. J. Krajewski, W. F. Peck, Jr., and L. W. Rupp. Jr., Physica C **165**, 419 (1990).

8. B. Bucher, J. Karpinski, E. Kaldis and P. Wachter, Physica C **157**, 478 (1989).

9. A. R. Bishop, R. L. Martin, K. A. Müller and Z. Tesanovic, Z. Phys. B **76**, 17 (1989).

10. E. Kaldis, P. Fischer, A. W. Hewat, E. A. Hewat, J. Karpinski and S. Rusiecki, Physica C **159**, 668 (1989).

11. C. Murayama, N. Môri, S. Yomo, H. Takagi, S. Uchida and Y. Tokura, Nature **339**, 293 (1989); J. T. Markert, J. Beille, J. J. Neumeier, E. A. Early, C. L. Seaman, T. Moran and M. B. Maple, Phys. Rev. Lett. **64**, 80 (1990).

12. a) Y. Ohta, T. Tohyama and S. Maekawa, Physica C **166**, 983 (1990), and b) preprint (1990).

13. S. D. Conradson and I. D. Raistrick, Science **243**, 1340 (1989); S. D. Conradson, I. D. Raistrick and A. R. Bishop, Science **248**, 1394 (1990).

14. R. Zamboni, G. Ruani, A. J. Pal and C. Taliani, Solid State Commun. **70**, 813 (1989).

15. L. Genzel, A. Wittlin, M. Bauer, M. Cardona, E. Schönherr and A. Simon, Phys. Rev. B **40**, 2170 (1989).

16. I. Batistic, A. R. Bishop, R. L. Martin and Z. Tesanovic, Phys. Rev. B **40**, 6896 (1989).

17. M. Kakihana, L. Börjesson, S. Eriksson, P. Svedlindh and P. Norling, Phys. Rev. B **40**, 6787 (1989) in their Raman study on substituting Co into $YBa_2Cu_3O_{7-\delta}$ to change the chain-plane charge transfer.

18. B. Renker, F. Gompf, E. Gering, G. Roth, W. Reichardt, D. Ewert, H. Rietschel and H. Mutka, Z. Phys. B **71**, 437 (1988); B. Renker, F. Gompf, E. Gering, D. Ewert, H. Rietschel and A. Dianoux, ibid. **72**, 309 (1989).

19. W. Reichardt, N. Pyka, L. Pintschovius, B. Hennion and G. Collin, Physica C **162–164**, 464 (1989).

20. K. A. Müller, Z. Phys. B **80**, 193 (1990).

21. a) I. Morgenstern, W. von der Linden, M. Frick and H. de Raedt, Physica B **163**, 641 (1990); b) M. Frick, W. von der Linden, I. Morgenstern and H. de Raedt, Z. Phys. B **81**, 327 (1990).

22. R. Brout, K. A. Müller and H. Thomas, Solid State Commun. **4**, 507 (1966).

23. H. Eschrig, Physica C **159**, 545 (1989); H. Eschrig and S. L. Dreschler, in Physics and Materials Science of High-Temperature Superconductors, Proc. NATO Advanced Study Institute, Bad Windesheim, Germany, August 1989, edited by R. Kossowsky, S. Methfessel and D. Wohlleben, pp. 151-160 (Kluwer, Dordrecht, 1990).

LOCAL PROBE METHODS

H. Rohrer

IBM Research Division
Zurich Research Laboratory
8803 Rüschlikon, Switzerland

Abstract - This lecture summarizes the essence, illustrates the capabilities and outlines the possibilities of local probe methods. Of the still increasing variety of methods, I have chosen approaches related to tunneling and to forces. They are at present the most and best implemented ones. The choice of the examples is a matter of convenience. Some topical STM and related areas are treated in more detail in the following contributions of this issue. For a broader coverage see recent proceedings[1-3] and reviews.[4]

The new local probe methods should be accepted and used for their capabilities different from those of conventional methods and for the new possibilities they open. At the beginning they lead, so to speak, an independent life and are pursued on their own. Once attention shifts away from instrumental and methodological aspects, they will find their place in close interdependence with existing approaches and new ones to come.

Highlights in Condensed Matter Physics and Future Prospects
Edited by L. Esaki, Plenum Press, New York, 1991

1. MOTIVATION FOR NANOMETER-SCALE SCIENCE AND TECHNOLOGY

Scanning tunneling microscopy (STM) and STM-derived local probe methods have given considerable impetus to nanometer-scale science and technology. They have created something like a casual relationship between us and individual atoms, molecules, clusters, and nanometer-sized regions of larger objects. In the following we call all these objects that range in size from 1 angstrom to, say, 100 nm nano-objects.

The motivation to work on a nanometer scale is manifold. By "work" we mean the observation of a single, individually selected nano-object, measuring and understanding its properties, manipulating it, modifying it, and ultimately the observation and control of its possible functions and related processes. The local probes assume the role of a kind of interface between our macroscopic world and the world of nano-objects, possibly in their natural environment. Observation and this microscopy is but an initial step in dealing with nano-objects. The "M" in STM, AFM (atomic force microscopy) or, generally, SXM (where X means some interaction between local probe and object) should therefore read "method" rather than "microscopy," in the following we use SPM for scanning probe methods.

The initial motivation of SPM was to observe and study local properties[5,6] - local in the sense of a preselected location on an object or a small object itself - possibly down to the atomic scale. The major attractions of local probe methods besides their resolution are their specific surface sensitivities, the large variety of accessible properties, and the various environments from UHV to electrolytes, and from cryogenic liquids to high temperatures. SPM as local characterization methods have already made their way into many scientific fields from surface science to electrochemistry. They prove especially useful for investigations of the growing variety of fine grain and composite materials, of grain boundaries and interfaces and of individual processes on an atomic or molecular scale. In all these cases, experimental contact with insight and procedures of numerically intensive computation can be established. Of great interest in an industrial setting with ever more demanding specifications are surface roughness standards on the nm scale, tribology, temperature distribution, in short, nanometer metrology. Furthermore, SPM is also expected to advance averaging analytical methods which might be handier and more convenient for various applications by correlation of nm scale features imaged by with more integral properties. Thus, SPM should be considered complementary to, rather than competitive with, existing averaging and other techniques.

A second motivation is the ability to work under extreme conditions. First, quantum effects become important or are dominant for the properties of extremely small objects of which the surface constitutes an appreciable or major part. Next, very high densities are attainable, e.g. 10 nano-amps dc through the end atom of a tunneling tip or a single molecule provide sufficient electrons for local electrical or optical experiments in the microsecond to nanosecond range. Likewise, such tips provide extremely high brightness for electron and ion beams when used as point sources.[7,8] Another interesting aspect is the high electric fields up to a volt per angstrom, anywhere on a flat or curved surface. We can even think of strong, local magnetic fields, in particular "effective" magnetic fields like exchange fields. Measurement and control of displacements and forces in the fm and pN range, respectively, with measuring times far below a millisecond become standard; improvements in specific cases by a few orders of magnitude are not unrealistic. At the same time, however, extremely high pressures in the GPa to TPa range are readily achieved. Finally, all these features will enable manipulation and modification of matter on the nm scale, an aspect of SPM treated in this issue by C. F. Quate.

The technologically inspired motivation is mainly toward miniaturization. We can think in terms of miniaturized tools, sensors or various technology components. Properties and features of components have to be observable and analyzable on a scale of one to two orders of magnitude smaller than the component itself which, for submicron size components, brings us into the nanometer range. Local probes with their nm scale capabilities are expected to play an important role in the continuing miniaturization of present day technologies as well as for emerging new solid state technologies. Miniaturization from mm to μm has been straightforward, through technologically challenging, and engineering ingenuity will carry it deep into the sub-μm regime. Nanometer-scale components, however, will require new concepts based on quantum effects, ballistic transport and so on. Thus miniaturization will probably not progress as fast as indicated by the extrapolation in Fig. 1; nevertheless, it will reach practical limits in the near future, be it in terms of compositional requirements, the minimum number of atoms necessary to form a functional unit, or energy dissipation. Miniaturization as a divi-

STORAGE: ONE BIT OF INFORMATION

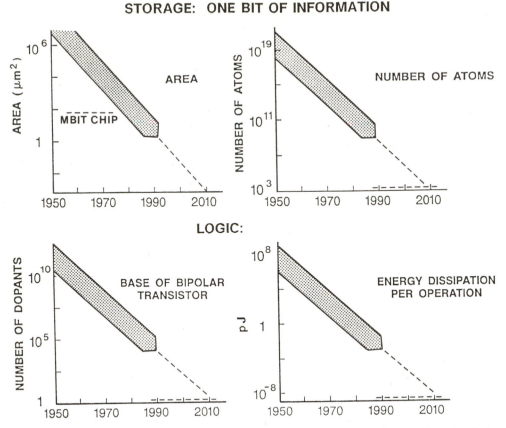

FIG. 1. Progress of miniaturization in information processing. The shaded beams comprise different types of devices or operations, e.g. storage contains disc file, magnetic bubble, optical disc and semiconductors. The horizontal, dashed lines give the practical limits. (Adapted from R. W. Keyes, Ref. 9).

sion into ever smaller and simpler blocks will be complemented by complex and intelligent assemblies built from molecular-sized functional units. "Molecular electronics," for instance, should be seen in a much broader sense than the simple replacement of semiconductor components in conventional circuits by molecules. The molecular assembly concert will sound much different, and it will hopefully be harmonious.

This brings us to another motivation, namely to work with selected, individual macromolecules. Of great interest to macromolecular chemistry and biology itself, this capability will be a crucial element for the building and control of molecular assemblies. This aspect is treated in this issue by B. Michel.

1.1. Methodology

The two central issues of using local probe methods at present are the probes themselves and the interactions used for probing or for working in general. The smallest nano-object observable with SPM is of the size of the local probe and of the distance between local probe and object. This means that nanometer-scale objects require nanometer-sized probes, and nanometer proximity of probe and object.

In the past decade, the control of probe and sample positions as well as of the probe-object distance became routine with various kinds of piezoelectric drives to below a tenth of a pm in a bandwidth of some tens of kHz. Changes in the interaction intensity, however, are monitored in the MHz range[10] and efforts for GHz operation are under way.[11,12]

The probes for tunneling and force methods are ground, etched or grown metallic or insulating tips. The end atoms of such tips serve as a natural, atomic size probe in tunneling and some force methods, as shown in Figs. 2a and 2b, provided the tip is reasonably structurally and chemically stable. So far most scientists have followed some recipe to obtain workable tips. In certain cases, especially when looking at larger objects or grooves and holes, tips with an overall small radius of curvature are preferred; for other applications where stability and maybe field homogeneity are important, tips with an overall large radius of curvature are required. Quite often, the front atoms or whole atom clusters are picked up from the object or object support. All these tips, of course, are neither chemically nor structurally well defined. This very often causes a certain ambiguity in the images such as tip-geometry-induced multi-tip[13] imaging or a tip-electronic structure dependent corrugation amplitude.[14] On the other hand, a virtue of "dirty" tips can be increased resolution[15] or better stability[16,17] due to reduced tip-object interactions. Well-defined pyramidal tips as shown in Figures 2c and 2d, even with apex atom or atoms different from the base atoms, have been fabricated[7,8] using procedures from field ion microscopy. So far, they have rarely been used as probe tips for SPM.[18] However, they have become significant in their own right as point sources for electrons and ions.[19] They exhibit a brightness several orders of magnitude higher than the existing field emission sources and produce beams with extraordinary properties. An alternative and promising approach for very well-defined probe tips is to fix an object to the apex of the tip and to use a specific interaction between this tip-object and the nano-object (see Fig. 2e). The local probe could then, e.g., be a selected molecular orbital, a magnetic moment or a specific molecular configuration.[20] Such "object-tips" should develop into the local probe of the future.

As to the interactions between probe and object, a large variety are in use, e.g. in tunneling microscopy the overlap of electronic wave functions of tip and object somewhere in the gap between them, and in force microscopy a variety of forces. since interactions depend on the distance between tip and object, imaging a property via a specific interaction requires independent control of the distance between probe and object. In a practical experiment, therefore, two interactions are usually required, the control interaction and the working or imaging interaction. The control interaction controls the position of probe with respect to the object. Scanning the tip at a constant control interaction strength gives a

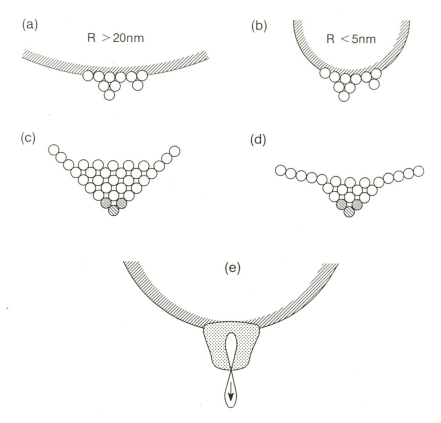

FIG. 2. Tips used in tunneling and force methods. a) and b) commonly used tips, mostly electrochemically etched before individual *in situ* treatment, with possible multitip action at edges. c) built-up tip and d) teton tip, both of well-defined shape and composition of the front atoms. e) tip of the future.

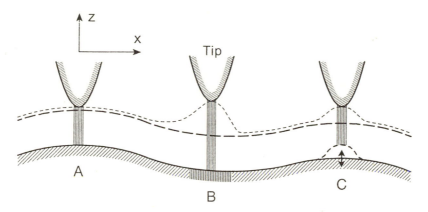

FIG. 3. Control of tip-object distance. Thick dashed line: tip-trace in the case of a laterally homogeneous control interaction; thin dashed line: tip trace due to inhomogeneities at B and C as explained in the text.

contour of the surface topography as shown in Fig. 3 at A, provided the control interaction is laterally homogenous. This is called the constant interaction mode or, in scanning tunneling microscopy, the constant current mode. However, in the case of an inhomogeneous control interaction, be it intrinsic or due to the close proximity of probe and object, the inhomogeneity is balanced by a changed probe-object distance as shown in Fig. 3 at B, where a locally stronger control interaction intensity is compensated by a larger distance. This causes an artifact in the object's topography, but of course also in the image of most other properties for which this inhomogeneous interaction was used as control interaction. A deformation of the object surface by probe-object forces, as shown at C, mimics a protrusion or indentation in the topography, but does not necessarily effect substantially the images of other properties. The art of local probe imaging or, quite generally, working with local probes is to find a control interaction which is laterally as homogeneous as possible. Stability of the position control further requires a control interaction which is monotonous in the probe-object distance in the range required for control.

2. SCANNING TUNNELING METHODS/MICROSCOPY

The probe-object interaction responsible for tunneling is the overlap of the electronic wave functions of probe and object in the gap between them. Tunneling, therefore, probes electronic properties at or outside the surface; subsurface and bulk properties are only sensed as far as they manifest themselves at the surface. The tunneling current, I, flowing between probe and object is a measure of this overlap which is in most cases also a measure of the local density of states at the surface. The electronic states contributing to the tunneling current are selected by the applied voltage V. For the theory of STM imaging, see Refs. 21-23 and references cited therein. The electronic property entering the tunneling current is, in addition to the local density of states, the transmittivity, exp (-2k s) where k is the inverse decay length of the wave functions in the tunnel gap of width s. k is related to the average, effective tunnel barrier height, ϕ_e, which these electrons experience, by $2k = \sqrt{2m\phi_e}\ /\ \hbar = 1.025\ \sqrt{\phi_e}$ if k is measured in Å^{-1}, ϕ_e in eV, and m is set equal to the free electron mass. The local density of states at energy $E = eV$ above or below the Fermi energy are obtained in tunneling spectroscopy from d(ln I)/d(ln V), and the tunnel barrier height from d(ln I)/ds. Various experimental procedures are used;[24] they essentially entail measuring I at various values of V or s, or by modulating V or s. The corresponding images are usually called spectroscopic images and work function profiles, respectively. "Work function profile" is insofar misleading as only at large gap width s, say $s > 15$ Å, is ϕ_e equal to the average work function, $\bar{\phi} = 1/2(\phi_p + \phi_o)$, of probe and object. At smaller distances, however, ϕ_e decreases rapidly. At small bias V,

$$- (\text{d ln I/ds}) = \sqrt{\phi_a} \approx \sqrt{\phi_e} + \frac{1}{2} (s/\sqrt{\phi_e})d\phi_e/ds + s^{-1} > \sqrt{\phi_e}\ .$$

ϕ_a is called the apparent barrier height and is not simply related to the effective barrier height. For details see Refs. 25-27.

The elastic response to the force interaction can also have drastic effects on the "work function profiles" or, better now, the "apparent barrier height" profiles.[27] Externally controlled is the probe position z rather than the tunnel gap width s. The measured response of the tunnel current to a position change is

$$\text{d ln I/dz} = \text{d ln I/ds}(1 + c\ \partial F/\partial s)^{-1},$$

where c is an effective compliance of the probe-sample system. The measured signal, d ln I/dz, is enhanced or reduced according to the sign of $\partial F/\partial s$, where $\partial F/\partial s < 0$ for increasing attractive forces with decreasing gap width and $\partial F/\partial s > 0$ otherwise. The latter case has been proposed as the main factor for the extremely small values of d ln I/dz usually obtained at ambient imaging, with some particle between tip and surface acting as the mediator of a strongly repulsive force.[28-30] The same mechanism - with or without additional force mediator - is also responsible for the giant corrugations observed on graphite and other materials of similar electronic structures.[29,30] Strong enhancements of d ln I/dz by attractive forces, on the other hand, have been observed on biological material.[31] Very often, the d ln I/dz image shows a much richer structure than the topography, in particular on biological materials.[31,32] Values of d ln I/dz considerably in excess of 2 $\mathring{A}^{-1}$, which translates into an effective tunnel barrier height of 4 eV, have to be attributed to such enhancement effects, and the images of d ln I/dz then reflect local force interactions and compliances rather than "work function profiles." They may well be as relevant as "work function profiles," in particular for biological materials.

STM provides images of electronic properties only and contact with the geometric properties is made with the distance dependence of the transmittivity. In most cases, scanning at constant overlap, i.e. constant tunnel current, means tracing contours of constant electronic state density in the gap. In the case of laterally homogeneous electronic properties, this trace reflects the surface atomic positions, and the constant current images are therefore referred to as surface topography. A crucial point in STM imaging is to find that contribution to I, which is laterally most homogeneous in order to be used as the control interaction. On clean metal surfaces, these are the conduction electrons at practically any energy. In most other cases, the connection of a constant current image with the object's topography can be quite involved. In GaAs, a simple example, the filled states of As and the empty states of Ga dominate a constant current image,[24,23] depending on sample polarity. An ac instead of a dc tunnel current, therefore, might be the appropriate control interaction. Another approach to deal with an inhomogeneous control interaction on relatively flat surfaces is to scan at an average control interaction, i.e. at an average height above the object surface. This is called the constant height mode, which also allows much higher scan speeds.[4] Local electronic or elastic responses to tip-sample interactions can complicate matters further. Therefore, great care is necessary when presenting the meaning of STM images. If done correctly, on the other hand, a rich variety of interesting local properties is obtained that are not yet accessible with other methods.

Figures 4 and 5 are two examples of constant imaging. Figure 4 shows the reconstructed (110) gold surface. The images represent contours of constant electron density at some distance from the surface given by the tunneling current and, therefore, a smoothed image of the gold atom positions. The upper image stems from the beginning of STM,[34] the lower one gives four snap shots at intervals of 15 minutes demonstrating appreciable diffusion of gold atoms. The progress lies not so much in the improved quality of the image - the upper image clearly exhibits all the salient features - but rather in the reproducibility, speed and ease of imaging. While atomic resolution was a happening at the time, it has now become standard.

Figure 5 shows a cross-sectional view of a cleaved GaAs-AlGaAs quantum well structure.[36] The upper left part is the GaAs, the lower right one the AlGaAs. The interface between the two layers appears atomically sharp. The apparent roughness of the AlGaAs is mainly an electronic and not a structural feature. It is caused by fluctuations of the state density induced by the fluctuations of the aluminum concentration. The high structures, too, are of electronic origin and are associated with oxygen atoms.[24,33] I-V characteristics taken at the same time show that the electronic properties also change quite abruptly at the interface.[37] The control interaction is laterally inhomogeneous, but homogeneous enough to obtain a set of significant I-V characteristics which can be viewed as a result of a working interaction. Figure 5 is also an example of combined instrumentation. Because of the strong reactivity of the cleaved AlGaAs surface the experiments have to be carried out within, say, an hour of cleaving. Therefore an UHV scanning electron microscope was used to find the interface region rapidly and to guide the tunnel tip to the interface.[38]

Spectroscopic imaging is one of the basic imaging techniques for semiconductor surfaces[24] (see Ph. Avouris in this issue). But it becomes also increasingly important in other fields as interest shifts from predominant structural properties to electronic ones. A nice example is the fluxoid imaging in superconductivity.[39]

There are many other local properties besides the corrugation of the density of states that can be investigated with STM. A first example is photon emission[40] as sketched on Fig. 6. The emitted photon can come from an inelastic tunneling process like that shown at A, from the decay

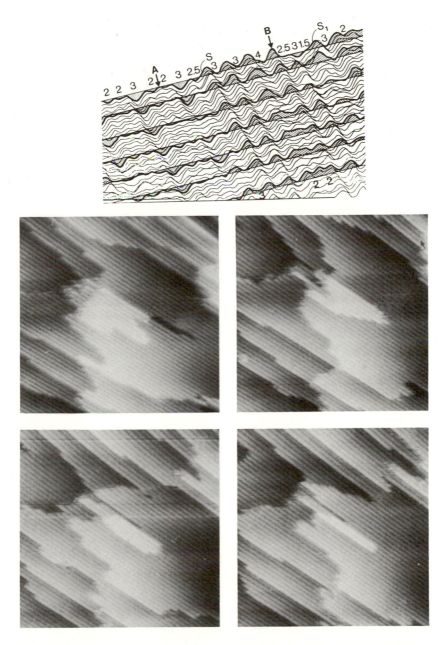

FIG. 4. Constant current images of reconstructed Au(110), showing the atomic rows in the [001] direction and illustrating the progress in STM imaging from the beginning (upper image, adapted from Ref. 34) to the present (lower four images, courtesy J. K. Gimzewski, see also Ref. 35).

FIG. 5. Constant current STM image of a cross-sectional GaAs-AlGaAs inter-face, where the AlGaAs was grown on the GaAs. Tip polarity is positive and therefore only the As atoms appear as protrusions. (Courtesy H. W. M. Salemink, see also Ref. 36).

of the excitation at B which was created at A, or from a recombination process at C of the electrons injected at A. Examples for processes like that at A are excitation of adsorbed molecules - which also can be detected with the so-called inelastic tunneling spectroscopy - or excitations of surface plasmons.[41] In the latter case the plasmon is created by the tunneling electron and the tip object configuration acts simultaneously as an antenna for emission of the photon when the plasmon decays. The information obtained thus relates both to the creation and the emission of the plasmon. If a plasmon travels to B before it decays, then the local aspect only comes into play via the excitation of

EMISSION   ABSORPTION

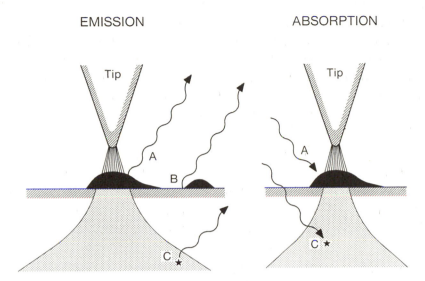

FIG. 6. Emission and absorption of photons in STM. In the case of emission, "local" is related to the creation of an excitation at A which emits a photon when decaying; the emission can occur at B and C, away from A. Absorption, on the other hand, should usually be specific to position A in order to provide a "local" absorption image.

the plasmon. In Fig. 7, a photon image is compared with a constant current image of a Cu(111) surface. In the photon image, most of the artificial protrusions appear dark, some, however, bright, reflecting an intrinsic property of the protrusions not evident in the topographic image. From recombination processes like that at C, information can be obtained regarding the electronic property at the position of the injection of the electron and its diffusion properties.[42] A luminescence image is shown in Fig. 8. Photon maps depend mostly on the number and energy of the injected electrons rather than on the probe-object distance; the control interaction is then simply the total tunnel-current at a given voltage. In the case of subsurface creation or decay of the excitation, photon maps also provide subsurface properties which are not observed at the surface. Photon emission in STM is in many ways a kind of local inverse photo emission, with the additional attractive features that the energy of the incoming electrons is not limited to values larger than the work function and that it can be performed in any transparent environment, e.g. water and electrolytes. Photon emission experiments with holes instead of electrons might also be feasible. Another class of photon experiments combines tunneling with absorbtion of photons.[43,44] Some approaches exploit an asymmetric, rectifying I-V tunnel characteristic as local photon detectors or frequency mixers, the imaging capability being in some cases of secondary importance.[43] Others use the contribution to the tunnel current of carriers created by the photons in the vicinity of the tunnel junction.

A second example is the imaging of local magnetic surface properties with spin polarized tunneling electrons. Various experiments[45] and theoretical studies[46] have treated the influence of polarization on tunneling in the classical metal-insulator metal sandwich tunnel junction. STM is in principle predestined for magnetic tunneling, since the tunnel barrier does not contain any magnetic moments which can destroy the

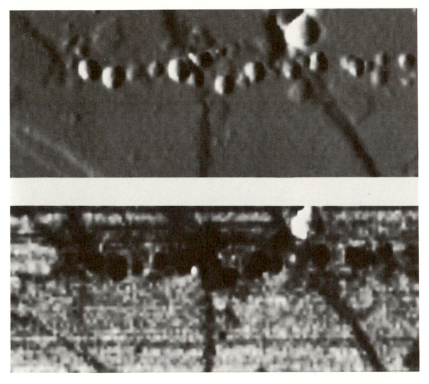

FIG. 7. Constant current (above) and photon (below) image of a Cu(111) surface exhibiting step lines and small artificial protrusions created by short current pulses. Differences are seen in the photon image which do not appear in the topographic image (400 Å × 200 Å). (Courtesy J. K. Gimzewski, see also Refs. 40, 41).

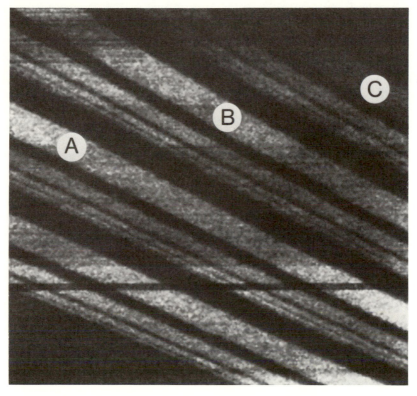

FIG. 8. Luminescence image of a GaAs (bright) - AlGaAs (dark) quantum well structure (2000 Å × 800 Å). (From Ref. 42, copyright 1991 American Vacuum Society).

spin memory of the tunneling electrons. Two experiments of "magnetic" STM have been reported so far. One dealt with the modulation of the tunneling current at some hundreds of MHz associated with the Larmor frequency of a single precessing magnetic moment.[10] This type of experiment does not require a magnetic probe. However, a magnetic moment at the tip apex would open interesting new possibilities. The other experiment[47] aimed at determining the surface magnetic polarization using the magnetic valve effect.[46] The contribution of polarized electrons to the tunneling current depends on the relative orientation of the polarized object and probe states. The parameters entering the magnetic valve effect are densities and decay lengths of "magnetic" and "nonmagnetic" object and tip electronic states, respectively, as well as the polarizations of the corresponding "magnetic" states and their relative direction. The general case of "magnetic" STM is thus a formidable task which makes "nonmagnetic" STM look easy by comparison. In the experiment conducted by the Basel group,[47] an elegant choice of the experimental conditions - including object and tip material - allows determination of the ratio of in-plain to out-of-plain magnetization components in Cr.[48] It is clear that well-defined magnetic properties of the probe tip, preferably even with external control, are essential for "magnetic" STM in the general case.

Other possibilities employing local tunneling include scanning potentiometry,[49] ballistic electron emission microscopy,[50] single electron tunneling,[51] standing wave microscopy,[52] thermal profilometry,[53] noise spectroscopy,[54] and various applications for position monitoring and control.

3. FORCE METHODS/MICROSCOPY

The atomic force microscope (AFM), invented by G. Binnig[55] and implemented by Binnig, Quate and Gerber,[56] uses forces as object-probe interactions. The motivation was to measure ultrasmall forces and to extend the STM topographic imaging capability in a stylus profilometer-type operation to the topography of insulators. The forces between

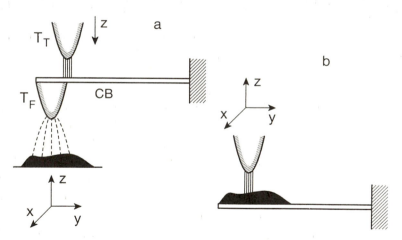

FIG. 9. Possible configurations for force microscopy. a) Principle of original instrument used in Ref. 56. The deflection of a cantilever beam (CB), due to the forces between sample and force tip (T_F) was sensed by the tunnel current flowing from the backside of CB to the tunnel tip T_T. The sample is approached to the force tip and scanned underneath it, contrary to STM where the tip is scanned. b) Configuration used in Ref. 57 for simultaneous tunneling and force microscopy.

object and probe - an atomically sharp tip like those in Figs. 2a and 2b, insulating or conducting - are sensed through the deflection of a submillimeter-sized cantilever beam, to which the probe tip is fixed. Sensitivity to small forces requires soft cantilever beams with a small mass and therefore small size allowing high-frequency operation which at the same time serves as "dynamic" protection against external low-frequency vibrations. In the original version, see Fig. 9a, the deflection was monitored with a tunnel current flowing from the back of the cantilever to a tunnel tip. Today it is usually measured optically with various optical methods.[58] Tunnel current detection, nevertheless, should still be very

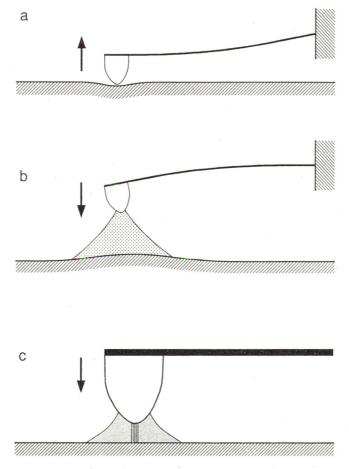

FIG. 10. Operating modes in force microscopy: a) net repulsive force, b) net weak attractive forces (shaded) at large distance, and c) attractive force (shaded) at tunneling distance.

attractive in low-temperature experiments for ultrasensitive force and deflection detection in the sub-fN range.[59] An alternative approach to force measurements makes use of the shift of the resonance farequency of the cantilever due to tip-object interaction.[57,58] This yields the derivative of the interaction force between object and probe, $dF/dz = \Delta f = 2 f \Delta\omega/\omega$, where dF/dz is the force gradient at object-probe separation s, ω and f are resonance frequency and stiffness of the cantilever beam at zero interaction, i.e. at $s = \infty$, respectively, and $\Delta\omega$ and Δf their respective changes from $s = \infty$ to a probe-object separation s. Important in this mode of operation is the Q value of the cantilever beam rather than its softness. In order to obtain the force, dF/dz has to be integrated and thus measured in the interval $s < z < \infty$. In many cases, however, dF/dz is sufficient, e.g. using dF/dz as control interaction.

Force microscopy is an ingeniously simple but effective method for topographic imaging of insulator surfaces down to the atomic level. Atomic resolution was first achieved on graphite[60] and has become routine in the "contact" mode (see Fig. 10a) provided the surface can withstand the very high, local perpendicular pressures and parallel stresses. Examples are shown for a layered material with strong in-plane bonds (Fig. 11) and the ionic crystal LiF (Fig. 12). Atomic resolution requires both an atomic-sized probe and a similar probe-object distance. In practice so far, this has meant contact between probe and object, i.e. repulsive forces. Since only the total force between probe and object is detected, an overall weakly repulsive force can mean quite a strong repulsive contact force, which has to overcompensate all the attractive forces. Under ambient conditions, strong capillary forces due to a liquid, e.g. water surface layer which the tip has to penetrate, can reach 10^{-7} N. Such forces are not tolerable when imaging soft materials or adsorbates. They can be avoided by emersing the entire cantilever system into a liquid,[63] appropriate when working with biological species, or by working in vacuum. Strong van der Waals, Coulomb and cohesive forces, however, can still remain a problem.[16,17] The van der Waals forces can also be reduced to an acceptable level in a liquid environment.[17] The other two forces should be taken care of by an appropriate choice of tip material. Very good results in imaging biological material in a liquid environment were obtained by the group at the University of California at Santa Barbara.[64] The repulsive imaging mode also attains great significance for metrology of insulating and conducting materials that are not too soft.

FIG. 11. STM (left) and AFM (right) images of 1T-TaSe$_2$. The STM image at low bias voltage V shows the Fermi level electron density corrugation. Both the long wave length charge density wave superlattice and the atomic lattice are observed. The AFM image, on the other hand, is dominated by the repulsive forces of all electrons interacting at contact and shows the atomic positions only. (Courtesy of the authors of Ref. 61).

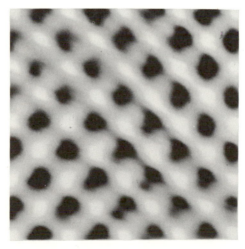

FIG. 12. AFM image of LiF. (Courtesy of the authors of Ref. 62).

Another class of force microscopy uses attractive forces such as van der Waals forces, F_V, magnetic forces, F_M, and Coulomb forces, F_C, at medium to large distances, say s ≥ 100 Å (see Fig. 10b). There, the force gradients are sufficiently small,[17] so that a very soft lever of the order of 0.1 to 0.01 N/m can be used, which in turn determines the sensitivity of the force measurement. Besides finding a laterally homogeneous control interaction, the various forces also have to be separated. In the case of magnetic imaging, one can usually find a distance s at which the total force is dominated by the magnetic forces, the force gradients, however, by van der Waals forces or Coulomb forces in the case of conducting object and tip. Magnetic force gradients saturate at a value of F_M/λ, where λ is the lateral extension of a magnetic domain, whereas the van der Waals and Coulomb force gradients diverge roughly as $F_{V,C}/s$. Therefore, we can find a distance s where $dF/ds \approx dF_{V,C}/s$. Taking dF/ds as control parameter, we obtain the sample topography provided dF/ds is laterally homogeneous. On the other hand, $F = F_M + F_{V,C} \approx F_M$ + const on the topography line dF/ds = const. The contrast in F then yields the magnetic image. In the case of conducting magnetic objects, it is convenient to work with Coulomb forces,[65] which can be controlled by an external voltage. In the experiment of Ref. 65, resolution reached about 100 Å. Note that a nonmagnetic control interaction is necessary for magnetic imaging since magnetic forces and force gradients can change sign on scanning.

Coulomb forces and van der Waals forces can in principle also be separated, since the van der Waals forces diverge stronger than the Coulomb forces, e.g. s^{-2} compared to s^{-1} for a tip-plane configuration.[17] Externally controllable Coulomb forces greatly facilitate separation. Fig. 13a presents the configuration used to detect single electron charges in a thin insulating layer which is on a conducting substrate[66] where even Coulomb forces of different origin are separated. Contributions to the total Coulomb force come from the charge trapped in the insulator, the contact potential and the applied voltage between substrate and tip. The total Coulomb force F is $\propto (\Delta\phi + V + \delta V)^2$, where $\Delta\phi$ is due to the contact potential and the trapped charge. V is an applied dc potential and δV is a small tickling voltage of frequency ω. The 2ω component, $\delta F_{2\omega} = g(s)\, \delta V^2$, is independent of $\Delta\phi$. Scanning at constant $\delta F_{2\omega}$ means scanning at a constant distance from the substrate provided that the thickness of the insulator does not vary too greatly. The total force, of course, contains the Coulomb forces both from the applied voltage, the

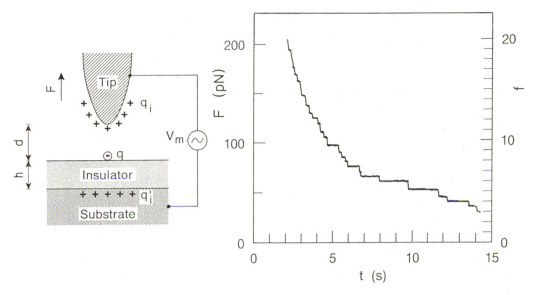

FIG. 13. Left: schematic of the force detection due to a charge q on an insulating film and its corresponding image charge q_i. In the experiment of Ref. 66, d, $h \approx 20$ nm, V_m is the applied voltage $V + \delta V$ (see text). Right: decay of F with time for a charged, 20-nm-thick Si_3N_4 film. The scale f on the right corresponds to the disappearance of one single electron charge from the film directly underneath the tip. (Courtesy Ch. Schönenberger and S. Alvarado, see also Ref. 66).

contact potential and the trapped charge. Their separation is more difficult. If, however, the contact potential is laterally homogeneous, then an applied voltage can compensate the contact potential, and the term F on contours $\delta F = $ const gives the Coulomb force due to the trapped charge. Note, that adjusting V for zero ω-component, $F_\omega \propto (\Delta\phi + V)\, \delta V = 0$, gives the total non-Coulomb force. Figure 13b shows the decay of charges injected into the insulator. It clearly demonstrates that force microscopy is sensitive enough to detect single electron charges over distances of 10 or more nanometers. This should be of interest for imaging charged molecules and slow charge transfer processes in biological functional units and control of such processes, as well as for memory applications.

Figure 10c corresponds to the measurement of attractive forces at close distance which would allow simultaneous tunneling and force microscopy. This is the most difficult case because the total attractive forces and force gradients at tunneling distance are not monotonous and can be quite significant.[16,17] Mechanical stability, therefore, requires a relatively stiff cantilever which in turn reduces the sensitivity of the force measurement. Actually, in the case of clean metal tips on clean metal surfaces, the attractive forces in the tunnel regime can be so large that tip and sample deformation lead to intrinsic instabilities[16] independent of the stiffness of the cantilever beam. In the case of an Ir tip on an Ir sample, cohesive Ir-Ir forces could be measured as function of distance[67] and even atomic features could be imaged[68] with them, using an instrument like that sketched in Fig. 9b and a tunneling current as control interaction. Other combinations of sample and tip material are more problematic.[69] Simultaneous performance of STM and force microscopy, however, should become most important in the imaging of soft materials, in particular biological materials, e.g. where local electron transfer through the material in question might be very inhomogeneous and may also depend strongly on tip-sample interactions, in particular on local strains.

An elegant variation of force microscopy is friction force microscopy, developed by the IBM Almaden group[70] and extended by O. Marti et al.[71] to a combined force and friction microscopy. To establish a contact between this atomic approach and friction and wear among larger objects of practical importance is crucial to the impact of these methods on general tribology.

4. PROSPECTS

The two examples of tunneling and force methods give an idea of the prospects of local probe methods. Known and sold still mainly for their imaging capability of surface structures with atomic resolution, their much wider interdisciplinary potential, which goes far beyond structures and imaging, is slowly being appreciated. The surface structure is a straightforward result of scanning at constant control interaction. The full potential of local probe methods, however, lies in the working interactions, which give access to the large variety of local properties and provide the means for manipulation, modification and control. This is convincingly demonstrated by contributions in this issue on surface chemistry, biology, and manipulation and modification.

Improvement of existing and invention of new instrumentation and methods will open many new avenues. Already mentioned in this context have been the probe tips like that shown in Fig. 3e where the tip end indeed becomes a multichoice tool, or instruments employing very different kinds of interactions like the combined STM/AFM or STM and photon emission/absorption. Dynamic studies like molecular processes and motion will greatly profit from an extension of the operation into the Gigahertz range not only electrically but possibly also mechanically. Large arrays of micromachined instruments, far less than one millimeter each,[72,73] for instance as read/write heads or other tools, with new ways of parallel operation could lead to new storage concepts or quite generally to new concepts of parallel processing beyond data processing. Multitip instruments with each tip working independently - maybe even separately controllable - at distances from each other down to, say, 5 nanometers could provide the basis for new experimentation. Other interactions, not widely used yet, like electromagnetic interactions in near-field optical methods,[74] and in certain types of thermal profilometry,[75] and interactions leading to transport of specific ions, atoms and molecules like in scanning ion conductance microscopy[76] and acoustic near-field microscopy[77] should prove very useful for selected problems, in spite of their so far limited resolution.

There are indeed a great many very promising avenues for local probe methods and certainly many new ones to come, the most interesting and rewarding of which will be singled out by scientists themselves. The prospects lie in the possibilities and opportunities we create and realize rather than in the predictions of what we now think will become important.

ACKNOWLEDGEMENTS

I appreciate and gratefully acknowledge many stimulating discussions with colleagues of the SPM community.

REFERENCES

1. Proceedings 4th Int'l. Conf. on Scanning Tunneling Microscopy/Spectroscopy, J. Vac. Sci. Techn. **A8**, (1990).

2. "Scanning Tunneling Microscopy and Related Methods" in NATO ASI Series **E 184**, (1990).

3. Proceedings of STM '90 and NANO I, to appear in J. Vac. Sci. Techn. (1991).

4. J. Tersoff and P. Hansma, J. Appl. Phys. 61 (1987) R1; H. K.

Wikramasinghe, Sci. Am. **261**, 74 (1989); H. Rohrer in Ref. 2, p. 1; D. Rugar and P. Hansma, Physics Today, 23 (October 1990).

5.   G. Binnig and H. Rohrer, Helv. Phys. Acta **55** 726 (1982).

6.   G. Binnig and H. Rohrer, Rev. Mod. Phys. **59**, 615 (1987).

7.   H. -W. Fink, Physica Scripta **38**, 260 (1988); idem in Ref. 2, p. 399.

8.   Vu Thien Binh, J. Microscopy **152**, 355 (1988); J. J. Saenz, N. Garcia, Vu Thien Binh and H. de Raedt in Ref. 2, p. 409.

9.   R. W. Keyes, IBM J. Res. Develop. **32**, 24 (1988).

10.   Y. Manassen, R. J. Hamers, J. E. Demuth and A. J. Castellano, Jr., Phys. Rev. Lett. **62**, 2531 (1989).

11.   G. P. Kochanski, Phys. Rev. Lett. **62**, 2285 (1989).

12.   H. J. Mamin, P. H. Guethner and D. Rugar, Phys. Rev. Lett. **65**, 2418 (1990).

13.   For STM see: Sang-il Park, J. Nogami and C. F. Quate, Phys. Rev. B **36**, 2863 (1987); H. A. Mizes, Sang-il Park and W. A. Harrison, Phys. Rev. B **36**, 4491 (1987). For AFM see. F. F. Abraham, I. P. Batra and S. Ciraci, Phys. Rev. Lett. **60**, 1314 (1988); F. F. Abraham and I. P. Batra, Surf. Sci. **209**, L125 (1989).

14.   J. Tersoff and N. D. Lang, Phys. Rev. Lett. **65**, 1132 (1990).

15.   E. P. Stoll, J. Phys. C **21**, L921 (1988).

16.   U. Landman, W. D. Luedtke, N. A. Burnham and R. J. Cotton, Science **248**, 454 (1990).

17.   F. O. Goodman and N. Garcia, Phys. Rev. B **42**, (1990, in press).

18.   Y. Kuk and P. J. Silverman, Appl. Phys. Lett. **48**, 1595 (1987); Y. Kuk, P. J. Silverman and H. Q. Nguyen, J. Vac. Sci. Techn. **A6**, 524 (1988); T. Hashizume, Y. Hasegawa, I. Kamiya, T. Ide, I. Sumita, S. Hyodo and T. Sakurai in Ref. 1, p. 233; T. Hashizume, I. Sumita, Y. Murata, S. Hyodo and T. Sakurai in Ref. 3; T. Sakurai et al., Progress in Surf. Sci. (1990 in press).

19.   W. Stocker, H.-W. Fink and R. Morin, Ultramicroscopy **31**, 379 (1989); H.-W. Fink, W. Stocker and H. Schmid, Phys. Rev. Lett. **65**, 1204 (1990).

20. J. Kaufman, Ch. Gerber and B. Michel, IBM Technical Disclosure Bulletin (1991).

21. J. Tersoff and D. R. Hamann, Phys. Rev. Lett. **50**, 1998 (1983); J. Tersoff in Ref. 2, p. 77 and references therein; J. Tersoff, Phys. Rev. B **39**, 1052 (1989); idem **40**, 11990 (1989) and **41**, 1235 (1990).

22. N. Garcia, C. Ocal and F. Flores, Phys. Rev. Lett. **50**, 2002 (1983); A. Baratoff, Physica B **127** 143 (1984); E. Stoll, A. Baratoff, A. Selloni and P. Carnevali, J. Phys. C. **17**, 2073 (1984).

23. N. Lang, Phys. Rev. Lett. **55**, 230 (1985); idem **56**, 1164 (1986); idem **57**, 45 (1987).

24. R. M. Feenstra in Ref. 2, p. 211; R. J. Hamers, R. M. Tromp and J. E. Demuth, Phys. Rev. Lett. **56**, 1972 (1986); N. D. Lang, Phys. Rev. B **34**, 5947 (1986); J. Ihm, Physica Scripta **38**, 269 (1988).

25. N. Lang, Phys. Rev. B **36**, 8173 (1987); R. Garcia, Phys. Rev. B **42** (1990); R. Garcia and J. J. Saenz, Surf. Sci. (1990 in press).

26. R. G. Forbes in Ref. 2, p. 163.

27. H. Rohrer in Ref. 2, p. 1.

28. J. H. Coombs and J. B. Pethica, IBM J. Res. Develop. **30**, 455 (1986).

29. J. M. Soler, A. M. Baró, N. Garcia and H. Rohrer, Phys. Rev. Lett. **57**, 444 (1986); H. J. Mamin, E. Ganz, D. W. Abraham, E. Thomson and J. Clarke, Phys. Rev. B **34**, 9015 (1986).

30. Y. Sugawara, T. Ishizaka and S. Morita, Jpn. J. Appl. Phys. **29**, 1533 (1990).

31. G. Travaglini, H. Rohrer, M. Amrein and H. Gross, Surf. Sci. **181**, 380 (1987); G. Travaglini, H. Rohrer, E. Stoll, M. Amrein, A. Stasiak, J. Sogo and H. Gross, Physica Scripta **38**, 309 (1988); G. Travaglini, M. Amrein, B. Michel and H. Gross in Ref. 2, p. 335.

32. A. Cricenti, S. Selci, A. C. Felici, R. Generosi, E. Gori, W. Djaczenko and G. Chiarotti in Ref. 1 and in Ref. 2, p. 359.

33. R. M. Feenstra, J. A. Stroscio, J. Tersoff and A. P. Fein, Phys. Rev. Lett. **58**, 1192 (1987); J. A. Stroscio, R. M. Feenstra and A. P. Fein, Phys. Rev. Lett. **58**, 1668 (1987).

34. G. Binnig, H. Rohrer, Ch. Gerber and E. Weibel, Surf. Sci. **131**, L379 (1983).

35. J. K. Gimzewski, R. Berndt and R. R. Schlitter in Ref. 3; idem Surf. Sci. (1991 in press).

36. O. Albrektsen, D. J. Arent, H. P. Meier and H. W. M. Salemink, Appl. Phys. Lett. **57**, 31 (1990).

37. O. Albrektsen, P. Koenraad and H. W. M. Salemink, to be published.

38. Ch. Gerber, G. Binnig, H. Fuchs, O. Marti and H. Rohrer, Rev. Sci. Instrum. **57**, 221 (1986).

39. H. F. Hess, R. B. Robinson and J. V. Warzczak, Phys. Rev. Lett. **64**, 2711 (1990).

40. J. K. Gimzewski, B. Reihl, J. H. Coombs and R. R. Schlitter, Z. Phys. B **72**, 497 (1988); R. Berndt, R. R. Schlitter and J. K. Gimzewski in Ref. 3.

41. R. Berndt, A. Baratoff and J. K. Gimzewski in Ref. 2, p. 269.

42. D. L. Abraham, A. Veider, Ch. Schönenberger, H. P. Meier, D. J. Arent and S. F. Alvarado, Appl. Phys. Lett. **56**, 1564 (1990); S. F. Alvarado, Ph. Renaud, D. L. Abraham, Ch. Schönenberger, D. J. Arent and H. P. Meier in Ref. 3.

43. T. E. Sullivan, Y. Kuk and P. Cutler, IEEE Trans. Electron. Devices **36**, 2659 (1989); H. Q. Nyguen, P. H. Cutler, T. E. Feuchtwang, Z. Huang, Y. Kuk, P. J. Silverman, A. A. Lucas and T. E. Sullivan, IEEE Trans. Electron. Devices **36**, 2671 (1989); W. Krieger, H. Koppermann, T. Suzuki and H. Walter, IEEE Trans. Instrum. Meas. (USA) **38**, 1019 (1989); W. Krieger, T. Suzuki, M. Volcker and H. Walter, Phys. Rev. B **41**, 10229 (1990).

44. R. J. Hamers and K. Market, Phys. Rev. Lett. **64**, 1051 (1990); Y. Kuk, R. S. Becker, P. J. Silverman and G. P. Kochanski, Phys. Rev. Lett. **65**, 456 (1990).

45. R. Meservey, T. Paraskevopoulos and P. M. Tedrow, J. Appl. Phys. **49**, 1405 (1978); M. Julliere, Phys. Lett. **54A**, 225 (1975); U. Gäfvert and S. Makawa, IEEE Trans. Magn. **18**, 707 (1982); Y.

Suezawa and Y. Gondo in <u>Proceedings of the Int'l. Symp. on Physics of Magnetic Materials</u>, 303, (World Scientific, Singapore, 1987).

46. J. C. Slonczewski, Phys. Rev. B **39** 6995 (1989) and references therein.

47. R. Wiesendanger, H. J. Güntherodt, G. Güntherodt, R. J. Gambino and R. Ruf, Phys. Rev. Lett. **65**, 247 (1990).

48. N. Garcia, private communication and comment in Phys. Rev. Lett. Ref. 47 (in press).

49. P. Muralt and D. W. Pohl, Appl. Phys. Lett. **48**, 514 (1986); A. D. Kent, I. Maggio-Aprile, Ph. Niederman, Ch. Renner and O. Fisher in Ref. 1, p. 459.

50. W. J. Kaiser and L. D. Bell, Phys. Rev. Lett. **60**, 1406 (1988); L. D. Bell, M. H. Hecht and W. J. Kaiser and L. C. Davis, Phys. Rev. Lett. **64**, 2679 (1990).

51. H. van Kempen in Ref. 2, p. 241.

52. J. H. Coombs and J. K. Gimzewski, J. Microscopy **152**, 841 (1988); J. A. Kubby, Y. R. Wang and W. J. Greene, Phys. Rev. Lett. (1990, in press).

53. J. M. R. Weaver, L. M. Walpita and H. K. Wickramasinghe, Nature **342**, 783 (1989).

54. R. Möller, A. Esslinger and B. Koslowski in Ref. 1, p. 590.

55. G. Binnig, US Patent 4 724 318, filed August 4, 1986, published February 9, 1988.

56. G. Binnig, C. F. Quate and Ch. Gerber, Phys. Rev. Lett. **56**, 930 (1986).

57. U. Dürig, O. Züger and D. W. Pohl, J. Microscopy **152**, 259 (1988).

58. G. M. McClelland, R. Erlandsson and S. Chiang in <u>Review of Progress in Quantitative Nondestructive Evaluation</u>, D. O. Thompson and D. E. Chimenti, Eds., (Plenum, New York 1987), Vol 6B, p. 307; Y. Martin, C. C. Williams and H. K. Wikramasinghe, J. Appl. Phys. **61**, 4723 (1987), G. Meyer and N.

M. Amer, Appl. Phys. Lett. **53**, 1045 (1988); for a review on force microscopy, see D. Rugar and P. K. Hansma in Ref. 4 and H. K. Wickramasinghe in Ref. 4.

59. M. F. Bocko, K. A. Stephenson and R. H. Koch, Phys. Rev. Lett. **61**, 726 (1988); idem Phys. Rev. A **40**, 6615 (1989).

60. G. Binnig, Ch. Gerber, E. Stoll, T. R. Albrecht and C. F. Quate, Europhys. Lett. **3**, 1281 (1987).

61. E. Meyer, R. Wiesendanger, D. Anselmetti, H. R. Hidber, H. J. Güntherodt, F. Lévy and H. Berger in Ref. 1, p. 495; E. Meyer, D. Anselmetti, R. Wiesendanger, H.-J. Güntherodt, F. Lévy and H. Berger, Europhys. Lett. **9**, 695 (1989).

62. E. Meyer, H. Heinzelman, H. Rudin and H. J. Güntherodt, Z. Phys. B - Cond. Matter, **79**, 3 (1990).

63. O. Marti, B. Drake and P. K. Hansma, Appl. Phys. Lett. **51**, 484 (1987).

64. D. Rugar and P. K. Hansma in Ref. 4.

65. Ch. Schönenberger, S. F. Alvarado, S. E. Lambert and I. L. Sanders, J. Appl. Phys. **67**, 7278 (1989); Ch. Schönenberger and S. F. Alvarado, Z. Phys. B **80**, 373 (1990).

66. Ch. Schönenberger and S. F. Alvarado, Phys. Rev. Lett. **65**, 3162 (1990).

67. U. Dürig, O. Züger and D. W. Pohl, Phys. Rev. Lett. **65**, 349 (1990).

68. U. Dürig and O. Züger, to be published.

69. U. Dürig, private communication.

70. C. M. Mate, G. M. McClelland, R. Erlandsson and S. Chiang, Phys. Rev. Lett. **59**, 1942 (1987); R. Erlandsson, S. Chiang, G. M. McClelland, C. M. Mate and G. Hadzioannou, J. Chem. Phys. **89**, 5190 (1988).

71. O. Marti, J. Colchero and J. Mlynek, Nanotechnology (1991 in press).

72. D. W. Pohl, U. T. Dürig and J. K. Gimzewski, European Patent Application, Publication No. 0 247 219, filed May 27, 1986, published December 2, 1987.

73. A first practical step in this direction is the "STM on a Chip," S. Akamine, T. R. Albrecht, M. J. Zdeblick and C. F. Quate, IEEE Electron. Device Lett. **10**, 490 (1989).

74. D. W. Pohl, W. Denk and M. Lanz, Appl. Phys. Lett. **44**, 651 (1984); A. Lewis, M. Isaacson, A. Harootunian and A. Murray, Ultramicroscopy **13**, 227 (1984); U. Fischer, J. Vac. Sci. Technol. **B3**, 386 (1985); U. Fischer in Ref. 2, p. 475; D. Courjon in Ref. 2, p. 497.

75. K. Dransfeld and J. Xu, J. Microscopy (UK) **152**, 35 (1988).

76. R. V. Coleman, W. W. McNairy, C. G. Slough, P. K. Hansma and B. Drake, Surf. Sci. **181**, 112 (1987); P. K. Hansma in Ref. 2, p. 299.

77. K. Dransfeld, Phys. Bl. (Germany) **46** 307 (1990).

BALLISTIC TRANSPORT IN NORMAL METALS

P. Wyder,[1] A. G. M. Jansen[1] and H. van Kempen[2]

[1] Max-Planck-Institut für Festkörperforschung
Hochfeld-Magnetlabor, 166X, F-38042 Grenoble
Cedex, France

[2] Res. Institute for Materials, Univ. of Nijmegen
Toernooiveld, NL-6525 Nijmegen, The Netherlands

Abstract - Using micro-contacts ("point contacts") between metals, conduction electrons can be injected ballistically into the metal; the relevant energy of these electrons can be tuned by varying the applied voltage. With the injected electrons it is possible to perform spectroscopy inside the metal just as it has usually been done with electrons, neutrons or photons from the outside. The observed deviations from Ohm's law in the tiny metallic contacts allow an energy-resolved spectroscopy of the interaction of the conduction electrons with elementary excitations (e.g. phonons, magnons, paramagnetic impurities, crystal-field levels, etc.) in a metal. In reasonable pure normal metals, the observed second derivative signal $d^2V/dI^2(V)$ is directly proportional to Eliashberg's form of the electron-phonon interaction $\alpha^2F(eV)$ with a slight modification due to a transport efficiency function. The point-contact method and its applications will be discussed. Specific experiments will show the usefulness of ballistic point-contacts for other purposes than just spectroscopy: high-frequency rectification with a point contact as a non-linear element, electron-focusing using a double point-contact setup, the study of interfaces (Andreev reflection), localization phenomena in a constriction, magnetoresistance of a point contact.

1. INTRODUCTION

Small metallic constrictions offer the possibility to study ballistic transport of conduction electrons within a metal. The smallness of the constriction is determined by the contact dimension (diameter d) with respect to the electron mean free path ℓ. By passing the contact in an applied electric field, the electrons are accelerated within a mean free path for clean contacts with d $\ll \ell$. This ballistic injection of the charge carriers in a point-contact geometry was recognized firstly by Sharvin[1] and has been used since then in various ways to study electrical transport properties in a metal under non-equilibrium conditions of the electrons.

Tuning the applied voltage V over the contact allows spectroscopic studies of the electron system with respect to the energy eV of the injected electrons. Because the electronic scattering processes in the contact area influence the contact resistance, the inelastic relaxation of the electrons yields nonlinear current-voltage characteristics for the metallic constriction with specific information on the energy-dependence of the interaction between the conduction electrons and other excitations in a metal. After the pioneering experiments by Yanson[2], this experimental technique, known as point-contact spectroscopy, has been applied very successfully in the determination of the Eliashberg function $\alpha^2 F$ for the electron-phonon interaction in normal metals. Also other interaction mechanisms with the electrons (e.g. magnons, paramagnetic impurities, crystal-field levels) have been studied as well using point-contact spectroscopy. For a point contact on a single crystal in an applied magnetic field, the injected electrons follow curved orbits on the Fermi sphere and can be focussed and collected on a second contact. Using this so called transverse-electron-focusing technique,[3] the fermiology of the metal and the reflection properties of the sample boundaries can be studied. For an interface between a normal metal and a superconductor, the double point-contact setup can be used for the observation of the retro-reflection of a quasi-particle at this interface (Andreev reflection).[4]

An introductory discussion of the point-contact method will be given with the important parameters (electron mean free path with respect to contact size) for the evaluation of the point-contact data, followed by a description of experiments with ballistic transport in metallic micro-contacts. For a detailed account of the point-contact technique, we refer to more extensive reviews on the subject.[5-8] Finally, experiments will be presented with applications of the point-contact technique: high-frequency rectification, transverse electron focusing, quantum interference phenomena in the electral transport in contacts with semi-metals and magnetic quantum oscillations in the point-contact magnetoresistance.

2. BASIC ELEMENTS IN POINT-CONTACT SPECTROSCOPY

The important parameters in the point-contact problem are the contact size (given by the contact radius a for a circular orifice) and the electron mean free paths with respect to elastic scattering (ℓ_i) and inelastic scattering ($\ell(\varepsilon)$). The condition for ballistic transport in a contact is given by $\ell_i, \ell(\varepsilon) >> a$. For the case of strong elastic scattering in the contact area ($\ell_i \leq a$), this condition can be relaxed to $\Lambda(\varepsilon) >> a$ for the inelastic diffusion length $\Lambda(\varepsilon) = (\ell_i \ell(\varepsilon))^{1/2}$. Although in this case the momentum distribution of the electrons is random, still the applied voltage V defines the shell of energies ε around the Fermi energy with

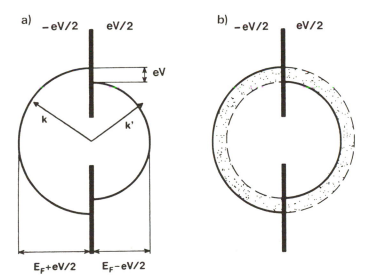

FIG. 1. Energy distribution for the electrons at the center of the constriction for a voltage V applied over ballistic contacts with $\ell_i, \ell(\varepsilon) >> a$ (a) and $\Lambda(\varepsilon) >> a \geq \ell_i$ (b). The arrows indicate a possible inelastic scattering event for a back-flow process.

non-equilibrium electrons. In Fig. 1 we have depicted schematically the distribution of energies for the electrons at the center of the contact.

The shown arrows in Fig. 1a illustrate a possible inelastic scattering event from state k to k′ with corresponding energies ε and ε', such that $\varepsilon = \varepsilon' + \hbar\omega$. For example, the energy loss $\hbar\omega$ could correspond to the energy of a spontaneously emitted phonon in the collision. Just this kind of scattering processes leads to negative corrections in the injected current, the so-called backflow current. By differentiating the current with respect to the applied voltage, the inelastic electron scattering can be investigated directly for the electrons with excess energy eV via the voltage dependence of the differential current-voltage characteristics.

For a formal understanding of the point-contact problem, the Boltzmann transport equation has to be solved with the appropriate boundary conditions for the contact geometry. In zeroth order without any collisions, the injection current I_0 can be calculated for the distribution of kinetic energies sketched in Fig. 1a, yielding

$$I_0 = \frac{V}{R_{Sh}} = V \frac{S_r S_k}{2\pi^2} \frac{e^2}{h} , \qquad (1)$$

where S_r is the cross section of the contact in real space and S_k the projection of the Fermi surface on the contact plane. For an isotropic free-electron metal with $S_k = \pi k_F^2$ and a circular orifice with $S_r = \pi a^2$, the Sharvin resistance R_{Sh} reduces to the commonly encountered expression $R_{Sh} = 4\rho\ell/3\pi a^2$, using a Drude model for the electrical resistivity ρ. The Sharvin resistance is independent of the electron scattering, but the Sharvin contact serves as an accelerator of the electrons within a mean free path inside the metal. The next order contribution I_1 to the current takes into account the backflow after inelastic scattering and can be written as

$$I_1 = -e\Omega_{eff}N_0 \int_0^{eV} \frac{d\varepsilon}{\tau(\varepsilon)}$$
$$= -\frac{2\pi e}{\hbar} \Omega_{eff}N_0 \int_0^{eV} d\varepsilon \int_0^{\varepsilon} d\varepsilon' \, S(\varepsilon - \varepsilon') \qquad (2)$$

N_0 is the density of states of the electrons at the Fermi level. Ω_{eff} (= $8a^3/3$ for a circular orifice) describes the volume near the contact where inelastic scattering is very effective to contribute to the backflow current I_1. It is readily seen that the current expansion goes like $I_1 \approx -I_0 a / \ell(\varepsilon)$, with $\ell(\varepsilon) = v_F \tau(\varepsilon)$ the inelastic relaxation length for an electron with energy ε above the Fermi level. According to Eq. (2), the inelastic relaxation time $\tau(\varepsilon)$ can be expressed in terms of the spectral function

$$
S(\varepsilon) = \frac{N_0}{32\pi^2} \int \frac{d^2k}{k^2} \int \frac{d^2k'}{k'^2} \, |g_{kk'}|^2
$$
$$
\times \, \eta(k, k') \delta(\varepsilon - \varepsilon_k + \varepsilon_{k'})
$$

(3)

for the interaction under study with matrix element $g_{kk'}$ between initial (k) and final (k') states integrated over the Fermi sphere. The efficiency function $\eta(k,k')$ for backflow reduces to $\eta(\theta) = (1 - \theta/tg\theta)/2$ by averaging over a constant angle θ between k and k', analogously to the function $(1 - \cos\theta)$ in the electrical transport problem. From Eq. (2) follows

$$
\frac{d^2I}{dV^2} = - \frac{2\pi e^3}{\hbar} \Omega_{eff} N_0 S(eV).
$$

(4)

Ballistic injection of electrons through a small constriction allows a direct measurement of the energy dependence of the spectral function for the concerned interaction with the electrons.

3. SPECTROSCOPIC POINT-CONTACT INVESTIGATIONS

In the very first explorative experiments dealing with point-contact spectroscopy[1], the investigated constriction consisted of a metallic short circuit in a MIM tunnel junction. Using a sharply etched whisker pressed on a flat surface, the point-contact technique could be applied to a large variety of materials including single crystals. Even two pieces of bulk material, pressed gently together, are useful point-contact devices. Although very easy to fabricate and to (re-)adjust, the exact contact area is not very well under control with these pressure-type contacts. Recently, with advanced micro-fabrication techniques three-dimensional nanobridges are produced, as small as 10 - 20 atoms across and of well defined geometry.[9]

Point-contact spectroscopy has been applied especially for the study of the electron-phonon interaction in metals. In fact, for normal metals it was the first method with a direct measurement of the electron-phonon interaction function. For superconductors, a comparison can be made of the point-contact spectra with superconducting tunneling data analyzed in the Rowell-McMillan scheme. In Fig. 2 we have given an example of the point-contact spectrum for Cu. The second derivative d^2V/dI^2 can be directly related to the spectral function for the electron-phonon interaction, the well known Eliashberg function α^2F. Because of the before discussed convolution of the point-contact spectral function with the efficiency function $\eta(k,k')$, the point-contact spectrum will be slightly modified with respect to α^2F. Comparing the point-contact data in Fig. 2 with the phonon density of states F obtained by inelastic neutron-scattering experiments, it is found that the electron-phonon interaction at the transverse phonon frequencies of Cu is strongly enhanced compared to the longitudinal ones. The reason for this lays in the importance of umklapp scattering processes in the noble metals. Besides, the backflow efficiency of umklapp scattering enhances this effect in the point-contact spectra.[10]

Nowadays, the phonon spectra have been studied with point-contact spectroscopy for many elements. Even to metallic compounds

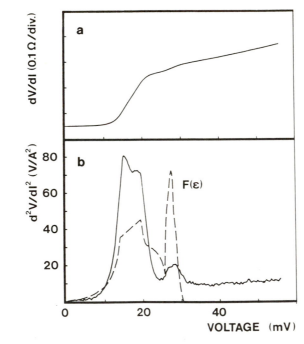

FIG. 2. Differential resistance dV/dI and second derivative d^2V/dI^2 for a Cu-Cu point contact with resistance 3.3 Ω at temperature 1.5 K. For comparison the phonon density of states $F(\omega)$ obtained by inelastic neutron scattering.

the point-contact method can be applied. Although the purity of these samples can be such that $1_i \leq a$ for the studied contacts, the contact can still be in a ballistic limit with respect to energy relaxation. For $\Lambda(\varepsilon) > a$, the applied voltage is still a well defined measure for the non-equilibrium distribution of the electrons in energy space (see Fig. 1b). However, in such a situation the measured spectra have a reduced intensity.[11]

Interaction mechanisms, other than the electron-phonon interaction, have been studied with point-contact spectroscopy. In experiments with rare-earth ferromagnetic metals, structure was found in the point-contact spectra at magnon energies, thus probing the electron-magnon interaction.[12] For noble metals with paramagnetic impurities dilutely dissolved (e.g. CuFe, AuMn), the point-contact resistance shows a maximum around zero bias voltage.[13] This phenomenon is a manifestation for the logarithmic divergence in the energy-dependent scattering rate (see Eq. (2)) due to the exchange interaction between conduction electrons and local moments in a Kondo system. In the following we will discuss the interaction with crystal-field levels in point contacts with $PrNi_5$.

In the hexagonal symmetry of the crystal field in $PrNi_5$, the ninefold (total angular momentum J = 4) degenerate multiplet of Pr^{3+} is split into three singlet and three doublet states. This crystal-field spitting has been measured by inelastic neutron scattering and point-contact spectroscopy. In Fig. 3 we have plotted the point-contact spectra for a $PrNi_5$ single crystal. The structure in the spectra is due to the spontaneous emission of phonons and to the excitation of the Pr^{3+} ion from the ground state by the ballistically injected electrons. The peak around 4 meV corresponds to the allowed transition with the lowest energy in the crystal-field-level scheme of Pr^{3+}. At higher energies both phonons and crystal-field transitions are observed.

In an applied magnetic field, point-contact spectroscopy has the unique possibility to measure the Zeeman splitting of the crystal-field levels in a direct way.[14] In Fig. 3 the spectra have been plotted in strong magnetic fields, applied parallel to the a-axis of the studied single crystal. Both the splitting of the 4-meV transition and new allowed transitions are observed in a magnetic field. The inset of Fig. 3 summarizes the observed transition energies as a function of the magnetic field for various point contacts. The experimental results for the other crystal orientations (magnetic field parallel to b and c axes) clearly show the anisotropy of the hexagonal symmetry in the problem. For the c axis

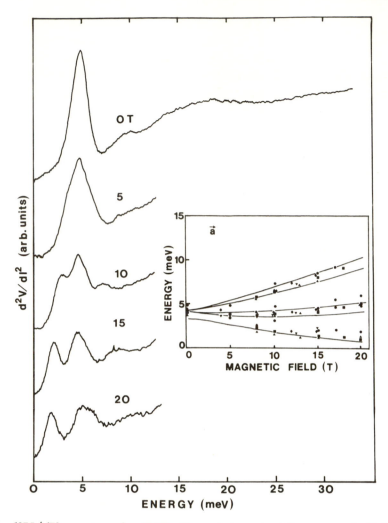

FIG. 3. d^2V/dI^2 spectra of a $PrNi_5$-Cu point contact for magnetic fields along the a axis of the orthohexagonal cell. The inset shows the data points for the crystal-field levels (different symbols for different contacts), together with the full curves as calculated for the Zeeman splitting in a magnetic field.

only the splitting of a doublet was observed and for the b axis the spreading in the field dependence of the levels was smaller. The crystal-field spitting can be calculated using the Hamiltonian $H = H_{CEF} + H_Z$ for Pr ions in $PrNi_5$, with H_{CEF} for the crystal-field Hamiltonian of the hexagonal symmetry of the lattice and H_Z for the Zeeman term in a molecular field approximation. The full curves in the inset of Fig. 3 show the solution for the eigenvalue problem of the Hamiltonian. This analysis yields in particular new information about the crystal-field levels to which the transition from the ground state is forbidden in zero magnetic field.[14]

4. HIGH-FREQUENCY RECTIFICATION

Usually the differential current-voltage characteristics of a point contact are measured by applying a low-frequency current to the contact and using phase-sensitive detection methods. In a high-frequency electromagnetic field the point-contact geometry (whisker against flat surface) serves as an antenna to induce high-frequency currents in the contact. Because we have a non-linear current-voltage characteristic, the electromagnetic radiation can be detected via the rectifying properties of the point-contact device. Fig. 4 gives an example of such an experiment. Radiation of a FIR laser, chopped at a low frequency, is applied to the contact. The signal over the contact, at this low frequency, is proportional to the second derivative d^2V/dI^2, as measured with a low-frequency current-modulation technique. The characteristic relaxation time τ for electrons at phonon energies is of the order of 10^{-14} - 10^{-13} s. At the used radiation frequency ω, the condition $\omega\tau < 1$ is fulfilled for the laser-detected signal to follow the non-linear DC current-voltage

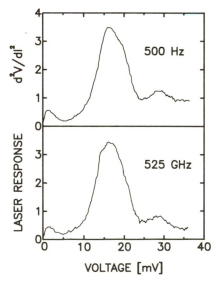

FIG. 4. d^2V/dI^2 spectrum for a Cu-Cu point contact measured with modulation techniques at 500 Hz, compared to the video-detected signal using an optically pumped FIR laser at 525 GHz.

characteristic. Hence, a good agreement is found between the two spectra measured at two totally different frequencies (see Fig. 4).[15]

In the measured point-contact spectra an additional signal is observed on which the electron-phonon interaction function is superimposed. This "background" signal extends above the Debye energy, where the phonon spectrum should go to zero. The background is explained in terms of additional scattering of the electrons with non-equilibrium phonons in the contact area. Compared to the low-frequency experiment, the observed background signal is much smaller in the video-response signal at FIR frequencies (see Fig. 4). The relaxation of non-equilibrium phonons does not follow the applied high-frequency current, thereby reducing the background intensity.[16]

5. LIMITATIONS OF POINT-CONTACT SPECTROSCOPY

The limitation, in the application of point contacts as a spectroscopic tool, is given by the very important condition for ballistic transport: inelastic scattering length $\ell(\varepsilon) > a$, or inelastic diffusion length $\Lambda(\varepsilon) > a$. In the dirty limit of a contact ($\ell(\varepsilon)$, $\Lambda(\varepsilon) < a$), the applied voltage defines no longer the excess energy of the electrons in a direct way. Because of the strong inelastic scattering inside the dirty contact, local contact heating of the contact area describes the non-Ohmic behavior of the point contact. In the dirty limit, the resistance of a circular contact is given by the Maxwell expression $R_M = \rho/2a$, now depending on the purity (i.e. resistivity) of the bulk sample. The voltage dependence of the contact resistance $R_M(V)$ resembles the temperature dependence of the resistivity $\rho(T)$. This phenomenon has been observed in point-contact experiments with ferromagnetic Co, Fe and Ni by tuning the temperature at the center of the contact through the Curie temperature by means of an applied voltage.[17]

Many experiments have been performed on mixed-valency and heavy-fermion systems, showing strong non-linearities in the point-contact characteristics.[18,19] Analogously to tunneling experiments, these experimental results have been explained by the structure in the density of states around the Fermi level in these materials. However, the proportionality between differential conductance $dI/dV(V)$ and density of states $N(eV)$ does not follow in a simple way from point-contact theory and a more elaborate analysis is needed.[20]

A different explanation of the experimental data in these compounds has been given in terms of local contact heating.[20,21] As reflected in the strong temperature dependence of the electrical resistivity, the electrons experience strong inelastic scattering (electron-electron interactions). At the applied voltages, the inelastic diffusion length $\Lambda(eV)$ can be smaller than the contact dimension. In Fig. 5 we give an example of the voltage-dependent contact resistance for the heavy-fermion compound CeB_6 in magnetic fields up to 20 T. In the applied magnetic field, a qualitative resemblance is seen between the contact resistance R(V) and the resistivity $\rho(T)$. However, a good quantitative agreement is not always possible in the heating model. For instance, the relative change in $\rho(T)$ is much larger than that in R(V). For a contact in the thermal limit, any inhomogeneity in the constriction will complicate an exact description of point-contact data with the model of local heating. A better agreement is already found by comparing the voltage-dependence of the contact resistance at low temperatures with the temperature-dependence of the contact resistance at zero voltage.[20] The asymmetry, with respect to the applied voltage, in the R(V) curves of Fig. 5 can be explained in terms of thermoelectric voltages over the heated contact between different materials (Cu against CeB_6 for the presented data).

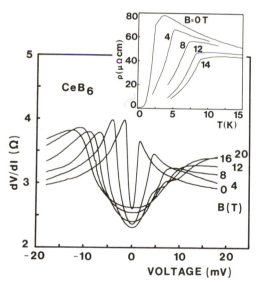

FIG. 5. Differential resistance dV/dI for a Cu-CeB_6 point contact in the indicated magnetic fields. For comparison, the inset shows the temperature dependence of the bulk resistivity.

6. TRANSVERSE ELECTRON FOCUSING

In an applied magnetic field, the isotropically injected electrons follow curved orbits and focus at specific places away from the injecting contact. For electron focusing occurring at the surface of the sample, the injected current can be detected by another point contact at the focusing spot. This effect has been demonstrated by Sharvin and Fisher[22] in a setup with two point contacts, the emitter and the collector, at both sides of a thin crystal lined up parallel to the applied magnetic field (longitudinal electron focusing). Tsoi[3] improved the technique by placing the two contacts at one side of the crystal with the field parallel to the crystal surface and perpendicular to the line connecting the two contacts (transverse electron focusing). The distance between the contacts should be smaller than the mean free path of the electrons, in order to maintain ballistic transport from emitter to collector. Transverse electron focusing can be used to study the fermiology of pure metals. At interfaces (sample boundary, superconductor) encountered by the ballistic electrons, the reflection properties of the electrons can be studied.

In Fig. 6 we have plotted the detected voltage over the collector for applied magnetic fields with an injected current through the emitter on a Ag surface. The observed peaks correspond to the focusing of electron trajectories along belly orbits (α_0) and four-cornered-rosette orbits (β_0). The rosette orbits enclose energy surfaces of higher energy and the electrons encircle these orbits in opposite sense, yielding signals at opposite fields compared to the belly orbits. Both the position and the shape of the peaks yield information about, respectively, the size and the shape of the extremal cross section of the Fermi surface perpendicular to the magnetic field. The signals α_1, β_1, β_2 result after one or two specular reflections of the electrons at the crystal surface, probed by the electrons from the inside of the metal. By tuning the applied voltage over the emitter, the signal at the collector contains information about the energy-dependence of the electron-phonon interaction along specific crystal orientations.[23]

Using transverse electron focusing, the Andreev reflection of electrons, at the boundary between a normal metal and a superconductor, can be observed directly.[24] If an electron passes from a metal into a superconductor, the electron condenses into the superconducting state forming a Cooper pair with the withdrawal of an extra electron from the metal, resulting in the retro-reflection of a hole (Andreev reflection). For a single crystal of Ag backed by a superconducting film in a magnetic

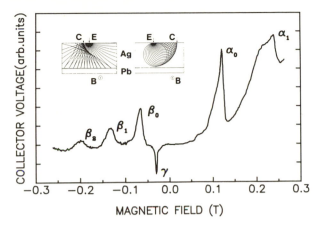

FIG. 6. Transverse-electron-focusing signal as a function of the applied magnetic field for point-contact emitter (E) and collector (C) on the (001) surface of a Ag crystal backed by a superconducting Pb film. The insets show schematically the focusing orbits of the quasi-particles in the Ag crystal (right inset) for the α_0-peak and after Andreev reflection at the Ag/Pb interface (left inset) for the γ-peak.

field, examples are given for the orbits of injected and retro-reflected quasiparticles in the inset of Fig. 6. Again focusing occurs at a specific magnetic field (peak γ). Because of the positive charge of the reflected "hole" a signal is observed opposite in sign compared to the electronic orbits. The observation of the Andreev reflection demonstrates in an elegant way the k,-k pairing in a BCS-type superconductor. For high-T_c superconductors, this type of experiment is of importance for a fundamental understanding of superconductivity in the copper-oxides.[25]

7. RECENT DEVELOPMENTS

A remarkable phenomenon is observed in the point-contact spectra of semi-metals. For example, in Fig. 7 we show the second derivative dV^2/dI^2 for a point contact with As. The measured electron-phonon-interaction spectrum corresponds to a decrease in the point-contact resistance at characteristic phonon energies. Usually in point-contact spectroscopy, the contact resistance shows a (metallic-like) increase due to phonon-mediated backflow corrections to the current (i.e. positive d^2V/dI^2 spectrum as in Fig. 2 for Cu). A similar phenomenon has been observed in point contacts with Sb.[26] In freshly prepared contacts it is possible to observe the usual positive spectrum, but re-adjusted contacts always show the negative spectra. An explanation for the phenomenon

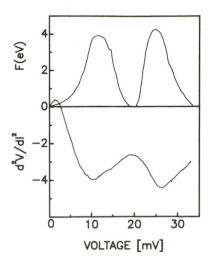

FIG. 7. Negative d^2V/dI^2 spectrum for a Cu-As point contact at 1.5 K. For comparison, the phonon density of states F(eV) for As obtained from inelastic-neutron-scattering experiments.

can be given in terms of weak localization. In a disordered system, interference effects between complementary trajectories of scattered electrons lead to enhanced backscattering.[27] The effect is seen as a precursor for localization. In a bulk system the resulting enhanced resistivity can be suppressed by increasing the temperature, destroying the phase coherence of scattered electrons by inelastic scattering events. Provided that the electron energy is conserved in the diffusional transport in the disordered contact region, also spontaneous emission of phonons destroys the phase coherence and the contact resistance will decrease at the phonon energies. The effect is especially important in semi metals because of the large de-Broglie wavelength ($\simeq 2$ nm for As) compared to the elastic mean-free-path in the contact region. Thus the possibility of point-contact spectroscopy is established in disordered materials, where quantum localization is of importance. Point-contact data in an applied magnetic field confirm the discussed interpretation for the observed negative d^2V/dI^2-spectra.[28]

In high magnetic fields, point-contact experiments show unexpected effects due to Landau quantization of the electrons.[29] In analogy to the Shubnikov-de Haas effect for the bulk resistivity, magnetic quantum oscillations are observed in the magnetoresistance of Al point contacts (see Fig. 8). The oscillations result from quantized electron orbits on pockets in the third zone of Al. In a simplified model for the Shubnikov-de Haas effect, the oscillating resistivity can be understood in terms of an oscillating scattering of the electrons via the Landau-level structure in the density of states (the electrons scatter between occupied and empty states near the Fermi level). The oscillation amplitude $\delta\rho$ for the resistivity is approximately given by the amplitude $\delta N(E_F)$ in the density of states at the Fermi level E_F

$$\delta\rho/\rho \simeq \delta N(E_F)/N(E_F) \simeq q^{-1/2}, \tag{5}$$

where $q = E_F/\hbar\omega_c$ for the cyclotron frequency $\omega_c = eB/m$. For the small pockets in the third zone of Al, the expected relative amplitudes in the resistivity are not more than $5 \ 10^{-3}$ for the investigated field range.[29] For the clean Al contacts, the scattering dependent part of the contact resistance $R_M = \rho/2a$ constitutes only a small fraction of the total resistance. Therefore, the observed oscillation amplitudes in the contact resistance are too large to be explained by a contribution according to Eq.(5).

The ballistic part of the contact resistance $R_{Sh} = 4\rho\ell/3\pi a^2$ is independent of the scattering of the electrons and, at first sight, no quantum oscillation is expected for this term. Calculating the Sharvin resistance for Landau-quantized electrons yields $\delta R_{Sh}/R_{Sh} \simeq q^{-1}$,[29] again too small to explain the observed amplitudes in point-contact experiments. For the understanding of the unexpectedly large magnetic quantum oscillations,

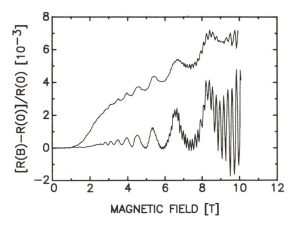

FIG. 8. Magnetoresistance of point contacts with Al, showing the magnetic quantum oscillations originating from extremal orbits on the third zone of the Fermi sphere.

we have to realize that the contact dimension is comparable in size to the magnetic length $\Lambda_B = (\hbar/eB)^{1/2}$. For a field B of a few tesla this length is of the order of 10 to 20 nm. The magnetic length determines the extent of the electronic wavefunctions perpendicular to the magnetic field. In such a situation diffraction of the electron at the contact has to be taken into account. An incoming electron can be diffracted into other states and the current is determined by the number of empty states an electron can be diffracted to. Instead of an ensemble of randomly distributed scatter centers in the case of a bulk metal, we have one center (the small orifice) in the case of a point contact. The amplitude for the Shubnikov-de Haas oscillations in the point-contact resistance is then again given by Eq. (5), but now for the variation of the total contact resistance $\delta R_{Sh}/R_{Sh}$. Using the model with diffraction of the electronic wavefunctions, the amplitudes of the observed quantum oscillations in the point-contact magnetoresistance can be understood.[29] In an alternative approach,[30] the local scattering of electrons, just inside the contact area, has been considered. After a scattering event at the constriction, the electron continues as a Landau-quantized electron in the pure banks of the point-contact electrodes. Such a model would also explain the enhanced Shubnikov-de Haas amplitudes in point-contact experiments compared to bulk experiments. Further experiments, with a better control of the sample purity at the constriction, are necessary to clarify these points.

REFERENCES

1. Yu. V. Sharvin, Zh. Eksp. Teor. Fiz. **48**, 984 (1965) [Sov. Phys. JETP 21, 655 (1965)].

2. I. K. Yanson, Zh. Eksp. Teor. Fiz. **66**, 1035 (1974) [Sov. Phys. JETP 39, 506 (1974)].

3. V. S. Tsoi, Pis'ma Zh. Eksp. Teor. Fiz. **19**, 114 (1974) [JETP Lett. 19, 70 (1974)].

4. P. A. M. Benistant, H. van Kempen and P. Wyder, Phys. Rev. Lett. **51**, 817 (1983).

5. A. G. M. Jansen, A. P. van Gelder and P. Wyder, J. Phys. C: Solid State 13, 6073 (1980).

6. I. K. Yanson, Fiz. Nizk. Temp. **9**, 676 (1983) [Sov. J. Low Temp. Phys. 9, 343 (1983)]; I. K. Yanson and O. I. Shklyarevskii, Fiz.

Nizk. Temp. 12, 899 (1986) [Sov. J. Low Temp. Phys. 12, 509 (1986)].

7. I. O. Kulik, A. N. Omel'yanchuk and R. I. Shekhter, Fiz. Nizk. Temp. **3**, 1543 (1977) [Sov. J. Low Temp. Phys. 3, 740 (1977)]; I. O. Kulik and I. K. Yanson, Fiz. Nizk. Temp. 4, 1267 (1978)[Sov. J. Low Temp. Phys. 4, 596 (1978)].

8. A. M. Duif, A. G. M. Jansen and P. Wyder, J. Phys.: Condens. Matter 1, 3157 (1989).

9. K. S. Ralls, R. A. Buhrmann and R. C. Tiberio, Appl. Phys. Lett. **55**, 2459 (1989).

10. M. J. G. Lee, J. Caro, D. G. de Groot and R. Griessen, Phys. Rev. **B31**, 8244 (1985).

11. A. A. Lysykh, I. K. Yanson, O. I. Shklyarevskii and Y.G. Naidyuk, Solid State Commun. **35**, 987 (1980).

12. A. I. Akimenko and I. K. Yanson, Pis'ma Zh. Eksp. Teor. Fiz. **31**, 209 (1980) [JETP Lett. 31, 191 (1980)].

13. A. G. M. Jansen, A. P. van Gelder, P. Wyder and S. Strässler, J. Phys. F: Metal Phys. **11**, L15 (1981).

14. M. Reiffers, Yu.G. Naidyuk, A. G. M. Jansen, P. Wyder, I. K. Yanson, D. Gignoux and D. Schmitt, Phys. Rev. Lett. **62**, 1560 (1989).

15. R. W. van der Heijden, A. G. M. Jansen, J. H. M. Stoelinga, H. M. Swartjes and P. Wyder, Appl. Phys. Lett. **37**, 245 (1980).

16. O. P. Balkashin, I. K. Yanson and Yu. A. Pilipenko, Fiz. Nizk. Temp. **13**, 389 (1987) [Sov. J. Low Temp. Phys. **13**, 222 (1987)].

17. B. I. Verkin, I. K. Yanson, I. O. Kulik, O. I. Shklyarevski, A. A. Lysykh and Yu. G. Naydyuk, Solid State Commun. **30**, 215 (1979).

18. B. Bussian, I. Frankowski and D. Wohlleben, Phys. Rev. Lett. **49**, 1026 (1982); E. Paulus and G. Voss, J. Magn. Magn. Mater. **47&48**, 539 (1985).

19. I. Frankowski and P. Wachter, Solid State Commun. **41**, 577 (1982); M. Moser, P. Wachter, F. Hulliger and J. R. Etourneau, Solid State Commun. **54**, 241 (1985).

20. A. A. Lysykh, A. M. Duif, A. G. M. Jansen and P. Wyder, Phys. Rev. **B38**, 1067 (1988).

21. Yu. G. Naidyuk, N. N. Gribov, A. A. Lysykh, I. K. Yanson, N. B. Brandt and V. V. Moshchalkov, Pis'ma Zh. Eksp. Teor. Fiz. **41**, 325 (1985) [JETP Lett. **41**, 399 1985)].

22. Yu. V. Sharvin and L. M. Fisher, Pis'ma Zh. Eksp. Teor. Fiz. 1, 54 (1965) [JETP Lett. **1**, 152 (1965)].

23. P. C. van Son, H. van Kempen and P. Wyder, Phys. Rev. Lett. **58** 1567 (1987).

24. P. A. M. Benistant, H. van Kempen and P. Wyder, Phys. Rev. Lett. **51**, 817 (1983).

25. H. F. C. Hoevers, P. J. M. van Bentum, L. E. C. van der Leemput, H. van Kempen, A. J. G. Schellingerhout and D. van der Marel, Physica C **152**, 105 (1988).

26. I. K. Yanson, N. N. Gribov and O. I. Shklyarevskii, Pis'ma Zh. Eksp. Teor. Fiz. **42**, 159 (1985) [JETP Lett. **42**, 195 (1985)].

27. P. A. Lee and T. V. Ramakrishnan, Rev. Mod. Phys. **57**, 287 (1985).

28. N. N. Gribov, P. Samuely, J. A. Kokkedee, A. G. M. Jansen, I. K. Yanson and P. Wyder, Physica **B166&167**, 917 (1990).

29. H. M. Swartjes, A. P. van Gelder, A. G. M. Jansen and P. Wyder, Phys. Rev. **B39**, 3086 (1989).

30. E. N. Bogachek and R. I. Shekhter, Fiz. Nizk. Temp. **14**, 810 (1988) [Sov. J. Low Temp. Phys. **14**, 445 (1988)].

ATOM-RESOLVED SURFACE CHEMISTRY WITH THE STM: THE RELATION BETWEEN REACTIVITY AND ELECTRONIC STRUCTURE

Phaedon Avouris

IBM Research Division, T. J. Watson Research Center
Yorktown Heights, NY 10598, U.S.A.

Abstract - We will discuss the application of scanning tunneling microscopy (STM) and spectroscopy (STS) to probe and induce surface chemistry on the atomic scale. First, we will show that by taking advantage of the change in the local density-of-states that takes place when reaction occurs at a particular site, one can determine atom-by-atom the spatial distribution of surface reactions. By simultaneously recording STS spectra we can relate chemical reactivity to local electronic structure. We will illustrate these capabilities using three examples from the chemistry of the Si(111)-7x7 surface: (a) the initial stages of the reaction with O_2, (b) the reaction with NH_3, (both being reactions that preserve the 7x7 reconstruction), and (c) the reaction with boron which destroys the 7x7 reconstruction. Topographic and spectroscopic studies, coupled with the results of electronic structure calculations allow the complete characterization of the above chemical processes. We will then consider the role of the tip chemical composition and electronic structure on the topography and spectroscopy of surfaces. Electronic structure calculations suggest that the "size" of adsorbed atoms in STM images can be a function of the chemical nature of the tip. Moreover, tips with narrow

Highlights in Condensed Matter Physics and Future Prospects
Edited by L. Esaki, Plenum Press, New York, 1991

peaks in their density-of-states lead to distorted STS spectra and possibly to the development of diode-like I-V characteristics.

We will discuss the use of the STS not only as a probe of chemistry but also as an active participant in it. Experiments involving processes such as site-resolved field-desorption will be used to illustrate this potential. Finally, we will discuss future prospects for the application of the STM in chemistry and materials science.

1. INTRODUCTION

The electronic structure of the surfaces of solids has been studied in the last couple of decades using powerful techniques such as photoemission and inverse-photoemission spectroscopies. These techniques average over an area of the surface defined by the size of the probe (photon or electron) beam. This area usually encompases more than 10^{12} atomic sites. Surface chemistry, on the other hand, is a localized phenomenon. Reactions may take place at different sites of a crystal surface, at defect sites, or may involve adsorbed foreign atoms. Furthermore, neighboring sites may interact and influence the reactivity of each other. It is clear that we need to know the spatial distribution of a reaction with atomic resolution. Moreover, to understand surface chemistry we need to relate the reactivity of a certain site to its local electronic structure. Thus, we also need an electronic structure probe with atomic resolution.

The technique of scanning tunneling microscopy (STM) developed by Binnig and Rohrer (1) allows the imaging with atomic resolution of the spatial distribution of the density of valence electrons with energies near the Fermi energy. In its spectroscopic version, scanning tunneling spectroscopy (STS), it allows the determination of the energies of shallow valence energy levels of the sample. It is precisely these loosely-bound electrons that are responsible for the chemistry of materials.

Here we will show that by taking advantage of the change in the local density-of-states that results when a chemical reaction takes place at a particular site, STM and STS can be used to determine the spatial distribution of surface reactions with atomic resolution. The reactivity of the different sites can then be related to the local electronic structure determined by STS.

As specific examples, we will use reactions of the complex

Si(111)-7x7 surface. We will first discuss the electronic structure of the clean surface as revealed by STS measurements. We will show that there are charge-transfer interactions between dangling-bond sites and that the direction of the charge-transfer changes as the reaction proceeds, leading to a time-dependent electronic structure and reactivity for unreacted sites. It will be shown that there is strong atomic site chemical selectivity which is different for different reactants. Even features such as sub-surface stacking faults can have a profound effect on the reactivity of the surface layer. We will outline how STM and STS results have been used to resolve long-standing problems such as the mechanism of the initial stages of Si(111) oxidation. We will show that STM can be valuable in the study of even more complex processes. As an example, we will consider briefly the high temperature doping reaction of boron with Si(111)-7x7, a reaction that destroys the 7x7 surface reconstruction.

We then will consider the role that the tip chemical composition and structure play in STM and STS. Electronic structure calculations will be discussed which suggest that the "size" of adsorbate images can be a function of the chemical nature of the tip. We also find that if the STM tip has localized states with narrow density-of-states (DOS) spectra, STS measurements will not provide a good measure of the sample valence electron spectrum. In this case, tunneling I-V curves obtained over localized sample sites may show areas of negative differential resistance (NDR). This NDR behavior is localized in areas of the surface with atomic dimensions (~1nm).

We then will show that the STM is not only a probe of chemical reactivity but can be an active participant inducing chemical processes and allowing the manipulation of individual atoms. Finally, we will discuss possible future applications of STM to chemical problems.

2. THE PRINCIPLE OF SCANNING TUNNELING MICROSCOPY AND SPECTROSCOPY

The principle and operation of STM is rather simple. A sharp metal tip, usually made of W or Pt-Ir alloy, is brought very close (e.g., 5-10 Å) to the sample. At this distance the wave functions of the sample and the tip, which are decaying exponentially in the vacuum barrier region, overlap. If a bias voltage is applied to the sample, an electron tunneling current flows between the tip and sample. The direction of electron flow depends on the sign of the bias applied to the sample; thus, the occupied or unoccupied states of the sample can be studied. The tip can be moved in three dimensions using, for example, three orthogonal piezoelectric translators labeled x, y, and z in Fig. 1. The z translator varies the tip-surface distance while the x and y translators scan the tip over the surface. The tunneling current depends exponentially on the tip-surface distance, typically varying by almost an order of magnitude for a change of only 1 Å in this distance. However, one usually performs the scan keeping the tunneling current constant. This is accomplished by the use of a feedback loop (see Fig. 1). By applying a voltage on the z piezo the tip-sample distance is changed so that the current remains constant. The STM topographs are maps of the voltage that has to be applied to the z piezo to keep the current constant. This mode of STM operation is preferred because it has been shown by Tersoff and Hamann (2) that if the tip wavefunction is an s-wave, the resulting STM topographs have a simple physical interpretation: they represent contours of constant local density-of-states (LDOS). The conditions under which this conclusion is valid will be discussed in more detail in section VI.

Thus, in general, the STM does not show the location of surface atoms, but rather the spatial distribution of valence electrons with ener-

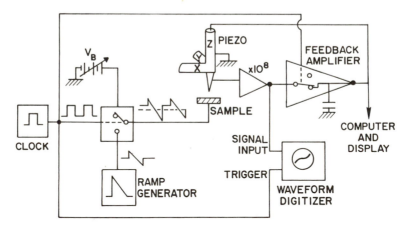

FIG. 1. Schematic diagram of the experimental set up. Constant-current topographs are obtained with a dc bias applied to the sample and an active feedback-loop. To obtain spectra (I-V curves) along with a topograph, a clock and a ramp generator are used to produce a waveform consisting of a dc part and a ramp. During the dc part, topographic information is obtained, while during the ramp the feedback loop is inative so that the tunneling current can be measured as a function of applied voltage.

gies not too far from the Fermi energy (E_F). For metals this distribution closely corresponds to the location of surface atoms, while, for example, in low-dimensional materials exemplifying charge-density waves there is little relation between the STM image and the atomic positions. For semiconductors this valence electron distribution is dominated by surface states. So, for example, images of the Si(100)-2x1 surface show the charge distributions of the Si-Si dimer π-bonds (negative sample bias) and π^*-antibonds (positive sample bias) (3). In GaAs (110) for negative sample bias the As atoms are imaged, while for positive bias the Ga atoms are imaged (4). Finally, in the case of the Si(111)-7x7 surface, on which we will concentrate here, the surface states are p_z-like orbitals of the surface Si atoms, so that STM images do show the location of surface atoms in this case.

One of the major new capabilities that STM brings to surface science is the ability to determine the energy of valence states with atomic resolution. Several different ways for doing such an experiment have been proposed (5). In our experiments to obtain spectroscopic information, instead of applying a DC bias to the sample, a waveform composed of a DC part and a voltage-ramp generated by a clock and a ramp-generator (Fig. 1) is applied to the sample (6). During the DC part, the feedback-loop is active, the tunneling current is kept constant and thus a topograph is obtained. When the ramp generator is triggered, the feedback-loop is inactivated, the tip-surface distance is fixed and the tunneling current is measured as a function of sample bias. The tunneling current, however, depends not only on the density-of-states of the sample (DOS_S) but also on the tip density-of-states (DOS_T) and on the electron transmission coefficient T, as shown in eq. 1.

$$I \propto \int_{E_F}^{E_F + eV} DOS_S(\vec{r}, E)\, DOS_T(\vec{r}, E - eV)\, T(\vec{r}, E, eV)\, dE \qquad (1)$$

It is usually assumed (see, however, section VI) that the DOS_T is structureless and thus does not distort the spectra. To remove the exponential dependence of the transmission coefficient on V, Stroscio et al (7) suggested plotting $(dI/dV)/(I/V)$ vs. eV. Such a plot corresponds to roughly an energy spectrum on top of a usually smoothly varying background. First-principles calculations by Lang (8) support this interpretation.

3. THE APPLICATION OF STM AND STS TO SURFACE CHEMISTRY

The chemistry of semiconductor surfaces at not too high temperatures is usually dominated by the chemistry of dangling-bonds (DB). Electrons in DB states have non-bonding or weakly bonding (see the case of Si(100)-2x1 surface (9)) character; thus it takes less energy to react at these sites since no bonds need to be broken for new ones to form. The DB sites are the chemically "active sites" of semiconductor surfaces. However, the reactivity of DBs is itself a function of the local environment. Charge-transfer between sites may affect their occupation and charge state, local strain considerations may affect their reactivity while cooperation between close-by DB sites may be needed to affect a chemical reaction. Because of the coupling between DB sites, reaction at

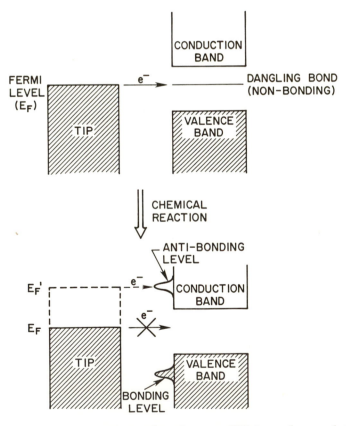

FIG. 2. Schematic diagram illustrating the way STM can be used to study the spatial distribution of surface reactions on semiconductor surfaces.

a particular site may affect (increase or decrease) the reactivity of neighboring sites.

Since DBs are directly imaged by the STM and reaction usually takes place at these sites, monitoring chemical reactions in an atom-resolved manner by the STM is, in principle, simple (6, 10). As is illustrated in Fig. 2, one adjusts the sample-tip voltage difference so as to image the semiconductor DB surface distribution. Reaction at a particular site will lead to a change in LDOS which, if we keep the sample bias the same, will appear as a change of contrast in a gray-scale topograph. In the chemical reaction the DB electrons and electrons from the reactant species form bonding and antibonding levels usually far removed from E_F, thus reacted sites are expected to appear darker than unreacted sites. In some cases the reactant itself may have energy levels in the same energy range as the DBs, while in other cases (see the oxidation of Si) the reaction may not take place at DB sites but may affect the DBs indirectly. As Fig. 2 suggests, by readjusting the sample bias we may be able to image the products (if more than one is formed) of the chemical reaction and thus determine their spatial distribution. In the case of DB sites that are not easily imaged topographically, as in the case of the second layer restatom sites of the Si(111)-7x7 surface, spectral maps can be used for the same purpose (6, 10). Finally, we note that analogous considerations can be used to study the chemistry of metal surfaces (11).

4. THE ELECTRONIC STRUCTURE AND TOPOGRAPHY OF THE Si(111)-7x7 SURFACE

The Si(111)-7x7 surface is ideal for the demonstration of the unique capabilities of the STM described above. This is because the 7x7 surface has DBs in several structurally inequivalent sites. Currently, the dimer-adatom-stacking fault (DAS) model proposed by Takayanagi et al. (12) for the reconstruction of this surface is generally accepted. The first STM studies of the Si(111)-7x7 surface by Binnig et al. (1) played an important role in the elucidation of the structure of this surface. A schematic diagram illustrating the structure is given in Fig. 3, which shows a top view (A) and a side view (B) along the long axis of the unit cell. In this picture, atoms at increasing depth from the surface are represented by circles of decreasing diameter. There are two triangular subunits surrounded by Si dimers. These subunits are rendered inequivalent by a stacking fault in the left subunit. A large reduction in the number of surface dangling bonds occurs when Si atoms ejected from the original

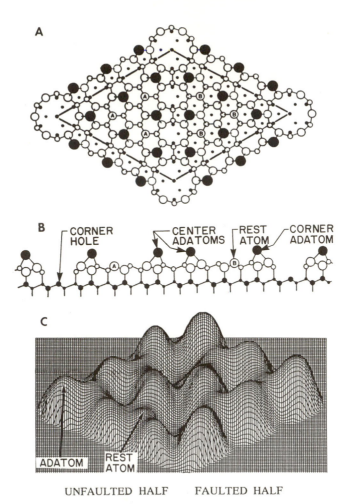

FIG. 3. (A) Top view of the dimer-adatom-stacking fault model of the Si(111)-7x7 reconstruction. The 7x7 unit cell containing the stacking fault (left side) is outlined. (B) Side view along the long diagonal of the unit cell. (C) Three dimensional topograph of the occupied states of Si(111)-7x7 unit cell (sample bias = -3V).

surface become adatoms on top of what is now the second atomic layer, each one of them eliminating three dangling bonds on that layer while introducing a new one. Finally, at the corners of the unit cell there are Si vacancies - so called corner holes. As a result of this reconstruction, from the 45 original dangling bonds only 19 survive in the 7x7 unit cell. Six of them are located on the now second layer, triply coordinated Si atoms (A and B in Fig. 3) - - the so-called rest atoms. Twelve are on the adatoms, and finally one dangling bond is located at the atom on the bottom of the corner vacancy, the so-called corner hole. For reasons that will become apparent later in the discussion, we further separate the adatoms in two groups: the six located near to a corner hole are called corner adatoms while the six not neighboring corner holes are called center adatoms.

In Fig. 3C, we show an experimental STM topograph of the occupied states of the 7x7 unit cell obtained with the sample biased at -3V. The two triangular subunits showing the presence of the stacking fault, the adatoms, and rest atoms are all clearly evident. We also note that the adatoms are more prominent on the faulted half of the unit cell.

We now examine the electronic structure of the DB states using STS. In Fig. 4(a) (top) we show a topograph of the unoccupied states of 7x7 and underneath it we show tunneling spectra obtained over DB sites. Spectrum A over second-layer restatom sites is characterized by a strong occupied state peak at about 0.8eV below E_F. The energy of this surface state implies that it is doubly occupied. The adatom spectra (B, C) show occupied and unoccupied states at ~ 0.3eV below and above E_F, respectively. These DB sites are less than half-filled, with the occupation of the corner-adatom DBs (spectrum B) being higher than those at center sites (spectrum C). As has been discussed elsewhere, these occupation patterns reflect an adatom-to-restatom charge-transfer process on the clean surface (6, 10, 13). The different charge state of restatoms and adatoms seen by STS is found to be reflected in some of their chemistry (14).

5. EXAMPLES OF SURFACE CHEMISTRY

5.1. The Reaction of Si(111)-7x7 With NH$_3$

As a first example which illustrates several of the STM capabilities discussed above, we use the reaction of Si(111)-7x7 with ammonia (NH$_3$) gas (6). NH$_3$ dissociates on the 7x7 surface and the resulting NH$_2$ and H species tie up the surface DBs (6, 15, 16). In Fig. 4(b) we show a topograph of the 7x7 surface after exposure to a couple of langmuirs (1L=10^{-6} torr.s) of NH$_3$. Reacted sites appear dark because of the reduction in LDOS that takes place when the DBs are saturated by the NH$_3$ fragments. Thus, such a topograph directly reflects the spatial distribution of the surface reaction. Important chemical information can be obtained by visual inspection. For example, by simple site counting we find that center-adatom sites are ~4 times more reactive than corner-sites (6). Spectra of dark, reacted sites, such as reacted restatoms (spectrum A) or adatoms (spectrum B, dashed line) show the absence of surface states. Moreover, systematic studies of such spectra show that the restatom DB states are eliminated by NH_3 faster than the corresponding adatom states. Most importantly, the electronic structure of even unreacted adatom sites is changed upon reaction of their neigh-

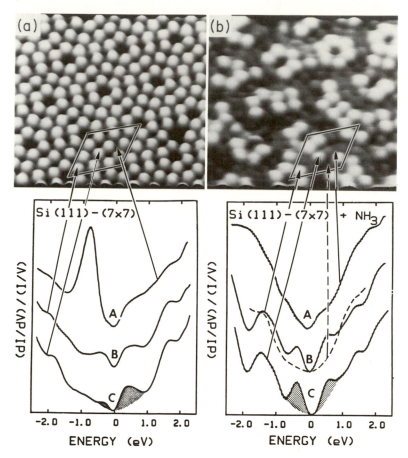

FIG. 4. (a) Topograph of the unoccupied states of the clean 7x7 surface (top) and atom-resolved tunneling spectra (below). Curves A, B and C give the spectra over restatom, corner-adatom, and center-adatom sites, respectively. (b) Topograph of the unoccupied states (top) and atom- resolved tunneling spectra (below) of an NH_3-exposed surface. Curve A gives the spectrum over a reacted restatom site, curve B (dashed) gives the spectrum over a reacted corner adatom, while curves B (solid line) and C give the spectra over unreacted corner and center adatoms, respectively.

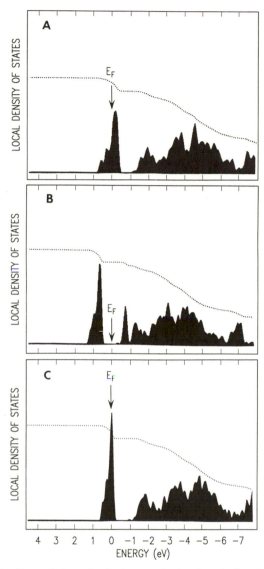

FIG. 5. Tight-binding slab calculations of the local density-of-states at a Si adatom site. (A) The unit cell contains only adatoms (2x2 layer), (B) the unit cell contains one adatom and one restatom, and (C) the restatom dangling-bond has been saturated by a hydrogen atom.

boring restatoms. Spectra B and C in Fig. 4(b) show that upon restatom reaction a reverse charge-transfer from restatoms to adatoms takes place which increases the adatom DB occupation (6). This observation is supported by the theoretical results shown in Fig. 5 (17). It is seen that in the absence of restatoms the adatom DB-state is half occupied (Fig. 5A). When the unit-cell contains one adatom and one restatom the adatom DB state is totally empty (Fig. 5B), (in the 7x7 unit-cell there are twice as many adatoms as there are restatoms, thus the adatom DB is not totally empty). Finally, when the restatom DB is saturated by hydrogen, a reverse charge-transfer takes place, repopulating the adatom DB state. Thus, it is found that the electronic structure and reactivity of individual unreacted sites varies during the course of a reaction. This coupling between sites has important implications for the spatial distribution of surface reactions. For example, it determines to a large extent whether reacted sites will cluster or disperse over the surface.

5.2. The Initial Stages of Si(111)-7x7 Oxidation

As a second example of the application of STM and STS in surface chemistry we will discuss the initial stages of Si(111)-7x7 oxidation (18, 19). This is an important problem which has attracted the interest of many workers. However, not only does the mechanism of the reaction remain unclear, but there is no agreement on even what structures are formed. Several different configurations have been proposed for the oxygen-containing sites. They involve oxygen atoms saturating the DBs of top layer Si atoms (20-25), oxygen atoms inserted in back-bonds but leaving the DBs intact (26-32), or molecular forms of oxygen bonded to surface atoms (33-35).

In Fig. 6 we show a topograph of the unoccupied states of a Si(111)7x7 surface after exposure to $\sim 0.2L$ of O_2 (19). Two oxygen-induced adatom sites are evident: (a) sites where a decrease in LDOS has taken place, as in the case of the NH_3 reaction, thus causing them to appear dark, and (b) sites where an apparent increase in LDOS has taken place so that these sites appear brighter than unreacted adatom sites. Previous STM studies (22, 23) of Si oxidation have observed only the dark sites. In only one other independent study (30) were bright sites observed. Both sites have strong surface site preferences. For example, there are about 4 times as many bright sites at corner-adatom sites than there are at center sites (a behavior opposite to that which we observed with NH_3). Moreover, there are 8 times more bright sites in the faulted half of the 7x7 unit-cell than there are in the unfaulted half. The

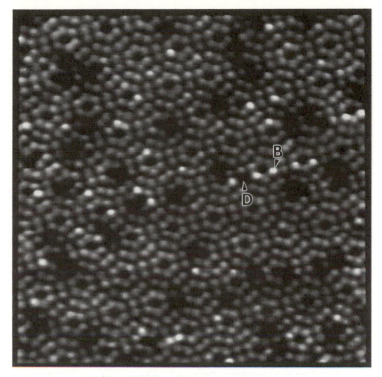

FIG. 6. STM topograph of the unoccupied states of a Si(111)-7x7 surface exposed to 0.2L of O_2 at 300K (sample bias voltage +2V). Oxygen-generated dark and bright adatom sites are indicated by D and B, respectively.

stacking fault has also been observed to affect the initial stages of metal deposition on Si(111)-7x7 (36, 37). It is truly remarkable that a rotation of atoms in the 5th atomic layer can have such an effect on the reactivity of the top layer atoms. Dark sites show the same selectivity trends but the selectivity ratios are much smaller, a fact that suggests that bright and dark sites are not two different stages in the oxidation of a particular Si atom, but are produced by two largely independent channels. This is further confirmed by the behavior of these two sites as a function of coverage. As the exposure is increased, dark sites become dominant. In Fig. 7 we show schematically the STM determined spatial distribution of the bright sites (black dots) and dark areas (unshaded islands) of a surface exposed to 0.2L of O_2, and of the dark areas (shaded islands) produced by exposure to 0.6L of O_2. As can be seen from this picture, there is little correlation between the spatial distributions of the bright sites and

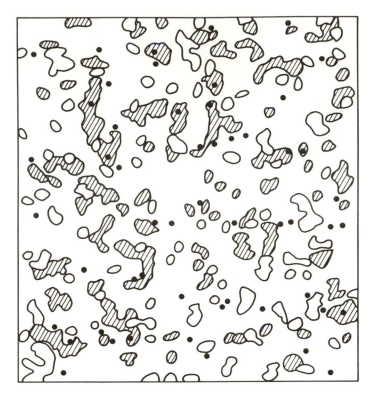

FIG. 7. Schematic representation of the distribution of the early oxidation reaction products on a Si(111)-7x7 surface. The dots indicate the positions of bright adatom sites induced by 0.2L O_2 exposure while islands indicate the dark areas produced by the same O_2 exposure. Shaded islands indicate the distribution of new dark areas produced by a 0.6L O_2 exposure.

the new dark sites. However, old dark areas do appear to act quite often as nucleation areas for the formation of new dark sites. This is a kind of autocatalysis, indicating that dark sites perturb the local electronic structure and facilitate the formation of more dark sites in their vicinity.

From the above STM evidence we can conclude that not only one but at least two different products are formed early on in the oxidation reaction. Moreover, STS can help identify their nature. In Fig. 8 we show spectra obtained over dark sites (A), bright sites (B) and perturbed adatom "gray" sites (C). Gray sites usually appear in the perimeter of large dark areas. Dark sites do not show the presence of surface states, a fact that suggests that the adatom DB has been saturated by oxygen. The bright site spectrum is characterized by strong occupied and unoccu-

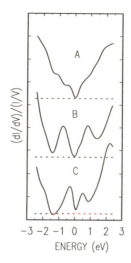

FIG. 8. Tunneling spectra obtained over oxygen-induced dark (spectrum A), bright (spectrum B) and perturbed (gray, spectrum C) adatom sites of the Si(111)-7x7 surface.

pied state features at ~0.8eV below and above E_F, respectively. Theoretical studies of the LDOS of possible structures (17) and photoemission spectra of molecularly adsorbed O_2 (19) show that in order to obtain a spectrum such as that shown in Fig. 8B the adatom DB must be intact but perturbed by nearby oxygen. The most likely stable structure to possess low-lying occupied and unoccupied states is shown in Fig. 9A. It involves an oxygen atom inserted in one of the back-bonds of an adatom. The unoccupied peak of spectrum 8A is correlated with LDOS peak **a** which has DB character. The occupied peak is correlated with LDOS peak **b** which has back-bond character. The bright appearance of such sites in topographs of unoccupied states is due in part to the fact that the adatom in structure 9A lies higher than normal adatoms and also because its DB is emptied by charge-transfer to the oxygen atom. Chemisorbed O_2 could also appear as an unstable bright site which upon dissociation turns dark. In this respect we note that we do observe bright sites turn dark during scanning without any further exposure to O_2. Turning to the dark sites we note that, in principle, either of the two structures in Fig. 9B and C could account for them since these structures do not possess low-lying occupied or unoccupied states (see STS spectrum 8A). Valence photoemission spectroscopy helps resolve this issue (19). Spectra obtained at 100K show the quenching of the adatom and

527

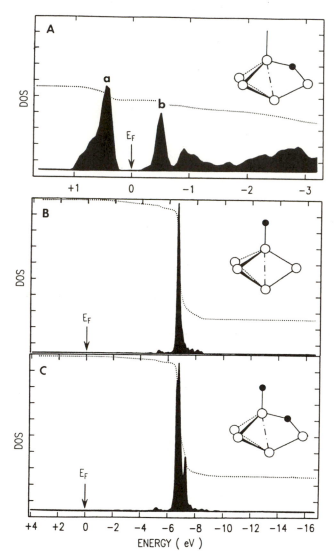

FIG. 9. Tight-binding calculations of the local density-of-states at the adatom site for various atomic oxygen adsorption configurations. The configurations are shown as insets, where a circle indicates a Si atom and a full circle an O atom. Positive energies indicate unoccupied states while negative energies indicate occupied states.

restatom surface states and the appearance of a new photoemission peak at ~4eV below E_F. This last feature is characteristic of molecularly chemisorbed O_2, a precursor to atomic chemisorption (31). Comparison of photoemission spectra with LDOS calculations (17) suggest that this species involves O_2 bonded to a single Si adatom. When the surface exposed to O_2 at 100K is annealed to higher temperatures the 4eV peak is eliminated, while peaks characteristic of atomic oxygen at ~7 and ~11eV grow in intensity. At the same time no significant change in the intensity of photoemission from DB states is observed (19). From the above observations one can conclude that the molecular precursor decomposes to insert one oxygen in one of the back-bonds while leaving the other oxygen atom tying up the adatom DB; i.e. it gives structure 9C. Both structures 9A and C have destabilized Si-Si back-bonds and calculations show (17) that another O_2 can directly insert one O atom into the remaining two back-bonds to give SiO_3 and SiO_4 dark sites, respectively.

Finally, we can address the issue of site selectivity. We know that the oxidation reaction proceeds via a molecular precusor (31). The large increase in the workfunction that accompanies the formation of this precursor (38) shows it to be negatively charged -- nominally O_2^-. Calculations (17) also show that O_2 adsorbed at various DB sites will be rather strongly bonded locally and will not diffuse on the surface. Thus, the distribution measured by the STM will be the one produced by the initial O_2 sticking process. Therefore, the observed site selectivity of the oxidation sites reflects a site selective O_2 sticking process where the faulted-half and corner-adatom sites of the 7x7 unit cell are preferred. What is special about these two sites? STM and STS provide the answer. STS spectra such as those in Fig. 8A show that the occupation of corner-adatom sites is higher than that of center sites. Moreover, the faulted half of the unit cell appears brighter in topographs of occupied states (see Fig. 3C). This could imply a higher occupation for the faulted half of the cell; that this is true is confirmed by electronic structure calculations (39, 40). We propose that the sticking of the O_2 to form the negatively charged precursor involves charge-transfer from the lowest binding energy adatom surface state and that this process favors the faulted-half and corner-adatom sites because of their higher electron density. As an O_2 molecule approaches the surface its electron affinity increases because of the image-like interaction of the resulting negative ion with the substrate. When the energy of the $2\pi^*$ affinity level crosses the Fermi level, charge-transfer takes place, the O_2^- is formed and is accelerated towards the surface, loses energy by exciting phonons and becomes trapped to form a precursor. This sticking mechanism involving charge-transfer is usually referred to as harpooning. The different precursors decompose to give the stable oxidation products. As we will show in the next section, the above model is further supported by experiments in which the occupation of the Si adatom DBs is altered through surface doping.

In the above examples we saw that STM and STS measurements with the help of photoemission and theory provide crucial new insights necessary to resolve complex problems such as the mechanism of the initial stages of silicon oxidation. The application of STM and STS to semiconductor surface chemistry is not, however, limited to the chemistry of DBs. Etching reactions such as the reaction of chlorine with Si(111)-7x7 have been studied (41). The removal of individual adatoms by chlorine can be viewed directly and adatoms bonded to different numbers of chlorine atoms can be imaged at higher bias. Even more complex processes which destroy the original surface reconstruction have been studied successfully. We discuss an example below.

5.3. The Surface Doping of Si(111)-7x7 with Boron

An interesting example of a process that destroys the original surface reconstruction is provided by the high temperature surface doping of Si(111)-7x7 with boron (42, 43). The boron is introduced in the form of decaborane molecules ($B_{10}H_{14}$). These molecules are adsorbed on the Si surface at room temperature and then the sample is heated to temperatures above 500°C to decompose the decaborane, desorb the hydrogen and dope the surface. In this case the 7x7 reconstruction is destroyed but the STM/STS results, in conjunction with theoretical calculations and results from conventional surface science techniques, allow the characterization of this complex reaction (42, 43). Only a few aspects will be considered. In Fig. 10 we show a region of a Si(111)-7x7 which has been exposed to 0.2L of decaborane then heated to 600°C. As an inset we show images of adsorbed decaborane molecules. The annealed surface shows large, intact 7x7 areas, along with areas where local decomposition of decaborane has resulted in the doping of the surface. In the upper part of the surface there was a line defect which acted as a trap for decaborane molecules. In this area we can see the creation of a new ($\sqrt{3} \times \sqrt{3}$) R30° phase along with globular clusters of displaced atoms. By adding more decaborane and annealing up to 1000°C we finally obtain a pure $\sqrt{3} \times \sqrt{3}$ phase which, however, contains a fair amount of substitutional and other types

FIG. 10. STM topograph of a Si(111)-7x7 surface exposed to 0.2 langmuir of decaborane and briefly annealed to 600°C. Inset: Image of decaborane molecules on the Si(111)-7x7 surface at 300K.

of defects (Fig. 11). Low energy ion-scattering studies of this surface (44) show that the top layer is composed of Si adatoms but photoemission and other techniques (44, 45) show that the boron is very close to the surface. STS shows that the surface is not metallic, as in the case of Si adatoms on the 7x7 surface, but is semiconducting. Most interestingly, a structure with the same characteristics is obtained by segregating boron from heavily bulk-doped Si samples (46, 47). All the measurements point to a structure where the boron is not at the surface but substitutes for a Si atom in the 3rd atomic layer under a Si adatom (see inset in Fig. 11). This is a unique adsorption site never encountered before. As in the case of the "bright" adsorption sites formed in the case of oxidation discussed earlier, the boron atom empties the Si adatom DB. In our discussions above, we stressed the relation between electronic

FIG. 11. Topograph of the B/Si(111)-($\sqrt{3}$ x$\sqrt{3}$) surface.

structure and reactivity, in particular, the role of DB occupation. There-fore, we should expect the chemistry of the Si adatoms on the B/Si(111)-$\sqrt{3}$ × $\sqrt{3}$ surface to be different from that of adatoms on the Si(111)-7x7 surface. Indeed, this is the case. For example, no O_2^- molecular precursor is observed in the low temperature photoemission spectra of O_2^--exposed $\sqrt{3}$ × $\sqrt{3}$ surface because the charge-transfer (harpooning-like) process cannot take place and correspondingly the oxidation process is significantly supressed (48). Also, NH_3 does not dissociate on the B-doped surface but only adsorbs molecularly at low temperature (43).

6. THE ROLE OF THE STRUCTURE AND COMPOSITION OF THE TIP

Up to this point we have assumed that the chemical nature and structure of the tip does not influence the topography or the tunneling spectra measured by STM and STS. These assumptions may be adequate in many cases, especially if one is only interested in qualitative aspects of surface topography. In general, however, the nature of the tip (49) is important, especially in studies such as those described above in which the tip is exposed to various chemicals. In the following we will explore these questions in more detail.

6.1. Tip Composition Effects on Topography

The possible effects of the chemical nature of the STM tip on the images of adsorbates measured by STM can be illustrated by the recent calculations of Walkup et al. (50). In these calculations, Lang's model (51) of an STM experiment was employed. In this model (see schematic diagram of Fig. 12), the sample is composed of a jellium surface (J_S) with an adsorbed atom (A_S) (silicon in the present case). The tip is modeled by another jellium surface (J_T) with an adsorbed atom (A_T) on it to represent the tip atom through which the tunneling takes place. The role of tip composition on STM topographs can be investigated by changing the chemical nature of this atom. Within the Bardeen approximation, it is sufficient to use wavefunctions that are separately calculated for the tip and the sample. Quite generally, the wavefunction for each side may be regarded as the sum of the wavefunction for bare jellium, plus the wavefunction that is due to the presence of the adatom. The tunneling current thus can be written as:

$$I_{TUN.} = I(J_T - J_S) + I(A_T - A_S) + I(A_T - J_S) + I(J_T - A_S). \quad (2)$$

The first term in equation 2 involves electrons tunneling between the two semi-infinite jellium substrates; it is clear that it has to be subtracted out

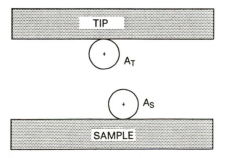

FIG. 12. Schematic diagram of the model used to represent the sample and tip in the local density functional calculations of tunneling.

of the total tunneling current (51). The 2nd and 3rd terms involve electrons tunneling out of the tip atom into the adsorbed sample atom and the surrounding sample surface, respectively. The last term represents tunneling from the jellium on the tip side to the adsorbed atom on the sample side. This last term is treated in two different ways: (a) it is subtracted from the net tunneling current, thus simulating the situation of a perfectly sharp tip where all the current flows through a single atom. We refer to this tip model as a "single atom tip". In case (b) the current flowing through the immediate area (e.g. a 10Å diameter) surrounding the tip atom is included so as to simulate a tip with an adatom adsorbed on a small flat plateau. We call this tip an "adatom tip." In Fig. 13A we

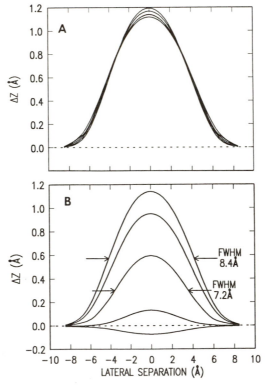

FIG. 13. (A) Calculated image of a Si atom adsorbed on a $r_s=2$ jellium using various single-atom tips: Na, Al, Si, C, Mo. (B) Calculated image of the same Si atom using (top to bottom): Na single atom tip, Na adatom tip, Si adatom tip, C adatom tip, F adatom tip.

show the predicted STM profile of a Si atom adsorbed on a $r_s=2$ jellium and probed by a single atom tip as a function of the nature of the tip atom. The tip atoms are selected to span a variety of different electronic (s, p, d) configurations. The constant-current image shown is for a tip atom to sample atom distance of 7Å (nucleus to nucleus) when the lateral separation is zero. It is clear from Fig. 13A that the Si atom image is not dependent on the chemical nature of the single atom tip. This result can be considered as a validation of the conclusions of Tersoff and Hamann (2). That is, when the tip has a well-defined center (in this case, the tip atom nucleus), the STM image is determined approximately by the local density-of-states at E_F evaluated at the center of the tip. However, when an adatom tip is used, it is seen (Fig. 13B) that the chemical nature of the tip atom is very important. The "size" of the Si adatom depends on the nature of the adatom tip. The nature of this dependence can be traced to the chemisorption-induced change in LDOS at E_F, i.e. to the effect of the tip atom on the surrounding jellium surface. Electropositive elements like Na lead to an increased LDOS and a strong STM signal. Electronegative elements, on the other hand, reduce the LDOS at E_F. In the extreme case of fluorine, the calculations suggest that the Si atom would appear as a dip rather than a bump in an STM topograph. Thus, the above model calculations suggest, under certain circumstances, a rather complex dependence of the amplitude of the STM signal on the chemical nature and atomic structure of the tip. The situation becomes more complex at closer tip-surface distances. Because of the breakdown of the Bardeen approximation (52) when the tip-surface interaction is strong, calculations are much more difficult to perform in this regime. However, one expects that higher l-waves (e.g. d-waves on the tip) may become important in the tunneling process (53). Moreover, the proximity of the tip to the surface can induce site-specific and laterally confined states (54, 55) leading to a site-dependent effective barrier. The tip images the corrugation of this barrier. It has been suggested that the formation of such localized states may, in fact, be responsible (56) for the anomalous corrugations observed for the nominally flat (111) surfaces of simple (57) and noble (58) metals. The influence of orbital symmetry on the overlap of tip and sample wavefunctions could lead to interesting contrast effects that may allow, with judicious choices of the tip material, the selective enhancement of the images of different adsorbates on a sample surface. Such experiments may confer a certain degree of chemical specificity to STM imaging.

6.2. The Role of the Tip in Tunneling Spectroscopy - Novel Electronic Effects

The composition and structure of the tip should also influence the results of scanning tunneling spectroscopy experiments. In general, the tunneling current depends on the electronic states of both sample and tip as indicated by eq. 1. When atomic resolution is achieved in STM images, tunneling should involve one or a couple of atoms at the apex of the tip (59). Under these conditions, the DOS of the bulk metal from which the tip is made should provide a poor representation of the DOS of the active area of the tip. One should expect a peaked distribution, most likely a Lorentzian DOS profile with a width that depends on the coupling of the active area to the rest of the tip. If this width is large, then a plot of (dI/dV)/(I/V) vs. sample voltage (V) would provide a reasonable representation of the DOS spectrum of the sample. On the other hand, tip DOS widths can be particularly narrow if the active area of the

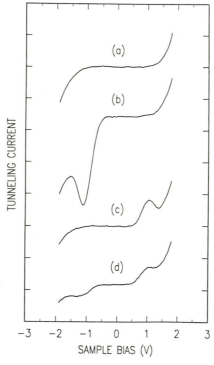

FIG. 14. Tunneling current versus sample bias (I-V) curves. (a) Normal I-V curve obtained over a majority site of the B/Si(111)-$\sqrt{3}$ x$\sqrt{3}$ surface. Curves (b) to (d) have been obtained over defect sites and illustrate the different types of negative differential resistance behavior observed in our studies.

tip is composed of an adatom or small cluster of atoms of different chemical composition from that of the rest of the tip. Foreign atoms on the tip may be present from the tip preparation process (e.g. metal oxides or carbides), from exposure of the sample to reactants, or through unintentional contact of the tip with the sample. Tips with narrow DOS can lead to distorted spectra, i.e. to new tip-related peaks and to shifts of the sample peaks. A most interesting behavior is observed when such a tip is used to study the I-V behavior of localized sample states (43,60). These localized states can, for example, be defect states such as the substitutional and other surface defects of the B/Si(111)-$\sqrt{3} \times \sqrt{3}$ surface (see Fig. 11). In Fig. 14 we show a collection of tunneling I-V curves obtained over majority (curve a) and various defect sites (curves b - d). While the perfect surface sites show normal I-V characteristics, the I-Vs of defects show regions of negative differential resistance (NDR). The corresponding (dI/dV)/(I/V) vs. V plots show regions of negative "DOS". Analogous observations were made independently by Bedrossian et al. (61). It is important to note that NDR is not simply a property of a site but also depends on the state of the tip. Contaminated tips are conducive to the appearance of NDR. In fact we have observed NDR over sites of the clean Si(111)-7x7 surface using such tips (60). The most interesting aspect of these findings is the extreme localization

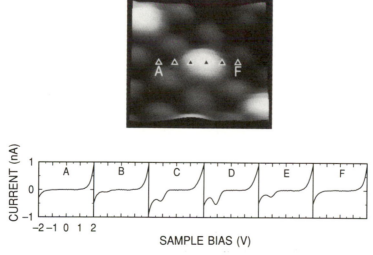

FIG. 15. Top: STM topograph showing a defect site whose I-V curve shows negative differential resistance. Bottom: I-V curves obtained at different points A to F across the above defect site. The points are spaced 3 Å apart.

of the NDR behavior. In Fig. 15 we show a defect site and I-V curves obtained 3Å apart from point A to F. It is seen that NDR behavior is localized to a region with a diameter of only ~10Å. In effect, the surface site acts as a quantum dot of atomic dimension, or viewed in another way, the surface site and the tip atom through which the current flows form a tunnel diode whose active region is just these two atoms.

Our experiments suggest that the essential ingredient for the development of atomic scale NDR is the presence on the sample and the tip of narrow DOS features at appropriate energies. This is illustrated by the computer simulations of Fig. 16A (41). On the left, the DOS of the sample (top) and of the tip (bottom) are shown, and on the right the resulting I-V curves. NDR is observed (curve C) only when tunneling involves relatively narrow peaks in the DOS of both sample and tip (curve C). NDR appears at the bias at which the DOS peaks of sample and tip have the same energy. This conclusion is in agreement with

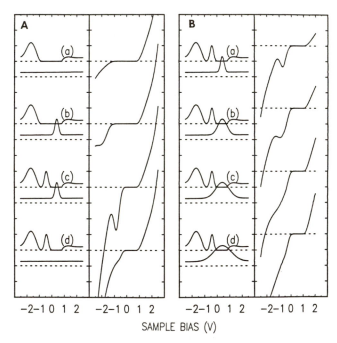

SAMPLE BIAS (V)

FIG. 16. Computer simulations of I-V characteristics. The barrier height and the tip-to-surface distance are 4eV and 8Å, respectively. (A) Left: DOS of sample (top) and tip (bottom). Right: Resulting I-V curves. (B) the role of the width of the tip DOS in the development of negative differential resistance.

observations based on first-principles electronic calculations by Lang (62). In Fig. 16B we see that NDR disappears as the width of the tip DOS is increased. I-V curves showing NDR for positive sample bias require a narrow peak in the unoccupied states DOS spectrum of the sample. DOS appears when the tip Fermi edge crosses this state. This NDR is, however, quite weak; a stronger NDR develops if the tip has a narrow DOS peak also. Finally, we should note that, in principle, NDR could arise by a Coulomb blockade mechanism (63). It is unlikely, however, that this mechanism is the source of the atomic scale NDR observed here (43).

7. THE STM TIP AS AN ACTIVE CHEMICAL PARTICIPANT

In the previous sections, we have seen that STM and STS provide powerful probes of the electronic structure and chemistry of surfaces on the atomic scale. However, STM can also be a unique tool which allows us to modify the local structure and chemistry of surfaces and manipulate individual atoms and molecules. The key element in these surface modifications is, of course, the STM tip. There are several different ways the tip can accomplish such surface modifications. It can act as a "scalpel" to dig trenches or produce indentations. In most cases the resulting features have dimensions in the order of hundred nm (64), but features with dimensions of a few nm have also been achieved (65-67). An exciting possibility will be to use the tip to "operate" on biomolecules such as DNA and proteins. Recently a most elegant use of the STM tip as a micromechanical device has been achieved (68). Instead of relying on a short-range repulsive interaction between tip and surface, the tip was made to gently approach noble-gas atoms adsorbed on a nickel surface at 4K. When the Van der Waals interaction between the tip and an individual noble gas atom exceeds the barrier for lateral motion of the atom, the tip can be moved at constant height and the atom will follow the tip motion. In this way, Eigler and co-workers (68) were able to arrange the noble gas atoms so as to form characters and write words. This new development opens up new possibilities for studying inter-adsorbate interactions. For example, different adsorbate arrangements can be produced and their stability and electronic structure can be studied by STM and STS.

Another manipulation scheme that can exhibit atomic resolution involves field desorption. Field desorption/evaporation has been studied using a field emitter tip for a long time (69). The STM's capability to image a surface and be able to place the tip over a particular atomic site of interest opens up the possibility of atom-selective field-desorption. The high electric fields (~1V/Å) needed can be generated by, for example, applying a voltage pulse when the tip is over the selected site. An even simpler scheme is illustrated by Fig. 17 (70). In this case, a Si(111)-7x7 surface was exposed to H_2O which then dissociates on it, with the H and OH fragments tying up surface dangling bonds. The dark area in the center of topograph 17(A) is a cluster of such reacted surface sites. The objective is to remove the adsorbates, regenerate the dangling bonds and thus restore local reactivity. This can be achieved by scanning

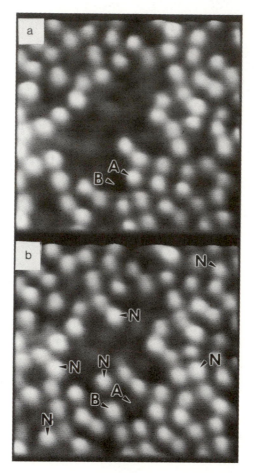

FIG. 17. (a) STM image of an area of H_2O-exposed Si(111)-7x7. Sample bias +2V. (b) STM image of the same surface area as in (a) after it has been scanned once with the sample biased at +3V and then imaged at +2V. New dangling-bond sites in (b) are denoted by N.

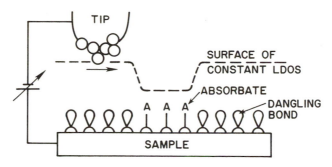

FIG. 18. Schematic diagram showing a constant current (i.e. constant LDOS) line-scan over a partially reacted surface.

the same area at a high positive sample bias voltage ($V_S \geq +3V$) while keeping the current constant. After a single high voltage scan the bias is lowered and the surface is imaged again (Fig. 17(B)). As we can see, this procedure regenerated six dangling bonds, all at their proper positions in the 7x7 unit cell, and, in addition, it stimulated the diffusion of the adsorbate on site (A) to site (B). These phenomena can be understood by refering to the Tersoff-Hamann theory (2). Since the high voltage scan was performed at constant current, the tip should follow a surface of approximately constant LDOS (see Fig. 18). Over unreacted dangling bonds, LDOS is high near E_F, while over reacted sites the LDOS is drastically reduced. Therefore the tip has to get very close to the reacted sites to maintain the same current. As a result, the field strength over reacted sites is high and field-desorption is possibly aided by tip-adsorbate chemical interactions. Schemes such as this can have nano-lithographic applications. For example, one could envision passivating a semiconductor surface, then field-desorbing adsorbates to form a pattern of dangling bonds. A new chemical can then be introduced to react with the dangling bonds and "develop" the pattern. The above results also show that there is a fine line separating the regime where the STM can be considered a non-perturbing probe of surface structure and the regime where it strongly perturbs that structure.

Single atom deposition also appears to be feasible using similar approaches. Thus, Becker et al. (71) have applied high bias (4V) to a tungsten tip contaminated with germanium and observed locally bumps on the Ge(111) substrate with a fwhm ~8Å, most likely due to desorption of individual Ge atoms from the tip.

8. THE FUTURE

In the above, we illustrated a few of the ways that STM can be used in the study of surface chemistry. These applications in no way exhaust the possible applications of STM in chemistry. In our opinion, the prospects for novel STM applications look particularly bright. We expect applications to involve: (a) new types of materials, (b) new processes and (c) new types of measurements.

Most of the published detailed STM/STS studies involve metal and semiconductor surfaces in ultra-high vacuum environments. However, there are now studies appearing on conducting organic crystals, liquid crystals, polymer surfaces, solid materials under liquids and electrochemical systems. Even thin insulating layers may, under the appropriate conditions, be studied by the STM. This can be accomplished by non-linear alternating-current STM, photo-excitation of carriers to the conduction band, direct ejection of electrons in the conduction band or non-linear optical effects.

The application of STM/STS to study surface chemistry need not be limited to thermal reactions. For example, the study of the surface-site dependence of photochemical reactions may give new insight into the factors that determine the efficiency of excited state quenching processes at surfaces.

It already is clear that STM is not just a surface science technique. In many studies the role of the substrate surface can be considered analogous to that of a "chemical bench" on which the system of interest is placed to be studied by STM. The problem of possible chemistry between the substrate (bench) and the molecular species to be studied can, in most cases, be eliminated by using a combination of inert substrates and cryogenic temperatures. Similarly, the problem of anchoring the species so that it does not diffuse during the STM measurement can be eliminated at sufficiently low temperatures.

STM and STS can be used not only to image and study adsorbed systems but also to modify and manipulate them. This can be accomplished in various ways. For example, the tip could be used as a scalpel to "operate" on macromolecules and cut off particular sections. It can also push and bring different molecular segments together and induce chemistry between them. One can envision breaking chemical bonds by using the field of tunneling or field emitted electrons to excite local chromophores to dissociative states. In other cases, reactions which are specifically induced by low-energy electrons, such as dissociative electron attachment, can be used to create locally free radicals and ions. As we discussed earlier, field-induced desorption and diffusion processes can be utilized to produce atomic scale modifications.

The STM also can be used to study dynamical aspects of surface phenomena. Fast scan STMs can allow the observation of physical processes such as diffusion and nucleation, and of chemistry following, for example, pulse dosing of the surface or excitation by a laser pulse. Monitoring changes in the tunneling current following photoexcitation with the tip stationary over predetermined sites would provide both high time- and spatial-resolution.

An important weakness of the STM is that it does not provide direct information on the chemical nature of surface species. This weakness can be largely offset by developing the capability to measure vibrationally inelastic tunneling. Vibrational spectra provide excellent "fingerprints" of chemical structures. STM vibrational spectroscopy requires low temperatures and exceptional vibrational isolation. However, these conditions can be achieved. Another scheme that can provide chemical identification with a few nm resolution involves using the STM tip in the field-emission mode to excited low-energy Auger transitions.

Finally, as the applications of STM grow and it is applied under a wider range of conditions, it will be imperative that detailed studies be performed to better understand the tip-surface interaction and the nature of the STM images. The Tersoff-Hamann interpretation of the STM topographs, on which we have relied in this paper, is valid in a relatively narrow range of operating conditions. It is necessary for experimental and theoretical studies to address the issue of the nature of STM images under strong tip-surface interactions and the perturbation of the surface electronic structure by local high electric fields. In the future it may be possible to utilize the chemical specificity of strong tip-surface interactions to achieve a measure of chemical specificity in STM imaging.

ACKNOWLEDGEMENTS

It is a pleasure to acknowledge useful discussions with I.-W. Lyo, N.D. Lang and R.E. Walkup.

REFERENCES

1. G. Binnig, H. Rohrer, Ch. Gerber and E. Weibel, Phys. Rev. Lett. **49**, 57 (1982); G. Binnig and H. Rohrer, Rev. Mod. Phys. **57**, 615 (1987).

2. J. Tersoff and D.R. Hamann, Phys. Rev. B **31**, 805 (1985).

3. R.M. Tromp, R.J. Hamers and J.E. Demuth, Phys. Rev. Lett. **55**, 1303 (1985).

4. R.M. Feenstra, J.A. Stroscio, J. Tersoff and A.P. Fein, Phys. Rev. Lett. **58**, 1192 (1987).

5. R.J. Hamers, Ann. Rev. Phys. Chem. **40**, 531 (1989).

6. Ph. Avouris and R. Wolkow, Phys. Rev. B **39**, 5081 (1989); Phys. Rev. Lett. **60**, 1049 (1988).

7. J.A. Stroscio, R.M. Feenstra and A.P. Fein, Phys. Rev. Lett. **57**, 2579 (1986).

8. N.D. Lang, Phys. Rev. B **34**, 5947 (1986); Comments Condens. Matter Phys. **14**, 253 (1989).

9. R.J. Hamers, Ph. Avouris and F. Bozso, Phys. Rev. Lett. **59**, 2071 (1987).

10. Ph. Avouris, J. Phys. Chem. **94**, 2247 (1990).

11. See for example: J.Wintterlin, H. Brune, H. Hoeffer and R.J. Behm, Appl. Phys. A **47**, 99 (1988); F.M. Chua, Y. Kuk and P.J. Silverman, Phys. Rev. Lett. **63**, 386 (1989).

12. K.Takayanagi, Y. Tanishiro, M. Takahashi, H. Motoyoshi and K. Yagi, J. Vac. Sci. Technol. A **3**, 1502 (1985).

13. J.E. Northrup, Phys. Rev. Lett. **57**, 154 (1986).

14. Ph. Avouris and F. Bozso, J. Phys. Chem. **94**, 2243 (1990).

15. L. Kubler, E.K. Hlil, D. Bolmont and G. Gewinner, Surf. Sci. **183**, 503 (1987).

16. F. Bozso and Ph. Avouris, Phys. Rev. B **38**, 3943 (1988).

17. B. Schubert, Ph. Avouris and R. Hoffmann, to be published.

18. I.-W. Lyo, Ph. Avouris, B. Schubert and R. Hoffmann, J. Phys. Chem. **94**, 4400 (1990).

19. Ph. Avouris, I.-W. Lyo and F. Bozso, J. Vac. Sci. Technol. B, April 1991.

20. R. Lukede and A. Koma, Phys. Rev. Lett. **34**, 1170 (1975).

21. G. Hollinger and F. Himpsel, Phys. Rev. B **28**, 3651 (1983).

22. A.J. Schell-Sorokin and J.E. Demuth, Surf. Sci. **157**, 273 (1985).

23. K. Uno, A. Namiki, S. Zaima, T. Nakamura, N. Ohtake, Surf. Sci. **193**, 321 (1988).

24. F.M. Liebsle, A. Samsavar, and T.-C. Chiang, Phys. Rev. B **38**, 5780 (1988).

25. H. Tokumoto, K. Miki, H. Murakami, H. Brando, M. Ono and K. Kajimura, J. Vac. Sci. Technol. A **8**, 255 (1990).

26. M. Green and K.H. Maxwell, J. Phys. Chem. Solids **13**, 145 (1960).

27. F.M. Meyer and J.J. Vrakking, Surf. Sci. **38**, 275 (1973).

28. H. Ibach, H.D. Bruchmann and H. Wagner, Appl. Phys. A **29**, 113 (1982).

29. S. Ciraci, S. Ellialtioglu, and S. Erkoc, Phys. Rev. B **26**, 5716 (1982).

30. P. Morgen, U. Hoefer, W. Wurth and E. Umbach, Phys. Rev. B **39**, 3720 (1989).

31. U. Hoefer, P. Morgen, W. Wurth and E. Umbach, Phys. Rev. B **40**, 1130 (1989).

32. J.P. Pelz and R.H. Koch, to be published.

33. M. Green and A. Lieberman, J. Phys. Chem. Solids **23**, 1407 (1962).

34. H. Ibach and J.E. Rowe, Phys. Rev. B **10**, 710 (1974).

35. W.A. Goddard III, A. Redondo and T.C. McGill, Solid State Commun. **18**, 981 (1976); A. Redondo, W.A. Goddard III, C.A. Swarts and T.C. McGill, J. Vac. Sci. Technol. **19**, 498 (1981).

36. V.K. Koehler, J.E. Demuth and R.J. Hamers, Phys. Rev. Lett. **60**, 2499 (1988).

37. St. Tosch and H. Neddermeyer, Phys. Rev. Lett. **61**, 349 (1988).

38. C. Silvestre and M. Shayegan, Phys. Rev. B **37**, 10432 (1988); C. Silvestre, J. Hladky and M. Shayegan, J. Vac. Sci. Technol. A **8**, 2743 (1990).

39. R.D. Meade and D. Vanderbilt, Phys. Rev. B **40**, 3905 (1989).

40. M. Fujita, H. Nagayoshi and A. Yoshimori, Surf. Sci., to be published.

41. J.S. Villarrubia and J. Boland, Phys. Rev. Lett. **63**, 306 (1989).

42. I.-W. Lyo, E. Kaxiras and Ph. Avouris, Phys. Rev. Lett. **63**, 1261 (1989).

43. Ph. Avouris, I.-W. Lyo, F. Bozso and E. Kaxiras, J. Vac. Sci. Technol. A **8**, 3405 (1990).

44. F. Bozso and Ph. Avouris, to be published.

45. A.B. McLean, L.J. Terminello, and F.J. Himpsel, Phys. Rev. B **41**, 7694 (1990).

46. R.K. Headrick, I.K. Robinson, E. Vlieg and L.C. Feldman, Phys. Rev. Lett. **63**, 1253 (1989).

47. P. Bedrossian, R.D. Meade, K. Mortensen, D.M. Chen and J.A. Golovchenko, Phys. Rev. Lett. **63**, 1257 (1989).

48. Ph. Avouris and F. Bozso, to be published.

49. R.M. Tromp, E.J. van Loenen, J.E. Demuth and N.D. Lang, Phys. Rev. B **37**, 9042 (1988).

50. R.E. Walkup, Ph. Avouris, L. Conner and N.D. Lang, to be published.

51. N.D. Lang, Phys. Rev. Lett, **55**, 230 (1985).

52. J. Bardeen, Phys. Rev. Lett. **6**, 57 (1961).

53. C.J. Chen, Phys. Rev. Lett. **65**, 448 (1990).

54. E. Tekman and S. Ciraci, Phys. Rev. B **40**, 10286 (1989).

55. S. Ciraci, A. Baratoff and I.P. Batra, Phys. Rev. B **41**, 2763 (1990).

56. E. Tekman and S. Ciraci, Phys. Rev. B **42**, 1860 (1990).

57. V.M. Hallmark, S. Chiang, J.F. Rabolt, J.D. Swalen and R.J. Wilson, Phys. Rev. Lett. **59**, 2879 (1987).

58. J. Winterlin, J. Wiechers, H. Brune, T. Grietsch, Hoefer, and R.J. Behm, Phys. Rev. Lett. **62**, 59 (1989).

59. Y. Kuk, P.J. Silverman and H.Q. Nguyen, J. Vac. Sci. Technol. A **6**, 524 (1988).

60. I.-W. Lyo and Ph. Avouris, Science **245**, 1369 (1989).

61. P. Bedrossian, D.M. Chen, K. Mortensen and J.A. Golovchenko, Nature, **432**, 258 (1989).

62. N.D. Lang, Phys. Rev. B **34**, 5747 (1986).

63. R.J. Hamers and R. Koch, The Physics and Chemistry of SiO_2 and $Si - SiO_2$ interface, edited by C.R. Helms and B.E. Deal (Plenum, New York, 1988).

64. For a review see: G.M. Shedd and P.E. Russell, Nanotechnology **1**, 67 (1990).

65. G. Ebert, M. Greenblatt, T. Gustafsson and S.H. Garofalini, Science **246**, 99 (1989).

66. Ph. Avouris and R. Wolkow, Mater. Res. Soc. Symp. Proc. **131**, 157 (1989).

67. E.J. van Loenen, D. Dijkkamp, A.J. Hoeven, J.M. Lenssinck and J. Dieleman, J. Vac. Sci. Technol. A **8**,574 (1990).

68. D.M. Eigler and E.K. Schweizer, Nature **344**, 524 (1990).

69. E.W. Muller and T.T. Tsong, Field Ion Microscopy, Principles and Applications, Elsevier, New York, 1969.

70. I.-W. Lyo and Ph. Avouris, J. Chem. Phys., **93**, 4479 (1990)

71. R.S. Becker, J.A. Golovchenko and B.S. Swartzentruber, Nature **325**, 419 (1987).

STM IN BIOLOGY

B. Michel

IBM Research Division
Zurich Research Laboratory
8803 Rüschlikon, Switzerland

Abstract - Scanning tunneling microscopy (STM) and local probe methods in general offer attractive capabilities for investigations of macromolecules: To observe and study individual objects or molecules in their natural environment, to follow and control molecular processes, and to modify matter on a molecular level. In a first step biological samples have been coated with a conducting film for STM investigations in order to make contact with established electron microscopy methods. Of special interest is the three-dimensional imaging of individual objects with sub-nanometer resolution. The substrates for the direct investigation of molecules used so far are cleaved pyrolytic graphite and flame annealed or epitaxially grown (111) surfaces of gold and platinum. Good imaging with STM and atomic force microscopy was achieved for a variety of molecules ranging from simple benzene rings to DNA and proteins. Surprisingly STM imaging was readily possible with very large structures of paraffins and progeins up to 100 nm in size. This implies sufficient electron transfer through these substances, many orders of magnitude larger than generally assumed. For STM investigations of macromolecules it is essential to strongly bind unmodified molecules to a well-characterized, atomically flat reactive surface. We functionalized an atomically flat gold (111) surface with various mercaptanes and disulfides. The molecular recognition of biotin-functionalized self-assembled monolayers by streptavidin yielded protein covered surfaces with different crystallinity due to different lateral mobility. The ability of the STM to study and modify molecules in real space on an atomic scale make this instrument an essential tool for the design of biosensors and of molecular devices. The final goal, the use of molecules to produce artificial structures, however, can only be achieved by a combined effort of techniques from biology, chemistry and physics.

1. INTRODUCTION

The past decades saw the emergence of new disciplines such as supramolecular chemistry, molecular biology and solid state nanometer scale science and technology. Seemingly unrelated at first glance, these fields, however, have common objectives: The investigation and control of matter at a molecular or supramolecular level.

In chemistry, which has dealt mainly with isolated molecular species of subnanometer size, supramolecular systems have entered the scene. In supramolecular chemistry, the functional properties of molecules are not only determined by their chemical structure but to an even larger extent by their environment. The process where isolated molecules in solution recognize a partner molecule before they dock to the recognized site and form a complex is called self-assembling. The new goal is to synthesize molecules that self-organize to form complex assemblies where molecules interact in a purpose-oriented manner, like parts in a machine. The organization of monomolecular assemblies at solid surfaces provides a rational approach for fabricating interfaces with well-defined composition, structure and thickness. Such assemblies provide a means to control the chemical and physical properties of interfaces for a variety of heterogeneous phenomena including catalysis, corrosion, lubrication and adhesion as well as for scanning tunneling microscopy (STM) investigations of molecules. When self-assembling is used to build more complex structures, surfaces become very important, since the recognition of a new molecule always takes place at the surface of the preassembled complex. This, and the many reactions at phase boundaries introduced in the past decade, have shifted the focus of chemistry to reactions at interfaces.

Biology and biochemistry have undergone a transition from describing taxonomic and functional properties of living organisms to the investigation of biological processes at a molecular scale. Moreover, they make use of such processes as molecular recognition, site-specific proteolysis by proteases, *in vitro* DNA transcription and translation, and site directed mutagenesis. Although it is a long process to sequence an entire genome, it is just a matter of time and costs until this, too, has been done. It is widely accepted that the translation of a gene to a protein via the RNA working copy determines all protein sequences of an organism and that the primary structure of the protein controls the generation of the secondary structure and thereby the function of the protein. From this fact one might think that with the knowledge of the genome, one knows everything about any organism. This is not true for several reasons; mathematical models that can predict the 3-D structure of a protein from the DNA or protein sequence do not yet exist. In addition, it is very difficult to deduce the function of a protein from the sequence and the 3-D structure alone. The genesis of an organism is a fragile interplay between the expression and the blocking of many genes. Our present methods do not allow the regulation of gene expression and the folding of proteins to be explained on a molecular level.

In solid state nanometer scale science and technology the size of artificially built structures is being reduced to dimensions considerably less than 100 nm. The properties of such nanoscopic structures differ markedly from those of the bulky material and hence new effects become important. Physics has laid the basis for understanding these new properties where quantum effects in particular play an important role and has also provided investigation methods such as electron microscopy, halography and scanning tunneling microscopy. Materials science provides the artificial materials such as ceramics for superconductors, and technology provides the processes for miniaturization.

2. MOTIVATION

The motivation to investigate biological systems from a general scientific standpoint is twofold: To discover biological concepts on a molecular scale as such and for later use of such concepts in human-built systems and applications of biological matter directly in systems of contemporary concept.

Many varieties of molecular recognition are used in living organisms, the lowest level being the recognition sequences on proteins that guide the transport to the different compartments of the cell. Once in the correct compartment, the pre-sequence is cleaved off and the proteins self-organize to structures like cytoskeleton, microtubuly etc. In fact the genesis of any organism is guided by self-organization. The only input to that organization is the information stored in the genes. The best known example for this genetically determined self-assembly is the virus bacteriophage T4. The biological system of information processing is a good example for the mass storage of information. Transcription and translation process of DNA involves many selective molecular recognition processes; this causes the very low error rate of a few errors in one billion copied or translated bases.

The fault tolerance, the process of learning and the efficient pattern recognition of our brain have led to the current interest in neural computing. The brain is a parallel distributed processing system that can adapt to new situations by modification of the interconnections. The basic unit of the brain, the neuron, uses principles that have been developed by millions of years of evolution. With instruments that can investigate processes at a molecular level this information can be made accessible for future use.

Biological matter is already applied in analytical kits that use the specificity of antibodies to detect traces of virus antigens in blood. At present the specificity of antibodies exceeds the specificity of chemical catalysts by many orders of magnitude. For that reason a new generation of detectors is being built that uses this specificity of antibodies or enzymes, the biosensors. To detect a macroscopic response, the antibodies or enzymes have to be immobilized, preferably as a monolayer on a surface. The response is detected, for example, by surface plasmons or by protein-sensitive silicon field effect transistors (FET). The concepts that are necessary to build biosensors can be extended to externally controllable complex functional units and biological mass storage. One ultimate goal is to build active components or systems entirely from molecules. Again these molecules have to be organized on a surface so that they can communicate with the outside world.

The STM has opened a new approach to molecules. They can be studied as individual objects in contrast to all other methods where properties of molecules can only be derived from the statistical behavior of many molecules. STM and other STM-derived methods with atomic resolution enable us to interface our macroscopic world with individual molecules or molecular structures. Once molecules are immobilized in their functional state, we can study their properties and functions. In a next step we can try to interfere with their functions and to control and modify them. Finally, the tip can then be used to modify molecular structures, to assemble them in a specifically ordered configuration and to have them perform a set of functions in a controlled sequence.

3. RESULTS

To elucidate the imaging mechanism on small adsorbed molecules several STM studies on adsorbed benzenes (Ohtani et al., 1988) and thiols (Hallmark et al., 1987) have been undertaken in UHV. It seems that the STM is able to visualize either lowest unoccupied or highest occupied orbitals of a molecule. The trifold symmetry found on a benzene ring in Fig. 1a is probably not due to the undisturbed orbitals but due to π or π^* orbitals of a benzene Rh π complex. The role of the coadsorbed CO is not clear, but it could contribute π orbitals to regularly saturate the Rh surface.

To achieve pure conditions and eliminate most unknown parameters due to contaminations, it is best to work with UHV. The easier interpretation in UHV has not prevented experiments under normal conditions whether in air, under protective gas, or under liquid. Good results have been obtained on smectic octylcyanobiphenyl liquid crystals adsorbed on highly oriented pyrolytic graphite (HOPG) (Foster & Frommer, 1987; Smith et al., 1990).

In Fig. 1b orientation and packing of the liquid crystal molecules are clearly visible. Moreover, it is possible to resolve different functional groups: The aromatically bonded carbon atoms of the tilted benzene rings are visible as bright spots, the cyano group appears less bright, and the aliphatic carbons are faintly visible. By decreasing the tunnel gap resistance, the underlying graphite substrate is imaged, allowing the registry of the adsorbed molecules with the graphite to be deduced. The electronic nature of the STM contrast is confirmed by the good agreement of the images with results of ab initio Hartree-Fock orbital calculations. STM is sensitive to electron density at the Fermi level which is close to the level of lowest unoccupied molecular orbital (LUMO) or highest occupied molecular orbital (HOMO) of aromatic carbons or of the nitrogen (Smith et al., 1990).

STM imaging of larger biological species was initially performed mainly on material coated with a conducting film, although it was recog-

a

⌶ 0.2 nm

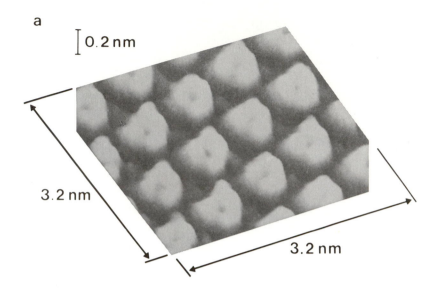

3.2 nm

3.2 nm

b

⌶ 0.5 nm

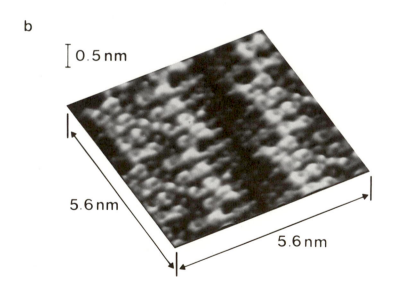

5.6 nm

5.6 nm

FIG. 1(a). Three-dimensional image of benzene and carbon monoxide coadsorbed as a 3 * 3 superlattice on Rh (111). Imaging conditions: voltage <0.5 V, current 4 nA. From Ohtani et al. (1988) (b) Three-dimensional images of octylcyanobiphenyl (8CB) liquid crystals on pyrolytic graphite. Imaging conditions: tungsten tip voltage 1.0 V, current 400 pA. From Smith et al. (1990).

554

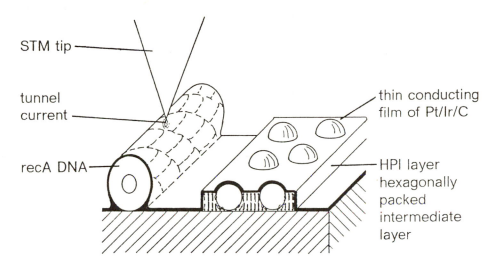

STM tip

tunnel
current

recA DNA

thin conducting
film of Pt/Ir/C

HPI layer
hexagonally
packed
intermediate
layer

atomically flat substrate

SCHEME 1. Imaging of coated objects. From Michel (1990).

nized quite early (Baro et al., 1985), that electron transfer through large
biological molecules might be sufficient for STM imaging of uncoated
material. The STM imaging of coated material profits from the preparation techniques developed for electron microscopic studies and should
complement those techniques with its atomic resolution capabilities. In
practice, the resolution is limited mainly by the roughness of the
substrate and the grain size of the coating. In the following example, the
molecules were adsorbed on a freshly cleaved mica surface that had been
pretreated with glow discharge. The sample was freeze-dried in high
vacuum and covered with 1.5 nm of Pt-Ir-C (Scheme 1, Travaglini et al.,
1987). Subsequently, STM imaging was performed. The sensitivity was
good enough to resolve the single helix of recA-DNA (Amrein et al.,
1988), HPI layer, a hexagonal lattice of proteins in cell membranes of
bacteria (Michel & Travaglini, 1988), and proteins that compose the head
of the virus bacteriophage T4 (Amrein et al., 1989b). Figure 2a shows
double-stranded DNA with a diameter of 2 nm partially complexed with
recA proteins. The complex appears as a right-handed single helix with a
diameter of 10 nm and a pitch of 10 nm. The handedness of molecules
was introduced by E. Fischer (1891). Ever since, the handedness of
molecules was based on his rule with no relation to real space. With this

a

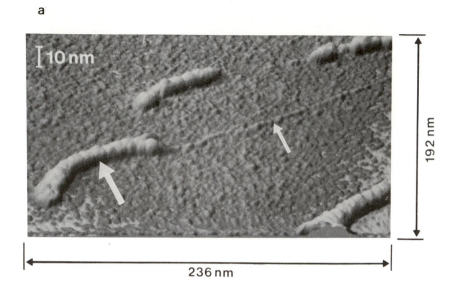

b

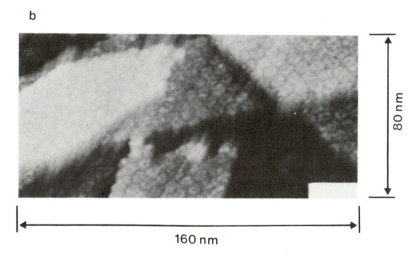

FIG. 2(a). Three-dimensional image of freeze-dried recA-DNA complexes coated with Pt-Ir-C film. Imaging conditions: gold tip, voltage 800 mV, current 500 pA. Large arrow points to recA-DNA complex, small arrow points to free DNA. (b) Top view of freeze-dried and Pt-Ir-C coated polyheads. The brighter parts of the polyheads lie on top of those being directly adsorbed to the support. Imaging conditions: gold tips, voltage 0.1 - 1 V, current 0.1 - 1 nA.

investigation on recA-DNA the handedness of molecules could be related
to real space. The right-handed helix of recA DNA confirmed the rule of
E. Fischer: D. symmetry corresponds to right-handed and L symmetry
to left-handed molecules. The head of bacteriophage T4 is composed of
a hexagonal lattice of proteins with a lattice size of 13 nm and a vertical
corrugation of 1 nm. The average capsomere morphology was determined by correlation averaging. When this is done, the sixfold symmetry
and the central depression of 0.5 nm of the proteins becomes clearly
visible (Amrein et al., 1989b). Owing to the high signal-to-noise ratio of
the tunneling data, only a few unit cells were needed to reveal a stable
average. With the highly reproducible sample, tip geometry could be
deduced for individual experiments.

The capability of STM imaging at ambient conditions and even in
electrolytes and the apparently generally sufficient conductivity of
biological materials prompted widespread STM work on bare molecules.
Langmuir-Blodgett (LB) films (Fuchs et al., 1987) and membranes
(Hörber et al., 1988) as well as proteins and DNA (Amrein et al.,
(1989a) were imaged. The dimensions of the recA-DNA complex
imaged on a MgAc-treated Pt-C surface in Fig. 3 show that the topography with and without coating is essentially the same. From this, we
concluded that direct imaging of biological molecules reveals the true
topography, even of large molecules.

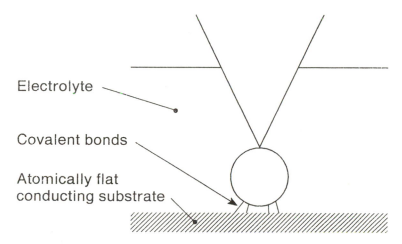

Electrolyte

Covalent bonds

Atomically flat
conducting substrate

SCHEME 2. Imaging of uncoated molecules. From Michel (1990).

a

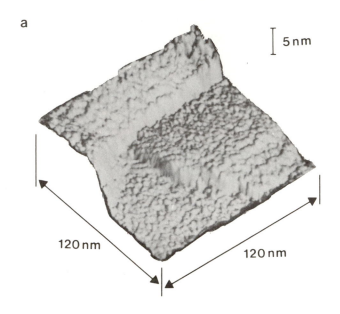

5 nm

120 nm

120 nm

b

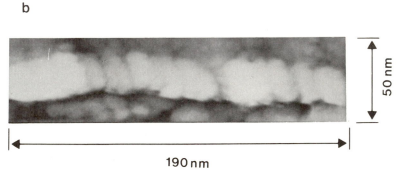

50 nm

190 nm

FIG. 3(a). Three-dimensional image of an air-dried uncoated multilayer config-uration of purple membranes adsorbed on Pt-C support. The step height at the membrane edges is about 4.5 nm for a single step. The topmost membrane shows periodicity typical for this membrane. (b) Top view of an individual uncoated recA-DNA complex adsorbed on a $MgAc_2$-treated Pt-C film. Each striation corresponding to a helical turn is composed of three to four repeating parts. From Amrein et al. (1989a).

The study of proteins in their native environment is an important field of application for STM. Examples of structures that have not been accessible with high resolution, not even with diffraction methods, are surfaces of membranes. Direct imaging is not subject to artifacts due to sample dehydration, and the imaging conditions can be selected to simulate native conditions for the molecules under investigation. The method can be applied to physisorbed molecules when molecules are densely packed on a surface, or when the shape of molecules ensures good adhesion. In this case the resolution is only limited by the molecular vibrations and elastic tip-molecule interactions. In other cases molecules are often displaced by the scanning tip so that no high resolution can be obtained.

The different forms of DNA (alpha, beta, Z) have been demonstrated by X-ray structure analysis, but it is not known whether other structures exist. It is only possible to determine the majority structure of DNA. Recently, several investigators could verify the structure of alpha, beta and Z DNA by STM (Arscott et al., 1989; Bebee et al., 1989; Dunlap & Bustamante, 1989; Driscoll et al., 1990; Lindsay et al., 1989; Travaglini & Rohrer, 1987). It is well known that DNA is extremely compacted in bacteria, in cell nuclei, sperm heads and virus capsids. Livolant et al. (1989) have shown with electron microscopy and X-ray diffraction a columnar longitudinal and a hexagonal lateral order with intermolecular distances from 2.8 to 4 nm for the highly concentrated phase of DNA molecules. Figure 4a shows the surface of such a concentrated DNA phase with large arrays or ordered tubes having a lateral periodicity of 14 nm with a similar appearance but a different periodicity as observed by Livolant et al. (1989). The individual DNA double strands that are part of the supercoiled structure and some individual strands of DNA double helix on the surface can be detected in Fig. 4. The periodicity along the supercoil is approximately 3.8 nm, which is in accordance with the packing density of DNA found by Livolant.

The local deviations from the helical structure are supposed to play an important role in the binding of promotors and repressors. Such unique structures are not accessible for statistical methods. Since the STM is able to image unique molecular structures it is a good candidate to resolve unique structures of DNA and thereby to study problems of gene regulation. For STM investigations the bases are sufficiently exposed only in single-stranded DNA. Single-stranded DNA, however, is structurally less stable than double-stranded DNA and, therefore, more

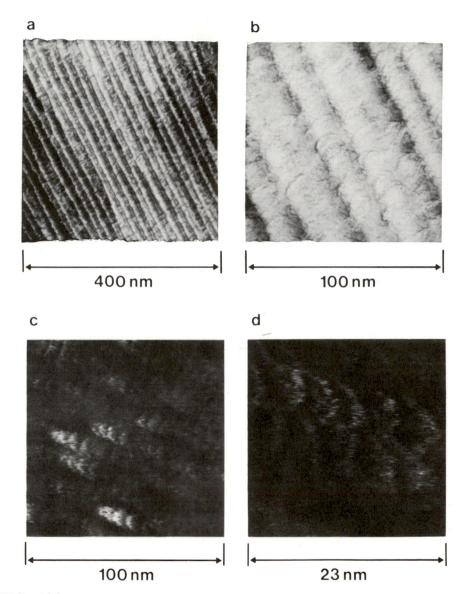

FIG. 4(a). Top view of the highly concentrated liquid-crystalline phase of sonicated salmon sperm double-stranded DNA on a Pt (111) surface. Imaging conditions: gold tip, voltage 300 mV, current 250 pA. (b) Close-up showing the 16 nm lateral and the 4 nm longitudinal spacing of the supercoiled structures. (c) Top view of a preparation of single-stranded DNA of bacteriophage Φ X-174 RF I on a Pt (111) surface. Imaging conditions as in (a). (d) Top view of one of the coils of (c). From Michel (1990).

difficult to study by STM. Figures 4c and d show a preparation of single-stranded DNA of bacteriophage Φ X-174 RF I on a Pt (111) surface. Under the selected conditions (10 mM Tris-HC1 pH 7.4, 10 mM NaC1), the single-stranded DNA seems not to be stable in the linear form. Figure 4c shows supercoiled structures with a lateral surface periodicity of 12 nm. The periodicity along the supercoil is approximately 4 nm. In some places a substructure with a periodicity of 1.2 nm is visible along an individual strand of the supercoil.

For the investigation of untreated molecules, the dominant technical problems concern preparation of suitable substrates which remain atomically flat under various ambient conditions, immobilization of molecules in their active state preferably on a reactive surface, and understanding the imaging mechanism.

The substrates used so far are cleaved pyrolytic graphite and flame annealed or epitaxially grown (111) surfaces of gold and platinum. Figure 5a shows an overview 4000 * 4000 nm in size on an epitaxially grown gold (111) surface consisting of crystal columns of 400 nm average width, ending in large atomically flat (111) terraces (See Fig. 5c). When the gold surface is disturbed with the tip it rearranges into newly formed, strongly terraced (111) surfaces (see Fig. 5b).

From a macroscopic point of view, most biological substances are poor conductors or even insulators. In the case of adsorbates or adsorbate layers of a thickness of less than one nanometer, the imaging contrast may arise from adsorbate-induced changes in tunneling probability and/or changes of the adsorbate electronic structure due to adsorbate/substrate interactions. For large molecules or thick adsorbate layers, however, direct tunneling through them can be excluded. Other electron transport processes from the surface of the adsorbate to the conducting substrate have therefore been proposed in order to allow STM imaging. They included band-type and hopping-type buld conductivity and surface conductivity (Amrein et al., 1989a).

To explore STM imaging in the absence of polar and charged groups which might induce a surface conductivity, we investigated linear alkanes or paraffins adsorbed to gold (111) and HOPG (Michel et al., 1989). Linear alkanes are the macromolecules with the most simple electronic structure; they are composed solely of C-C and C-H sigma bonds. They readily form orthorhombic and monoclinic prismatic layered crystals which have been extensively studied by electron diffraction methods (Dorset, 1978). The thickness of the layered crystals was found to be

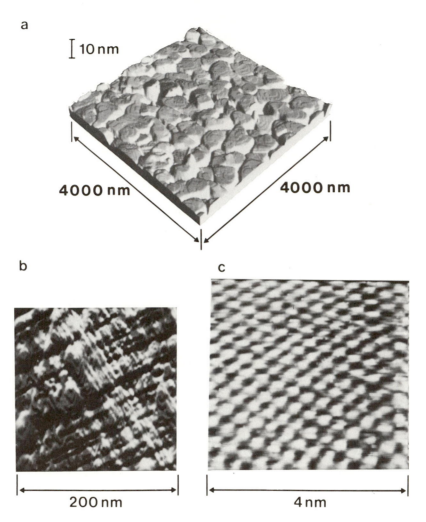

FIG. 5(a). Three-dimensional image of a gold (111) surface. Imaging conditions: Iridium tip, current 100 pA, voltage 300 mV. The monoatomic gold steps on the column terraces are clearly visible. (b) Top view of slope $dz(x,y)/dx$ of a gold (111) surface reformed after disturbance with tip. Imaging conditions as in (a). (c) High resolution image showing 0.01 nm corrugation of gold (111) packing. Imaging conditions: Tungsten tip, voltage 500 mV, current 100 pA. (C. Gerber, 1990, unpublished).

a

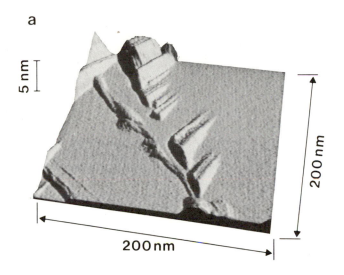

b

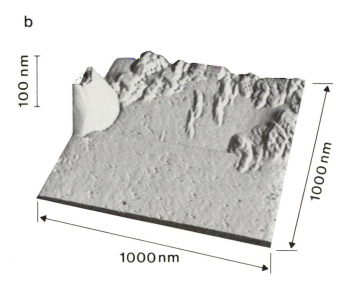

FIG. 6(a). Three-dimensional image of *n*-heptadecane in HOPG. Imaging conditions: Gold tip, current 150 pA, voltage 300 mV. (b) Three-dimensional image of *n*-hexatriacontane on gold (111), imaging conditions as in (a).

equal to the length of the chains (2.1 nm for heptadecane (arrows in Fig. 6a), 2.5 nm for octadecane, and 4.5 nm for hexatriacontane). Most of the crystallites were 2 to 10 nm high but in some places screw dislocation caused much larger structures. Figure 6b shows such a tower-like structure with a height of 100 nm, above a gold (111) plane. On other parts of the (111) plane, chain-like structures were resolved indicating at least a monolayer coverage with hexatriacontane molecules. Other parts of this image are covered with smaller, tilted, layered crystallites as in Fig. 6a. In the case of the low melting alkanes the large objects disappeared as soon as the entire STM reached the melting temperature of the respective alkane. Moreover, pre-melting processes were observed when the temperature was slightly below the melting temperature. In this temperature range the formerly stable crystallites became much more mobile. Layers of alkanes could be seen that grew on surfaces as well as structures which disappeared. These melting experiments served simultaneously as independent identification of the observed structures as *n*-octadecane and *n*-heptadecane. Finally, the liquid alkanes appeared to be non-conducting, since the flat surroundings of large alkane structures did not change upon melting.

Initially, biological molecules were physisorbed to a surface or membranes were transferred with the LB technique prior to the STM investigations. These molecules were often not stable enough to allow detailed reproducible STM investigations. Chemisorption on the other hand tightly immobilizes the molecules. Direct covalent bonding of molecules results in sufficient immobilization. Heckl et al., 1989) but can impair their functional state. Immobilization, therefore, has to be achieved by activating a surface via a bifunctional molecule. One functional group reacts with the solid surface and the other functional group binds the desired molecules (Scheme 3). Different second functional groups can be used for a given first functional group. This facilitates a wide range of molecules to be reacted with the surface. Once the desired molecules are chemically crosslinked, they can be subjected to a prolonged and reproducible investigation with a local probe method.

Much of the interest in organic monolayers stems from their relationship to biological membranes and from their potential use as building elements of artificial systems. Kuhn and co-workers have demonstrated that planned structures may be assembled by successive deposition of LB films (Kuhn, 1989). Adsoprtion of amphiphatic molecules on solid surfaces has been known to lead to a formation of closely packed monomolecular films (Nuzzo & Allara, 1983; Sagiv, 1980; Bain &

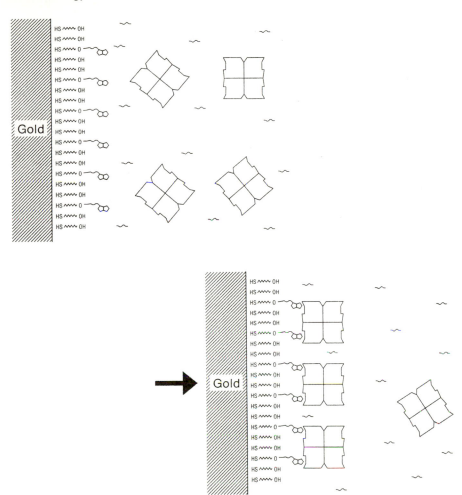

SCHEME 3. Streptavidin adsorbed to self-assembled monolayers. From Häussling et al. (1990).

Whitesides, 1988). The experimental evidence points to a similarity between the structure of these self-assembled monolayers and of LB films.

If the adsorption of such a film is followed by a chemisorption like that for mercaptanes on gold, the stability is increased by many orders of magnitude. Such films can easily be studied with local probe methods with no risk of displacing molecules (Häussling et al., 1990). In addition, the properties of the film can be altered by changing either the spacing of the head or tail groups. Figure 7 shows a film of mercaptoundecanol on gold. The typical appearance of such layers is seen in Fig. 7a: a smooth surface with randomly distributed depressions. The smooth parts covering approximately 85% - 95% of the surface exhibit the same terraced structure as the uncovered gold films with the same height and comparable sharpness of the steps (Fig. 7b and bisection in Fig. 7c). The depressions have lateral dimensions of 3.5 to 5 nm and a maximum depth of 1.1 nm. From this and additional experiments we concluded that the gold substrate is covered by a densely packed monolayer of mercaptoundecanol except for the holes, which are either empty or filled with loosely packed molecules. We attribute this to the van der Waals

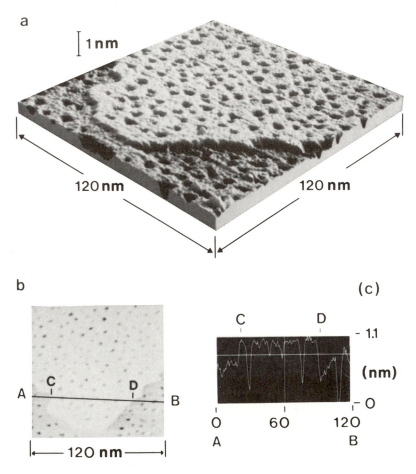

FIG. 7(a). Three-dimensional image with a chemisorbed self-assembled monolayer of mercaptoundecanol on a gold (111) surface. Imaging conditions: Tungsten tip, 200 mV, and 20 pA. (b) Top view of (a) with 1.5 nm from black to white. (c) Cross section along line A-B. From Häussling et al. (1990).

radii of the head groups or to the lattice given by the underlying solid being larger than the van der Waals radii of the tail groups. The bending of the tail groups due to their van der Waals interaction might then cause the randomly distributed "holes" in the film.

The shape and distribution of the depressions observed on the mercaptane films can vary slightly in a series of images when the current is set to a level larger than 1 nA. Then the force exerted on the film might be large enough to change the tilt of the chains and redistribute the holes. The electrical resistance of a single molecule is experimentally derived from that of an extended monolayer. A classical example is the LB film where the resistance associated with one molecule is 10^{17} ohms. In an STM experiment with the tip directly on top of a molecule the resistance of the molecule has to be smaller than the tunnel resistance. Experimentally the resistances are less than 10^9 ohms. Many explanations have been proposed to deal with this difference of eight orders of magnitude but none of them, so far, does justice to all the experimental findings. The I-V characteristics of the self-assembling monolayer shows a small gap at large tip - gold substrate distances. This gap, however, vanishes when the tip is closer to the substrate. We interpret this linear ohms regime with a valence band coming close to the Fermi level as the tip - substrate distance is reduced (Dürig et al., in preparation).

The next step was to adsorb streptavidin on biotin functionalized self-assembled monolayers. Unlike Langmuir monolayers on water, chemisorbed monolayers on solids exhibit only limited lateral and no vertical mobility. The recognition of biotin functionalized gold (111) surfaces by streptavidin is very slow and leads to randomly distributed proteins on the surface. Since the recognition is very slow, crystallization is sometimes favored over recognition, which leads to needle-shaped crystals. The molecular recognition of biotinlipids spread on a buffer surface leads to the well-known "H"-shaped structure of the streptavidin crystals (Ahlers et al., 1989). Such structures were imaged with the STM after they had been transferred from the water surface to a gold (111) surface with the Langmuir-Schäfer technique.

4. CONCLUSION

We have seen that the STM and the derived local probe methods offer a wide range of novel investigations of macromolecules.

STM imaging of objects coated with a conducting film is a reliable and quick method to obtain sub-nanometer resolution on any structure irrespective of the conductivity. The method is suitable for making contact to electron microscopy studies, since the coating technique so far was very similar. However, more gentle coating methods for STM might be developed in the future thanks to simpler requirements of STM imaging. Of more impact, however, we believe, is the direct *in vivo* imaging with local probe methods. In the case of STM, this requires good electron transfer through the objects. The generally low resistance of molecules adsorbed to surfaces as experienced in an STM experiment is still an open question. Such and other aspects of tunneling microscopy have been discussed elsewhere (Behm et al. eds., 1990). The reversible use of the STM tip for modification and investigation of paraffin structures leads us to believe that many organic and biological materials are accessible to STM investigations - more than originally anticipated and more than many still want to believe.

A great deal of STM efforts in biology concentrate on imaging biological structures with the intention of being able to sequence DNA. To a non-expert the simple and straightforward STM technique seems to be an ideal approach. The STM should eventually be able to read a single DNA molecule over its entire length with no need for restriction enzymes, cloning etc. The problems, however, start with the preparation of single-stranded DNA on a surface exposing all bases such that they are accessible to the tip. The properties of the four bases which might be readily distinguishable by a local probe method and thus allow an acceptable throughput have yet to be found. Possible parallel imaging of local probe instruments might greatly facilitate DNA sequencing. Alternative local probe methods, however, are at least as promising. As mentioned above, the most interesting topics that are not accessible to other methods are the binding of promotors and repressors to DNA. STM or related methods can indeed be envisaged to directly study the DNA/RNA transcription at a DNA/RNA polymerase or the RNA-protein translation at a ribosome.

Alternative methods for sequencing DNA which would read the DNA code directly at a DNA polymerase look equally promising. Moreover, novel procedures along such lines should be the basis for a wider application of local probe methods as an interface to molecular processes. In this sense, the local probe method would serve as an interface to the biological world rather than as a microscopic tool. In an ultimate step the local probe method for interfacing might be in a fixed array and the molecules would organize themselves accordingly. Self-assembly of structures on surfaces can make the connection between lithography and the self-assembly commonly found in biological systems.

The experiments with self-assembled monolayers have also shown that methods like grazing angle infrared and contact angle microscopy can predict an average coverage even of an inhomogeneous surface. On a molecular basis, however, the structure of a given film can vary widely. This is even more pronounced when enzymes are immobilized on biosensor surfaces. Here, the enzymic activity is taken as a measure of determining coverage. The enzymes quite often assemble to larger complexes, leaving most of the surface uncovered, but the total activity can still be close to that of a monolayer.

Another local probe method, atomic force microscopy (AFM) has been developed into a promising tool for the investigation of macromolecules (Drake et al., 1989; Rugar & Hansma, 1990). With AFMs, molecules can be studied on insulating substrates and in liquids irrespective of an electron transfer mechanism of the molecule or its surrounding liquid.

Most interesting, however, might be the combination of force and tunneling microscope. Such an instrument will enable us to deconvolute distortions of images present with either method from the true topography. Moreover, forces could be an effective means to investigate and influence molecular processes and thereby provide an alternative interface of our macroscopic world to molecules, in particular since mechanical methods can also be extended to the GHz or THz range in the molecular world.

We are confident that the final goal - the use of molecules to produce artificial structures - can be achieved by a combined effort of techniques from chemistry, biology and physics. With these techniques we will be able to gain access to the experience and expertise of nature, a vast database which has been collected in the course of billions of years of molecular evolution and is stored in the genome of organisms. With some adaptations we can make use of this information to induce molecules to self-assemble to human-designed molecular devices.

ACKNOWLEDGEMENTS

I would like to thank H. Rohrer for valuable discussions and Ch. Gerber for his collaboration and for providing images of atomic resolution on gold (111).

REFERENCES

1. M. Ahlers, R. Blankenburg, D. W. Grainger, P. Meller, H. Ringsdorf and C. Salesse, Thin Solid Films **180**, 93 (1989).

2. M. Amrein, R. Dürr, S. Stasiak, H. Gross and G. Travaglini, Science **243**, 1708 (1989a).

3. M. Amrein, R. Dürr, H. Winkler, G. Travaglini, R. Wepf and H. Gross, J. Ultrastruct. Mol. Struct. Res. **102**, 170 (1989b).

4. M. Amrein, A. Stasiak, H. Gross, E. Stoll and G. Travaglini, Science **240**, 514 (1988).

5. P. G. Arscott, G. Lee, V. A. Bloomfield and D. F. Evans, Nature **339**, 484 (1989).

6. C. D. Bain and G. M. Whitesides, Science **240**, 62 (1988).

7. A. M. Baro, R. Miranda, J. Alarnan, N. Garcia, G. Binnig, H. Rohrer, Ch. Gerber and J. L. Carrascosa, Nature **315** 253 (1985).

8. T. P. Beebe, T. E. Wilson, D. F. Ogletree, J. E. Katz, R. Balhorn, M. B. Salmeron and W. J. Siekhaus, Science **243** 370 (1989).

9. R. J. Behm, N. Garcia and H. Rohrer (Eds.), Proceedings of the NATO Meeting, Basic Concepts and Applications of Scanning Tunneling Microscopy (STM) and Related Techniques, Erice, Italy, April 17-29, 1989, Vol. 184 (1990), Kluwer Academic Publishers.

10. D. L. Dorset, Z. Naturforsch. **33a**, 964 (1978).

11. P. Drake, C. B. Prater, A. L. Weisenhorn, S. A. C. Gould, T. R. Albrecht, C. F. Quate, D. S. Channell, H. G. Hansma and P. K. Hansma, Science **243**, 1586 (1989).

12. R. J. Driscoll, M. G. Youngquist and J. D. Baldeschwieler, Nature **346**, 294 (1990).

13. D. D. Dunlap and C. Bustamante, Nature **342**, 204 (1989).

14. U. Dürig, C. Joachim, L. Häussling and B. Michel, in preparation.

15. E. Fischer, Ber. **24**, 2683 (1891).

16. J. S. Foster and J. E. Frommer, Nature **333**, 542 (1987).

17. H. Fuchs, W. Schrepp and H. Rohrer, Surf. Sci. **181**, 391 (1987).

18. V. M. Hallmark, S. Chiang, J. F. Rabolt, J. D. Swalen and R. J. Wilson, Phys. Rev. Lett. **59**, 2879 (1987).

19. L. Häussling, B. Michel, H. Ringsdorf and H. Rohrer, submitted to Angewandte Chemie (Sept. 1990).

20. W. M. Heckl, K. M. R. Kallury, M. Thompson, C. Gerber, H. J. K. Hörber and G. Binnig, J. Amer. Chem. Soc. **5**, 1433 (1989).

21. H. J. K. Hörber, C. A. Lang, T. W. Hänsch, W. M. Heckl and H. Möhwald, Chem. Phys. Lett. **145**, 151 (1988).

22. H. Kuhn, Molecular Electronics, F. T. Hong (Ed.), Plenum Publishing Co., (1989).

23. S. M. Lindsay, T. Thundat, L. Nagahara, U. Knipping and R. L. Rill, Science **244**, 1063 (1989).

24. F. Livolant, A. M. Levelut, J. Doucet and J. P. Benoit, Nature **339**, 724 (1989).

25. B. Michel and G. Travaglini, J. Microscopy **152**, 681 (1988).

26. B. Michel, G. Travaglini, H. Rohrer, C. Joachim and M. Amrein, Z. Phys. B **76**, 99 (1989).

27. B. Michel, Proc. 4th Toyota Conf. "Automation in Biotechnology", Aichi-ken, Japan, October 21-24, 1990, North-Holland, in press (1990).

28. R. G. Nuzzo and D. L. Allara, J. Am. Chem. Soc. **105**, 4481 (1983).

29. H. Ohtani, R. J. Wilson, S. Chiang and C. M. Mate, Phys. Rev. Lett. **60**, 2398 (1988).

30. D. Rugar and P. Hansma, Physics Today **23**, (1990).

31. J. P. Ruppersberg, J. H. K. Hörber, Ch. Gerber and G. Binnig, FEBS Lett. **257**, 460 (1989).

32. J. Sagiv, J. Am. Chem. Soc. **102**, 92 (1980).

33. D. P. E. Smith, J. H. K. Hörber, G. Binnig and H. Nejoh, Nature **344**, 641 (1990).

34. G. Travaglini, H. Rohrer, M. Amrein and H. Gross, Surf. Sci., **181**, 514 (1987).

35. G. Travaglini and H. Rohrer, Spektrum d. Wissenschaft **No. 8**, 14 (August 1987); in English: Scientific American **257**, 30 (November 1987).

MANIPULATION AND MODIFICATION OF NANOMETER SCALE OBJECTS WITH THE STM

C. F. Quate

Edward L. Ginzton Laboratory
Stanford University, Stanford, CA and
Xerox Palo Alto Research Center, Palo Alto, CA

1. INTRODUCTION

The scanning probes associated with the STM are used in the fields of surface physics, surface chemistry and molecular biology to record images that display electronic, atomic and molecular structure.

In this presentation we are not concerned with imaging. We are concerned with the generation of structure, the modification of surfaces and the manipulation of objects on these surfaces. We are interested in the "nano-scale" where structures have dimensions of nanometers.

In the summer of 1985, Gerd Binnig and Heini Rohrer convened a small meeting in Oberlech, Austria. At that meeting they commented on the diversity of the STM. We cannot improve on their description. They said[1] -- "This local probing capability, ..., make it increasingly attractive for use in diverse areas of science and technology, reaching beyond its initial use in atomic-scale imaging. In many applications, its primary role ... is used primarily to select and define the location of the experiment. In other applications, ..., imaging might not be used at all."

Highlights in Condensed Matter Physics and Future Prospects
Edited by L. Esaki, Plenum Press, New York, 1991

The illustration of Fig. 1 - a scratch in a gold film made with the tip of an STM - is a vivid reminder that the instrument is more than a device for imaging; it is a multi-function device with a power that extends beyond imaging.

The scanning probe, illustrated in Fig. 2, with its sharp tip and intense E-fields, can deposit, remove and arrange surface atoms. It can affect the direction of magnetization in thin magnetic films. The energy in the tunneling electrons can be exploited to dissociate molecules, to

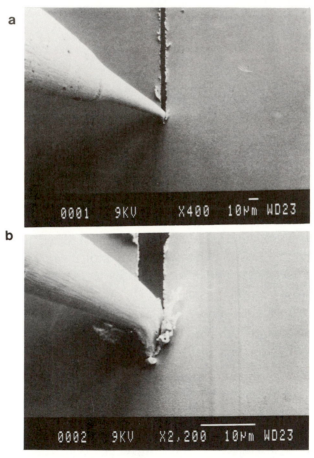

FIG. 1. (a) Scratch generated with the STM in a gold film 300 Å thick vacuum deposited on graphite
(b) Enlarged area indicating that the graphite substrate is still intact. (From H. Fuchs and R. Laschinski, "Surface investigations with a combined scanning electron-scanning tunneling microscope," *Scanning*, **12**, 126-132 (1990).)

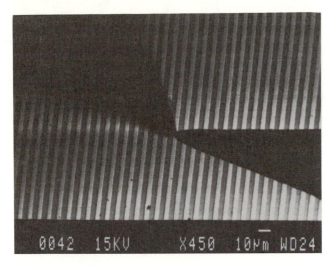

FIG. 2. The tungsten tip of the STM hovering over a grating etched in silicon. (Courtesy of A. de Lozanne, University of Texas.)

attach molecules to substrates, to detach molecules from substrates, to raise the temperature above the melting point for materials with low thermal conductivity, and to enhance the reaction on the substrate for selected etchants. The forces associated with the tip can be used to move atoms along the surface of the substrate and to transfer atoms between the tip and substrate. The modification, and manipulation, takes place in a well controlled fashion over regions with nanometer dimensions. These probes are tools for performing experiments on the "local scale" in the small region at the end of the tip.

The modifications are examined with the scanning tip, itself. It is only necessary to operate the instrument in the imaging mode where the tip voltage is reduced to a minimum.

A report on this field was prepared for the NATO workshop held in Erice in the spring of 1989.[2] The results that have appeared in the intervening months are quite remarkable. In that short interval, devices - albeit primitive - have actually been fabricated.

We begin our discussion with the early attempts to modify surface structure by applying high voltage pulses to the tip positioned at precise locations on the sample.[3]

2. GOLD SURFACES -- INITIAL RESULTS

At the meeting in Oberlech, a group from Berkeley[4] discussed the modification of a gold surface with a tungsten tip of the STM. We quote from their article - "By lowering the tip to the surface we have been able to produce indentations typically 100 Å across; ... By operating the microscope at high tunneling current (...) and reduced gap spacing, we have been able to deposit material on the surface to form hillocks roughly 300 to 400 Å across and 12 Å high. In both surface-modification processes, surface diffusion was observed with a characteristic time of the order of minutes." This succinct statement, summarizing the important properties, established the ground for much that followed.

Their report did not generate the interest that it deserved. The exciting element, the focus of our primary interest at the Oberlech conference, was the atomic structure in many of the images. The interest in the holes and gold hillock of Fig. 3 was limited. We will learn that this view was very short-sighted when we return to this subject in a later section.

Surface modification on the atomic scale was first demonstrated by Becker et al.[5] when they "wrote" on the surface of Ge(111) with the STM by increasing the bias on the tip to 4 volts. They discovered that this procedure would create atomic-scale hillocks on the germanium surface. The hillocks were approximately 1 Å in height and 8 Å in diameter. Their success rate for transforming the surface of germanium was improved when they moved the tungsten tip to a remote location on the Ge surface and deliberately touched the tip to the sample. They inferred from this that Ge atoms, adhering to the tip after touching, were transferred to the sample during the "writing" cycle. In effect, a field-induced transfer of atomic species across the gap from the tip to sample.

The report from Becker stands as the pioneering effort in "writing bits" on the atomic scale, but it was not the first attempt to modify surfaces with the scanning probes. That was done earlier by Ringger et al.[6] and McCord et al.[7] Ringger succeeded in "scratching" silicon by scanning while the tip was in virtual contact with the silicon surface. The recorded pattern was a series of parallel linear lines scribed in the hydrocarbon films covering their surfaces.

Hydrocarbon films are always present in those systems that use oil diffusion pumps for the vacuum. These films are easily polymerized when exposed to electron beams. The first report came from Stewart[8] working with electrons at 200eV..."In an evacuated tube in which the slightest traces of organic vapors may occur,...insulating layers are formed on surfaces subject to electron...bombardment. These layers may be attributed to carbon compounds and their formation is related to the polymerization of organic vapors..."

Broers[9] demonstrated the utility of such a system for lithography in his work with the SEM.

In direct analogy with this, McCord et al. used the electrons from a scanning tip to polymerize hydrocarbon films. They used the scanning tip in the field emission mode with a large spacing between tip and sample. The STM, modified in this way, resembles the scanning electron

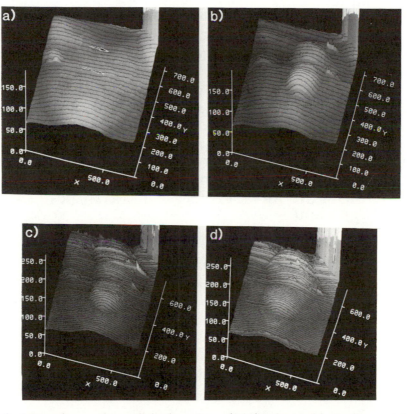

FIG. 3. Mounds on the polished surface (210) of a gold single crystal. The hillocks were produced by increasing the tunneling current to 1 μA with the tip positively biased at 5 mV. (From D. W. Abraham, H. J. Mamin, E. Ganz and J. Clarke, "Surface Modification With the Scanning Tunneling Microscope," *IBM J. Res. Develop* **30**, 492-499 (September, 1986).)

microscope. The tip was an intense source of electrons - "without space charge or aberrations." With a 10 volt bias on the tip they were able to form lines 1000 Å in width.

In a second experiment, they formed monolayers of docosenoic acid (CH_3 $(CH_2)_9$: $(CH_2)_{12}$ $COOH$) in a Langmuir-Blodgett trough and transferred the monolayers onto Al films (100 Å thick) vacuum deposited on silicon substrates. Docosenoic acid was chosen since it was known from prior work[10] that these films are easily exposed with electrons. McCord used electrons from the scanning tip at 25 volts to expose the monolayers. The Al in the unexposed areas was removed with wet etchants. The Al lines formed in this way were 1000 Å in width.

The field emission mode was also used by Marrian and Colton[11] to expose a film of polydiacetylene. Their films were prepared on silicon substrates in a Langmuir-Blodgett trough. Their tip was positioned close to the substrate within tunneling range. When they ramped the tip voltage to 10-20 volts they formed patterns on the substrate that were different from that reported by McCord. Their process occasionally produced depressions in the organic film but more often hillocks were formed and it was the hillocks that attracted most of their attention. They were 80 Å in height and 200-600 Å in diameter.

3. MONO-MOLECULAR LAYERS -- LANGMUIR-BLODGETT AND SELF-ASSEMBLY

Organic films as monomolecular layers are easily formed on substrates. The established techniques[12] include: the Langmuir-Blodgett trough (LB), self-assembly (SA), depositing molecules from solution, and ordered layers formed at the interface between the substrate and a liquid-crystal. These films are obvious and natural candidates for modification with the STM at the local level, but the initial results are ambiguous and inconclusive. Nevertheless, there is a widespread perception that the potential in this field is very strong. The continuing effort to exploit phenomena in this arena warrants attention in our review.

3.1. Langmuir-Blodgett Films

In the LB technique, the molecules are spread on the surface of a liquid, compressed to form a contiguous film, and transferred to the substrate by dipping it through the compressed film. Bilayers of

Cadmium-Arachidate, transferred to graphite, were the first organic molecules studied with the STM.[13] The molecules were arranged perpendicular to the substrate in a periodic array. The individual molecules in the STM images stood out from the background with strong contrast.

These films must be formed on a substrate that is smooth and flat on the atomic scale. Graphite has been extensively used for this purpose but it is not ideal; the surface is hydrophobic and inert. Film adhesion is problematical. Epitaxial gold grown on mica has often been used, but a preferred substrate, recently revealed by Fuchs, et al.[14], is WSe_2. The surface of Tungsten Diselenide is hydrophilic and molecular adhesion to the substrate is very good. Stable films of 22-tricosenoic acid covering large areas of the substrate are easily prepared. The ($CH = CH-(CH_2)_{20}$ $COOH$) molecules, 32 Å in length, can be imaged more than two months after the films are laid down.

There is a problem with linear molecules arranged perpendicular to the substrate. The mechanism that gives rise to the current along the length of the molecules is not understood. The path along the linear chain is much too long for tunneling.

For parallel alignment, where the molecules lie parallel to the surface, the current can be explained with tunneling electrons moving across the narrow dimension of the molecule. An example of this comes from the work of Kuroda et al.[15] They used the LB trough to form a bilayer of behenic acid ($CH_3 (CH_2)_{20} COOH$) on graphite. The images, recorded with a negative bias on the sample, indicate that the molecules are in registration with the graphite lattice.

Various polymer films [poly(methyl methacrylate) (PMMA) and poly(methyl pentene sulphone) (PMPS)] on graphite, prepared as thin films in a LB trough, have been studied by Dovek et al.[16,17] They used the STM to image and modify the structure of the molecules. Their images showed arrays of polymer fibrils lying parallel to the surface. There was no evidence for molecular order. These films are easily modified by applying a negative voltage pulse to the tip, 4 volts in amplitude and 0.1 microseconds in duration.

The modification of graphite can proceed without the overlying organic film. Albrecht[18] observed etch pits beneath the tip on clean graphite bathed in a vapor of water molecules. Rabe et al.[19] in his study of octylcyanobiphenyl on graphite reports that etch pits were created in graphite when the tip bias is raised above 2.5 V. They found that the pits were formed in air, in water vapor (10 mbar of pressure), and silicone oil as well as octylcyanobiphenyl. We will return to this subject in a later section for a more complete discussion.

3.2. Liquid-Crystals

Foster and Frommer[20] have described a simple process for organizing molecular arrays on the surface of graphite. A small drop of liquid crystal, with cyanobiphenyl compounds as the major component, was placed on a heated graphite surface. The molecules are physi-absorbed onto the surface over a period of 1-2 hours. They discovered that the interaction at the interface changes the liquid-crystal molecules on the substrate described as "a two-dimensional solid crystal" by Mizutani et al.[21] The two benzene rings of each molecule were clearly displayed in many of the images. Their beautiful images of the molecules arranged in ordered patterns were easily simulated with molecular modeling.

A further study of these molecules by Smith et al.[22] confirmed the work of the Foster group. Smith went on to analyze the structural properties of a class of molecules labeled as $4'$-n-alkyl-4-cyanobiphenyl (mCB, where $m = 8,10, 12$ indicates the number of carbons in the alkyl group). The molecular order, the components of the individual molecules, and the registry with the substrate lattice, were all examined in great detail.

In a second paper,[23] Foster's group raised the question of manipulating individual molecules. For that study they covered a graphite surface with di(2-ethylhexyl)phthalate and moved the tip toward the substrate into the tunneling range. The familiar atomic structure of graphite appeared in the image when the surface was scanned with the tip biased at 30 mV. When the tungsten tip was pulsed to a voltage of 4 volts for 1 microsecond a sizeable structure appeared on the surface. They thought that the structure was a molecule of phthalate chemically bonded to the substrate. The object could be removed by pulsing the tip a second time with the tip positioned directly above the object. The voltage threshold for both processes was 3.5 volts.

This system with the pinning and depinning of individual molecules is an attractive method for modifying surfaces. The energy from the tunneling electrons is sufficient to either form, or break, the chemical bond. However, this process is confused by questions of the type raised in the work described below.

3.3. Alkanes Adsorbed on Graphite

Bernhardt et al.[24] took up the study of fabricating "nano-scale patterns" with the STM. They examined the system introduced by Foster and moved on to study ordered monomolecular layers of the n-alkanes. There is a great deal of background material in the literature relating to the adsorption on graphite of n-alkanes with chain lengths of 2 to 40 atoms. They used this information to adsorb a uniform ordered layer of n-$C_{32}H_{66}$ on graphite from a solution of n-$C_{10}H_{22}$ and 2-2-4 trimethylpentane (isooctane). The patterns were imaged with a Pt/Ir tip in the STM. The images, with fine detail in the molecular chains, indicated that the molecules were in registration with the graphite substrate.

They followed this with a series of experiments designed to produce the nanometer-scale features described by Foster's group. The work is best summarized with a sequence of four images of Fig. 4.

For the first image, they pulsed the Pt/Ir tip to 4 volts in air for 0.2 microseconds and found a hillock on the substrate with a height of 5 Å, a length of 25 Å and a width of 17 Å. For the second image, Bernhardt used a tungsten tip immersed in decane ($C_{10}H_{22}$) and pulsed at 4 volts for 1 microsecond. The hillock measured 6.5 Å in height, 102 Å in length, and 17 Å in width. For the third image, a tungsten tip was immersed in dimethylphthalate and pulsed at 4 volts for 0.4 microseconds. The hillock was 8 Å in height, 21 Å in length, and 9 Å in width; similar in many respects to the hillock formed by Foster's group. There are "some striking similarities between the formation of hillocks in air and in liquids. In air, a likely mechanism for hillock formation is the transfer of material from tip to surface."

On several occasions the voltage pulses produced depressions in the graphite rather than a hillock. For the image of Fig. 4(d), the Pt/Ir tip, immersed in decane, was pulsed to 4 volts for 0.2 microseconds. This procedure produced a hole with a depth of 80 Å and a diameter of 260 Å.

These four examples illustrate the dilemma. The nature and source of the hillocks are undefined. If the hillocks were all made of n-alkane molecular clusters it would seem reasonable to see some evidence of molecular structure in the ordered molecular arrays - but so far the hillocks are featureless. There is no signature that can be used to identify the chemical nature of the attached species. The large depressions, or craters, are inconsistent with molecular pinning. More likely they are related to the enhanced etching of graphite.

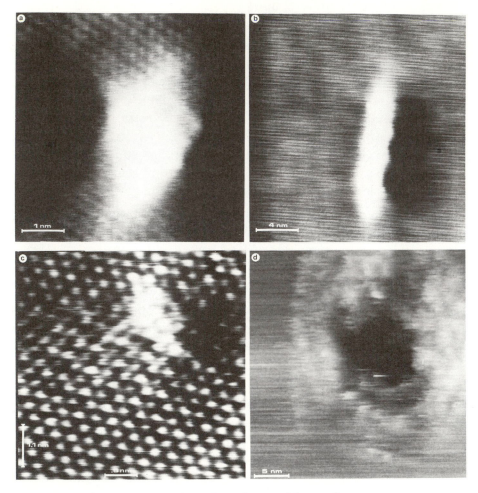

FIG. 4. Four views of the modification of a graphite surface
(a) Hillock formed in air with a pinning pulse of 4 V for 200 ns. (Pt/Ir tip)
(b) Hillock formed in decane with a pinning pulse of 4 V for 1 μs. (cold worked tungsten tip)
(c) Hillock formed in dimethyl phthalate with a pulse of 4 V for 400 ns. (etched, ion-milled tungsten tip)
(d) Depression in Highly Oriented Pyrolytic Graphite (HOPG) in decane with a pulse of 4.5 V for 200 ns. (Pt/Ir tip)
(From R. H. Bernhardt, G. C. McGonigal, R. Schneider and D. J. Thomson, "Mechanisms for the Deposition of Nanometer-Sized Structures From Organic Fluids Using the Scanning Tunneling Microscope," *J. Vac. Sci. Technol. A* **8**, 667-671 (Jan/Feb, 1990).)

The inconsistency in size, the change from hillocks to depressions, and the differing environments all contribute to a state of confusion in this field. Reactive processes which erode the graphite surface beneath the tip must be a factor in some of this. We will return to this subject in a later section.

4. LIQUID-SOLID INTERFACE -- ELECTROCHEMISTRY AND LIQUID CRYSTALS

There is a substantial body of work concerning the operation of the STM in liquid environments. Surface reactions are observed in electrochemical cells with the tip and sample immersed in various liquids; polar solvents,[25] non-polar solvents,[26] and conductive aqueous solutions.[27] In these solutions, where the current is carried by ions, a Heaviside double layer forms at the interface between the electrode and the liquid. Much of the potential drop is across the double layer. We direct our attention to the electrochemical reactions which occur in the region between the tip and sample.

Lin[28] and his colleagues in the laboratory of Bard in Texas, have demonstrated that the current from a scanning tip can control the etching rate associated with the light induced decomposition of III-V compounds. It is based on well-known work on photo-oxidation and the anisotropic etching of III-V semiconductors via the photoelectrochemical technique (PEC).[29-31] The incident light generates minority carriers in the depletion layer at the interface between the semiconductor and the electrolyte. Photons with energy greater than the band-gap participate in the generation of carriers and this permits one to use incoherent light. The lattice bonds, weakened in the presence of the minority carriers, are attacked by agents in the electrolyte.[32] The ratio of light to dark etch rates is greater than 100:1.

In the innovation introduced by Lin et al. the etching rate was enhanced by the current from the tip of the STM, modified to operate in an electrochemical cell. The tip, at the end of a wire sheathed in glass, was exposed to the electrolyte through a narrow opening in the glass tubing. It was spaced from the substrate by 1 micron and biased to -3 volts. The aqueous solution covering the sample consisted of 5 mM of NaOH and 1 mM of EDTA. When the sample was irradiated with a tungsten-halogen lamp they found that the "ratio of the local reaction rate across the substrate to that immediately under the tip was controlled by the current density distribution at the substrate surface." In the end, they were able to move the tip along a predetermined path and etch a line 0.3 microns in width on the GaAs surface.

Nagahara[33] and his colleagues working with Lindsay in Arizona, determined that they could use the photoelectrochemical technique in a buffered solution of $KAu(CN)_2$ to deposit gold atoms onto a substrate of p-GaAs. They worked with a Pt/Ir probe coated with Apiezon wax[34] to minimize the faradaic leakage currents. There was a small opening at the tip to allow for the tunneling current. The tip was spaced approximately 100 Å from the substrate and biased at +2 volts with respect to the solution. The region was simultaneously irradiated with light.

Gold lines approximately 1000 Å in width were deposited on the substrate beneath the tip, as shown in Fig. 5. The light creates electron-

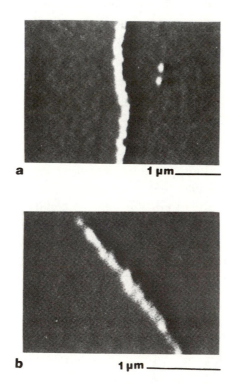

FIG. 5. Images from the STM of gold lines formed directly on p-type GaAs with the STM tip by photoelectrodeposition. The lines were written with a tip bias of +2 V with respect to the AG/AgCl electrode. (Courtesy of L. A. Nagahara, Arizona State University, Private Communication

hole pairs in this process and electrons move into the depletion layer to neutralize the gold ions in the solution. The neutral metal atoms are subsequently deposited on the substrate.

In this work, with the large gap between tip and sample, the line width is controlled by the lateral diffusion of the carriers, rather than the dimensions of the tip. It doesn't quite meet the spirit of the "local labo-

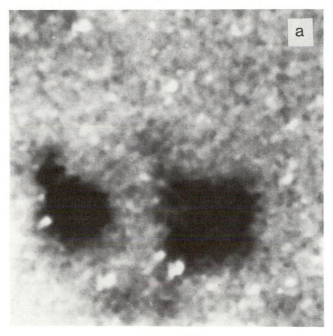

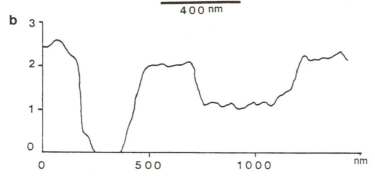

FIG. 6. (a) STM image of holes etched in silicon (100) in 0.05% HF solution. The tip was biased at 1.4 V with 1 nA of current

(b) Profile of the two etched pits showing the depth of the depression. Courtesy of L. A. Nagahara, Arizona State University, Private Communication

ratory." In a second experiment, Nagahara and his colleagues[35] moved the tip within a few Angstroms of the sample and reproducibly etched lines in Si(100) with the STM tip. Light was excluded from the sample and the surface was covered with a dilute (0.05%) of HF solution. In this situation, where the Heaviside double layers at the tip and substrate overlap, the potential of the solution is not relevant. The process is controlled by the potential between tip and substrate. They biased the tip at +4 volts to etch the desired pattern. The minimum line width was 200 Å. The etching did not occur when they immersed the surface in other electrolytes, nor when the tip spacing was increased to 500 Å. They deduced that the etching was a field induced electrochemical process. A typical result is shown in Fig. 6.

5. SILICON

The group at Philips with van Loenen,[36] in their study of the

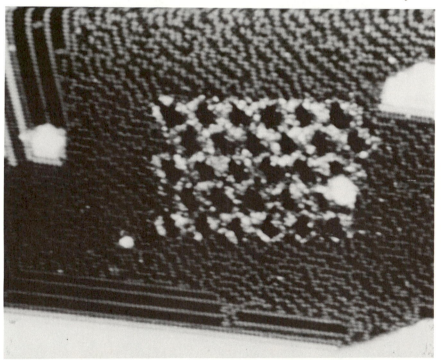

FIG. 7. Writing on silicon with the STM tip. (From E. J. van Loenen, D. Dijkkamp, A. J. Hoeven, J. M. Lenssinck and J. Dieleman, "Direct Writing in Si With a Scanning Tunneling Microscope," *Appl. Phys. Lett.* **55**, 1312-1314 (September 25, 1989).)

surface of silicon, found that atoms could be easily removed from the Si(110) surface with a negative tip bias of 1.25 V. In their process, lateral scanning was interrupted and the z-piezo was extended to advance the tip toward the substrate by 20 Å. It was then retracted back to the original height. The time for this cycle was 8 ms. Subsequent imaging revealed that a hole was formed in the substrate 20 - 50 Å in diameter and 7 Å in depth. The cycle was controlled and easily reproduced. They were able to "write" continuous lines on the silicon as narrow as 25 Å.

They examined the field dependence of this process and found that the holes still formed in the substrate when the field was turned off. The pattern of Fig. 7 is typical.

Iwatsuki et al.[37] at JOEL, in their study of silicon, connected the lateral motion of the tip to a "joy-stick" so as to move the tip under manual control. They found that silicon atoms on the Si(111) 7x7 surface could be continuously removed when the tip was moved across the surface with a bias of 10 volts. The "joy-stick" was used to 'write' lines in any desired pattern.

A group at Aarhus with Besenbacher and Mortensen,[38] have also used the technique of advancing the tip toward the substrate by extending the z-piezo while the tip is held stationary over a selected pixel. Their results are shown in Fig. 8.

In a similar way, Avouris[39] has written on Silicon (111) with the results shown in Fig. 9.

5.1. H-Passivated Silicon

The surface of silicon is chemically modified with the diffusion of oxygen into the surface. It is known that the diffusion rate is increased when the surface is illuminated with electrons from an SEM.[40]

Dagata et al.[41] had this in mind when they speculated that diffusion of oxygen into a silicon surface might be enhanced with the E-field associated with the tip of the STM. Furthermore, if it could be done in air (or ambient oxygen), the vacuum environment of the SEM would not be required. As a first step, it would be necessary to passivate the surface. A system for doing this was disclosed several years ago in the work of Grunthaner and Grunthaner.[42] In that process the silicon wafer was cleaned and dipped in dilute hydrofluoric acid to remove the residual oxide. Predominantly, the dangling bonds of silicon are terminated with hydrogen, rather than fluorine, and it is this termination that passivates

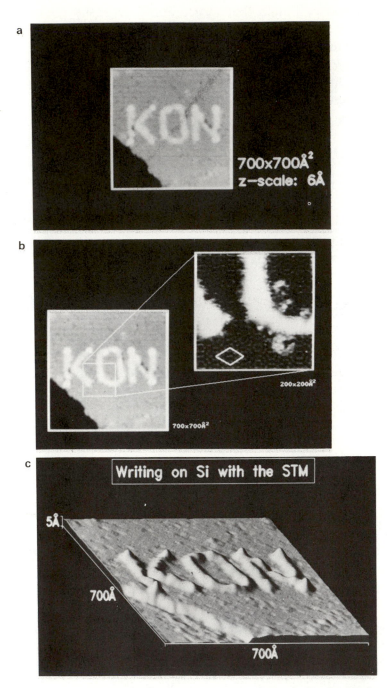

FIG. 8. Writing on silicon (111) 7x7 in honor of Prof. K. O. Nielsen. Tip biased at 2 V used in the manner described in the text. (Courtesy of K. Mortensen and F. Besenbacher, Aarhus University, Denmark.)

FIG. 9. Writing on Silicon (111). (Courtesy of Ph. Avouris, IBM.)

the surface. The subject has been revisited by Fenner et al.[43] They state that "H termination reduces the rate of uptake of impurities, at low coverage, on the Si(100) surface by at least 10 orders of magnitude relative to the unterminated Si(100) surface." Surfaces passivated with this technique will survive in air for about 30 minutes before hydrocarbon contaminants clutter the surface and impede further processing. This is not a long interval of time, but it is most useful.

With this information in mind, the group with Dagata passivated silicon with hydrogen and scanned the tip over a preselected pattern. The tip voltage was fixed at 3.5 volts. When the oxygen molecule is dissociated, or ionized, by the intense E-field of the tip, the increased reactivity of the new species enhances the diffusion into the substrate. The process is labeled "diffusion enhanced oxidation."

In one instance they "wrote" a series of parallel lines - 200 nm in width, 3 microns in length, and spaced apart by 300 nm. The minimum line width was limited by the residual surface roughness of the silicon wafer. The surface texture of their wafers was 10-50 nm after passivation, but as they point out, this can be reduced with oxidation annealing prior to passivation.[44] A smoother surface would allow them to "write" finer lines. The scanned regions appeared as depressions in the STM images, but this is misleading. It was not topography, it was a change in the conductivity. When the conductivity of the surface region decreases the tip moves toward the sample to maintain a constant current. This movement appears as a depression in the image. The scanned area was examined in a SIMS apparatus and the examination revealed that the oxygen content was incorporated into the region to a depth of 1-20 nm.

In the final step, Dagata demonstrated that the patterned region was chemically distinct from the unmodified regions since the oxidized features could be selectively etched to a depth of 1 micron. The results are shown in Fig. 10.

6. PHASE TRANSITIONS

Phase transitions from the amorphous to crystalline state are interesting to many investigators. The lateral extent of the transition can be controlled if the material is heated above the transition temperature with a focused laser beam. The electron beam from the STM tip is a natural candidate for heating on the "local scale," but there is a problem. The temperature rise with a point source for heating is a function only of the thermal conductivity and the mean free path of the electrons. In crystalline materials, where the thermal conductivity is of the order of 100 W/mk and the mean free path is of the order of 10 nm, the temperature rise is only a fraction of a degree. However, in amorphous materials, where these values are reduced by a factor as large as 100, the heating produced by the tunneling electrons in the STM is substantial. We will discuss the results from the groups working in this area.

Staufer et al.[45] in Basel, examined the properties of metallic glasses. They found that this material can be smoothed to a surface roughness less than 1 Å by ion sputtering (Ar+). A substrate with this degree of smoothness is suitable for nanometer-scale surface modification. With the tip in position over a selected point on the substrate, they increased the voltage to 2 volts (sample positive). This was followed by an increase in the current to 300 nA. The onset of instabilities in the current indicated that dramatic changes were taking place in the

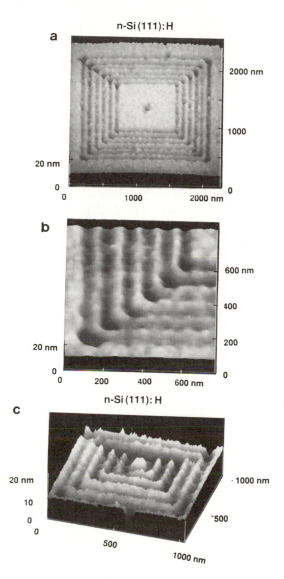

FIG. 10. The etched features on H-passivated, n-Si(111), $\rho = 10$ ohm-cm, by an STM operating in air

(a) Lines as written with a bias voltage of 3 V (tip positive)

(b) Detail of corner of the image of (a)

(c) Patterned features written on n-Si(111), with +3 V on the STM tip.

(From J. A. Dagata, J. Schneir, H. H. Harary, J. Bennett and W. Tseng, "Pattern generation on semiconductor surfaces by a scanning tunneling microscope operating in air," J. Vac. Sci. Technol. **B9**, 1384-1388 (Mar/Apr 1991).

substrate material. Indeed, the evidence indicates that the temperature rise exceeds the melting temperature and a small pool of liquid formed under the tip. The molten glass was drawn toward the tip by the strong electrostatic forces to form a "Taylor cone." The current was lowered while the voltage was maintained at 2 volts. The cone froze into a hillock 300 Å in diameter and 150 Å in height. The size was dependent on the tip voltage - lower voltages produce smaller hillocks. It is easy to fabricate hillocks in the form of arrays, as shown in Fig. 11.

The group at Basel is proceeding to adapt their technique to media used in thermo-magnetic recording.[46] In that system, a ferromagnetic film is biased with a uniform magnetic field and heated over a small "local" region to a temperature above the Curie temperature. The direction of the magnetization in the small heated region will change to match the direction of the bias field. The bias field is aligned with the bulk magnetization for writing, or against for erasing. Staufer et al. studied a Co-Tb alloy in the form of a metallic glass, $Co_{35}Tb_{65}$. A sample bias of 1.2 volts and a current of 50 nA was sufficient to melt the film and raise the "Taylor cone." They were able to "write" arrays of cones 50 Å in

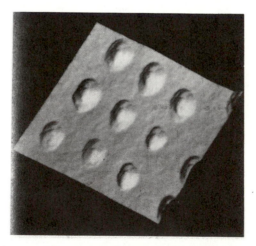

FIG. 11. STM image of cones written on the surface of $Rh_{25}Zr_{75}$. (From U. Staufer, L. Scandella and R. Wiesendanger, "Direct Writing of Nanometer Scale Structures on Glassy Metals by the Scanning Tunneling Microscope," *Z. Physik B- Condensed Matter* 77, 281-286 (1989).)

diameter and 5 Å in height in the presence of a magnetic biasing field. Information on the state of magnetization within the altered regions can be measured with magnetic tips, but this information is not yet available.

Hydrogenated amorphous silicon (a-Si:H) is another material with low thermal conductivity and short mean path for the electrons (10 Å). Jahanmir et al.[47] have studied a 200 Å film of amorphous silicon deposited on a silicon substrate. They pulsed the tip to 10 volts for 35 microseconds. The current during the pulse reached values as high as 100 microamps. They suggest that the increase in current elevates the temperature above the transition temperature and changes the film to the crystalline state. Lines were created by scanning the tip through distances as large as 1 micron. The lines were 150 Å in height and 1400 Å in width. The minimum size of these protrusions was limited by the inherent roughness of the surface.[48]

Hartmann,[49] in the group of Koch in Munich, has also studied the properties of amorphous silicon irradiated with the e-beam from an STM. They worked with n-type layers grown on p-type crystalline silicon substrates. The amorphous layers doped with hydrogen were 600 Å in thickness. They were able to modify the α:silicon layer with a negative bias of 10 volts on the tip. The modification, in the form of lines, is shown in Fig. 12(a) and (b). The linear features do not represent topography. It is a local change in conductivity induced by the high bias on the tip. The lines in Fig. 12 are 600 Å in width and 3500 Å in length. They were produced with a writing speed of 200 Å/sec. Unfortunately, the written features are not stable. The relaxation time is measured in hours.

The mechanism is not yet understood. But the pn-junction at the interface between the n-type amorphous layer and the p-type silicon substrate must play a role. The phenomena does not occur with layers deposited on n-type substrates. It should, also, be kept in mind that the 10 volt bias is a very high value. It is in the range where atoms can be transferred between tip and substrate if the spacing is close.

Foster et al.[50] have considered the transition from the amorphous state to the crystalline state in a material such as GeTe. They propose to alter and convert discrete amorphous regions by heating to a temperature above the transition temperature with the STM. In the amorphous state the material is a semiconductor with a bandgap of 0.8 volts, whereas in

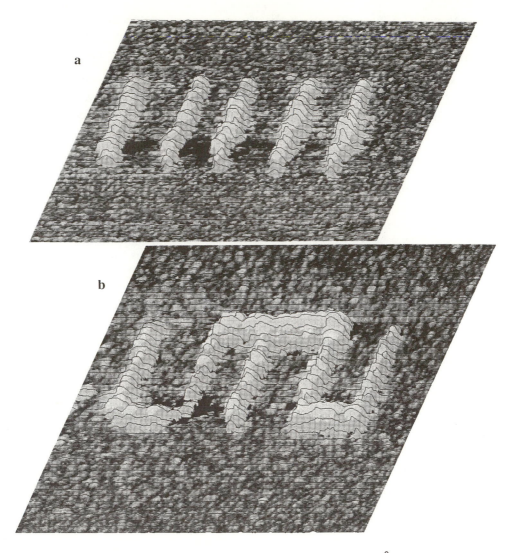

FIG. 12. (a) Hydrogenated amorphous n-type silicon (600 Å thick) deposited on a p-type silicon substrate. The linear features, 600 Å in width and 3500 Å in length, are written with a positive bias on the substrate of 10 volts. The images are recorded with a tip bias of 3-4 volts.

(b) Similar to (a) with the addition of the horizontal lines. In both (a) and (b) the linear features represent a change in the conductivity. The topography is not modified. (Courtesy of E. Hartmann, G. Krötz, G. Müller, R. J. Behm and F. Koch, University of Munich.)

594

the crystalline state it is nearly metallic with a bandgap of 0.1 volts. They estimate that the change from the amorphous state to the crystalline state can be induced in 50 nanoseconds with 8 volts on the tip. The reverse transition from crystalline to amorphous state might be induced with a similar voltage applied to the tip for 1 millisecond. The large change in the electrical properties should be easily detected with the STM in the imaging mode. These materials are highly reactive in air and a protective layer might be necessary to passivate the surface. Such a layer would impede the tunneling current and restrict the utility of this system.

7. CONVENTIONAL METHODS FOR "READ-WRITE" STORAGE

There are two conventional systems for storage of digital information where the information can be "written" and "erased" with electrical signals. The first relies on the polarization of magnetization in small magnetic domains and the second relies on charge trapped in dielectric layers overlying a semiconducting substrate.

7.1. Magnetic Recording with a Modified Tip in the STM

Moreland and Rice[51] have replaced the rigid tip of their STM with a compliant magnetized tunneling tip to study the field patterns on a magnetic hard disk. The magnetic probe was made from an iron foil, 5 microns thick, shaped in triangular form. The apex of the triangle served as the tip. The probe was attached directly to the piezo-tube of the STM. The disk, taken from a computer drive, was aluminum with a magnetic coating of 1 micron film of CoCrTa. In one instance, the disk was overcoated with 1000 Å of gold.

In the writing cycle, the iron tip was pressed against the substrate with a force that was estimated to be 2×10^{-9} N. The work was done at room temperature. They found that the magnetization in the film was altered over the full extent of the scanned area. They were able to "write" bits that measured 5000 Å across. The bits, shown in Fig. 13, could be imaged using the same tip with a resolution of 200 Å.

Their observation of recording and imaging magnetic regions in a controlled manner is an event that is central to the program of surface

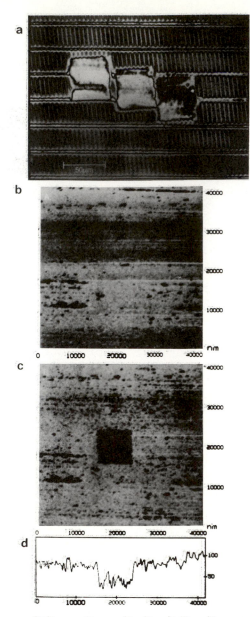

FIG. 13. Optical image of the surface of a hard disc decorated with ferrofluid showing the three squares with magnetization induced by the magnetic tip
(a) Optical image of the surface of a hard disk decorated with ferrofluid showing the three squares with magnetization induced by the magnetic tip
(b) The magnetic image of the hard disk before writing
(c) A magnetic square written with the magnetic tip
(d) The profile of the square of (b)
(From J. Moreland and P. Rice, "High-resolution, tunneling-stabilized magnetic imaging and recording," *Appl. Phys. Lett.* **57**, 310-312 (July 16, 1990)

modification. Our understanding of the physical process underlying this phenomena will increase and it is probable that the bit size will shrink from the scale of microns to the scale of nanometers.

7.2. Charge Storage - MNOS

Charge storage is the second area where the information can be written and erased in micron-scale structures. In 1981, Iwamura et al.[52] published a description of a system under the title "Rotating MNOS Disk Memory Device." It should be easy to adapt this to STM technology and create nanometer-scale structures. Their device was based on the Metal-Nitride-Oxide-Semiconductor technology used in conventional memory systems.

MNOS memory devices consist of a silicon substrate coated with a thin layer of oxide, about 20 Å in thickness. A second layer on silicon nitride with a thickness of 100 Å (or more) is placed on top of the oxide layer. A metal film is deposited over the nitride layer to complete the structure. In operation, a voltage on the metal film, positive with respect to the silicon, will cause electrons to tunnel from the silicon substrate through the oxide layer into the nitride where it is trapped. The trapped charge creates a depletion layer at the silicon interface. The stored charge can be erased with a voltage pulse of the opposite polarity on the metal electrode.

The reliability for long-term storage in MNOS devices has been extensively studied. The process is well understood[53-56] and it gives us a firm foundation for future work with moving probes.

In the work reported by Iwamura, the deposited metal electrode was replaced with a movable gate electrode in the form of a rotating stylus 10 microns in diameter. The stylus moved in contact with the nitride layer - analogous to the motion of the flying head in magnetic hard disks. The stored information was read by sensing the depletion layer with the moving probe.

It is straightforward to adapt this technology to nanometer-scale structures by replacing the moving stylus with the narrow tip of the STM[57] to provide a new "read-write" system for high density storage.

8. ORGANOMETALLIC DEPOSITION ALA E-BEAM CVD

Deposition of atoms onto a substrate from a gas is known as chemical vapor deposition - CVD. When the gas molecules are irradiated with energetic particles they dissociate into volatile and non-volatile components. The volatile components move off and the non-volatile species deposit on the substrate. The energy for dissociation is furnished by various energetic beams that illuminate the region where the reaction takes place. Laser beams are used in laser-beam CVD,[58] and electron beams in e-beam CVD.[59] Metals, semiconductors and inorganic materials can be deposited with CVD. The feature size from laser CVD is limited to micron-scale structures, but e-beam CVD with the TEM, the SEM, or the STM can generate nanometer-scale features. We will introduce the discussion with the results from the TEM.

Matsui and Ichihashi[60] have used the "electron-beam-induced surface reaction" to dissociate WF_6 and deposit tungsten atoms onto a silicon substrate. They could watch the growth process in situ with their TEM. When the gas was irradiated with the electron beam, small tungsten clusters were formed. The initial clusters were quite mobile. A layer of tungsten was formed by the continuous coalescence of the clusters.

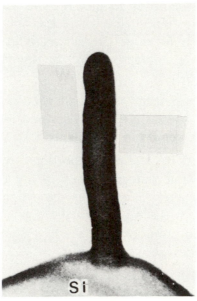

FIG. 14. A tungsten rod fabricated with a 120 keV electron beam via e-beam CVD. (From S. Matsui and T. Ichihashi, "*In Situ* Observation on Electron-Beam-Induced Chemical Vapor Deposition by Transmission Electron Microscopy," *Appl. Phys. Lett.* **53**, 842-844 (September 5, 1988).)

In this system for direct writing, the features are formed in place by deposition of metal atoms onto the substrate.

In a separate experiment, they fabricated the tungsten rod shown in Fig. 14. It has a diameter of 150 Å and a length of 1000 Å.

Matsui[61] also used a modified SEM to study electron-induced surface reactions as a method for etching micron-scale patterns in poly(methyl methacrylate) (PMMA). The lithography was carried out on the surface of PMMA in a vapor of ClF_3. In the absence of e-beam irradiation, the PMMA film was inert and immune to etching. In the presence of e-beam (10 keV, 100 pA) irradiation, the ClF_3 molecules dissociated and etched the PMMA to a depth proportional to the electron dose.

The technique for writing on PMMA with the SEM should lead to a similar system for lithography with the STM. The STM has not yet been used for etching PMMA, but it has been used to deposit metal atoms with e-beam CVD.

Electrons tunneling from a tungsten tip of the STM to a silicon substrate were used as the energetic source for molecular dissociation in the work of Silver et al.[62] Dimethyl cadmium (DMCd) molecules, with a dissociation energy of 3.14 eV, were adsorbed on a silicon substrate. The dissociation was then carried out with the STM operating at 11 V with a current of 500 nA. They deposited cadmium on the substrate with a feature size as small as 200 Å.

McCord, Kern and Chang[63] operated an STM in the field emission mode (5-40 V) to deposit tungsten atoms on a silicon substrate with feature size as small as 100 Å. Two organometallic gases, tungsten hexacarbonyl and dimethyl-gold-acetylacetonate, were used as the vapor. These materials are easy to handle since they are solids that sublime at room temperature. In a vapor of $W(CO)_6$ with 16 mTorr of gas pressure, metallic lines were deposited on a silicon wafer with a central ridge 100 Å wide. The writing was done with a 30 V, 10 nA beam with the tip moving at 0.25 microns/sec.

In another instance, they wrote tungsten dots in a square array with a pitch of 0.25 microns. A single dot 300 Å in diameter was written in 0.5 seconds. A tungsten post standing 2800 Å high with a diameter of 250 Å was fabricated in 2 seconds.

This demonstration of tungsten lines, dots and posts, will impact nanometer-scale devices where the dimensions of the structures are comparable to the wavelength of the electrons. If we ask about small ferromagnetic systems where the magnetic response is dominated by quantum mechanics, we turn to the work of Awschalom et al.[64] They report on a system with "...nanometer-scale magnets configured in regular arrays."

The nanometer-scale magnets were formed by dissociating pentacarbonyl [Fe(CO)] molecules with the STM. The magnetic structures were 250-1000 Å in height and 100-300 Å in diameter. The content of each cluster was estimated to be 60% FE and 40% Carbon. The magnetic particles, which are assumed to be amorphous, were grown under a pressure of 160 mTorr of Fe(CO) in about 100 msecs. The dimensions of the clusters of Fig. 15 are small compared to the thickness of the domain-wall in crystalline ferromagnets.

020205 25KV X60.0K 0.50um

FIG. 15. Fe clusters formed by e-beam CVD with the STM. (Courtesy of M. A. McCord, IBM; also, D. D. Awschalom, M. A. McCord and G. Grinstein, "Observation of Macroscopic Spin Phenomena in Nanometer-Scale Magnets," Phys. Rev. Lett. 65, 783-786 (August 6, 1990).)

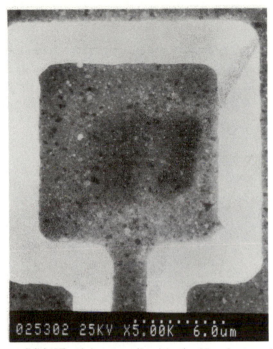

FIG. 16. A planar SQUID pick-up coil with the Fe clusters in place. (Courtesy of M. A. McCord, IBM; also, D. D. Awschalom, M. A. McCord and G. Grinstein, "Observation of Macroscopic Spin Phenomena in Nanometer-Scale Magnets," *Phys. Rev. Lett.* **65**, 783-786 (August 6, 1990).)

The important aspect of this work is that these small particles were fabricated in place on a special substrate which contained previously fabricated structures. The particles were placed on a sapphire substrate carrying the planar coils of a dc SQUID micro-susceptometer described by Awschalom[65] and shown in Fig. 16. The pick-up loop consisted of a pair of superconducting loops of lead, 25 microns on a side. One of the loops was covered with a gold film 800 Å thick. The magnetic particles were deposited in place on this loop with e-beam CVD, using the STM. They measured well-defined resonant peaks in the magnetic susceptibility at 500 Hz as the temperature was lowered below 0.2°K. This experiment is of great interest, even though their findings are not yet reconciled with the theoretical picture. It is a prime example of using the STM to fabricate an experimental device.

9. XENON ON NICKEL

A strategy for manipulating atoms on the surface is contained in the report from Eigler and Schweizer.[66] They propose to position xenon atoms by sliding them along a nickel surface with the tip of the STM. In a system cooled to 4° K they first adsorb a few xenon atoms on the nickel surface and then use the STM at 10 mV and 1 nA to image the atoms scattered randomly over the nickel surface. In the image, they select a given atom, position the tip over that atom and adjust the electronic system to increase the tunneling current. This advances the tip toward the atom and increases the force between the tip and the atom. They find that the force can be adjusted (by fine control of the spacing) to the point where the atom will slide and follow the tip as it is translated along the surface.

By sliding xenon atoms along a given row on the Ni(110) surface they form a well defined array of xenon atoms. The arrangement is stable with a spacing between xenon atoms of 5 Å. An example of a ring structure is shown in Fig. 17.

The fundamental character of these forces, the balance between the binding and translating forces, is independent of temperature. In the

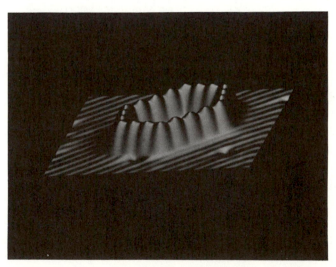

FIG. 17. Atomic structure arranged by sliding xenon atoms on nickel with the tip of an STM. (Courtesy of D. M. Eigler, IBM; also, D. M. Eigler and E. K. Schweizer, "Positioning Single Atoms With a Scanning Tunneling Microscope," *Nature*, **344**, 524-526 (April 5, 1990).)

near future, we will discover a system where the atoms are assembled at room temperature by translation along the atomic rows of a smooth surface.

10. GRAPHITE

The surface of graphite is easily modified with the STM when the tip, in close approach to the surface, is pulsed to a voltage above 4 volts. At first, this is quite surprising. Graphite is a layered structure where bonds between the carbon atoms in a given layer are very strong. The surface is inert and impervious. But, it should not be surprising, since the combustion and gasification of graphite proceeds via chemical erosion of the surface when the graphite is heated in a gaseous atmosphere. In the initial stages of combustion the surface erosion is similar in some respects to the surface modifications that are brought about in the STM. We will first review the combustion process and then review examples of electron enhanced etching.

In the process of chemical erosion, adsorbed particles, such as molecules and ions, react to form a new species with a lower binding energy. The new species desorb from the surface and leave a vacancy at the site of the erosion. The erosion rate increases when the energy of the adsorbed molecules is increased. The molecular energy is increased either by heating, or by illuminating the surface with secondary particles, such as electrons and photons.

In the combustion of graphite the material is heated in a gaseous atmosphere. In the initial stage, chemical erosion removes a monolayer of carbon over a small circular region 100 Å in diameter. The eroded regions appear as small etch pits with a depth of one atomic layer. A study of the characteristics of the etch pits leads to an understanding of the combustion process. More recently, it has been determined that etch pits of a similar nature can be created with a voltage pulse on the tip of an STM. Surface modification carried out in this way is reliable and well behaved. It is the point of interest for our report. Before proceeding, we will present background material on the combustion and erosion processes.

The reaction, central to processes of combustion and gasification of solid fuels, takes place when carbon is heated to 650° C in an atmosphere of gases such as H_2O, H_2, O_2, or CO_2. The reaction begins with the creation of the small pits. The pits are enlargements of vacancies, and other defects, where an edge is exposed in the uppermost layer. The

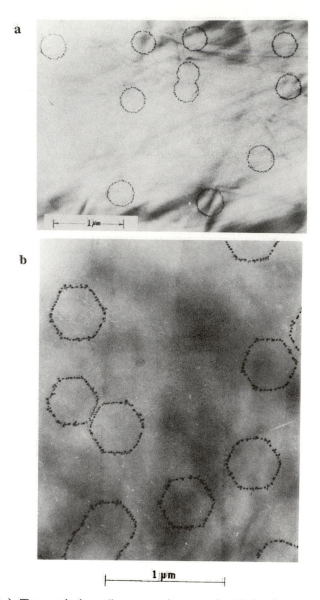

FIG. 18. (a) Transmission electron micrograph of the basal plane surface of graphite after oxidation for 25 min at 650° C with 0.2 atm O_2 followed by gold decoration.
(b) TEM micrograph of gold-decorated monolayer etch pits on the basal plane formed by 23 torr H_2O (in 1 atm N_2) at 600° C for 14 h.
(From R. T. Yang, "Etch-Decoration Electron Microscopy Studies of the Gas-Carbon Reactions," in *Chemistry and Physics of Carbon*, vol. 19, pp. 163-210, P. A. Thrower, ed. Marcel Dekker, Inc. New York (1984).)

reaction is initiated at the site of the sp^2 electrons. The rate of removal of atoms from an edge site in the layer plane is faster by ten orders of magnitude than the rate of removal of atoms from the surface of a perfect layer. Elementary rate constants, mechanisms for catalyzed reactions, and information on the earliest stages of combustion are revealed in the study of the etched pits.

The traditional method for studying etch pits is with the TEM using a technique labeled "etch-decoration transmission electron microscopy - ED-TEM." The technique was introduced in the late 50's by Hennig.[67] He found that gold particles deposited on carbon migrate to the edge of the pits and outline the circumference in a way that is easily recognized in the TEM. The shape and size of the pits are evident in the TEM images of Fig. 18. The depth is not available in the TEM images, it must be inferred from indirect evidence.

Yang[68] tells us that the technique for decorating with gold particles is "...difficult for beginners - in fact, temperamental..." This difficulty is circumvented if the STM is used to image the etch pits. Atomic detail with information on both shape and depth stand out in the STM images with high contrast; decoration with gold particles is unnecessary. Chang and Bard[69] have studied the etch pits formed by heating HOPG graphite to 650° C in an air furnace for several minutes. Their results are illustrated in Fig. 19. They propose the STM as a "powerful technique for studying the early stages of the gasification of graphite."

The chemical erosion, or deterioration, of graphite used as moderators in reactors has been dealt with in the work of Ashby et al.[70] They simulated the reactor environment by bombarding the graphite surface with atomic hydrogen. When they analyzed the volatile molecules in a mass spectrometer they found that methane (CH_4) was the major component in the desorbed species. They observed a "major reactivity enhancement" when the surface was simultaneously bombarded with electrons. The rate of methane production was increased by a factor of twenty.

Haasz[71] was able to confirm their results when he carried out similar experiments in a station with moderate vacuum, but he found no evidence for electron enhanced etching when he improved the vacuum with prolonged baking in a UHV station. Haasz found that in systems with a moderate vacuum methane molecules are adsorbed on the sur-

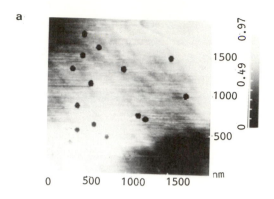

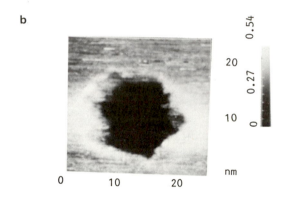

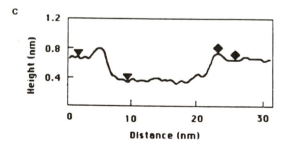

FIG. 19. (a) Image from the STM of etch pits in graphite after heating in air for several minutes at 650° C

(b) An enlarged view of the etch pits in (a)

(c) A profile of the etch pit shown in (b)

(From H. Chang and A. J. Bard, "Formation of Monolayer Pits of Controlled Nanometer Size on Highly Oriented Pyrolytic Graphite by Gasification Reactions as Studied by Scanning Tunneling Microscopy," *J. Am. Chem. Soc.* **112**, 4598-4599 (1990).)

faces. He suggests that e-beam irradiation increases the methane production by releasing the adsorbed molecules.

Erosion of graphite at room temperature is, also, observed when the surface is bombarded with electrons. Enhanced etching with electrons is evident in the early work of Hennig and Monet.[72] They determined that single vacancies were created in perfect layers of graphite when the surface was irradiated with electrons (0.14 to 0.26 keV).

There is clear evidence[73] that "electron enhanced etching" is frequently encountered in the field of electron microscopy where carbon films are often used as supports for the specimens. It is common to find that these films are thinned in the region exposed to the electron beam. In one model for this process it is assumed that the water molecules remaining on the film are ionized with electron beams. The reactivity of the ionized species increases to the level where they can react with the carbon atoms to form CO. The volatile hydrocarbon removes carbon from the surface and creates the etched pits. Original work on this system is contained in the paper by Egerton and Rossouw.[74]

"Electron enhanced etching" of graphite has been reported by several groups working with the STM. Albrecht et al.[18] report that the graphite surface can be etched with a single pulse on the tip of the STM. Their STM was operating in air with high relative humidity. The pit, shown in Fig. 20, was formed by pulsing the tip to 5 volts for a few microseconds. The size and shape of the pits were identical to those observed by Chang and Bard in their thermal etching system. When the Albrecht group placed the samples in a vacuum station, thereby removing the H_2O molecules, they found that it was not possible to mark the graphite surface, regardless of the strength of the voltage pulse. In an atmosphere of N_2, the graphite was equally impervious to damage. However, the pits were readily formed with the voltage pulse when they reintroduced H_2O molecules. A typical pattern is shown in Fig. 21.

Terashima et al.[75] found similar results in their work on surface modification. Mizutani et al.[76] used this process to scribe structures with a special shape in the uppermost layer of graphite. They positioned the tip at successive points along the circumference of a circle 100 Å in diameter. At each point they pulsed the tip voltage to form small overlapping pits as a circular ditch around the circumference. The monolayer

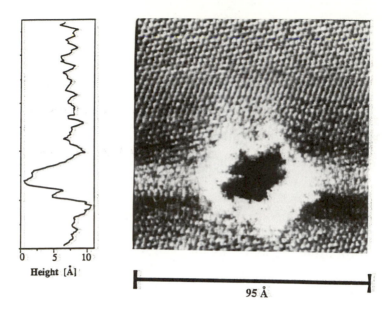

Height [Å]

95 Å

FIG. 20. Etch pits in graphite formed by pulsing the STM tip to 4 V for several μs in humid air. (From T. R. Albrecht, M. M. Dovek, M. D. Kirk, C. A. Lang and C. F. Quate, "Nanometer-Scale Hole Formation on Graphite Using a Scanning Tunneling Microscope," *Appl. Phys. Lett.* **55**, 1727-1729 (October 23, 1989).)

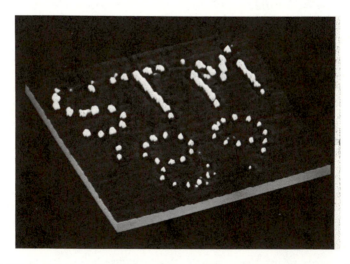

FIG. 21. Etch pits in graphite written with voltage pulses on the STM tip. (Courtesy of T. R. Albrecht and M. D. Kirk, Stanford University.)

disk inside the circle was exfoliated leaving the circular hole, conforming to the scribed outline, shown in Fig. 22.

Shen et al.[77] have observed similar pits when the tip of their STM was pulsed to 100 volts for 1 μs. The tip was spaced 100 Å from the substrate. It is likely that with this large voltage and large spacing, the molecules are ionized in the gap before they reach the substrate.

We mentioned in a previous section, the related experiment where Bernhardt[78] and Rabe[79] have successfully etched pits in graphite in ambient air.

More recent work has been reported by Penner et al[80] with an STM immersed in water. Domes, rather than pits, were formed on the graphite surface when the tip was pulsed to 4 volts for 20 μs. The domes were 7 Å in diameter and 1 Å high. The domes were converted to the monolayer pits 30-40 Å in diameter by positioning the tip over the dome and pulsing the tip to 200 mV. The moderate voltage for conversion suggests that the domes are metastable intermediates in the path to the formation of pits.

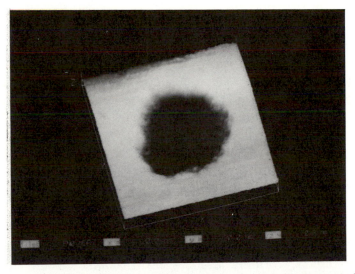

FIG. 22. A circular feature formed in graphite by fabricating overlapping etch pits around the circumference. (From W. Mizutani, J. Inukai and M. Ono, "Making a Monolayer Hole in a Graphite Surface by Means of a Scanning Tunneling Microscope," *Jpn. J. Appl. Phys.* **29**, L815-L817 (May, 1990)

11. EMISSION OF ATOMS -- ATOMIC EMISSION

A theme that occurs repeatedly in the work on surface modification is the creation of hillocks and craters. Several authors attributed their results to the transfer of material between the tip and substrate, but the mechanism for transfer was obscure. The situation remained cloudy until Mamin, Guethner and Rugar[81] did a careful study of atomic emission from a gold tip in close proximity to a gold substrate. They demonstrated that atoms can be transferred across this gap in a controlled fashion, consistently and reliably, with a threshold field of 0.1 V/Å. This is an order of magnitude lower than the field required to evaporate atoms from a tip in free space as in the Field-Ion Microscope (FIM).[82] This changes the character of the problem. Atomic emission should be considered in all of the events associated with the creation of hillocks and craters.

Before continuing with the Mamin-Guethner-Rugar report, we will discuss the various results on gold surfaces that followed the initial report from Berkeley.

11.1. Gold -- Revisited

The tip of the STM has been used to alter the surface of gold and silver in several case studies. Gimzewski and Möller[83] examined the modification of thin silver films deposited on a Silicon (111) surface. They worked in UHV with a Platinum/Iridium tip biased in the positive direction with respect to the silver surface. They found that they could create hillocks 30-100 Å in diameter and 20 Å in height. The "...smaller structures tended to anneal out over a period of up to several minutes,..." Different conditioning of the tips, "contaminated tips," would produce structures that appeared as indentations in the surface. Gimzewski and Moller suggested that "the tip itself may have contained a thin overlayer of silver previously deposited by field-emission cleaning. This may behave rather similar like a liquid-metal source reforming directly after contact.." They envisioned that the liquid metal formed a "neck of material which breaks, forming a local protrusion..." when the tip was retracted.

Rabe and Bucholz[84] reported on similar events for the surface of silver films grown as epitaxial layers on cleaved mica.[85] They used a conventional STM with a Pt/Ir tip operating in ambient air. Holes appeared in the film when the pulsed the tip above 5 volts. On occasion, they observed mounds on the substrate when the E-field between tip and film approached 1 V/Å. Their results were attributed to field evaporation of atoms.

Jaklevic and Elie[86] studied the surface of a single crystal of gold. They deliberately advanced the tip toward the surface by increasing the voltage on the piezo element controlling the spacing. Surface craters were formed in this operation but they changed shape and filled within minutes. This was attributed to surface diffusion of mobile atoms. Jaklevic[87] reports that diffusion is very high on some surfaces with the

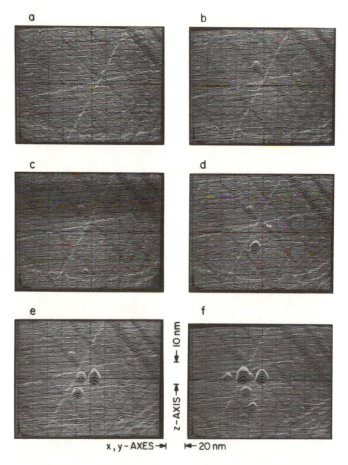

FIG. 23. A series of mounds written on gold (111) facet with 0.7 V on the tip of an STM with the surface covered with fluorocarbon grease. (From J. Schneir, R. Sonnenfeld, O. Marti, P. K. Hansma, J. E. Demuth and R. J. Hamers, "Tunneling Microscopy, Lithography and Surface Diffusion on an Easily Prepared, Atomically Flat Gold Surface," *J. Appl. Phys.* **63**, 717-721 (1988).)

cleanest surfaces exhibiting the highest diffusion rates. On other surfaces the diffusion rate was small, or non-existent.

Schneir et al.[88] took up the study with surfaces formed on the (111) facets of small gold balls. The balls were prepared by heating a gold wire to the melting temperature in the flame of an oxyacetylene torch, following a recipe described by Hsu and Cowley.[89] The facets, covered with fluorocarbon grease, were scanned with the STM tip immersed in the grease. They pulsed the tip voltage (biased negatively) to a value in excess of 3 volts to create both craters and mounds. The mounds were easily reproduced and in one instance the letter "T," as shown in Fig. 23, was written on the facet of their gold sample. The most reproducible mounds were generated "with the tips that had previously been pushed 100 nm into the gold surface with the z-axis piezoelectric element. (The tip was moved to a new region of the sample at least 20 microns from the site at which this procedure was performed.)" They found that the profiles flattened in a matter of hours with the gold diffusing outward from the mounds. Attempts to reproduce these results with the tip scanning in air were not successful.

Li et al.[90] report that the smooth (111) facet of a gold ball prepared by melting the end of a gold wire[91] can be "electroetched," a term they coined to describe the cratering of the surface when the STM tip voltage is suddenly increased to 2.7 volts. The work was done in air with a negative bias on the tungsten tip. The craters were 20-80 Å wide and 8-14 Å deep. They found no evidence of surface diffusion. The pattern shown in Fig. 24 is representative of what was done.

Emch et al.[92] studied the surface of a gold film epitaxially grown on mica.[93] They observed the formation of holes, and occasionally bumps, when the tip (biased positively) was pulsed to 3 volts for a time of 10-30 nanoseconds. The holes, 50-100 Å in diameter, were created on freshly prepared surfaces. On those surfaces where the diffusion rate was high, some of the step edges moved with a velocity of 5 Å/second.

The confusing aspect of this body of work is the different values for the diffusion rates. The time scales associated with a change in the features range from minutes to days and even months. The diffusion rates appear to depend on the nature of the atmosphere during evaporation of the gold atoms. Jaklevic tells us that "... a third element must be involved."[94]

One informative study that bears on the problem of diffusion rates

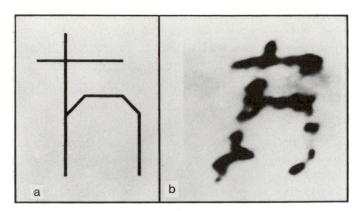

FIG. 24. (a) A computer generated template of the symbol ℏ The template is comprised of 201 coordinate points out of a possible 10,000. At each specified coordinate point, a voltage pulse is applied to a tungsten tip executing an STM scan.

(b) A subsequent scan of the STM tip over the same region reveals the STM image which is the ℏ symbol electroetched into a flat Au substrate. The ℏ symbol is ~ 65 nm high, ~ 40 nm wide and ~ 2 nm deep. (Courtesy of R. Reifenberger, Purdue University; also, Y. Z. Li, L. Vazquez, R. Piner, R. P. Andres and R. Reifenberger, "Writing Nanometer-Scale Symbols in Gold Using the Scanning Tunneling Microscope," *Appl. Phys. Lett.* **54**, 1424-1426 (April 10, 1989).)

was carried out by Francois et al.[95] Their evidence proves that the character of deposited gold films is dependent on the environment during evaporation. Francois studied the deposition of gold clusters onto a silicon substrate from an LMIS. (LMIS is an ion source described in the next section.) They used a strong magnetic field between source and substrate to spatially separate species with different charge to mass ratios. They found subtle differences in films deposited in the presence of energetic ions and those deposited in the absence of energetic ions. If energetic ions were present the film morphology showed a granular character with good integrity and strong adhesion to the substrate. It was inferred from the shape of the balls on the surface that there was little surface diffusion. The shape of the balls reproduced the shape of the spherical droplets emanating from the LMIS source. When the energetic ions were removed by deflection in the magnetic field the film morphology changed, the film integrity was lost and the adhesion to the substrate was degraded.

The work of Francois is enlightening for it tells us that the surface of gold is ubiquitous. The stability of nanostructures - however they are created - depends on the environment during creation. With this evidence in mind, it should not be surprising that gold hillocks deposited from the STM tip exhibit different behavior when they are formed in UHV, in air, and in other environments. The process must involve the transfer of material between tip and sample but the mechanism for transfer is unclear. The situation was clarified in the work of Mamin and Rugar, but before we inquire into that we want to discuss two classical systems for extracting atoms from condensed systems: the Liquid Metal Ion Source (LMIS), with a threshold field of 0.1 V/Å, and the Atom-Probe, with a threshold field of 1 V/Å.

11.2. Emission From LMIS

The Liquid Metal Ion Source (LMIS) is a prolific source of ions used for applications in ion beam microlithography,[96] and micromachining.[97] The conventional LMIS employs a heated tungsten needle with a shank 2 mm in diameter tapered to an apex with a radius of 2 microns. The shank is immersed in liquid metal and capillary action coats the apex with a thin layer of the molten metal. The liquid metal, wetting the needle, forms into a cone when a high voltage (10 kV) is applied.[98] It is known from the work of Taylor[99] that a liquid surface under the influence of a strong electric field will form into a stable cone where the stress of the electric field tending to stretch and elongate the cone is balanced by the surface tension acting to contract the cone. In "Taylor cones" with a half-angle of 49.3 degrees these two counter forces balance each other over the entire surface. Gallium, with a surface tension of 0.72 joules/mtr^2, and melting temperature near 30° C, is the preferred metal, but gold with a surface tension of 1.1 joules/mtr^2, and a melting temperature near 1000° C, has been employed in many instances. Other metals using LMIS[100] include: Ga, In, Al, Au, Ag, AuSi, Pd_4B_6, and Pd_2As.

Copious amounts of ions are emitted from the liquid "Taylor cones" when the field reaches the threshold. The threshold field for atomic evaporation is above 1 V/Å, but the field applied to the tip is much less (0.1 V/Å). The reduction in applied field comes about in the following way. The maximum field at the apex of the cone produces a finger-like protrusion from the apex with a small radius of curvature and this protrusion increases the field strength to the threshold for atomic evaporation.[101]

The LMIS is a unique ion source with the following properties: (1) high angular beam intensities, (2) moderate energy spreads, (3) small source size, and (4) high stability with little movement of the cone over long periods. The stability makes it feasible to use focused beams from the LMIS for the repair of microcircuits. These attributes and applications are discussed by Swanson et al.[102] The characteristics of gold ions from an LMIS are discussed by Wagner and Hall.[103]

In variation of the standard version of the LMIS, Bell et al.[104] turned to a configuration similar to the STM. They found stable emission with low voltages using gallium coated tungsten needles when the spacing between the apex of the cone and the substrate was reduced to a small value. Their experiment was complicated by the extension of the liquid surface when the cone was formed. The extension can be as large as 1 micron. In spite of this, Bell and his colleagues were able to place the LMIS source 1000 Å from the substrate and deposit lines with their tips biased at 100 volts. These lines, 1000 Å in width, stood as a prime achievement until Mamin and Rugar demonstrated that atoms could be directly emitted from the solid gold tip. In this circumstance, where there is no molten metal at the tip, it is an easy matter to place the tip in close proximity to the surface.

11.3. Emission From Atom Probes

The Field-Ion Microscope (FIM)[105] which provides a view of individual atoms on metal surfaces can be combined with a mass spectrometer. This combination is used to identify the chemical species of evaporated atoms. It is known as the Atom-Probe FIM.[106,107]

The process for removing atoms from the metal tip of the FIM is known as field evaporation,[108] or field desorption.[109,110] In this process, a 3 kV voltage pulse is applied to the FIM for several nanoseconds. This pulse generates a field in excess of 1 V/Å at the surface of the metal tip.[111] Several monolayers can be removed in a few nanoseconds.[112] Multiple charged ions, neutral atoms, and atomic clusters, are frequently observed in the emitted beams.

It is a simple and surprisingly efficient process for removing atoms from the top layer from a solid metal tip.

11.4. Gold Surfaces -- Final Results

We return to our final topic - the demonstration by Mamin,

Guethner and Rugar of field evaporation from solid metal tips. Their advance over previous work was the discovery that the barrier for field evaporation is reduced when a conducting surface is in close proximity to the tip. The barrier is lowered to the point where they get atomic emission with a threshold voltage of 3-5 volts between tip and sample.

Ultra-High Density Storage with the STM

a

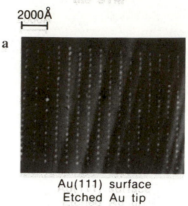

Au(111) surface
Etched Au tip

Pulse parameters: −4.4 V
                    300 nsec

Density ~10^{12} bits/in^2

Direct Deposition From A Gold STM Tip

b

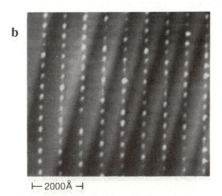

⊢ 2000Å ⊣

FIG. 25. Hillocks on gold formed with a negative pulse on the STM tip. The mounds are stable over periods in excess of 15 days. (From H. J. Mamin, S. Chiang, H. Birk, P. H. Guethner and D. Rugar, "Gold Deposition From a Scanning Tunneling Microscope Tip," J. Vac. Sci. Technol. **B9**, 1398-1402 (Mar/Apr 1991).

The evidence for lowering of the barrier comes from the experimental measurements. The measured threshold for a tip in close proximity to a metal surface is near 0.1 V/Å compared to a value in excess of 1 V/Å for a tip in free space.

The model for field desorption from a tip in free space has been worked out by Müller et al.[82] Mamin proposes that the height of the barrier in Müller's model is reduced by the overlap of the atomic potentials when a metal surface is brought close to the tip. The reduction in barrier height is significant for atomic emission is achieved with modest voltages.

Mamin et al.[81] used a gold tip formed at the end of a wire 250 microns in diameter electrochemically etched to a radius less than 1 micron. A single voltage pulse of 3.6 volts (sample positive with respect to the tip), 600 nanoseconds in length was sufficient to evaporate atoms and form mounds on the substrate. The procedure can be repeated without end. The tips appear to be self-annealing. This suggests that the gold atoms are drawn from the shank toward the tip to replenish the evaporated atoms. The mounds are 80-100 Å across at the base and 20-30 Å in height. The process is fast and reproducible. Mounds can be written with pulses as short as 10 nanoseconds. This is consistent with the data quoted earlier when 10 monolayers are evaporated in a few nanoseconds from the Atom-Probe. Furthermore, there is no diffusion when the gold atoms are deposited in air.

STM OPENS UP A NEW WORLD

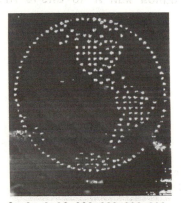

Scale 1:10,000,000,000,000

FIG. 26. Logo written on a smooth gold surface with atoms emitted from a gold tip of an STM operating in air. (Courtesy of H. J. Mamin, IBM.)

The printed dots are remarkably stable. The stability is illustrated in Fig. 25. The pattern of dots remains in place after many hours at elevated temperatures.

In closing, we present the logo of Fig. 26. It is a representative sample of the level of maturity that now exists in this field. The work of Mamin, Guethner and Rugar has clarified our thinking and resolved a number of conflicting issues. With gold tips and gold substrates they have given us a clear demonstration of the transfer of atoms from the tip to substrate, or from the substrate to the tip. This striking advance in surface modification opens a path for fabricating nano-scale structures.

ACKNOWLEDGEMENT

This work was supported by the Defense Advanced Research Projects Agency, the Joint Services Electronics Program and E. I. du Pont de Nemours and Company, and the National Science Founcation.

REFERENCES

1. G. Binnig and H. Rohrer, "Scanning Tunneling Microscopy," IBM J. Res. Develop. **30**, 355-369 (July 1986).

2. C. F. Quate, "Surface Modification With the STM and the AFM," in Scanning Tunneling Microscopy and Related Methods, pp. 281-297, R. J. Behm, N. Garcia and H. Rohrer, eds., Series E: Applied Sciences, Vol. 184, Kluwer Academic Publishers, Dordrecht, The Netherlands (1990).

3. G. M. Shedd and P. E. Russell, "The Scanning Tunneling Microscope as a Tool for Nanofabrication," Nanotechnology, **1**, 67-80 (1990).

4. D. W. Abraham, H. J. Mamin, E. Ganz and J. Clarke, "Surface Modification With the Scanning Tunneling Microscope," IBM J. Res. Develop. **30**, 492-499 (September 1986).

5. R. S. Becker, J. A. Golovchenko and B. S. Swartzentruber, "Atomic-Scale Surface Modifications Using a Tunneling Microscope," Nature, **325**, 419-421 (January 29, 1987).

6. M. Ringger, H. R. Hidber, R. Schlögl, P. Oelhafen and H.-J. Güntherodt, "Nanometer Lithography With the Scanning

Tunneling Microscope," Appl. Phys. Lett. **46**, 832-834 (May 1, 1985).

7. M. A. McCord and R. F. W. Pease, "Lithography With the Scanning Tunneling Microscope," J. Vac. Sci. Technol. **B4**, 86-88 (Jan/Feb, 1986).

8. R. L. Stewart, "Insulating Films Formed Under Electron and Ion Bombardment," Phys. Rev. **45**, 488-490 (April 1, 1934).

9. A. N. Broers, W. W. Molzen, J. J. Cuomo and N. D. Wittels, "Electron-Beam Fabrication of 80-Å Metal Structures," Appl. Phys. Lett. **29**, 596-598 (November 1, 1976).

10. A. N. Broers and M. Pomerantz, "Rapid Writing of Fine Lines in Langmuir-Blodgett Films Using Electron Beams," Thin Solid Films, **99** 323-329 (1983); also, G. G. Roberts, "Langmuir-Blodgett Films on Semiconductors," pp. 56-67, Part II, Thin Insulating Films, in Insulating Films on Semiconductors, M. Schulz and G. Pensl, eds. (Springer-Verlag, Berlin 1981).

11. C. R. K. Marrian and R. J. Colton, "Low-Voltage Electron Beam Lithography With a Scanning Tunneling Microscope," Appl. Phys. Lett. **56**, 755-757 (February 19, 1990).

12. See, for example, J. D. Swalen, D. L. Allara, J. D. Andrade, E. A. Chandross, S. Garoff, J. Israelachvili, T. J. McCarthy, R. Murray, R. F. Pease, J. F. Rabolt, K. J. Wynne and H. Yu, "Molecular Monolayers and Films," Langmuir, **3**, 932-950 (1987).

13. D. P. E. Smith, A. Bryant, C. F. Quate, J. P. Rabe, Ch. Gerber and J. D. Swalen, "Images of a Lipid Bilayer at Molecular Resolution by Scanning Tunneling Microscopy," Proc. Natl. Acad. Sci. USA **84**, 969-972 (February, 1987); also, C. A. Lang, J. K. H. Horber, T. W. Hansch, W. M. Heckl and H. Mohwald, "Scanning Tunneling Microscopy of Langmuir-Blodgett Films on Graphite," J. Vac. Sci. Technol. A **6**, 368-370 (1988); also, H. G. Braun, H. Fuchs and W. Schrepp, "Surface Structure Investigation of Langmuir-Blodgett Films," Thin Solid Films, **159**, 301-314 (1988); also, W. Mizutani, M. Shigeno, K. Saito, K. Watanabe, M. Sugi, M. Ono and K. Kajimura, "Observation of Langmuir-Blodgett Films by Scanning Tunneling Microscopy," Jpn. J. Appl. Phys., **27**, 1803-1807 (1988).

14. H. Fuchs, S. Akari and K. Dransfeld, "Molecular Resolution of Langmuir-Blodgett Monolayers on Tungsten Diselenide by Scanning Tunneling Microscopy," Z. Physik, B-Condensed Matter **80**, 389-392 (1990).

15. R. Kuroda, E. Kishi, A. Yamano, K. Hatanaka, H. Matsuda, K. Eguchi and T. Nakagiri, "Scanning Tunneling Microscope Images of Fatty Acid Langmuir Blodgett Bilayers," J. Vac. Sci. Technol. **B9**, 1180-1183 (Mar/Apr 1991).

16. M. M. Dovek, T. R. Albrecht, S. W. J. Kuan, C. A. Lang, R. Emch, P. Grütter, C. W. Frank, R. F. W. Pease and C. F. Quate, "Observation and Manipulation of Polymers by Scanning Tunneling and Atomic Force Microscopy," J. Microsc. **152**, 229-236 (October, 1988).

17. H. Zhang, L. S. Hordon, S. W. J. Kuan, P. Maccagno and R. F. W. Pease, "Exposure of Ultrathin Polymer Resists With the Scanning Tunneling Microscope," J. Vac. Sci. Technol. B **7**, 1717-1722 (Nov/Dec, 1989).

18. T. R. Albrecht, M. M. Dovek, M. D. Kirk, C. A. Lang, C. F. Quate and D. P. E. Smith, "Nanometer-Scale Hole Formation on Graphite Using a Scanning Tunneling Microscope," Appl. Phys. Lett. **55**, 1727-1729 (October 23, 1989).

19. J. P. Rabe, S. Buchholz and A. M. Ritcey, "Reactive Graphite Etch and the Structure of an Adsorbed Organic Monolayer - a Scanning Tunneling Microscopy Study," J. Vac. Sci. Technol. A **8**, 679-683 (Jan/Feb, 1990).

20. J. S. Foster and J. E. Frommer, "Imaging of Liquid Crystals Using a Tunneling Microscope," Nature, **333**, 542-545 (June 9, 1988); also, J. K. Spong, H. A. Mizes, L. J. La Comb, Jr. M. M. Dovek, J. E. Frommer and J. S. Foster, "Contrast Mechanism for Resolving Organic Molecules With Tunneling Microscopy," Nature, **338**, 137-139 (March 9, 1989).

21. W. Mizutani, M. Shigeno, M. Ono and K. Kajimura, "Voltage-Dependent Scanning Tunneling Microscopy Images of Liquid Crystals on Graphite," Appl. Phys. Lett., **56**, 1974-1976 (May 14, 1990).

22. D. P. E. Smith, H. Hörber, Ch. Gerber and G. Binnig, "Smectic Liquid Crystal Monolayers on Graphite Observed by Scanning Tunneling Microscopy," Science, **245**, 43-45 (July 7, 1989); also, D. P. E. Smith, J. K. H. Hörber, G. Binnig and H. Nejoh, "Structure, Registry and Imaging Mechanism of Alkylcyanobiphenyl Molecules by Tunneling Microscopy," Nature **334**, 641-644 (April 12, 1990).

23. J. S. Foster, J. E. Frommer and P. C. Arnett, "Molecular Manipulation using a Tunneling Microscope," Nature **331**, 324-326 (January 28, 1988).

24. R. H. Bernhardt, G. C. McGonigal, R. Schneider and D. J. Thomson, "Mechanisms for the Deposition of Nanometer-Sized Structures From Organic Fluids Using the Scanning Tunneling Microscope," J. Vac. Sci. Technol. A **8**, 667-671 (Jan/Feb, 1990); also, G. C. McGonigal, R. H. Bernhardt and D. J. Thomson, "Scanning Tunneling Microscopy of Ordered Alkane Layers Adsorbed on Graphite," Appl. Phys. Lett. **57**, 28-30 (1990).

25. R. Sonnenfeld and P. K. Hansma, "Atomic-Resolution Microscopy in Water," Science, **232**, 211-213 (1986).

26. J. Schneir and P. K. Hansma, "Scanning Tunneling Microscopy and Lithography of Solid Surfaces Covered With Nonpolar Liquids," Langmuir, **3**, 1025-1027 (1987).

27. R. Sonnenfeld and B. C. Schardt, "Tunneling Microscopy in an Electrochemical Cell: Images of Ag Plating," Appl. Phys. Lett. **49**, 1172-1174 (November 3, 1986).

28. C. W. Lin, F-R. F. Fan, and A. J. Bard, "High Resolution Photoelectrochemical Etching of n-GaAs With the Scanning Electrochemical and Tunneling Microscope," J. Electrochem. Soc., **134**, 1038-1039 (April, 1987).

29. P. A. Kohl, C. Wolowodiuk and F. W. Ostermayer, Jr., "The Photoelectrochemical Oxidation of (100), (111) and (111) n-InP and n-GaAs," J. Electrochem. Soc: Solid-State Sci. Technol. **130**, 2288-2293 (November, 1983).

30. F. W. Ostermayer, Jr. and P. A. Kohl, "Photoelectrochemical Etching of p-GaAs," Appl. Phys. Lett. **39**, 76-78 (July 1, 1981).

31. R. M. Lum, A. M. Glass, F. W. Ostermayer, Jr., P. A. Kohl, A. A. Ballman and R. A. Logan, "Holographic Photoelectrochemical Etching of Diffraction Gratings in n-InP and n-GalnAsP for Distributed Feedback Lasers," J. Appl. Phys. **57**, 39-44 (January 1, 1985).

32. J. J. Ritsko, "Laser Etching," Chap 6, 363-383, in <u>Laser Microfabrication</u>, D. J. Ehrlich and J. Y. Tsao, eds. (Academic Press, Inc. New York 1989).

33. L. A. Nagahara, private communication.

34. L. A. Nagahara, T. Thundat and S. M. Lindsay, "Preparation and Characterization of STM Tips for Electrochemical Studies," Rev. Sci. Instrum. **60**, 3128-3130 (October, 1989).

35. L. A. Nagahara, T. Thundat and S. M. Lindsay, "Nanolithography on Semiconductor Surfaces Under an Etching Solution," Appl. Phys. Lett. **57**, 270-272 (July 16, 1990).

36. E. J. van Loenen, D. Dijkkamp, A. J. Hoeven, J. M. Lenssinck and J. Dieleman, "Direct Writing in Si With a Scanning Tunneling Microscope," Appl. Phys. Lett. **55**, 1312-1314 (September 25, 1989); also, E. J. van Loenen, D. Dijkkamp, A. J. Hoeven, J. M. Lenssinck and J. Dieleman, "Nanometer Scale Structuring of Silicon by Direct Indentation," J. Vac. Sci. Technol. A **8**, 574-576 (Jan/Feb, 1990).

37. M. Iwatsuki, private communication; also, M. Iwatsuki, S. Kitamura and A. Mogami, "Development of the UHV-STM," in Proceedings XIIth Intl. Congress for Elec. Microscopy, pp. 322-323 (San Francisco Press, Inc., San Francisco, 1990).

38. F. Besenbacher and K. Mortensen, private communication.

39. Ph. Avouris, private communication.

40. R. E. Kirby and D. Lichtman, "Electron Beam Induced Effects on Gas Absorption Utilizing Auger Electron Spectroscopy: CO and O_2 on Si," Surf. Sci. **41**, 447-466 (1974); also, J. P. Coad, H. E. Bishop and J. C. Riviere, "Electron-Beam Assisted Adsorption on the Si(111) Surface," Surf. Sci. **21**, 253-264 (1970).

41. J. A. Dagata, J. Schneir, H. H. Harary, C. J. Evans, M. T. Postek

and J. Bennett, "Modification of Hydrogen-Passivated Silicon by a Scanning Tunneling Microscope Operating in Air," Appl. Phys. Lett. **56**, 2001-2003 (May 14, 1990); also, J. A. Dagata, J. Schneir, H. H. Harary, J. Bennett and W. Tseng, "Pattern Generation on Semiconductor Surfaces by a Scanning Tunneling Microscope Operating in Air," J. Vac. Sci. Technol. **B9**, 1384-1388 (Mar/Apr 1991).

42. F. J. Grunthaner and P. J. Grunthaner, "Chemical and Electronic Structure of the SiO_2/Si Interface," Matl. Sci. Rep. **1**, 65-160 (1986).

43. D. B. Fenner, D. K. Biegelsen and R. D. Bringans, "Silicon Surface Passivation by Hydrogen Termination: a Comparative Study of Preparation Methods," J. Appl. Phys. **66**, 419-424 (1989).

44. W. J. Kaiser, L. D. Bell, M. H. Hecht and F. J. Grunthaner, "Scanning Tunneling Microscopy Characterization of the Geometric and Electronic Structure of Hydrogen-Terminated Surfaces," J. Vac. Sci. Technol. **A6**, 519-523 (1988).

45. U. Staufer, R. Wiesendanger, L. Eng, L. Rosenthaler, H. R. Hidber, H.-J. Güntherodt and N. Garcia, "Nanometer Scale Structure Fabrication With The Scanning Tunneling Microscope," Appl. Phys. Lett. **51**, 244-246 (1987); also, U. Staufer, R. Wiesendanger, L. Eng, L. Rosenthaler, H. R. Hidber, H.-J. Güntherodt and N. Garcia "Surface Modification in the Nanometer Range by the Scanning Tunneling Microscope," J. Vac. Sci. Technol. A **6**, 537-539 (Mar/Apr, 1988); also U. Staufer, L. Scandella and R. Wiesendanger, "Direct Writing of Nanometer Scale Structures on Glassy Metals by the Scanning Tunneling Microscope," Z. Physik B - Condensed Matter, **77**, 281-286 (1989).

46. P. Chaudhari, J. J. Cuomo and R. J. Gambino, "Amorphous Metallic Films for Magneto-Optic Applications," Appl. Phys. Lett. **22**, 337-339 (April 1, 1973); also, J. C. Suits, D. Rugar and C. J. Lin, "Thermomagnetic Writing in Tb-Fe:Modeling and Comparison with Experiment," J. Appl. Phys. **64**, 252-261 (July 1, 1988).

47. J. Jahanmir, P. E. West, S. Hsieh and T. N. Rhodin, "Surface Modification of a-Si:H With a Scanning Tunneling Microscope Operated in Air," J. Appl. Phys. **65**, 2064-2068 (March 1, 1989).

48. R. Wiesendanger, L. Rosenthaler, H. R. Hidber, H.-J. Güntherodt, A. W. McKinnon and W. E. Spear, "Hydrogenated Amorphous Silicon Studied by Scanning Tunneling Microscope," J. Appl. Phys. **63**, 4515-4517 (May 1, 1988).

49. E. Hartmann and F. Koch, private communication.

50. J. S. Foster, K. A. Rubin and D. Rugar, "Data Storage Method Using State Transformable Materials," US Patent No. 4,916,688 (April 10, 1990).

51. J. Moreland and P. Rice, "High-Resolution, Tunneling-Stabilized Magnetic Imaging and Recording," Appl. Phys. Lett. **57**, 310-312 (July 16, 1990).

52. S. Iwamura, Y. Nishida and K. Hashimoto, "Rotating MNOS Disk Memory Device," IEEE Trans. Electron Devices, ED-28, 854-860 (July, 1981).

53. J. J. Chang, "Nonvolatile Semiconductor Memory Devices," Proc. IEEE, **64**, 1039-1059 (July, 1976).

54. M. M. E. Beguwala and T. L. Gunckel, II, "An Improved Model for the Charging Characteristics of a Dual-Dielectric (MNOS) Nonvolatile Memory Device," IEEE Trans. Electron Devices, **ED–25**, 1023-1030 (August, 1978).

55. J. E. Brewer, "MNOS Density Parameters," IEEE Trans. Electron Devices, **ED–24**, 618-625 (May, 1977).

56. R. S. Withers, R. W. Ralston and E. Stern, "Nonvolatile Analog Memory in MNOS Capacitors," IEEE Electron Dev. Lett. **EDL-1**, 42-45 (March, 1980).

57. Work relating to this project has been reported by J. E. Stern, B. D. Terris, H. J. Mamin and D. Rugar, "Deposition and Imaging of Localized Charge on Insulator Surfaces Using a Force Microscope," Appl. Phys. Lett. **53**, 2717-2719 (December 26, 1988); also, Y. Martin, D. W. Abraham and H. K. Wickramasinghe, "High-Resolution Capacitance Measurement and Potentiometry by Force Microscopy," Appl. Phys. Lett. **52**, 1103-1105 (March 28, 1988).

58. D. J. Erlich, R. M. Osgood, Jr. and T. F. Deutsch,

"Photodeposition of Metal Films With Ultraviolet Laser Light," J. Vac. Sci. Technol. **21**, 23-32 (May/June, 1982); also, D. V. Podlesnik, H. H. Gilgen and R. M. Osgood, Jr., "Waveguiding Effects in Laser-Induced Aqueous Etching of Semiconductors," Appl. Phys. Lett. **48**, 496-498 (February 17, 1986); also, H. S. Cole, Y. S. Liu, J. W. Rose and R. Guida, "Laser-Induced Selective Copper Deposition on Polyimide," Appl. Phys. Lett. **53**, 2111-2113 (November 21, 1988); also, R. J. von Gutfeld, "Laser-Enhanced Patterning Using Photothermal Effects: Maskless Plating and Etching," J. Opt. Soc. Am. B **4**, 272-279 (February, 1987).

59. S. Matsui and K. Mori, "New-Selective Deposition Technology by Electron Beam Induced Surface Reaction," J. Vac. Sci. Technol. B **4**, 299-304 (Jan/Feb, 1986).

60. S. Matsui and T. Ichihashi, "*In Situ* Observation on Electron-Beam-Induced Chemical Vapor Deposition by Transmission Electron Microscopy," Appl. Phys. Lett. **53**, 842-844 (September 5, 1988).

61. S. Matsui, "Electron Beam Lithography Using Surface Reactions With C1F$_3$," Appl. Phys. Lett. **55**, 134-136 (July 10, 1989).

62. R. M. Silver, E. E. Ehrichs and A. L. de Lozanne, "Direct Writing of Submicron Metallic Features With a Scanning Tunneling Microscope," Appl. Phys. Lett. **51**, 247-249 (July 27, 1987); also, E. E. Ehrichs, R. M. Silver and A. L. de Lozanne, "Direct Writing With the Scanning Tunneling Microscope," J. Vac. Sci. Technol. A **6**, 540-543 (Mar/Apr, 1988).

63. M. A. McCord, D. P. Kern and T. H. P. Chang, "Direct Deposition of 10-nm Metallic Features With the Scanning Tunneling Microscope," J. Vac. Sci. Technol. **B6**, 1877-1880 (Nov/Dec, 1988).

64. D. D. Awschalom, M. A. McCord and G. Grinstein, "Observation of Macroscopic Spin Phenomena in Nanometer-Scale Magnets," Phys. Rev. Lett. **65**, 783-786 (August 6, 1990).

65. D. D. Awschalom, J. R. Rozen, M. B. Ketchen, W. J. Gallagher, A. W. Kleinsasser, R. L. Sandstrom and B. Bumble, "Low-Noise Modular Microsusceptometer Using Nearly Quantum Limited dc SQUIDs," Appl. Phys. Lett. **53**, 2108-2110 (November 21, 1988).

66. D. M. Eigler and E. K. Schweizer, "Positioning Single Atoms With a Scanning Tunneling Microscope," Nature, **344**, 524-526 (April 5, 1990).

67. G. R. Hennig, "Electron Microscopy of Reactivity Changes Near Lattice Defects in Graphite," in Chemistry and Physics of Carbon, vol. 2, P. L. Walker, Jr., ed. pp. 1-49, Marcel Dekker, Inc., New York (1966); also, R. T. Yang, "Etch-Decoration Electron Microscopy Studies of the Gas-Carbon Reactions," in Chemistry and Physics of Carbon, vol 19, P. A. Thrower, ed. pp. 163-210, Marcel Dekker, Inc., New York (1984).

68. R. T. Yang, loc. cit.

69. H. Chang and A. J. Bard, "Formation of Monolayer Pits of Controlled Nanometer Size on Highly Oriented Pyrolytic Graphite by Gassification Reactions as Studied by Scanning Tunneling Microscopy," J. Am. Chem. Soc. **112**, 4598-4599 (1990).

70. C. I. H. Ashby and R. R. Rye, "Electron Enhanced Hydrogen Attack on First Wall Materials," J. Nucl. Matl. **103 & 104**, 489-492 (1981).

71. A. A. Haasz, P. C. Stangeby and O. Auciello, "Methane Production From Graphite and TiC Under Electron and Atomic Hydrogen Impact," J. Nucl. Matl. **111 & 112**, 757-762 (1982).

72. G. R. Hennig, loc. cit.

73. J. J. Hren, "Barriers to AEM: Contamination and Etching," in Analytical Electron Microscopy, J. J. Hren, D. Joy and J. I. Goldstein, eds. Chap. 18, pp. 481-505, Plenum Press, New York (1985).

74. R. F. Egerton and C. J. Rossouw, "Direct Measurement of Contamination and Etching Rates in an Electron Beam," J. Phys. D: Appl. Phys. **9**, 659-663 (1976).

75. K. Terashima, M. Kondoh and T. Yoshida, "Fabrication of Nucleation Sites for Nanometer Size Selective Deposition by Scanning Tunneling Microscope," J. Vac. Sci. Technol. A **8**, 581-584 (Jan/Feb, 1990).

76. W. Mizutani, J. Inukai and M. Ono, "Making a Monolayer Hole in

a Graphite Surface by Means of a Scanning Tunneling Microscope," Japn. J. Appl. Phys. **29**, L815-L817 (May, 1990).

77. T. C. Shen, R. T. Brockenbrough, J. S. Hubacek, J. R. Tucker, and J. W. Lyding, "Ion Irradiation Effects on Graphite With the Scanning Tunneling Microscope," J. Vac. Sci. Technol. **B9**, 1376-1379 (Mar/Apr 1991).

78. R. H. Bernhardt, loc. cit.

79. J. P. Rabe, loc. cit.

80. R. M. Penner, M. J. Heben, N. S. Lewis and C. F. Quate, "Mechanistic Investigations of Nanometer-Scale Lithography at Liquid-Covered Graphite Surfaces," Appl. Phys. Lett. **58**, 1389-1391 (Arpil 1, 1991).

81. H. J. Mamin, S. Chiang, H. Birk, P. H. Guethner and D. Rugar, "Gold Deposition From a Scanning Tunneling Microscope Tip," J. Vac. Sci. Technol. **B9**, 1398-1402 (Mar/Apr 1991); also, H. J. Mamin, P. H. Guethner and D. Rugar, "Atomic Emission From a Gold Scanning-Tunneling-Microscope Tip," Phys. Rev. Lett. **65**, 2418-2421 (November 5, 1990).

82. E. W. Müller and T. T. Tsong, in <u>Field Ion Microscopy, Principles and Applications</u>, Elsevier, New York (1969).

83. J. K. Gimzewski and R. Möller, "Transition From the Tunneling Regime to Point Contact Studied Using Scanning Tunneling Microscopy," Phys. Rev. **B36**, 1284-1287 (July 15, 1987).

84. J. P. Rabe and S. Buchholz, "Fast Nanoscale Modification of Ag(111) Using a Scanning Tunneling Microscope," Appl. Phys. Lett. **58**, 702-704 (1991).

85. S. Buchholz, H. Fuchs and J. P. Rabe, "Surface Structure of Thin Metallic Films on Mica as Seen by Scanning Tunneling Microscopy, Scanning Electron Microscopy and Low-Energy Electron Diffraction" J. Vac. Sci. Technol. **B9**, 857-861 (Mar/Apr 1991).

86. R. C. Jaklevic and L. Elie, "Scanning-Tunneling-Microscope Observation of Surface Diffusion on an Atomic Scale: Au on Au(111)," Phys. Rev. Lett. **60**, 120-123 (January 11, 1988).

627

87. R. C. Jaklevic, private communication.

88. J. Schneir, R. Sonnenfeld, O. Marti, P. K. Hansma, J. E. Demuth and R. J. Hamers, "Tunneling Microscopy, Lithography and Surface Diffusion on an Easily Prepared, Atomically Flat Gold Surface, " J. Appl. Phys. **63**, 717-721 (1988); also J. Schneir and P. K. Hansma, "Scanning Tunneling Microscopy and Lithography of Solid Surfaces Covered With Nonpolar Liquids," Langmuir, **3**, 1025-1027 (1987).

89. T. Hsu and J. M. Cowley, "Reflection Electron Microscopy (REM) of fcc Metals," Ultramicroscopy, **11**, 239-250 (1983).

90. Y. Z. Li, L. Vazquez, R. Piner, R. P. Andres and R. Reifenberger, "Writing Nanometer-Scale Symbols in Gold Using the Scanning Tunneling Microscope," Appl. Phys. Lett. **54**, 1424-1426 (April 10, 1989); also, US Patent No. 4,896,044 (January 23, 1990).

91. T. Hsu and J. M. Cowley, loc. cit.

92. R. Emch, J. Nogami, M. M. Dovek, C. A. Lang and C. F. Quate, "Characterization of Gold Surfaces for Use as Substrates in Scanning Tunneling Microscopy Studies," J. Appl. Phys. **65**, 79-84 (January 1, 1989).

93. K. Reichelt and H. O. Lutz, "Hetero-Epitaxial Growth of Vacuum Evaporated Silver and Gold," J. Crystal Growth, **10**, 103-107 (1971); also, V. M. Hallmark, S. Chiang, J. F. Rabolt, J. D. Swalen and R. J. Wilson, "Observation of Atomic Corrugation on Au(111) by Scanning Tunneling Microscopy," Phys. Rev. Lett. **59**, 2879-2882 (December 21, 1987); also, E. Holland-Moritz, J. Gordon II, G. Borges and R. Sonnenfeld, "Motion of Atomic Steps of Au(111) Films on Mica," Langmuir **7**, 301-306 (February 1991).

94. R. C. Jaklevic, private communication.

95. M. Francois, K. Pourrezaei, A. Bahasadri and D. Nayak, "Investigation of the Liquid Metal Ion Source Cluster Beam Constituents and Their Role in the Properties of the Deposited Film," J. Vac. Sci. Technol. **B5**, 178-183 (Jan/Feb, 1987).

96. R. L. Seliger, J. W. Ward, V. Wang and R. L. Kubena, "A High-Intensity Scanning Ion Probe With Submicrometer Spot Size," Appl. Phys. Lett. **34**, 310-312 (March 1, 1979).

97. J. Puretz, R. K. DeFreez, R. A. Elliott and J. Orloff, "Focused-Ion-Beam Micromachined AlGaAs Semiconductor Laser Mirrors," Electron. Lett. **22**, 700-702 (June 19, 1986).

98. R. Gomer, "On the Mechanism of Liquid Metal Electron and Ion Sources," Appl. Phys. **19**, 365-375 (1979).

99. G. I. Taylor, "Disintegration of Water Drops in an Electric Field," Proc. Roy. Soc. London, Ser. A, **280**, 383-397 (1964).

100. A. E. Bell, private communication.

101. G. Ben Assayag and P. Sudraud, "LMIS Energy Broadening Interpretation Supported by HV-TEM Observations," J. de Phys. Colloque **C9**, Tome 45, C9-223-226 (December, 1984).

102. L. W. Swanson, G. A. Schwind, A. E. Bell and J. E. Brady, "Emission Characteristics of Gallium and Bismuth Liquid Metal Field Ion Sources," J. Vac. Sci. Technol. **16**, 1864-1867 (Nov/Dec, 1979).

103. A. Wagner and T. M. Hall, "Liquid Gold Ion Source," J. Vac. Sci. Technol. **16**, 1871-1874 (Nov/Dec, 1979); also, A. Wagner, "The Hydrodynamics of Liquid Metal Ion Sources," Appl. Phys. Lett. **40**, 440-442 (March 1, 1982).

104. A. E. Bell, K. Rao and L. W. Swanson, "Scanning Tunneling Microscope Liquid-Metal Ion Source for Microfabrication," J. Vac. Sci. Technol. **B6**, 306-310 (Jan/Feb, 1988); also, G. Ben Assayag, P. Sudraud and L. W. Swanson, "Close-Spaced Ion Emission From Gold and Gallium Liquid Metal Ion Source," Surf. Sci. **181**, 362-369 (1987).

105. T. T. Tsong, in Atom-Probe Field Ion Microscopy, Cambridge Univ. Press, Cambridge (1990).

106. E. W. Müller, J. A. Panitz and S. B. McLane, "The Atom-Probe Field Ion Microscope," Rev. Sci. Instrum. **39**, 83-86 (January, 1968).

107. J. A. Panitz, "The 10 cm Atom Probe," Rev. Sci. Instrum. **44**, 1034-1038 (August, 1973).

108. E. W. Müller, "Field Ionization and Field Ion Microscopy," in Advances in Electronics and Electron Physics, vol. XIII, pp. 83-179 (Academic Press Inc. New York, 1960).

109. E. W. Müller, "Field Desorption," Phys. Rev. **102**, 618-624 (May 1, 1956).

110. R. Gomer and L. W. Swanson, "Theory of Field Desorption," J. Chem. Phys. **38**, 1613-1629 (April 1, 1963).

111. T. T. Tsong, "Studies of Solid Surfaces at Atomic Resolution-Atom-Probe and Field Ion Microscopy," Surf. Sci. Rep. **8**, 127-209 (North-Holland, Amsterdam, 1988).

112. "One meter of atoms per second" is the rate of removal of atoms from a surface in the Atom Probe, so it is not surprising to observe atomic layers removed in nanoseconds; J. Panitz, private communication.

THE PHYSICS OF TIP-SURFACE APPROACHING:
SPECULATIONS AND OPEN ISSUES

E. Tosatti*

International School for Advanced Studies
Via Beirut 2, 34014 Trieste, Italy
and
International Center for Theoretical Physics
P.O. Box 586, 34014 Trieste, Italy

Abstract - This lecture describes the current state of understanding of some physical issues raised by the STM/AFM work of the 80's, as seen by an interested outsider. Some tentative ideas about new things which could be done, and which seem of interest to a surface theorist, are presented as digressions. In the rest of the lecture I try very briefly to focus on problematic and unresolved questions, rather than on the many well-established, well understood pieces of physics, for which excellent reviews can already be found.

1. THE PHYSICS OF TIP-SURFACE APPROACHING AND TOUCHING

The physics of two approaching bodies, and particularly the forces between them, have received much attention in the general area of adhesion.[1] This field developed quietly over many decades, until the Scanning

---

* Work done partly in collaboration with O. Tomagnini, F. Ancilotto, A. Selloni, S. Iarlori and F. Ercolessi.

Tunneling Microscope (STM) broke into the scene in the early 80's,[2,3] soon followed by the Atomic Force Microscope (AFM).[4] These inventions, along with a host of related more recent devices, have made it possible to sense both electrically and mechanically the tip-surface interaction, to an unprecedented level of resolution and detail.[5] With their help, we have learned a great deal about both morphology and electronic structure of substrates in very broad areas like surface physics, biology, technology etc. Correspondingly, our microscopic understanding of the tip-surface interaction has also increased.

I shall not endeavor here to offer a comprehensive review of the results which have emerged in this vast field. Rather, in line with the spirit of this meeting, I will qualitatively comment on the main aspects of tip-surface interaction, with the purpose of pin-pointing at least some issues which appear to be interesting and potentially rich of development in the 90's. This will also partly be done through digressions onto sidetracks, whose choice is by necessity highly arbitrary. The review material is taken from the literature. Most of the digression material and ideas presented are unpublished, and are offered obviously as tentative rather than definitive results. Although I do of course hope this lecture does not contain too many mistakes and omissions, freedom of speculation rather than precision of account will be my main goal.

The classical description of the interaction between two approaching bodies had been historically mostly based on macroscopic concepts. With the discovery of STM, the level suddenly became fully microscopic. However, the emphasis switched from mechanical interaction to the onset of electrical conduction, due to electron tunneling at very close proximity. For several years in the early and mid-80's, attempts at understanding tunneling conditions and the resulting I-V characteristics were detached from any references to mechanical tip-surface interaction. The development of AFM has subsequently underlined the need for a new microscopic understanding of mechanical interactions as well. Very recently, Dürig et al.[6] have shown that it is possible to measure simultaneously both current *and* forces as a function of the tip-surface distance. As will be discussed in more detail later, this combined approach appears to be extremely instructive, and is very likely to bear more fruit in the future.

In the end, even when they are not measured together, forces and currents should be understood together. In the following I shall review briefly how forces and tunneling currents are currently understood in the various regime of tip-surface approach. The discussion will be broken by digressions and speculations over side tracks, as announced earlier.

2. THREE REGIMES OF TIP-SURFACE APPROACHING

Between the two extreme situations of very large distance (where tip and surface do not "see" each other) and of zero distance (point contact), the changes are mostly gradual. Gradually, as the tip-surface distance decreases, weak van der Waals (vdW) attractive forces grow between the two bodies. At the same time, weak tunneling processes become possible, and some current can be passed across. This is the large distance regime, which I shall review first. In this regime, the theory for the force (van der Waals attraction) and that for the current (Bardeen tunneling) are both long-established, and totally unrelated to one another. At closer distance, the mutual modifications which tip and surface cause to each other can no longer be ignored. Electronic wavefunctions overlap strongly, and this affects both force and current. This is the medium distance regime. At even closer quarters, forces acquire a dominant role, mechanical deformations become crucial and must be well understood before any understanding of currents can even be considered. This is the third and final regime of shortest approach, just preceding and leading to the point contact. Leaving aside the point contact itself (whose physics is also very interesting, as discussed separately in Wyder's lecture[7], large, medium and short-distances are the three regimes I shall separately address below.

To be sure, it is not always so clear to what regime a given experimental situation corresponds. Even theoretically, the tip-surface distance is not unambiguously defined. We shall conventionally use the closest distance between the outermost ion cores of the substrate and those of the tip as our definition of the tip-surface distance, s. Very loosely speaking, and subject to revision by the experts of any particular case, large distance means $s > 5\text{Å}$, medium means $3\text{Å} < s < 5\text{Å}$, and short distance $s < 3\text{Å}$. Here, 5Å is taken as representative of a distance where most tip-surface wavefunction overlap becomes exponentially small and practically irrelevant, while 3Å is in the order of the interatomic spacing of a metal. Let us begin, now.

3. LARGE DISTANCE REGIME: VAN DER WAALS FORCES

The van der Waals, or dispersion force arising from virtual photon exchange gives rise to a potential between two parallel semi-infinite media a distance s apart, given by

$$U_{vdW}(S) = -\frac{A \, x(area)}{12\pi s^2} \,,$$ (1)

where conventionally A is called the Hamaker constant[1] and is related to the dielectric functions $\varepsilon_1(\omega)$ and $\varepsilon_2(\omega)$ of the two media by the Lifshitz formula[8]

$$A \simeq \frac{3\hbar}{4\pi} \int_o^\infty d\omega \, \frac{\varepsilon_1(i\omega) - 1}{\varepsilon_1(i\omega) + 1} \, \frac{\varepsilon_2(i\omega) - 1}{\varepsilon_2(i\omega) + 1}$$ (2)

The Hamaker constant A is typically in the eV range. For instance, it can be estimated to be 1.6 eV for a Si - Si interface, and 3.4 eV for an Au -Au interface.[9] While this constant itself is unique for a given pair of materials, the law of attraction is obviously geometry-dependent. For example, for a sphere of radius R approaching a flat surface at a distance s the law of attraction (1) is to be replaced by

$$Uvdw = -\frac{A}{6}\left(\frac{R}{s}\right),$$ (3)

and other similar laws can be obtained for other geometries.[1]

As an interesting gauge of the absolute magnitude of these attractive potentials, we can consider by comparison the adhesion energy E_{adh}, which is the minimum value which U(s) will reach for an optimal s, when tip and surface adhere, or "weld" to form a single body. In that case, if the two media are not too different, one may estimate an adhesion energy simply as the sum of the two surface free energies γ_1 and γ_2

$$E_{adh} = U(s_{min}) = (\gamma_1 + \gamma_2) \times (area)$$ (4)

For an Au-Au contact, $E_{adh} \simeq 0.18$ ev/$\mathring{A}^2$, while $U_{vdw}(s_{min}=2.4\mathring{A}) \simeq 0.016$ev/$\mathring{A}^2$, one order of magnitude smaller (2.4 $\mathring{A}$ is the Au-Au distance in crystalline gold).

The conclusion which this comparison suggests, namely that van der Waals forces dominate at large distances but become utterly negligible at reasonably close tip-surface distances, although correct in the particular case considered, is however geometry-dependent, and thus not particularly safe. Already for a spherical tip on a flat surface, or for a truncated conical tip, the importance of van der Waals can be much larger. This is due to the fact that the microscopic adhesion area is strongly reduced in these geometries, while the strength of the van der Waals attraction is reduced much less, due to its long-range nature. Essentially based on this argument, the point can be made[10] that for certain interesting tip shapes, particularly those where a sharp atomic-size cluster is deposited on top of a large blunt tip support, the importance of van der Waals forces could be dominant even at mechanical contact.

What consequences could this have? The total tip-surface interaction energy

$$U(s) = U_{vdw}(s) + U_{short\ range}(s) \tag{5}$$

would in the case of those particular tips be essentially identical to $U_{vdw}(s)$ at all distances outside a repulsive hard core, provided by $U_{short-range}$. Contributions to $U_{short\ range}$ would in that case come from a strong compressional deformation of the first few atomic layers in mechanical contract. Therefore, the use of a blunt support tip with a small cluster "nipple" on top (named teton tips in ref. 11) can realize a situation where the first few tip layers are strongly pressed against the substrate, even when the overall tip-surface potential is still in the attractive region ($dU/ds > 0$).

This point is relevant to the short-distance regime, to be discussed further below. In the intermediate and large distance regime, this kind of tip geometry would also produce peculiar results. Because of the very gradual behavior of van der Waals forces with distance, the magnitude of the total attractive force dU/ds is expected to be relatively small in the intermediate distance regime (where usually dU/ds is large), and yet relatively large in the large distance regime (where usually dU/ds is very small).

On the experimental side, there appears to be very little microscopic work which deals with these aspects of van der Waals forces yet. The experimental study of these different force regimes in relation to tip geometry, and also to work backwards from forces to geometry, is thus vastly unexplored, and represents an open area for the future.

4. DIGRESSION NO. 1: INFLUENCE OF VAN DER WAALS FORCES ON SURFACE MELTING.

Even though at large distance the macroscopic van der Waals estimate of the tip-surface interaction is essentially exact, the actual force is small. Using for example the Hamaker constant for Au-Au as typical, we would obtain, between a spherical tip of radius 50Å and a flat surface placed at $s = 10\text{Å}$ away on attractive force given by $dU/ds = AR/6s^2 \simeq 0.5$ times $10^{-9}N$, which is about one order of magnitude smaller than values one can typically measure.[6] While of course, as stressed earlier, larger forces can be attained by particular geometries, it is interesting to point out the existence of surface phenomena which would be influenced in an important way even by such a weak force. One such phenomenon of this kind is expected to be surface melting.

Surface melting consists of the gradual disordering of certain crystal surfaces (in fact of most stable crystal surfaces, since only exceptionally one finds in practice surface "non-melting") with formation of a thin liquid, or "quasi-liquid" layer. This layer is in equilibrium between the solid and the vapor already somewhat below the melting temperature T_m of the solid.[12] The thickness 1 of the liquid layer grows in a critical fashion when T approaches T_m,

$$1 \sim 1_o(1 - T/T_m)^{-\nu} \tag{6}$$

where the exponent ν is related to the power-law decay exponent n of long-range forces in the solid $V(r) \sim r^{-n}$, in the simple form $\nu = 1/(n-3)$. Hence, for dispersion forces (which dominate in nonpolar insulators as well as in metals) $n = 6$, and $\nu = 1/3$. The feature which is somewhat surprising, is that even a very small change of interparticle potential, so small as to affect very little the bulk crystal properties, can modify dramatically surface melting. For example, it was demonstrated by Trayanov and Tosatti[13] that by modifying only the very weak tail of the interparticle potential of a Lennard-Jones system one could very strongly modify, or even suppress surface melting. This extreme sensitivity of surface melting to long range forces is the crucial feature which makes it interesting in the present context.

Suppose we take a solid and bring it to a temperature T close, but below its melting point T_m. Let us focus on a particular surface, choosing one which is prone to surface melting. The details of its surface melting behavior, in particular the development of a quasi-liquid layer of thickness $1(T)$ will be determined solely by long-range forces inside the solid (since the vapor above has negligible density). Next, let us consider an approaching tip, for example idealized as a sphere of radius R, held a distance s away from the surface. How will the additional van der Waals forces between tip and surface modify the surface melting behavior?

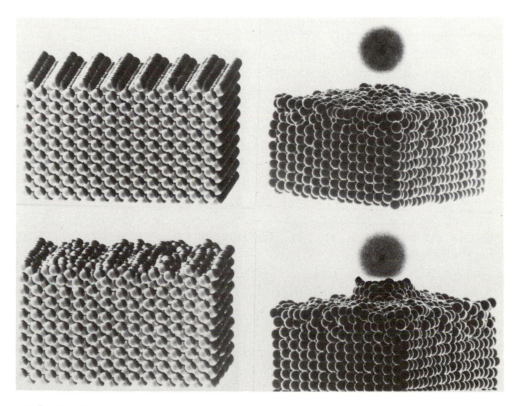

FIG. 1. Molecular dynamics simulation of the effect of a spherical Au tip approaching a Au(110) surface, which undergoes surface melting (Ref. 15). Tip and surface atoms interact via a van der Waals force. (a) The 1x2 reconstructed surface at low temperature without tip. Note the characteristic "missing-row" reconstruction; (b) Snapshot of the same surface, when heated at T=1000 K. Thermal disorder has removed the missing-row reconstruction, but there is no surface melting yet; (c) Snapshot at T=1250K (roughly 95% of the melting temperature) with the spherical tip (R=50 A) hovering at s=7 A above the nominal surface level. At this temperature, the first two-three layers are melted, and a small bulge, barely visible, begins to form under the tip; (d) Snapshot at T=1250 K, with s=5 A. The quasi-liquid layer has snapped out to wet the tip. (Courtesy of O. Tomagnini).

The precise and detailed answer to this question, which seems of practical interest for current and future experimental studies, will be presented elsewhere[14], but we can already anticipate the main effects based on reasoning, and on preliminary molecular dynamics results.[15] First of all there is an outwards bulging of the liquid film, even if very thin, towards the tip. Secondly, the crystalline order also "bulges" outwards under the tip, so that the quasi-liquid film does not locally thicken much. The liquid layer protrudes very dramatically as the tip approaches, until quite suddenly a liquid neck is formed, and at some critical distance s_c the tip is wetted. Pictures taken from a simulation of a 50Å Au spherical tip onto a melting Au (110) surface (Fig. 1) illustrate the situation described above, and yield $s_c \simeq 6$Å. Hence at such large distances as 7Å, where the tip-surface force is purely van der Waals and amounts to only about ~ 1nN, the effect of the tip on a melting surface is quite dramatic. Moreover, motion of the tip couples to important mechanical and thermodynamic changes like mass transport and melting/crystallization under the tip. Hence, a vibrating tip such as in the setup of Ref. (6) would undergo both a frequency shift and an increase of damping when coupled to such a melting surface. Experiments on a low-melting point metals and semimetals, e.g., Hg, Ga and Pb, seem entirely feasible and highly interesting.[16]

5. LARGE DISTANCE REGIME: BARDEEN TUNNELING

In this regime, the tip and surface wavefunctions overlap only very marginally, and the simple first-order perturbation theory first applied to tunneling by Bardeen,[17] and specialized to tip-surface tunneling by Tersoff and Hamann,[18] can be applied. In that approximation, assuming a spherical tip with s-wave wavefunction and featureless density of states, and with other reservations discussed in detail, e.g., in a recent review by Tersoff,[19] the differential tunnel conductance is simply proportional to the Fermi level surface electronic density of states, measured at the tip center, r_0,

$$(dJ/dV) \sim \rho (r_0, E = E_F) = \sum_{nk} |\Psi_{nk}(\vec{r}_0)|^2 \delta(E_{nk} - E_F) \qquad (7)$$

The $|\Psi_{nk}|^2$ decay exponentially outside the surface, roughly like $\exp\left(-2z\sqrt{2m\Phi}/\hbar\right)$, and since the work function Φ is typically 3 -4 eV, this implies that the exponential decay of the tunnel conductance with tip-surface distance is expected to occur roughly at a rate e^{-s/a_B}, where a_B is the Bohr radius. The lateral variations of conductance map out the contours of the important wavefunctions at the Fermi level. This map has the full symmetry of the surface lattice, and can be used as a surface topograph to the extent to which the wavefunctions at the Fermi level are sufficiently representative of the full set of occupied states. This may be the case in a regular metal, but is clearly not so for a semi-metal and even less for a semiconductor. Historically, Selloni et al[20] were the first to exemplify the large difference between an STM map and the actual electronic charge corrugation map in a semimetal, by explicitly calculating dJ/dV for graphite. They also pointed out very early that the conductance at finite tunneling voltage V could have provided a tool for Scanning Tunneling Spectroscopy (STS), and proposed that approximation (5.1) could be extended to finite V in the form

$$dJ/dV \sim \rho(r_0, E = E_F + V) \, T(V) \tag{8}$$

where the density of states is now sampled at an energy V away from the Fermi level, with an additional barrier tunneling factor T(V) whose WKB expression is

$$T(V) = \exp \left\{ - s \frac{4\sqrt{2m}}{\hbar V} [\Phi^{3/2} - (\Phi - V)^{3/2}] \right.$$
$$\left. + s \frac{2\sqrt{2m}}{\hbar} (\Phi - V)^{1/2}] \right\} \tag{9}$$

What Eq. (8) really says, is that tunneling of electrons into the surface will exhibit a peak when the tip negative voltage -V is such to give them an energy which matches exactly that of a prominent empty state of the surface. Or, conversely, that electron tunneling out of the surface will have a peak when the positive tip voltage +V matches the energy position below E_F of an important filled state of the surface. Important features, such as the π^* states of graphite, predicted to yield a peak through approximation (8), were indeed found by STS.[21] Considerable interest and new understanding has since developed in these issues. Lang[22] has shown that the generalization of (8) for STS is considerably less accurate than (7) is for STM. Tersoff[23] has qualitatively re-analyzed the graphite results. Graphite has in the meanwhile independently become a very popular STM substrate, where several other issues have come under debate, including different explanations for large observed corrugations.[24,25] Yet, in spite of its crudeness, an approximation of the type (8) still remains the simplest way to provide a first guess for both the STS spectrum and the STM map of a surface, at least in the large distance regime. To be sure, there have been several interesting formulations, including those of Flores and his group,[26] of Doyen et al[27] and of Noguera[28] which appear very promising in particular situations. The construction of a universal alternative algorithm which should both quantitative and computationally viable for most situations of practical interest remain as one of the challenges for the future.

6. DIGRESSION NO. 2: SEARCHING FOR "NEGATIVE HUBBARD U" STATES ON SEMICONDUCTOR SURFACES

Electrons in semiconductor surface states (which essentially consist of narrow-band forming dangling bonds), are strongly correlated. A prototype lattice model for strong electron correlations is the well-known Hubbard model, which is obtained by adding to a regular narrow band electron propagation an additional on-site interaction U,

$$H = \Sigma_{i\sigma}\varepsilon_{i\sigma}c_{i\sigma}^+c_{i\sigma} - t\Sigma_{ij\sigma} c_{i\sigma}^+c_{j\sigma} + U\Sigma_i n_i\!\uparrow n_i\!\downarrow \tag{10}$$

where i, j denotes sites, ε_i are site energies, σ denotes spin and t is a first neighbor hopping energy. For real electrons moving on a rigid lattice, U is simply a mimic of the intra-atomic repulsive interaction, and is therefore positive. If the lattice instead is mobile and can readjust depending upon site occupancy, then in principle situations become possible where the combined effect of electron and lattice may turn out to work like a "negative Hubbard U."[29] In a system with $U < 0$, one thing which will tend to happen, is that two absolutely equivalent atoms carrying an electron each may prefer to become inequivalent, one of them finally acquiring both electrons and the other being left with none. Dangling bond states are ideal candidates for this type of phenomenon, because the on-site one electron energy ε_i depends very strongly upon hybridization, which in turn is rigidly tied to bond angles with the neighboring sites. Thus, for example, on a (111) surface the dangling bond one-electron energy ε_i will rise or fall by as much as 1 eV (which is of the order of the bandwidth), relative to the bulk states, when the atom is respectively depressed onto the second layer or lifted out of it, by just a fraction of an Å. Due to essentially this mechanism, negative Hubbard U behavior has been demonstrated earlier for bulk Si vacancy,[30] which is nothing but four interacting dangling bonds.

The stable state of most clean semiconductor surfaces is a state based on suppression of dangling bonds. Dangling bonds have been either eliminated altogether (by some kind of rebonding, or by saturation with adatoms), or else they remain, but become empty or doubly filled. The latter situation is in a way already a manifestation of negative Hubbard U, and can be recognized to play a role in the Si(111)7 × 7[31] and Ge(111)c(2×8)[32] reconstructed states. In both cases, there is electron transfer from the dangling bond of an adatom (which has been rendered p_z like by depression) to that of a rest atom (which is more s-like due to lifting). However these dangling bonds are not really equivalent from the start, and there is no true symmetry breaking associated with this sort of negative U behavior. The case of Si(100)2×1 "buckled dimer"[33] is a clearer realization of negative U, since the buckling takes two otherwise absolutely identical surface atoms and makes them inequivalent, with a change transfer from the depressed to the lifted one.

Having noted that negative Hubbard U is rather universal on the semiconductor surfaces, the point of this digression is to suggest that it will apply also to new situations like a surface vacancy, and moreover that the resulting negative Hubbard U states will be very easy to pick up by STM/STS.

A recent study by the Car-Parrinello method[34] of the Si(111)2×1 π-bonded chain state by Ancilotto et al,[35] has shown that very reliable results can be obtained by this type of first-principles molecular dynamics, concerning both atomic and electronic structure of surfaces. This work has now been extended to include the full structure of a vacancy on the π-bonded Si(111)2×1 surface.[36] A negative Hubbard U type of effect can be found already on the perfect 2×1 reconstructed surface, which exhibits a symmetry-breaking "buckling" of the π-bonded chains. However, a much stronger negative U effect is predicted for a vacancy along the π-bonded chain. Fig. 2 gives a qualitative picture of the change of atomic positions which occur near the vacancy. The two previously equivalent vacancy nearest-neighbor atoms along the chain have become inequivalent, one, H being raised, the other, L, lowered. Simultaneously, the lowered atom engages in a weak bond with the atom C next nearest to the vacancy. Fig. 3 presents the total change density contour along a vertical plane which contains the H and L atoms, showing that the negative Hubbard-U-related electron charge transfer is

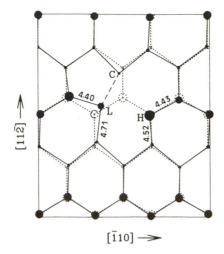

FIG. 2. A top-layer vacancy in the π-bonded chain (2x1)-reconstructed Si(111) surface (top view). Larger dots indicate atoms which stick out more than smaller dots. An undefected π-bonded chain is visible at the bottom of the figure. Dotted contours indicate atomic positions prior to removing an "up" atom of the (slightly buckled) π-bonded chain. There is charge transfer between atoms L and H, which have spontaneously become inequivalent, by moving respectively down and up (negative Hubbard U). Atom L also develops a weak bond (dashed) with atom C. From Ref. 36).

quite noticeable. The voltage-dependence of the STM conductance can be estimated from the local density of states as in (8) and predicts a shift of 0,7 eV between similar electronic features of atoms H and L[36].

The possibility to pursue the discovery and study of these, and other similar interesting negative U states by combined use of theory and tunneling spectroscopy seems a point of some interest for the future.

7. MEDIUM DISTANCE REGIME: STRONG TIP-SURFACE INTERACTION, AND BARRIER COLLAPSE

By the time the tip and surface wavefunctions overlap and thus electrons begin, so to speak, to mix, the previously separate electronic structures of the two systems merge into a single one (except for the important caveat that their link is narrow, i.e., it has nearly zero lateral extension). Electron tunneling is replaced by electron flow across the narrow link, whose detailed transport properties are much less clear. At the same time, van der Waals forces give way to much stronger adhesion forces. These forces result essentially from a gradual loss of the surface

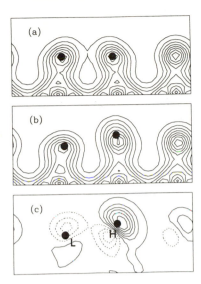

FIG. 3. Total ground state electronic density contours calculated on a vertical plane crossing atoms L (left dot) and H (right dot), of Fig. 2. Panel (a): vacancy ground state obtained by requiring L and H to remain equivalent. Panel (b): vacancy ground state obtained without requirements, and letting thus L and H free to move independently. This state is about 0.5 eV lower than that in (a), and is the true ground state. Panel (c): difference between electron density of (b) and of (a), showing electron transfer from L (dashed contours) to H (solid contours). From ref. 36.

free energy for those tip and surface portions which are brought into contact as the tip-surface distance s.decreases. At a more basic physical level, the short-range tip-surface attraction originates from a gain of kinetic energy of the electrons which cross the link.

Apart from these generic considerations, there is unfortunately no simple, ready-made theory, such as that for van der Waals attraction or Bardeen tunneling, available for us to borrow for a description of the tip-surface properties in this regime. Brave early attempts at a microscopic understanding of tip-surface current transport are due chiefly to Lang.[22,37,38] A microscopic theory of adhesion forces has previously been put forward, among others, by Rose, Ferrante and Smith.[39,40] More recently, computational approaches such as that of Ciraci, Baratoff and Batra[25,41] address simultaneously both currents and forces. All these theoretical approaches are based on the local-density approximation (LDA) to the electronic structure and/or total tip-surface energy. Although one could easily imagine strongly correlated situations where this kind of approximation does not apply very well, LDA is by far the best we have for this problem, and we shall probably learn much more from these approaches.

On the experimental side, Gimzewski and Möller have shown, already several years ago,[42] how the exponential decrease of the tip-surface resistance R for decreasing s gets gradually less steep at smaller s until eventually the resistance levels off in a range of values not too different from the Sharvin point contact limit $h/2e^2 = 12.9k\Omega$. If one insists in assuming that resistance is still due to tunneling across an effective barrier of thickness s and height Φ_{eff}, one obtains $\Phi_{eff} = (\hbar^2/2m)(dlnR/ds)^2$. The gradual leveling off of R for small s implies that the barrier ϕ_{eff} collapses from its large s value (essentially the work function), to zero, just before point contact. Barrier collapse between tip and surface seems to be a well-established concept. Fig. 4 illustrates the barrier collapse obtained theoretically in an idealized Na-atom tip-jellium surface contact.[38]

Ciraci, Baratoff and Batra[41] have very recently shown, in an Al-tip, Al-surface system, that the barrier collapse is a) very local, giving rise to a narrow parabolic collapsed-barrier channel b) very position-dependent, varying from a broad to narrow channel as a function of relative lateral tip-surface position. As a consequence of a), electron transit through the narrow hole due to local barrier collapse is impeded by quantum effects, i.e., the electron must still cross an effective barrier, whose height is the zero-point energy in the narrow hole. As a consequence of b), additional lateral resolution is predicted. This fact is particularly interesting, since there is fairly clear evidence[43] that even on regular metals, the lateral STM resolution is definitely better than purely topographic resolution.

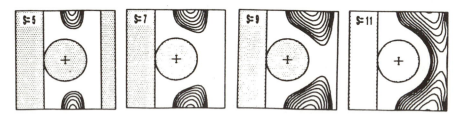

FIG. 4. Contour map of the effective potential seen by an electron, calculated for two plane jellium "electrodes" one of which carries a single Na atom "tip," at various mutual distances s (in a.u.). The repulsive tunneling barrier, present at large distances (right) collapses as the distance decreases (left). (From Ref. 38).

At present, it is not clear what might cause this unexpectedly high resolution. A different possibility which has been put forward as an explanation for this very high resolution is a high apparent corrugation, due in turn to deformation caused by strong tip-surface forces.[44,45] This possibility seems plausible, at least in cases where one can prove or surmise a strong short-range tip-surface repulsion. The reason is that in case of repulsion, the true tip-surface gap modulation is in reality smaller than the nominal one (because the surface deforms to avoid the tip) which in turn simulates a larger corrugation. This situation is a likely one for STM on dirty surfaces, e.g. surfaces covered by an insulating layer which acts as a kind of mechanical gasket between tip and surface.[45] The state of affairs is much less clear for clean surface-tip contacts. Both common sense (see section 2), and detailed local density calculations[39-41] indicate a strong attractive regime, before the optimal tip-surface distance is reached, and followed by repulsion. It would seem reasonable, in presence of short-range forces only, to identify the optimal spacing roughly with the electrical point contact onset. If this were true, then most STM data would be taken when the tip and surface are still attracting each other. The STM/AFM results of Dürig et al for an Ir tip on an Ir surface shown on Fig. 5, demonstrate precisely that.[6] If there is attraction, however, the surface deforms exactly out of phase with the tip. This deformation acts to reduce the apparent corrugation, as pointed out also by Ciraci et al,[41] and this cannot explain the high resolution. Unfortunately, STM/AFM data are not generally available in an arbitrary tip-surface combination. Therefore, we do not know whether the very high resolution found, e.g. on Au(111) was obtained in the attractive or repulsive regime. An important role of tip-surface repulsion for clean surfaces has anyway been claimed in the context of several STM studies,[43-46] including very clean surfaces.

An interesting possibility is the combined interplay of long-range attraction and short-range tip nipple-surface repulsion, already discussed in section 3, and in Refs. (25, 10). In that case, overall attraction could still prevail, while local surface deformation would be of the repulsive type, which enhances the apparent corrugation, and STM resolution with it.

These and related questions are very much under present debate, and consensus is scarce. Experimentally, it is clear that we could learn a lot from extended STM/AFM measurements done with tips of controlled nature and shape. Theoretically, we need tip-surface simulations which should include self-consistently electronic structure and atomic deformations, and where the tip-surface resistance should be quantum-mechanically calculable. Although conceptually feasible practical implementation of this program remains very much a challenge for the coming decade.

8. DIGRESSION NO. 3: THE MYSTERY OF TUNNELING INTO INSULATING PARAFFINS

So far we have discussed tunneling from a metallic tip into a metallic surface. Semiconductors surfaces can also be studied very well, since either doping, or temperature, or both, make them conducting enough. With strong bulk insulators, on the other hand, STM cannot be used: electrons remain trapped at the surface, and charging prevents any further current. There appear however to be exceptions, which are as striking as little understood. The main exception has to do with some organic insulators, namely paraffins. Paraffin molecules, also called n-alkanes, consist of long saturated nonpolar chains $(CH_3) - (CH_2)_n - (CH_3)$, with $n > 10$. They solidify into orthorombic or monoclinic layered crystals, which are highly insulating in the bulk ($\rho \sim 10^{18} \Omega$ cm). The surprise[47] comes from STM: paraffin films as thick as 200Å, sometimes even 1000Å, deposited on a conducting substrate (gold, graphite) can be imaged by tunneling in full detail with quite ordinary currents (0.5nA) and quite low voltages (0.1 - 0.3V). This seems to imply that tunneling current can somehow be conducted through the thick film, which thus does not at all behave as an insulator, against expectations. A similar surprise had been found a while earlier on Langmuir-Blodgett films, also organic and insulating, and also well-imaged by STM.[48,49] A second piece of the puzzle is that the paraffin film becomes totally invisible to STM (very probably meaning non-conducting) when it melts, but it promptly reappears at recrystallization. Yet, bulk paraffins do not change much their bulk conductivity at melting.

What is the solution of this puzzle? Various possibilities have been discussed so far.

First, it is well-known, and it has been repeatedly stressed that, once an electron is injected inside a long chain of organic bonds, it can move very well, and will easily be conducted away even at large distance. Hence bond conduction could explain the puzzle if one could figure how to inject the electron into this state. But this is the biggest problem, because the affinity level (or conduction band, or LUMO,...) of the molecule presumably lies several eV above the metal Fermi level, and not hundreds of millivolts, as used in STM. Again, only tunneling remains; but a barrier of the order of the eV and a thickness of hundreds of Å is the closest thing to a hard wall one can imagine.

Other groups[51] have emphasized the role of a strong tip-applied pressure. One imagines the possibility of resonant tunneling via some kind of empty state, shifted by pressure to an energetic position close to the Fermi level. After all, it is noted, tunneling transmittance at resonance is exactly one for any barrier thickness. Also, this explanation would fit with the disappearance of current at melting, since the liquid will not support tip pressure. The main trouble with this explanation is that it seems too specific, and thus not robust at all. Imagine the tip was in a configuration where the process goes as supposed. Since the barrier is very thick, one can easily convince oneself that the resonant level must be millivolts away from E_F, or else the transmittance would drop from large to very close to zero (transmittance *does* depend dramatically on thickness away from exact resonance). As soon as the tip is scanned, it is hard to see by what kind of miracle should there always be a helpful resonant level, consistently located within millivolts of E_F. Unless there were consistently hundreds or thousands of such states, forming a kind of continuum. This would amount to injection into a band of tip-induced interface states, whose existence is entirely hypothetical, and should be proven before marching on along this line.

A third possibility involves the existence of well-defined band of surface states, which should be located inside the insulating gap, and conduct current laterally, without any tunneling or injection inside the bulk. It has been noted[47] that such surface states do exist, at least on Longmuir-Blodgett films. Although nothing is known about surface states of paraffins, and even less about their change of behavior, or energetic location, with melting, this way of explaining the puzzle does not seem totally implausible. And it seems at least relatively easy to imagine experiments aimed at establishing whether electron conduction takes place through the bulk or the surface of the film, (maybe by engineering suitable substrates). The puzzle is open, and the answer will require some work. This work is however well worthwhile since, as STM has underlined, our understanding of subtle conduction processes involving the boundaries of these insulators is inadequate, and more physics hides there than was suspected hitherto.

9. SHORT-DISTANCE REGIME, AND THE REALIZATION OF POINT CONTACT

Eventually, as the approach distance s decreases, the tip is rammed into the surface, and a point contact of some kind is established. Before that happens, intimate adhesion can be approached, at least conceptually, in a number of different ways. Tip and surface may simply approach, and begin to merge together uneventfully, i.e., without too much reciprocal deformation. Alternatively, one may instead imagine that tip and surface could at some point jump together[51] since by doing that a large adhesion energy can be suddenly gained, while the elastic cost of stretching the rest of the tip and substrate can be macroscopically distributed and reduced. This has also been called an adhesion avalanche.[52] A different kind of jumping together can take place if the substrate is plastic or liquid, and if it wets the tip. In that case, bulging of the surface towards the tip is followed by sudden formation of a wetting neck, very much as in the surface melting situation, Sect. 4.

When tip and surface jump together in this way all useful STM/AFM scanning probably terminates, since the nature of both tip and surface is irreversibly changed. Therefore, useful situations appear to be, loosely speaking, of the non-wetting type, since in that case tip and surface will avoid merging as much as possible.

What orders of magnitude are the tip-surface forces at real short distance, and what effect do they cause? The adhesive force can be expected to reach $10^{-9} \div 10^{-8}$N before decreasing again to zero and turning repulsive. A force of this value will generally cause only a minor elastic surface deformation, except on a liquid, as explained earlier. If the tip is further pushed into the surface, the repulsive force will initially cause elastic deformations, for magnitudes below $\sim 10^{-8}$N, but is expected to cause irreversible plastic deformations when larger.[53] These numbers apply for a metal surface, and should be reduced one to two orders of magnitude for a soft molecular substrate. Returning to metals, we note that the pressure which 10^{-7}N exert locally on area of, say, 10×10 Å is in face 1 Mbar, which should be disruptive to many materials. Visual examples of tip-induced plastic deformations have been provided recently, through molecular dynamics computer simulation.[54] Given all this, it is often surprising how good atomic resolutions are sometimes obtained in AFM with extremely strong applied pressures, which should instead disrupt the surface. These phenomena remain at present rather poorly understood.

The electrical properties of the link at very short tip-surface distance can of course only be expected to be strongly connected to the mechanical aspects just discussed. Because of the collapse of the repulsive barrier for electron motion; conductivity in this regime becomes less and less distance-dependent, as s decreases. The upwards conductance jump[42] often observed to mark the onset of point contact (see e.g. Fig. 5), has been attributed to mechanical jumping together of tip and surface.[45] Electron transport at this point becomes ballistic[55] through the contact. Further plastic deformations are probably responsible[56] for other conductivity features, or oscillations, generally observed at further approaching of tip and surface.

In summary, it is clear from the above that a detailed understanding of forces and conductivity in this regime is by necessity very system-dependent. The mechanical aspects will generally require a molecular dynamics simulation of some kind. Only after that does it make sense to worry, at least theoretically, about tip-surface conductivity. In such a non-universal scenario, the physics to be learned from STM/AFM will

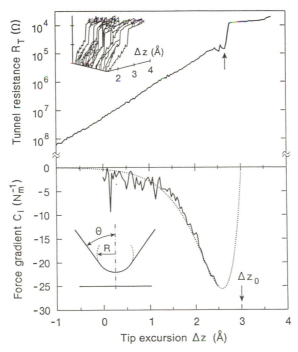

FIG. 5. Simultaneous measurement of tunnel current and of force by combined STM/AFM, for an Ir tip on an Ir surface. (Courtesy of U. Dürig; see also Ref. 6).

be strongly connected with the particular understanding of one specific case, rather than with any general property of this regime.

10. CONCLUDING REMARKS

We have given a bird's-eye overview of some of the physics addressed by the STM/AFM. Large, medium and small distance regimes have been discussed in sequence, and clearly represent a scale of increasing complication. The large distance regime is the easiest to describe, and to interpret. A role for attractive van der Waals forces which is probably more important than usually acknowledged is emphasized. The dramatic effect that tip-surface forces can have even when very weak, in a marginal phenomenon such as surface melting, has been demonstrated. For what concerns conductivity, stretching the Bardeen-Tersoff-Hamann tunneling theory beyond its range of validity, although basically incorrect, is all the same very handy in providing at least a hunch for what to expect for the voltage-dependent tunneling spectrum. Searching for negative Hubbard U at a silicon surface vacancy is a typical game one could play by STM, based on such a hunch. The medium distance regime is undoubtedly more difficult. Here the tunneling barrier collapses, and is likely to be replaced by a subtler quantum-mechanical barrier. Attractive adhesion forces begin to play a role, but cannot explain the exceedingly high resolutions often attained in this regime, which remain somewhat mysterious. Another independent mystery is the demonstrated possibility to image thick insulating paraffin films by tunneling, as if they were conductors. The short distance regime may be characterized by jump instabilities and/or by strong tip-surface deformations, and clearly seems the least universal, where each case is a world to itself.

Although this is only a bird's eye projection of a very vast field, I hope I have been able to convey the feeling for some of the challenges in this area of condensed matter, an area which probably still exhibits the smallest ratio of theorists to experimentalists engaged at present. There is ample room for this ratio to grow in the 90's.

ACKNOWLEDGEMENTS

I am deeply indebted first of all to Orio Tomagnini, Francesco Ancilotto, Annabella Selloni, Simonetta Iarlori and Furio Ercolessi, who are my collaborators in some of the digressions presented here, as well as in general surface physics. My appreciation goes also to Alexis Baratoff,

Salim Ciraci, Urs Durig, Jim Gimzewski, Bruno Michel, Dieter Pohl and Heini Rohrer who not only discussed with me their results, but also guided me through the literature. Finally, I wish to thank Leo Esaki for asking me to contribute to this discussion. Surely, without his tenacious insistence, this paper would never have been put together.

REFERENCES

1. See eg. J. N. Israelachvili, Intermolecular and Surface Forces, Academic Press (1985).

2. G. Binnig, H. Rohrer, C. Gerber and E. Weibel, Phys. Rev. Lett. **50**, 120 (1983).

3. For a good early review, see J. A. Golovchenko, Science, **232**, 48 (1986).

4. G. Binnig, C. F. Quate and C. Gerber, Phys. Rev. Lett. **56**, 930 (1986).

5. For a recent broad review, see e.g., Scanning Tunneling Microscopy and Related Methods, ed by R. J. Behm, N. Garcia and H. Rohrer, Kluwer Academic Publishers (1990).

6. U. Dürig, O. Züger and D. W. Pohl, Phys. Rev. Lett. **65**, 349 (1990).

7. P. Wyder, lecture presented at this conference.

8. E. M. Lifshitz, Sov. Phys. JETP **2**, 73 (1956); I. E. Dzyaloshinskii, E. M. Lifshitz and L. P. Pitaevskii, Adv. Phys. **10**, 165 (1961)

9. H. Krupp, Adv. Colloid Interf. Sci. **1**, 79 (1967); J. Visser in Surface and Colloid Science, Vol. 9, ed. E. Matejavec, Wiley (1976), p.1.

10. See also F. O. Goodman and N. Garcia, Phys. Rev **B43**, 4714 (1991).

11. Vu Thien Binh, J. Microscopy **152**, 355 (1988).

12. For a review, see E. Tosatti, in The Structure of Surfaces II, ed by J. F. van der Veen and M. A. van Hove, Springer (1987) p. 535.

13. A. Trayanov and E. Tosatti, Phys. Rev. Lett. **59**, 2207 (1987).

14. O. Tomagnini, F. Ercolessi, S. Iarlori, E. Tosatti, work in progress.

15. O. Tomagnini, unpublished.

16. One possible warning for the case of Ga could be the possibility that the higher conductivity of the liquid and/or surface close packing, might prevent surface melting. For a discussion of this point, see X. J. Chen, A. C. Levi and E. Tosatti, Il Nuovo Cimento B (to appear).

17. J. Bardeen, Phys. Rev. Lett. **6**, 57 (1961).

18. J. Tersoff and D. R. Hamann, Phys. Rev. **B31**, 805 (1985); Phys. Rev. Lett. **50**, 1988 (1983).

19. J. Tersoff, in Ref. (5) page 77.

20. A. Selloni, P. Carnevali, E. Tosatti and C. D. Chen, Phys. Rev. **B31**, 2602 (1985).

21. G. Binnig, H. Fuchs, C. Gerber, H. Rohrer, E. Stoll and E. Tosatti, Europhys. Lett. **1**, 31 (1986); B. Reihl, J. K. Gimzewski, J. M. Nicholls and E. Tosatti, Phys. Rev. **B33**, 5770 (1986); H. Fuchs and E. Tosatti, Europhys. Lett. **3**, 745 (1986).

22. N. D. Lang, Phys. Rev. **B34**, 5947 (1986).

23. J. Tersoff, Phys. Rev. Lett. **57**, 440 (1986).

24. J. M. Soler, A. M. Baro, N. Garcia and H. Rohrer, Phys. Rev. Lett. **57**, 444 (1986).

25. S. Ciraci, A. Baratoff and I. P. Batra, Phys. Rev. **B38**, 10047 (1988).

26. A. Martin-Rodero, F. Flores and N. H. March, Phys. Rev. **B38**, 10047 (1988).

27. G. Doyen, E. Koetter, J. Barth and D. Drakova, in Ref. (5) p. 7.

28. C. Noguera, Phys. Rev. **B42**, 1629 (1990).

29. P. W. Anderson, Phys. Rev. Lett. **34**, 953 (1975).

30. G. A. Baraff, E. O. Kane and M. Schlüter, Phys. Rev. **B25**, 548 (1982).

31. K. Takayanagi, Y. Tanishiro, S. Takahashi and M. Takahashi, Surf. Sci. **164**, 367 (1985).

32. R. D. Meade and D. Vanderbilt, Phys. Rev. **B40**, 3905 (1989).

33. D. J. Chadi, Phys. Rev. Lett. **43**, 43 (1979).

34. R. Car and M. Parrinello, Phys. Rev. Lett. **55**, 2471 (1985).

35. F. Ancilotto, W. Andreoni, A. Selloni, R. Car, M. Parrinello, Phys. Rev. Lett. **65**, 3148 (1990).

36. F. Ancilotto, A. Selloni and E. Tosatti, Phys. Rev. **B43**, 5180 (1991).

37. N. D. Lang, Phys. Rev. Lett. **55**, 230 (1985); Phys Rev. Lett. **56**, 1164 (1986); Phys. Rev. **B34**, 5947 (1986); Phys. Rev. **B36**, 8173 (1987); Phys. Rev. Lett. **58**, 45 (1987).

38. N. D. Lang, Phys. Rev. **B37**, 10395 (1988); Comm. Cond. Matter Phys. **14**, 253 (1988).

39. J. H. Rose, J. Ferrante and J. R. Smith, Phys. Rev. Lett. **47**, 675 (1985); Phys. Rev. **B28**, 1835 (1983); J. R. Smith and A. Banerjea, Phys. Rev. Lett. **59**, 2451 (1987).

40. J. Ferrante and J. R. Smith, Phys. Rev. **B31**, 3427 (1985).

41. S. Ciraci, A. Baratoff and I. P. Batra, Phys. Rev. **B42**, 7618 (1990).

42. J. K. Gimzewski and R. Möller, Phys. Rev. **B36**, 1284 (1987).

43. See, e.g., J. Wintterlin, J. Wiechers, H. Brune, T. Gritsch, H. Höfer and R. J. Behm, Phys. Rev. Lett. **62**, 54 (1989).

44. J. M. Soler, A. M. Baro, N. Garcia and H. Rohrer, Phys. Rev. Lett. **57**, 444 (1986).

45. J. B. Pethica and W. C. Oliver, Phys. Scr. **T19**, 61 (1987).

46. C. J. Chen and R. J. Hamers, preprint (1990).

47. B. Michel, G. Travaglini, H. Rohrer, C. Joachim and A. Amrein, Z. Phys. **B76**, 99 (1989).

48. C. A. Lang, J. H. K. Hörber, T. W. Hänsch, W. M. Heckl and H. Möhwald, J. Vac. Sci. Technol. **16**, 386 (1988).

49. H. Fuchs, Phys. Scr. **38**, 264 (1988).

50. S. M. Lindsay, O. F. Sankey, Y. Li, C. Herbst and A. Rupprecht, J. Phys. Chem. **94**, 4655 (1990).

51. J. B. Pethica and A. P. Sutton, J. Vac. Sci. Technol. **A6**, 2494 (1988).

52. J. R. Smith, G. Bozzolo, A. Banerjea and J. Ferrante, Phys. Rev. Lett. **63**, 1269 (1989).

53. See, e.g., H. Heinzelmann, E. Meyer, H. Rudin and H. J. Güntherodt, in Ref. (5), p. 443, for a broad review of forces and their effects.

54. U. Landman, W. D. Luedtke, N. A. Burnham and R. J. Colton, Science, **248**, 454 (1990).

55. H. van Kempen, in Ref. (5), p. 241.

56. S. Ciraci, in Ref. (5), p. 113 and references therein.

A NEW SPECTROSCOPY OF CARRIER SCATTERING

W. J. Kaiser,[1] L. D. Bell,[1] M. H. Hecht,[1]
and L. C. Davis[2]

[1] Center for Space Microelectronics Technology
Jet Propulsion Laboratory
California Institute of Technology, Pasadena, CA 91109

[2] Ford Motor Co, Research Staff, Dearborn, MI

Carrier transport properties of metal and semiconductor materials have been of central importance both in fundamental problems in condensed matter physics and in device technology. The early history of carrier transport theory had its origins in the Drüde model for electron relaxation,[1] developed directly after the discovery of the electron at the end of the last century. The experimental investigation of carrier transport was initially limited to conductivity measurements which probe carrier transport at the Fermi energy. More recently internal photoemission[2] and ballistic carrier[3-6] spectroscopies have provided methods for extracting detailed transport parameters for a wide range of systems. However, many of the most fundamental questions regarding the nature of carrier transport and scattering are still unanswered. In particular, the relative contributions of elastic and inelastic processes in scattering have not been directly probed. Further, experimental investigations have historically lacked an energy-resolved probe of scattering. Progress in this area is of increasing importance due to rapid advances in device technology based on ballistic carrier transport for transistor ampli-

Highlights in Condensed Matter Physics and Future Prospects
Edited by L. Esaki, Plenum Press, New York, 1991

fiers and novel radiation detectors. This paper describes a new experimental spectroscopy which directly measures the energy-dependence of carrier scattering and detects the products of carrier scattering events.[7]

The transport phenomena which are emphasized here are carrier-carrier scattering processes. The processes considered involve an incident carrier having an energy and wave vector describing a state outside the Fermi surface. The incident carrier scatters to a new empty state outside the Fermi surface while producing an electron-hole pair. A complementary process is also considered where an incident hole within the Fermi surface scatters with an electron while producing an electron-hole pair. Electron-hole pair creation by these and other processes has a primary influence on both the electronic and optical properties of materials. It is noteworthy that conventional measurements detect carrier scattering but do not resolve the subsequent propagation of both the electron and hole of the excited pair.

External photoemission spectroscopy of multilayer systems may provide a measure of hot electron attenuation in a variety of materials. However, the application of this technique to carrier transport is limited to large carrier energy since only electrons at an energy greater than the work function may escape and be detected. Internal photoemission extends optical excitation methods to low carrier energy by providing a low energy barrier to separate and detect excited carriers.[2] The application of internal photoemission to thin film metal/semiconductor diode structures has been used to extract an effective electron attenuation length in metal films. The electron attenuation length is a composite of all scattering processes including carrier-carrier scattering.

Advanced materials technology has enabled the fabrication of ballistic electron devices based on both the metal base transistor[4] and semiconductor heterostructures.[4-6] These devices offer important potential advantages over those based on diffusive transport. In addition, the investigation of these structures has allowed both electron[4-5] and hole[6] transport spectroscopies.

Several limitations exist for the conventional experimental methods described above. While both electron and hole attenuation may be separately measured, the contributions of carriers created by scattering may not be resolved.

Also, for some methods, the spectroscopy is limited to a narrow energy range. Further, the internal photoemission and ballistic electron spectroscopies must be applied to large area structures where the contributions of microscopic defects and material inhomogeneity may distort measured properties.

A new method for probing carrier transport and scattering has recently been implemented using the Ballistic-Electron-Emission

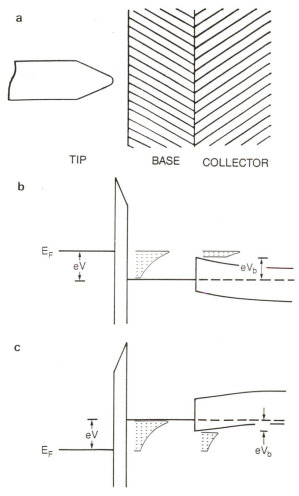

FIG. 1. (a) The experimental configuration for BEEM spectroscopy of ballistic carrier transport through materials and subsurface interfaces. The STM tunnel tip is shown with a metal/semiconductor interface structure. (b) The electronic structure for BEEM spectroscopy of ballistic electron transport with an n-type semiconductor substrate, showing the metal and semiconductor Fermi energy and the semiconductor conduction and valence bands. (c) The electronic structure for BEEM spectroscopy of ballistic hole transport with a p-type semiconductor substrate.

Microscopy (BEEM) technique.[8] BEEM is a three-terminal Scanning Tunneling Microscopy (STM) technique which enables imaging and electron spectroscopy of buried interfaces with nanometer spatial resolution. As shown in Figure 1, BEEM employs electrons injected by a tunnel tip which propagate ballistically through a structure before sampling a buried interface. The control of ballistic electron transport through interfaces allows many direct measurements of critical interface properties including Schottky barrier height, semiconductor heterostructure band offset, and spectroscopy of interface band structure.[9] BEEM also provides a hole transport spectroscopy, shown in Figure 1B, where electron tunneling is used to create a nonequilibrium hole distribution.[10] Interface properties which were previously inaccessible to experimental methods may now be imaged accurately with BEEM at nanometer spatial resolution.

In addition to interface investigation, BEEM provides a measurement method for transport. Specifically, inelastic scattering events which reduce incident carrier energy attenuate collector current. BEEM, therefore, may be applied to carrier transport investigation by measuring the dependence of the collector current spectrum on base thickness. However, BEEM allows an additional complementary and direct scattering spectroscopy method.

BEEM scattering spectroscopy, shown in Figure 2, is based on the injection of nonequilibrium carriers into a heterostructure, followed by the direct detection of carriers created by scattering. The investigation of electron scattering in a thin metal base layer is accomplished with a Schottky barrier structure prepared on a p-type substrate as shown in Figure 2a. For this system, incident electrons scatter and produce an electron-hole pair which continues to propagate. The p-type collector prevents the recovery both of electrons from the incident flux and those created by scattering. A preliminary consideration of this experiment would indicate that a collector current would not be observed. However, holes created by pair-production during scattering are allowed for transport into the collector for hole energies less than the valence band to Fermi energy Schottky barrier. The Schottky barrier provides, therefore, an energy filter which enables a spectroscopy of carrier scattering. A distinguishing feature of this experiment is that the sign of the collected current, due to holes, is opposite from the injected electron tunnel current. The complementary experiment for hole scattering, shown in Figure 2b, involves the decay of a nonequilibrium hole by pair creation

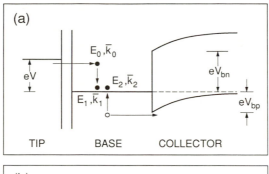

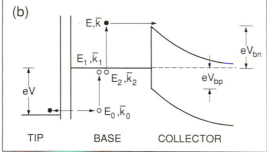

FIG. 2. (a) The electronic structure for BEEM spectroscopy of electron scattering with a p-type collector. The hole created below the Fermi energy is shown propagating through the interface and occupying a state in the semiconductor collector valence band. The labeling of the electron and hole states is described in the text. The Schottky barrier height is labeled as eV_{bp}. (b) The electronic structure for the BEEM complementary spectroscopy of hole scattering with an n-type collector. The electron created above the Fermi energy is shown propagating through the interface and occupying a state in the semiconductor collector conduction band. The labeling of the electron and hole states is described in the text. The Schottky barrier height is labeled as eV_{bn}.

and the detection of the excited electron in for n-type collector. In contrast to conventional measurements which measure the attenuation of an unscattered carrier flux, this unique method measures only scattering events.

Scattering spectroscopy is performed by measuring the carrier current created by scattering and transmitted through the interface into the collector as a function of tunnel bias, V, applied between tip and base.

Since the scattered carrier current measured between the collector and base terminals is typically less than 10 pA, a sensitive, low-noise preamplifier system must be employed for current detection. The current amplifier noise decreases with increasing impedance of the base collector junction. Since junction impedance depends on thermally activated processes, for low energy-barrier interface systems it is necessary to perform BEEM scattering spectroscopy measurements at low temperature to obtain large impedance and low-noise spectra. In addition, reduction in the BEEM system operating temperature results in reduced smearing of the tunnel tip Fermi distribution and therefore improved spectral energy resolution. The experimental results reported here were obtained with a BEEM apparatus employing a scanning tunneling microscope system operating immersed in liquid nitrogen. The entire apparatus including the liquid reservoir was contained in a flowing nitrogen gas environment. Scattering spectra are obtained by sweeping tunnel bias, v, holding tunnel current constant (at 1.0 nA) under feedback-control, and measuring collector current. The Au/Si(100) Schottky barrier interfaces investigated here were prepared by evaporation of Au in ultrahigh vacuum to a thickness of 100Å on chemically prepared Si wafer substrates (n-type; $n = 2 \times 10^{15} cm^{-3}$ and p-type; $p = 3 \times 10^{15} cm^{-3}$).[8]

The electron scattering spectrum for the Au/Si(100) system is shown in Figure 3a. This spectrum (corresponding to the experiment of Figure 2a) is compared with a ballistic hole transmission spectrum. The combined spectra show a region of zero observed collector current bounded by two thresholds in the current. The ballistic hole spectrum is obtained for positive tip bias, as shown in Figure 1c. The hole spectroscopy threshold for positive tip bias simply indicates the transmission threshold at the valence band maximum at $eV_{bp} = 0.35$ eV.[10] For negative bias, electrons created by tunneling are allowed to decay by scattering, create electron-hole pairs, and a fraction of the holes are allowed for collection. It is important to note that the threshold for the negative bias spectrum, the electron scattering spectrum, is well below the bias value (electron energy) corresponding to the energy barrier formed by the Fermi level to conduction band minimum separation. Processes involving the semiconductor conduction band may, therefore, be ignored for this energy. Indeed, the scattering spectrum is observed to have a threshold at an electron energy value equal to the valence band maximum to Fermi level Schottky barrier. This threshold simply marks the energy required for interface transport of the holes created by scattering. The comparison of the hole and scattering spectra also clearly shows the characteristic feature of this spectroscopy - the polarity of the

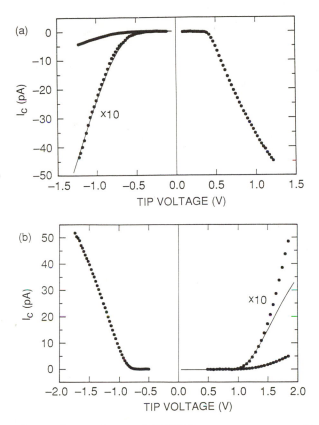

FIG. 3. Comparison of experimental BEEM spectroscopy results for electron and hole ballistic transport with electron and hole scattering. These spectra for the Au/Si(100) interface structure were measured at a tunnel current of 1nA. (a) The experimental ballistic hole transmission spectrum is obtained for positive tip bias. The Schottky barrier height, eV_{bp} (see Figure 2a), extracted from this spectrum is 0.35 eV. The experimental electron scattering spectrum (dots) for negative tip bias is compared with theory (solid line). A magnified plot of the electron scattering spectrum is also shown. (b) The experimental ballistic electron transmission spectrum is obtained for negative tip bias. The Schottky barrier height, eV_{bn} (see Figure 2b), extracted from this spectrum is 0.82 eV. The experimental hole scattering spectrum (dots) obtained for positive tip bias is compared with theory (solid line). A magnified plot of the hole scattering spectrum is also shown.

negative bias scattering spectrum is the same as that for the positive bias ballistic hole spectroscopy spectrum.

The complementary hole scattering spectrum for the Au/Si(100) system is shown in Figure 3b. The hole scattering spectrum (corresponding to the experiment of Figure 2b) is compared with a ballistic electron transmission spectrum in Figure 3b. As for the previous experiment the combined spectra show two thresholds in the collector current. The ballistic electron spectrum, for negative tip bias indicates the conduction band minimum at $eV_{bn} = 0.82$ eV. For positive bias, holes created by electron tunneling propagate through the base before scattering, and a fraction of the electrons created by pair production are collected. Both the scattering spectrum and the ballistic electron transmission spectrum show thresholds corresponding to the Fermi level to conduction band minimum separation. Both spectroscopy bias polarities yield an electron current in the collector.

The direct measure of the energy-dependent pair-production rate by scattering allows an investigation of fundamental transport processes. A treatment combining the energy-dependent scattering rate with interface transmission has been successful for treating experimental results for the Au/Si(100) system.[7] The incident carrier produced by tunneling occupies a state outside the Fermi surface with energy, E_o, and wave vector, k_o The incident carrier scatters to a state (E_1, k_1), also outside the Fermi surface. Scattering produces an electron-hole pair with an electron at (E_2, k_2) and a hole at (E, k). Both energy and momentum conservation conditions may be applied to determine the energy dependent phase space available for scattering. Energy conservation requires that

$$E_o - E_1 = E_2 - E. \tag{1}$$

Further, momentum conservation requires that

$$k_o - k_1 = k_2 - k. \tag{2}$$

The energy dependence of phase space available for scattering is combined with a matrix element for scattering, M. The Golden Rule gives the rate, $R(E, k)$, for the creation of holes at energy E and momentum $\mathbf{k}$:

$$R(E,\mathbf{K}) = 2\pi/h \sum_{\substack{\mathbf{k}_1, \mathbf{k}_2 \\ E_1, E_2 > E_F}} |M|^2 \delta(E_o + E - E_1 - E_2)\delta_{\mathbf{k}_o + \mathbf{k}, \mathbf{k}_1 + \mathbf{k}_2} \qquad (3)$$

For treatment of the Au/Si(100) results the matrix element for scattering has been taken to be energy and momentum independent. The implications of this approximation may be tested by comparing the predictions of this model with fundamental results. First, Equation 3 enables the calculation of the total electron- electron scattering rate for low energy electrons within a free-electron metal. Now, the states for available for hole creation are confined by energy conservation to a thin shell of thickness $\delta E = E_o - E_F$, at the Fermi surface. The volume of the allowed shell is proportional to δE, for $\delta E \ll E_F$. In addition, the states available for the incident electron to enter by scattering also lie within a shell of thickness δE at the Fermi surface. Therefore, for constant M, the electron-electron scattering rate for energies near E_F simply scales as the product of these two phase space factors; δE^2.[11] This result appears in classical transport phenomena as the well-known T^2 resistivity contribution due to electron-electron scattering in metals at low temperature. It is clear that BEEM scattering spectroscopy probes fundamental properties of electron transport since the scattering rate R(k) directly determines the measured collector current.

The current measured in BEEM scattering spectroscopy has been calculated by combining the carrier generation rate R(k) with the phase-space limited interface transport probability.[7] BEEM spectra for ballistic electron or hole transmission, see Figure 3, display a current threshold increasing as $(eV - eV_b)^2$, as a consequence of the phase space restrictions on interface transport. However, the square-law energy dependence of the electron scattering rate combined with interface transport produces a current threshold increasing as $(eV - eV_b)^4$. Figure 3 clearly shows this qualitative difference between the spectral shapes of the ballistic carrier and scattering spectroscopies.

The BEEM scattering spectroscopy theory is compared to the experimental results in Figure 3. The theoretical spectra are calculated by adjusting only a scale factor for the magnitude of the collector current.[7] The magnitude of the bias threshold taken for the calculated spectra is equal to that measured directly in the ballistic carrier spectroscopy mode. The agreement between theoretical and experimental spectra is accurate near threshold for both electron and hole scattering. Several phenomena are now under investigation to determine the discrepancy appearing at large bias between theoretical and experimental hole scattering spectra. The agreement between theoretical and experimental spectral shapes is significant since the calculated spectral shape is independent of the adjustable parameters. BEEM scattering spectra clearly show, therefore, the fundamental energy-dependence of carrier scattering.

In summary, carrier scattering in subsurface interface structures has been directly measured using a new energy-resolved spectroscopy. Theoretical spectra show detailed agreement with the experimental results. Applications for this new method exist in many areas of transport physics. The unique capabilities of an energy-resolved probe combined with a detection of scattering products are important for determining the role of various scattering processes on transport in a material or structure. The nature of the BEEM technique also allows spectra to be spatially resolved with nanometer resolution, thereby revealing the contributions of defects to scattering. This method is also applicable to a wide variety of structures comprised of metals, semiconductors, and thin insulating barriers. Finally, for certain structures the four BEEM spectroscopies for electron and hole ballistic transport may be combined with electron and hole scattering for comprehensive measurements on a microscopic scale.

ACKNOWLEDGEMENTS

The research described in this paper was performed at the Center for Space Microelectronics Technology, Jet Propulsion Laboratory, California Institute of Technology and was jointly supported in parts by the office of Naval Research and the Strategic Defense Initiative Organization/Innovative Science and Technology Office through an agreement with the National Aeronautics and Space Administration.

REFERENCES

1. P. Drude, Annalen der Physik 1, 566 (1900).

2. S. M. Sze, <u>Physics of Semiconductor Devices,</u> (Wiley, 1981) and references therein.

3. For a review of early work see M. Heiblum, Solid State Electronics **24**, 343 (1981), and references therein.

4. M. Heiblum, M. I. Nathan, D. C. Thomas, and C. M. Knoedler, Phys. Rev. Lett. **55**, 2200 (1985).

5. J. R. Hayes, A. F. J. Levi and W. Wiegmann, Phys. Rev. Lett. **54**, 1570 (1985).

6. M. Heiblum, K. Seo, H. P. Meier and T. W. Hickmott, Phys. Rev. Lett. **60**, 828 (1988).

7. L. D. Bell, M. H. Hecht, W. J. Kaiser, and L. C. Davis, Phys. Rev. Lett. **64**, 2679 (1990).

8. W. J. Kaiser and L. D. Bell, Phys. Rev. Lett. **60**, 1406 (1988).

9. L. D. Bell and W. J. Kaiser, Phys. Rev. Lett. **61**, 2368 (1988).

10. M. H. Hecht, L. D. Bell, W. J. Kaiser and L. C. Davis, Phys. Rev. B **42**, 7663 (1990).

11. N. W. Ashcroft and N. D. Mermin, Solid State Physics, (Holt, Rinehart, Winston, New York, 1976) p. 346.

ROLE OF CONDENSED MATTER PHYSICS IN INFORMATION MANAGEMENT AND MOVEMENT – PAST, PRESENT AND FUTURE

C. K. N. Patel

AT&T Bell Laboratories
Murray Hill, N.J.

1. INTRODUCTION

The invention of the transistor in 1947 launched the long relationship between condensed matter physics and the information management and movement (IM&M) industry which continues unabated. Significant recent discoveries and inventions in semiconductor and other materials include the laser and low loss optical fibers which have continued the impact of condensed matter physics on IM&M. I define IM&M very broadly to include all products and systems activities enabling the movement (transmission) and management (computing, storage, switching, etc.) of information. In this very broad definition I also include the consumer electronics arena. To describe all the major events that revolutionized the IM&M industry since the invention of the transistor is an impossible task. Therefore, I will select a few areas representative of the electronic and photonic industries to highlight both the progress as well as the future opportunities.

Condensed matter physics has advanced rapidly because of a close connection between physics and materials science activities which are

Highlights in Condensed Matter Physics and Future Prospects
Edited by L. Esaki, Plenum Press, New York, 1991

summarized in Figure 1. These strong inter-relationships are character-istics of not only the area of semiconductors but also of superconductivity, magnetism and insulators. Without new materials, new physics would have remained impotent indeed. Furthermore, it is not only new materials which have contributed to the advances, but also new processing techniques such as molecular beam epitaxy[1] (MBE) have had an enormous impact on both condensed matter physics and its applications.

In this paper I would like to cover four different areas where condensed matter physics has made significant impact on the technological progress in information management and movement. Even though the title indicates that I'm going to talk about past, present and future, I will concentrate mostly on present and the future and will not discuss the past accomplishments and the past progress in any significant detail. The

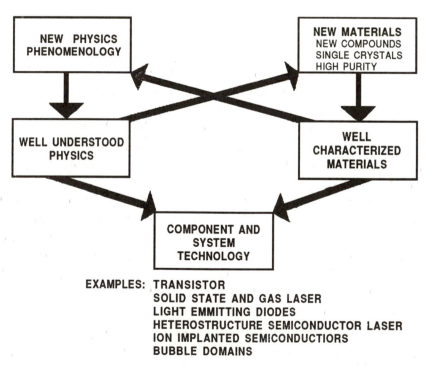

FIG. 1. Leading-edge component and system manufacture is the result of a broad-knowledge base in and coupling of physics materials science.

areas that I wish to cover are electronics and photonics, as they apply to both devices as well as transmission and switching and far out future research areas which may have significant impact over the long term from the standpoint of IM&M technology.

2. ELECTRONICS

2.1. Silicon

The invention of the transistor in the late 40's and the invention of integrated circuits in the late 60's and early 70's have revolutionized the technology of electronics completely. Needless to say, silicon is the primary vehicle for electronic circuits today. Compound semiconductors such as the III-V materials appear to be attractive from the point of view of both speed as well as lower power consumption. However, so far the III-V materials have remained a technology of tomorrow where complex integrated circuits are concerned. The principal and dominant trend in the silicon integrated circuits is the unabated reduction in the minimum feature width that one uses in an individual circuit. The minimum feature width, in some sense can be defined as the smallest transistor or other components that one is able to fabricate on silicon. Figure 2 shows the minimum feature width as a function of time starting from something

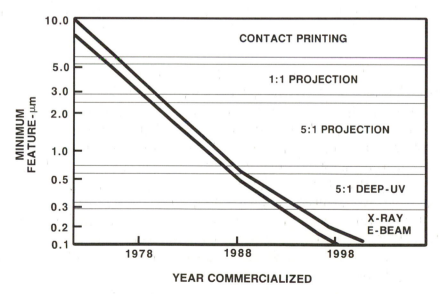

FIG. 2. Minimum future size of leading-edge DRAM devices as a function of year of commercial introduction.

FABRICATION LINE COST AS A FUNCTION OF DESIGN RULE

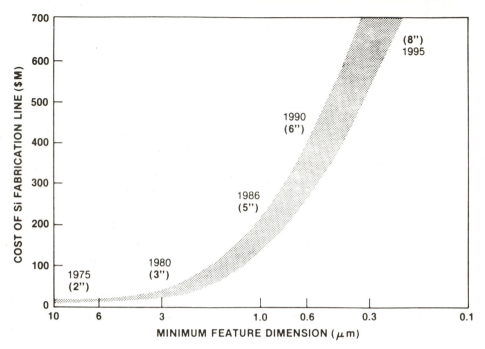

FIG. 3. The cost (capital requirements) of a high volume fabrication line is increasing at a phenomenal rate and will approach one billion dollars by the year 2000.

close to 10 or 20 micrometers down to the current minimum dimension of about 0.8 μm, and it is still getting smaller. The almost exponential reduction in the size is expected to continue in the coming decade with the expected minimum device dimension on a silicon integrated circuit by the year 2000 to be of the order of 0.25 μm. Along with the reduction in the minimum feature width there is an increase in the number of components or transistors that one fabricates on an integrated circuit. This number at present stands at about 2-10 million components and is likely to go up by at least a factor of 10 by the end of the decade. The number of components have gone up approximately by a factor of 2 every 18 to 24 months ever since the field of integrated silicon circuits began some twenty years ago.

The reduction in the minimum feature width in the coming decade will require use of different types of lithographic techniques. It will also require a more stringent control on the gate width and oxide thicknesses.

The present use of visible radiation for projection lithography is giving way to the use of deep ultraviolet light, and as we proceed towards the next century all indications point towards the use of x-rays as the radiation sources.

It has also been found that the cost of a silicon integrated circuit fabrication line has been going up exponentially as the minimum feature dimension has shrunk over the last 10 years or so. Figure 3 shows an approximate variation of this cost as a function of time. In order to make such a fabrication line cost economically competitive the wafer sizes have gone up. As the wafer size goes up, the conventional batch process becomes less than optimum for achieving the uniformity and yield required. Nonetheless, the expected cost of the fabrication line of about a billion dollars by the turn of the century indicates that only the very large silicon integrated circuit manufacturers will be able to stay at the forefront of technology. Clearly condensed matter physics and the material science technologies have a very important challenge - how to break the exponential escalation of the cost of the fabrication line.

One proposal which has been put forward is to use what is called an integrated manufacturing facility, sometimes also called controlled environment manufacturing of integrated circuits. Traditionally a silicon fabrication line consists of a variety of processing machines or equipment all of which are located inside a very clean environment. Inside this clean environment fabrication line workers move around operating the equipment as well as transporting the processed wafers from one step to the next. The drawback of this approach is that humans are the primary contaminators of the clean environment. The controlled environment processing philosophy attempts to invert the concept of the clean room by putting the cleanliness inside the machines and keeping the people out in only moderately a clean environment.

Figure 4 shows such a proposed setup. It consists of a set of equipment and processes which are necessary for taking a silicon starting wafer and processing it to the point where the completed integrated circuit emerges. Once the wafer is introduced in such a processing machine, it never emerges from either ultra high vacuum or a totally benign environment out into the open, where it could possibly either collect dust or react with atmospheric contaminants including oxygen and other naturally occurring gasses. Such an integrated process will allow processing of one wafer at a time and will assure uniformity of processing across an entire wafer by having appropriate on-line monitors as well as on-line process control techniques. It is clear that the uniformity of processing across the wafer is going to be exceedingly important for obtaining high yield of devices as we move towards the future when silicon substrate wafers of 12 inches diameter may be used.

The process cluster of equipment requires that all the processes be vacuum compatible. Currently this is not so. For example, the photoresists which are used for transferring the pattern from a mask onto a silicon wafer through the projection lithographic steps are liquids and still very often liquid etching steps are included in silicon integrated circuit manufacture. Clearly an integrated manufacturing process described in Figure 4 will require us to move towards all vacuum compatible lithographic resists as well as etching technologies. Condensed matter physics and material science have a very important role to play in making such a processing technology feasible.

While it is difficult to make precise cost estimates and cost comparisons between a traditional fabrication line whose costs are empirically projected to be of the order of a billion dollars by the year 2000, with the

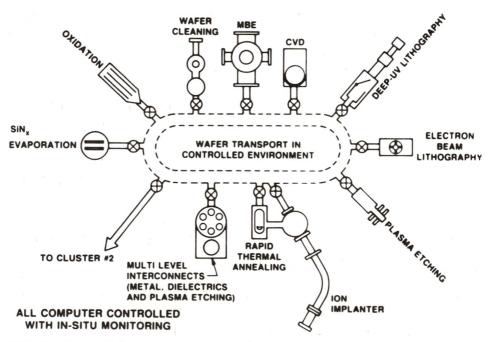

FIG. 4. Controlled environment processing as schematically illustrated here attempts to forego the cleanliness requirements of the processing room by keeping the wafers inside the processing machine. In-situ monitoring assures enhanced uniformity and yield.

cost of a cluster machine shown in Figure 4, we believe that such cluster machines will clearly be cost competitive with a traditional fabrication line. The throughput of each one of these integrated manufacturing clusters will be small but so will the cost. If integrated manufacturing is scaled up by having a large number of clusters it will be possible to obtain the same product as that which one would get from a traditional fabrication line at a cost which will be comparable to or lower than the cost of the traditional semiconductor IC fabrication line. An additional advantage which arises from such cluster philosophy is the fact that one can add processing capability in small steps as needed rather than having to build a new fabrication line even when, for example, all that is needed is a ten-percent greater capacity. Such a notion is a radical departure from the traditional way of silicon integrated circuit processing. Today, silicon IC manufacturing is a batch processing technique. We should be able to take advantage of the ideas that have arisen out of the chemical manufacturing industry, where continuous processing has yielded not only lower cost but also faster turn around times and improved quality through the use of on-line process monitors and on-line process control.

It is my belief that such a continuous processing environment will eventually become standard for silicon integrated circuit processing. However, the introduction of piece parts of the total scheme shown in Figure 4 is likely to occur even in the traditional IC fab line long before the complete structure is used. The subcluster machines are being fabricated even today and the cluster sizes will grow as we understand which optimization algorithms are needed for maximum throughput.

2.2. III-V Compound Semiconductors

Earlier I had mentioned that the compound semiconductor technology for electronic circuits is still the technology of tomorrow. This in some sense is a generalization. There are specific instances where the capabilities of III-V materials are being investigated in actual devices and circuits. One of the most significant advances of the last 20 years in the area of III-V materials is the invention of molecular beam epitaxy (Figure 5) which allows the deposition of III-V material heterostructures on a molecular level. It was shown[2,3] that one can grow multilayers of gallium arsenide and aluminum arsenide where each one of the layers is exactly two atomic layers thick and one can have layers as many as 100 thousand ultimate layers fabricated using the molecular beam epitaxy scheme.

This highly precise control of layer thickness has allowed a variety of new physics and new devices using molecular beam epitaxy grown

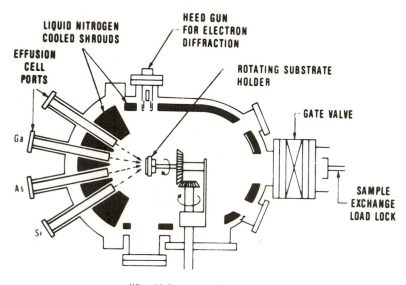

FIG. 5. Schematic diagram of the molecular beam epitaxy (MBE) system. The substrate is rotated during deposition to achieve a uniformity with variation less than 1% over a 3-inch diameter.

material. Table 1 shows a list of the new physics and devices that have become possible. Of these, let me point out one specific device which is a multistate resonant tunneling transistor[4] schematically shown in Figure 6. Such a tunneling transistor has multiple operating states which in principle allows one to replace a large number of traditional two-state transistors by one single multistate transistor. Specifically two applications have been investigated. One is a frequency multiplier and the other one is a four bit parity checking circuit. The four bit parity checking circuit, which uses only one of these multistate resonant tunneling transistors is able to carry out functions of 27 traditional 2-state transistors for checking parity of four signals. Speed of operation of such a parity checking multistate resonant tunneling transistor circuit has been measured and it's found to be in excess of 20 gigahertz. These switching phenomena, which rely upon the tunneling ideas coming from condensed matter physics and the materials growth technology based on molecular beam epitaxy, are likely to revolutionize device concepts for the next decade by allowing us to put more functionality in single integrated

TABLE 1. NEW PHYSICS AND DEVICES USING MBE GROWN MATERIALS

PHYSICS

- Modulation doping
- Bandgap engineering
- Strained layer superlattices
- Fractional quantized Hall effect

ELECTRONICS

- High speed modulation doped FETs
- Resonant tunneling transistors
- Heterojunction bipolar transistors

PHOTONICS

- Lasers
  - High uniformity and reproducibility
  - Ultralow threshold using quantum wells
  - Surface emitting lasers
- Detectors
  - Low noise APDs
  - Resonant tunneling long wavelength detectors

circuit devices compared to what was possible with the traditional 2-state devices.

In a second example of the impact of III-V materials on electronics is the indium gallium arsenide/indium phosphide bipolar heterostructure transistor. The heterojunction bipolar transistor energy level diagram is shown in Figure 7, where we see that electrons are injected from a large bandgap material, in this case indium phosphide, into the smaller bandgap material, which is an indium gallium arsenide across a very thin base layer. Because of the high energy with which the electrons are injected into the base layer, they can traverse the base without significant scattering. Speed measurements indicate a high frequency cutoff at ≥ 165 GHz (fig. 8).[5]

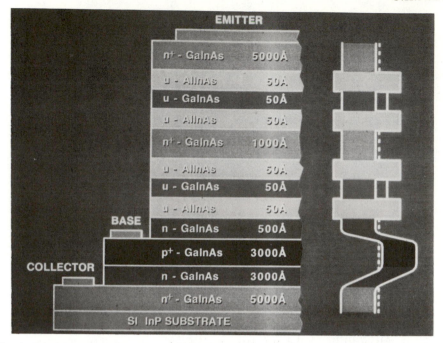

FIG. 6. Multistate resonant tunneling transistors, fabricated using molecular beam epitaxy, can replace a number of conventional two-state transistors. The schematic shows the multilayer structure and associated band diagram.

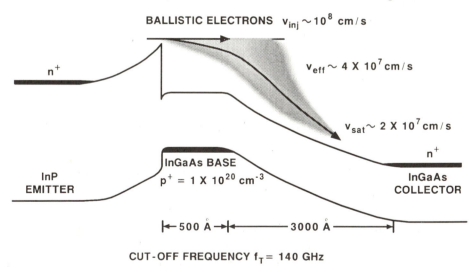

FIG. 7. Energy band diagram of a heterostructure bipolar transistor when the device is under bias. A wide-bandgap n-type emitter injects electrons with excess kinetic energy into the p-type base region.

676

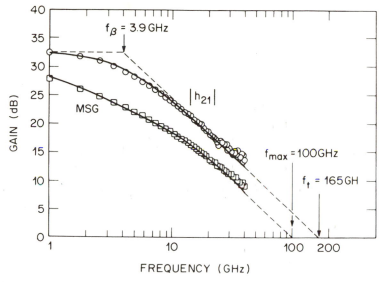

FIG. 8. Frequency dependence of the current gain, h_{21} and maximum available stable power gain (MSG) for the $Ga_{0.47}In_{0.53}As/InP$ bipolar transistor grown by MBE.

3. PHOTONICS

3.1. Devices

One of the more significant advances in semiconductor lasers to come along in recent years is the surface-emitting laser (SEL) geometry, first discussed by Iga.[6] Figure 9 shows a schematic of a SEL where the laser is formed by MBE grown alternate layers of GaAs and GaAlAs.[7] With proper design, high reflectivities can be obtained and lasing occurs in the same direction as the current flow. The current path is defined by lithography and therefore the position of the light output is precisely known. Coupling of such a laser output to an optical fiber is easily accomplished.[8] Further, because the cavity length is very short, typically of the order of 1 - 3 μm, such a laser naturally operates in a single longitudinal mode (Figure 10). Since the lasing regions are lithographically defined, it is now easy to fabricate individually addressable laser arrays[9] as seen in Figure 11 for a 4 by 4 example. There are several processing, operational and applications related advantages unique to SEL, and these are listed in Table 2.

677

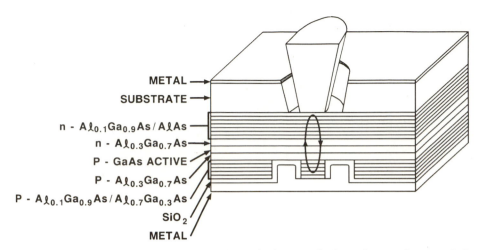

FIG. 9. Cross-sectional representation of the vertical-cavity surface-emitting laser consisting of n- and p-type distributed Bragg reflecting mirrors.

TABLE 2. SURFACE EMITTING LASERS

PROCESSING OPPORTUNITIES

- Single step processing
- Integral mirrors
- No regrowth
- Laser areas defined by lithography
- Potentially very high yield

OPERATIONAL OPPORTUNITIES

- Naturally single mode output
- Output beam defined by processing
- Coupling into fibers easy (mechanical assembly)

APPLICATIONS OPPORTUNITIES

- Inexpensive laser transmitter for fiber-to-the-home
- Arrays for high powers
- Sources in optical switching
- Optical interconnect elements on Si

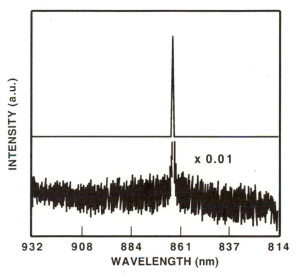

FIG. 10. Room termperature cw emission spectrum of a typical surface emitting laser diode, showing the single mode emission characteristic of this device.

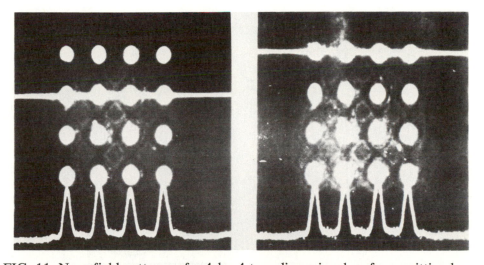

FIG. 11. Near-field patterns of a 4-by-4 two-dimensional surface emitting laser array. The geometry of the SEL easily allows arbitrary placement of laser outputs. As many as 1 million SELs have been fabricated on one square centimeter.

3.2. Transmission

Use of photonics technology in information movement requires semiconductor devices - lasers and detectors - and a reliable low loss transmission medium, optical fiber. Progress in silica optical fibers has been such that now fibers with optical transmission losses ≤ 0.19 dB/km are routinely available commercially. This progress in loss reduction has occurred over a period of some 10 years and required an in-depth understanding of the role of impurities and removal of impurities from silica. Figure 12 shows three snapshots of optical loss spectra of fibers in 1973, 1976 and 1983, respectively.[10] Along with the reduction in the total losses, the wavelength for the loss minimum has shifted to longer wavelengths. The rising loss at shorter wavelengths originates from scattering which has a λ^{-4} dependence. The rising loss at longer wavelengths, beyond ~ 1.6 μm, comes from overtone phonon wings corresponding to the Si-O stretch vibration. It is apparent that the minimum loss of 0.155 dB/km for a pure silica fiber at 1.55 μm is very close to what would be predicted theoretically. The loss peak at ~ 1.4 μm arises from an overtone of O-H vibration (~ 0.7 dB/km). At this level of absorption at 1.4 μm, O-H concentration in the silica core is at or below sub-ppB level. The 1.4 μm peak serves as a critical guide to monitor the level of OH in the fiber. The OH absorption will be "enhanced," making the fiber less

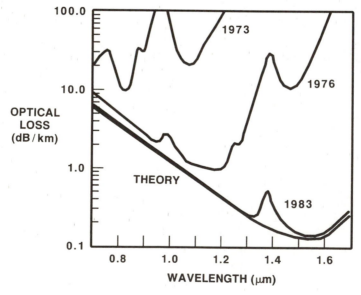

FIG. 12. Optical fiber loss as a function of wavelength. Dramatic progress in loss reduction has occurred over a period of 10 years.

desirable for use in information transmission in the long-haul arena if hydrogen were to enter into the fiber after manufacture. Water ingress serves to decrease the physical strength of the fiber, and therefore both hydrogen and water must be kept out of the fiber. Various hermetic coatings have been proposed and two are available. AT&T, so far, has used amorphous carbon coating while Corning has used titanium diffusion. The carbon-coated hermetic fiber which is now in production provides the long-term hermeticity against water and hydrogen ingress.[11]

The routine low loss of ~ 0.19 dB/km at 1.55 μm allows long unrepeatered spans. However, eventually the signal becomes too weak to detect with acceptable signal-to-noise ratio. Thus, in long-haul information transmission, repeaters are required at 50 - 150 mile separations to restore the optical signal to a high value. Traditional repeaters carry out three principal functions (see Figure 13): detection of the optical bit stream and converting it into an electrical bit stream, amplification,

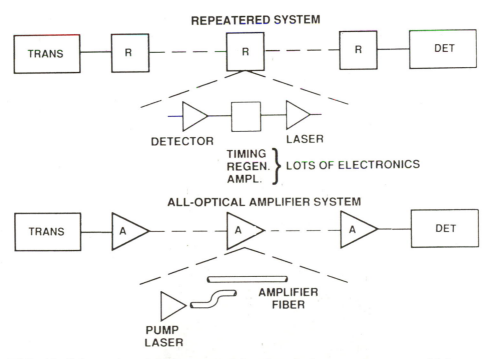

FIG. 13. Schematics of (A) a conventional optical repeater system which consists of a photodetector, an amplifier, a timing circuit and a laser; (B) an all-optical amplifier system which consists of a laser and an Erbium-doped fiber with directional couplers.

retiming and regeneration of the electrical signal, and finally driving a semiconductor laser to convert the electronic signals back into an optical bit stream. This scheme is universally utilized to date but as we go to higher and higher bit rates for information transmission, the electronic part of the system becomes decreasingly cost-effective. An "electronic-bottleneck" appears at bit rates $\gtrsim$ 5-10 Gb/s. Further, electronics capability and its cost effectiveness is continually improving. But once a repeater system is installed, such as the TAT-8 transatlantic cable, it is prohibitively expensive to replace the repeaters with improved electronics. In addition, wavelength division multiplexing (WDM) is not very cost effective using conventional repeaters, since each repeater now must demultiplex the WDM bit stream, and then carry out all the photonic-electronic-photonic operations on each of the wavelengths, and finally remultiplex the different wavelengths. What is needed is a repeater which is "transparent" to the bit rate or wavelength division multiplexing imposed on the system. Such a system can then be continually upgraded by improving only the end terminals.

A recent advance, optical amplifiers, is making such a bit-rate independent repeater possible (Figure 13). In the early days of optically pumped solid-state lasers, a variety of laser ions and hosts were investigated. Of these, erbium doped silica fiber is seen to provide optical gain at ~ 1.55 μm when optically pumped at either 0.98 μm or 1.48 μm (Figure 14). With pump power of ~ 10 mW at 1.48 μm gains as high as 30 dB have been reported for amplifier fiber length of ~ 50-70m.[12] Thus,

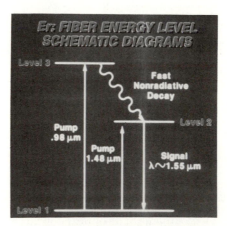

FIG. 14. Energy level diagram of Erbium ions in a silica fiber host showing that optical gain at 1.55 μm can be achieved with optical pumping at 1.48 μm or 0.98 μm.

it is possible to amplify the weakened optical signal without going through photonic-electronic-photonic conversion. The gain-bandwidth of the erbium-doped fiber amplifier (EDFA) is ~ 20 nm and hence poses no barrier even to the highest bit rates contemplated in the foreseeable future. Figure 15 shows results of a laboratory experiment which contained three EDFA appropriately pumped through directional couplers. Signal decay due to fiber loss and the restoration of the optical signal strength through amplification is demonstrated. The desired "transparent pipe" is now at hand and the data stream and wavelength division multiplexing which can be accommodated is now determined only by the end terminals.

The limitation set on distance between repeaters arises not only from the fiber loss. A further limitation arises from dispersion which results in pulse broadening. Figure 16 shows the dispersion caused pulse broadening which would arise in a single mode fiber as a function of wavelength for normal fiber and two other geometries. The fiber dispersion is zero near 1.3 μm. However, the loss minimum is near 1.55 μm (and EDFA are available, at present, only for 1.55 μm). At 1.55 μm, in a normal fiber we would expect a dispersion caused pulse broadening of ~ 15 ps/nm/km. Assuming only a Fourier spectral broadened pulse,

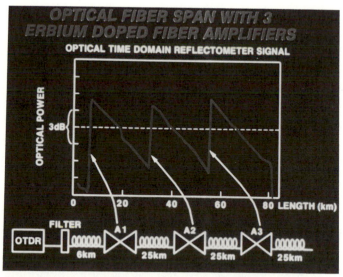

FIG. 15. Laboratory demonstration of signal decay due to fiber loss and the restoration of the optical signal strength through amplification by three Er-doped fiber amplifiers.

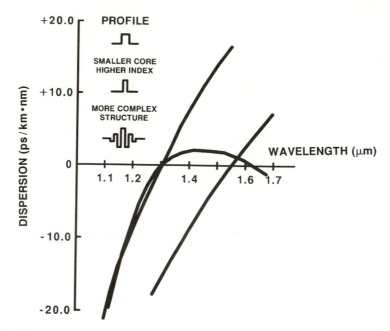

FIG. 16. The optical pulse broadening in three different single mode fibers as a function of wavelength.

a 10,000 kM span (using EDFA to compensate for the fiber loss) would lead to a maximum data rate of $\lesssim 1$ GHz! If the distributed feedback laser did not operate in a Fourier spectrum limited mode, due to chirp for example, the data rate would be even smaller. However, it is known that one can take advantage of the nonlinearity of the refractive index of the optical fiber to propagate pulses which retain their shape and width if the fiber loss can be overcome through the use of optical amplifiers described above.

The refractive index of silica fibers can be written as

$$n = n_0 + n_2 I + ... \tag{1}$$

The intensity dependent term $n_2 \approx -3/2 \times 10^{-16}$ cm^{-2} W^{-1} is small, but the small diameter of the single mode fiber ~ 6.8 μm results in significant optical intensity. For example, a 1 mW signal would correspond to I $\approx 10^4$ W cm^{-2}. The propagation of a pulse in such a fiber requires solution of the nonlinear Schroedinger's equation. It is seen that stable operating pulse shape is a hyperbolic secant and propagation over distances as long as $10\text{-}20 \times 10^3$ km is possible without significant distortion or broadening.

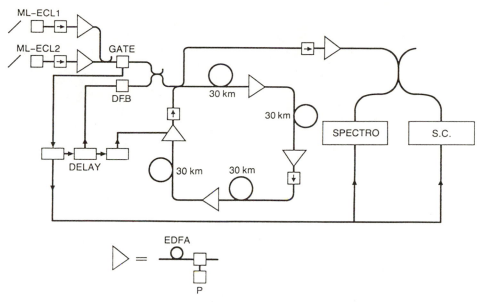

FIG. 17. Soliton transmission in fibers is a method recently demonstrated to overcome pulse broadening. Shown is an experimental setup for ultra-long distance propagation of solitons in a fiber loop. Solitons are generated in two mode-locked external cavity semiconductor lasers (ML-ECL) and gated into the 120 km long fiber loop containing four Erbium Doped Fiber Amplifiers (EDFA). The output is analyzed with a spectrometer and a streak camera (S.C.).

Such a pulse is called an optical soliton and was first predicted by Hasegawa and Tappert[13] in 1973 and experimentally demonstrated first by Mollenauer[14] in 1980.

Laboratory systems experimetns[15] have been carried out using a closed loop in which a pulse propagates around for many times to obtain an equivalent propagation distance of 10,000 to 20,000 km (Figure 17). The pulse shape is sampled periodically to obtain the pulse shapes at 3000, 5000 and 10,000 km as shown in Figure 18, confirming the expected soliton propagation behavior. It should be noticed that to maintain the soliton pulse shape and width, the intensity of the pulse has to be maintained by using appropriate optical amplifiers.

While experimental results have confirmed optical soliton propagation, it is instructive to look at an intuitive example constructed by Stephen Evangelides and L. F. Mollenauer showing a pack of runners on a mattress (Figure 19). The collective weight makes a depression in the

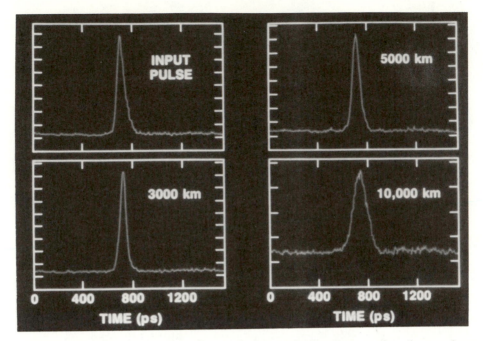

FIG. 18. Streak camera recordings of the optical soliton pulse shape after a propagation distance of 0.3000 km, 5000 km and 10,000 km in the loop of Figure 17.

SOLITONS: The bunching of high-intensity light. The light pulse creates a moving "valley" (of higher dielectric constant material).

RUNNING ON A MATTRESS: The moving valley pulls along slower runners and retards the faster ones.

FIG. 19. An analog of soliton propagation to a pack of runners on a mattress; the higher dielectric constant in the fiber resulting from the high-intensity light tends to keep the pulse shape from spreading.

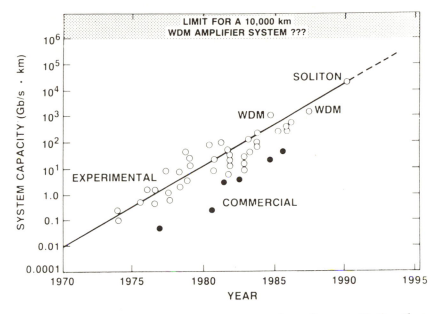

FIG. 20. Lightwave system capacity as a function of year. Notice that soliton propagation has significantly increased the performance.

mattress. The faster runner, thus, ends up running uphill and slows down; the slower runner is running downhill and speeds up, keeping the pack of runners together, as it were, in a soliton pulse. Absent the depression in the mattress (or equivalently the intensity dependent refractive index of the fiber) the pack of runners would spread out (as would an optical pulse described earlier).

Soliton propagation is now sufficiently well understood to assure us that at some time in the future long-haul communications systems would use this mechanism. A figure of merit of the optical fiber communication system has been the product of the bit rate and the maximum propagation distance without using traditional repeaters. The soliton propagation significantly increases the previous performance which has shown exponential growth since 1983 as seen in Figure 20. Finally, the gain bandwidth of the EDFA is such that four different wavelengths can be multiplexed to give a significant increase in the information capacity.

I should point out another potential advantage of the optical amplifier (as opposed to traditional repeatered systems). This is its ability to support bidirectional transmission of optical signals thus doubling the actual capacity of a fiber. Further, the counter propagating signals can be wavelength multiplexed and can be solitons. Soliton pulses pass through each other without any degradation.

687

3.3. Switching

Switching and computing make up the second half of information movement and management. While the progress of photonics in information transmission has grown exponentially, similar replacement of electronics by photonics has not occurred in switching and computing. Nonetheless there has been technical progress which needs to be noted. Reasons for and barriers to greater photonics in switching and computing

TABLE 3. ADVANTAGES OF OPTICS IN SWITCHING AND COMPUTING

- Noise immunity
- Absence of ground loops
- Handle signals already in the optical domain
- Compatible with high bandwidth, low loss fibers
- Massively parallel interconnects (free space)
- Potential for very high throughput

TABLE 4. SWITCHING PARALLELS

	ELECTRONICS	OPTICS
DEVICE	Transistor	Optical Switch (S-SEED)
OUTPUT	Electrical Signal Transmission	Optical Reflection
CONTROL	Electrical Input Signal	Second Beam of Light
STATE	Memory Configuration Possible	Memory Configuration Natural
SPEED	~180 GHz	~1 GHz

are based in cost and capability. In switching, both cost and capability are providing the needed incentive for photonics introduction since it would eliminate the electronic/photonic conversion at either end of the switch. There are a number of excellent reasons, as seen in Table 3, why increasing amounts of optical switching is desirable in a communications system. There are also parallels between electronic and optical switching where the switching is completely optical (as opposed to hybrid where in $LiNbO_3$, e.g., the control is electronic. These parallels are listed below in Table 4. For broadband switching of information, such as that coming down optical fibers, switching of destination can be provided through the use of $LiNbO_3$ guided-wave switches, a schematic of which is shown in Figure 21. The electrooptic effect in $LiNbO_3$ allows the coupling of energy in the two waveguides. Application of appropriate electric field results in complete switching of optical power from one guide to the other. Thus 2x2 switch can be multiplexed to a larger nxn switch. Figures 22 and 23 show a sketch of the architecture and a photograph of a 8x8 switch, respectively.

There are two issues associated with $LiNbO_3$ 2x2 switch and its integration in the higher level architecture. The first is that heretofore, the insertion loss of each 2x2 $LiNbO_3$ switch is quite high ~ 1-3 dB. For a 8x8 switch the total insertion loss can be as high as 7-10 dB. This loss,

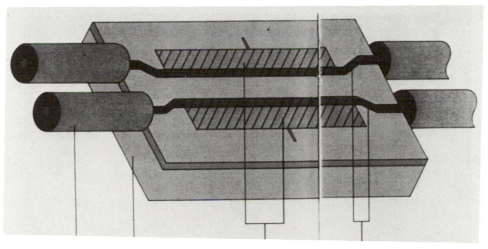

Optical Fiber Lithium Niobate   Electrodes   Waveguides

FIG. 21. Schematic of a $LiNbO_3$ guided-wave electrooptic effect switch. This switch is an example of switching signals in photonic form eliminating the need of photonic/electronic conversion.

in principle, can be compensated through the use of appropriate optical amplifiers, e.g., EDFA. The second issue is the physical dimension of the switch and the $LiNbO_3$ real estate needed for a large switch. The physical dimension needed for efficient transfer of energy from one guide to the next is such that a 512x512 switch, say, using such a technology is

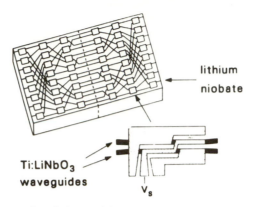

FIG. 22. Schematic of the architecture for a 8x8 photonic switch.

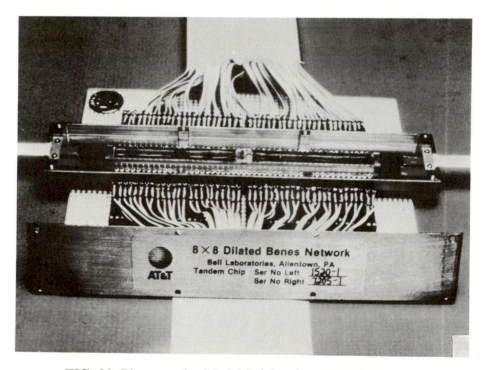

FIG. 23. Photograph of 8x8 $LiNbO_3$ photonic switch.

difficult. Lastly, the materials uniformity is not sufficiently high to assure identical switching voltages for each of the cascaded 2x2 switches making up the nxn switch. The result is that the switching voltage for each of the switches has to be individually adjusted adding to the cost of the switch. For a really large switch, such as that used in central switching offices, needing interconnection between ~ 50,000x50,000 lines, the $LiNbO_3$ 2x2 switch does not appear to be practical.

Another switching scheme[16] utilizes free space propagation and SEEDs (self electrooptic device). The SEED technology, though not quite as mature as the $LiNbO_3$ technology, is nonetheless better suited for high level multiplexing because the semiconductor technology used for fabrication allows manufacture of highly uniform devices. SEEDs which are primarily designed for switching at ~ 0.8 μm use GaAs/AlGaAs structures and rely on the electric field induced change in the absorption at the exciton peak (Figure 24). For operation of the SEED, the semiconductor material is sandwiched between high reflectivity multilayer mirror stacks so that when the material is non-abosrbing, we have a high transmitting high Q Fabry-Perot cavity. When the exciton peak shifts to the wavelength of bit stream, the Q of the cavity is destroyed and the incident steam is reflected by the front mirror.

To obtain a better switching capability, a newer structure - symmetric SEED, has been designed, and requires two control light beams

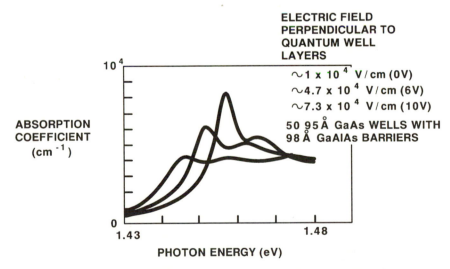

FIG. 24. Absorption coefficient as a function of photon energy (or wavelength) for different bias conditions. Notice the shift of the exciton peak for different bias voltages.

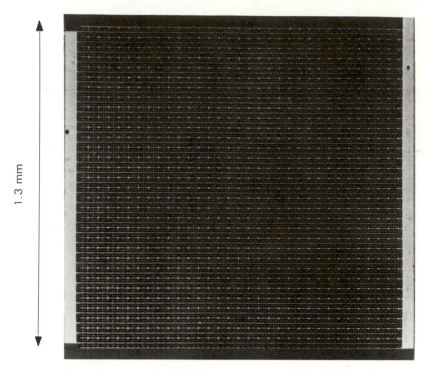

FIG. 25. Photomicrograph of a 64x32 symmetric self-electrooptic-effect device (S-SEED) developed at AT&T.

and switches two streams of light in a complementary fashion. Figure 25 shows a 32 × 64 array of AlGaAs/GaAs S-SEEDs which were used in a recently published optical computer experiment[17] A binary counter and a decoder Boolean circuit were demonstrated. In computing or switching applications, the free space propagation provides a way of very high level of parallelism for constructing a large scale switch. Nonetheless, both optical computing and optical switching using S-SEEDs have a long path ahead.

4. FUTURE

The discussion heretofore has focussed on the potential impact of condensed matter physics and materials science that is already reasonably well-understood. However, it is also important to look at the current forefront science activity in condensed matter physics and materials science which may form the basis of technology of the future. In the remainder of this paper, I will touch upon two exciting areas.

4.1. Clusters

The progress in electronics and photonics technology is predicated on making device sizes smaller and smaller. In this situation it is worth asking the question "At what minimum device size does the traditional condensed matter physics understanding begin to blur with the atomic physics ideas?" The work on semiconductor clusters is providing experimental information to obtain in-depth understanding of the bridge between atomic and condensed matter physics.

Semiconductor crystallite clusters of 100 to 10,000 atoms, ($\sim$ 15 to 60 A diameter) are too small to exhibit a bulk bandgap, even though they have the same unit cell and structure as the bulk semiconductor. These small crystallites, sometimes called *quantum dots*, have discrete, molecule-like excited electronic states that shift to higher energy with decreasing size, instead of the continuous electronic bands characteristic of macroscopic semiconductors. This phenomenon is a three dimensional electronic quantum size effect.

Controlled size clusters of semiconductors have been fabricated using three techniques, laser evaporation, chemical synthesis, and biological technique. The laser evaporation and subsequent separation using time-of-flight techniques yields the most monodisperse clusters of small sizes. However, the amount of cluster material produced is generally too small to be useful for measurements which require substantial quantities, e.g., optical absorption studies. On the other hand, the chemical and the biological techniques do not produce truly monodisperse distribution of cluster sizes, but yield semiconductor clusters in quantities large enough for detailed optical studies. Regardless of the technique used for fabricating clusters, there is a need for surface passivation of the clusters to prevent unanticipated coalescing into large uncontrolled crystallites. Steigerwald and his colleagues have invented synthetic methods to chemically modify the surface of the clusters, in order to be able to make macroscopic amounts of pure, reasonably monodisperse clusters with controlled reactivity and solubility properties.[18] For example, bonding phenyl organic groups to surface Se atoms on CdSe clusters changes the clusters from being hydrophilic to hydrophobic, and allows one to isolate and purify the clusters without spontaneous fusion of clusters into bulk CdSe. These *capped* clusters have intense nonlinear optical properties which are being investigated at present.

In order to nucleate, grow and chemically modify the surface of nanoscale semiconductor clusters, Brus et al. employ microscopically organized reaction environments. For example, inverse micelle solutions contain water pools (of controllable 10 - 40 Å diameter) coated with a surfactant monolayer, in a bulk hydrocarbon liquid. CdSe clusters can be precipitated inside the water pools if the appropriate reagent sources of Cd and Se atoms are added. These reactive clusters are stabilized against fusion with each other by adsorption of surfactant. Yet the adsorbed surfactant is sufficiently labile that additionally introduced molecular reactants can react with the clusters. Thus, a small CdSe "seed" can be grown larger by sequential addition of the Se and Cd reagents. It is also possible to grow a concentric layer of ZnS on a CdSe seed by addition of the appropriate reagents, or to bond individual paramagnetic transition metal ions such as Tm to the surface, while the cluster is stabilized in solution. In a final step, the S or Se atoms on the cluster surface can be bonded to organic groups, using different organometallic reagents, and the final "capped" particle recovered from the solvent as a pure chemical material.

Brus et al. have additionally discovered that clusters of very high crystalline quality can be grown at elevated temperatures in certain neat molecular liquids. These solvents have electron donating, Lewis base amine or phosphine groups that appear to catalyze the assembly of organized crystallites from monomer reagents. These moieties have the ability to form weak chemical bonds to exposed metal atoms. For example, nearly perfect 35 Å diameter wurtzite CdSe clusters can be grown near 200 C in a refluxing mixture of tributyl phosphine and tributyl phosphine oxide, starting with ~ 10 Å diameter seeds and a feedstock or organometallic Cd and Se reagents.

Brus and others have also discovered that some living cells (i.e., certain common yeasts), have the ability to nucleate and grow clusters of CdS crystallites capped with short peptides, in response to Cd introduced into their environment.[19] These biological semiconductor clusters can also be purified and isolated, and show a quite narrow size distribution near about 20 Å diameter, in one case quantitatively investigated. This observation establishes a new branch in the field of biomineralization, and has stimulated a search for additional unexpected natural processes of this sort. These CdS clusters are the first true examples of *biochips*.

With regard to electronic properties, nanoscale semiconductor clusters in principle form a bridge between the discrete electronic states of small molecules, and the continuous bands of bulk semiconductors.

Simple theory predicts that the electronic spectrum should be a dense manifold of discrete states, with the lowest state shifting to higher energy than the bulk bandgap as the diameter decreases. This shift is experimentally observed (Figure 26), however, the lowest state is spectrally broad and the higher states seem continuous.

A major source of broadening is due to the cluster size and shape distribution, even in samples where the diameter standard deviation is only 10%, because the spectra are a sensitive function of size. This result is observed in both absorption and luminescence spectra. Nevertheless, Brus et al have found spectroscopic techniques to remove this inhomogeneous broadening, to obtain the optical spectra of monodisperse CdSe clusters.[20] The excitation spectra of narrow sections of the overall luminescence reveal the absorption spectra of those monodisperse crystallites that emit at one specific wavelength. Additionally, single frequency laser holeburning experiments in the absorption spectra reveal the optical autocorrelation function of the monodisperse crystallites that absorb at that wavelength. Table 5 lists some of the key conclusions.

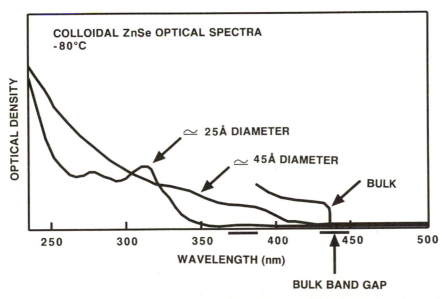

FIG. 26. Optical absorption spectra of ZnSe crystallites as a function of size. With decreasing diameter, the apparent bandgap becomes larger. The spectrum also develops partially discrete excited electronic states (adapted from Reference 21).

TABLE 5.

SEMICONDUCTOR SIZE EFFECTS:

TRANSFORMATION FROM MOLECULAR TO SOLID-STATE
BEHAVIOR

A)	~ 50 atoms ~ 15 A diameter	• Molecular structure • Molecular reactivity • Electron pairing in covalent bonds
B)	~ 100 - 200 atoms ~ 25 A diameter	• Unit cell of bulk structure • Bulk-like phonons • Band structure not yet developed
C)	~ 10^3 - 10^5 atoms	•Bulk gap, bulk conductivity, and bulk band structure slowly developing

The single size spectra thus obtained, coupled with dynamical measurements of bleaching and luminescence, show that the lowest electron-hole state exhibits substantial lifetime broadening due to surface localization within 150 fsec. It appears that the hole, initially created in a 1s orbital inside the cluster, localized on a surface Se atom. Higher electronic states show an even greater lifetime broadening, due to "hot carrier" relaxation down to the lowest state. Research continues into surface chemical structures that could control these relaxation processes.

4.2. Superconductivity

There is no better example than superconductivity for demonstrating the strong interplay between materials science and condensed matter physics described earlier in this paper (see also Figure 1). The field of high temperature superconductivity is dependent on new and improved materials both for understanding the mechanisms responsible for the phenomenon and for potential devices. This conference has dealt with this subject fully so I will not elaborate upon the scientific and technical progress except to point out that while the long-term prognosis of the impact of high temperature superconducting materials on the field of information management and movement is bright, many years of hard work remains to be devoted to improving material and related processing technologies. In the area of IM&M, a variety of potential applications exist and these include high speed electronics comprising of Josephson junction circuits, and superconducting interconnects on and between conventional semiconductor integrated circuits (Si and III-V). Other applications which involve simultaneous exposure of the superconducting material to both high current density and high magnetic fields such as those encountered in NMR and MRI magnets, etc., would require even greater amount of progress in bulk superconducting properties.

5. CONCLUSION

In this brief survey, I have described a few examples where condensed matter physics and materials science have played a crucial role in the almost exponential progress in a variety of technologies underlying the information management and movement activities. Needless to say, the simultaneous progress in condensed matter physics and materials science and associated processing technologies will continue to have similar impact on the area in the future.

ACKNOWLEDGEMENTS

I would like to thank Drs. L. E. Brus, A. Y. Cho and N. A. Olsson for critical comments on the manuscript.

REFERENCES

1. A. Y. Cho and J. R. Arthur, Progress in Solid State Chemistry **10**, G. Somorjai and J. McCaldin, Eds., New York, Pergamon Press, 1975, p. 157.

2. L. Esaki and L. L. Chang, Thin Solid Film **36**, 285 (1976).

3. M. B. Panish, H. Temkin, R. A. Hamm and S. N. G. Chu, Appl. Phys. Lett. **49**, 164 (1986).

4. F. Capasso, S. Sen, F. Beltram, L. M. Lunardi, A. S. Vengurlekar, P. R. Smith, N. J. Shah, R. J. Malik, and A. Y. Cho, IEEE Transactions on Electron Devices **36**, 2065 (1989).

5. R. A. Hamm, M. B. Panish, R. N. Nottenburg, Y. K. Chen and D. A. Humphrey, Appl. Phys. Lett. **54**, 2586 (1989).

6. K. Iga, S. Ishikawa, S. Ohkouchi and T. Nishimura, Appl. Phys. Lett. **45**, 1949 (1989).

7. K. Tai, R. J. Fischer, C. W. Seabury, N. A. Olsson, T.-C.D. Huo, Y. Ota and A. Y. Cho, Appl. Phys. Lett. **55**, 2473 (1989).

8. M. Orenstein, A. C. von Lehmen, C. Chang-Hasnain, N. G. Stoffel, J. P. Harbison, L. T. Forez, E. Clausen and J. E. Jewell, Appl. Phys. Lett. **56**, 2384 (1990).

9. H. -J. Yoo, A. Scherer, J. P. Harbison, L. T. Florez, E. G. Paek, B. P. vander Gaag, J. R. Hayes, A. von Lehmen, E. Kapon and Y.-S. Kwon, Appl. Phys. Lett. **198**, (1990).

10. S. R. Nagel, Proceedings of SPIE **1085**, (1989).

11. R. G. Huff, F. V. DiMarcello and A. C. Hart, Technical Digest for Optical Fiber Communications Conference, New Orleans, LA, 1988, Paper No. Tu-G2.

12. J. L. Zyskind, D. J. DiGiovanni, J. W. Sulhoff, P. C. Becker, and C. H. Brito Cruz, 1990 Technical Digest Series, Paper No. PDP6, **13**, Monterey, CA, August 6-8, 1990.

13. A. Hasegawa and F. Tappert, Appl. Phys. Lett. **23**, 171 (1973).

14. L. F. Mollenauer, R. H. Stolen and J. P. Gordon, Phys. Rev. Lett. **45**, 1095 (1980).

15. P. A. Andrekson, N. A. Olsson, J. R. Simpson, T. Tanbun-Ek, R. A. Logan and K. A. Wecht, 16th European Conference on Optical Communication, Amsterdam, 1990.

16. D. A. B. Miller, Optical and Quantum Electronics **22**, S61 (1990).

17. M. E. Prise, critical Review Series, **CR35, SPIE**, 3 (1990).

18. A. R. Kortan, R. Hull, R. L. Opila, M. G. Bawendi, M. L. Steigerwald, P. J. Carroll, and L. E. Brus, J. Am. Chem. Soc. **112**, 1327 (1990).

19. C. T. Dameron, R. N. Reese, R. K. Mehra, A. R. Kortan, P. J. Carroll, M. L. Steigerwald, L. E. Brus and D. R. Winge, Nature **338**, 596 (1989).

20. M. G. Bawendi, W. L. Wilson, L. Rothberg, P. J. Carroll, T. M. Jedju, M. L. Steigerwald and L. E. Brus, Phys. Rev. Lett. **65**, 1623 (1990).

21. N. Chesnoy, R. Hull and L. E. Brus, J. Chem. Phys. **85**, 2237 (1986).

INDEX

AC Stark effect, 304
adatom, 520
adiabatic transport, 19, 20, 260
Aharanov-Bohm
    effect, 45, 105
    phase, 46
Airy function, 69
Ambegaokar-Baratoff junction, 406
amorphous silicon, 593
Anderson's resonating valence bond
    (RVB), 371
Andreev reflection, 496
angular resolved photoemission
    spectroscopy (ARUPS), 386
angular resolved inverse
    photoemission spectroscopy
    (ARIPES), 387
anharmonic double-well
    potential, 460
anomalous magnetic moment, 8, 9
antiferromagnetic (AFM)
    insulators, 378, 434
    phase, 437
    spin fluctuations, 440, 445
antiferromagnetism, 365, 378
anyon, 28, 46, 433, 448
Arrhenius plot, 425
artificial
    heterostructures, 455
    superlattices, 416

atomic force microscopy
    (AFM), 466, 549, 569, 632
Auger analyzer, 86, 89

backscattering geometry, 193
band bending, 59, 60
band offset, 59, 118, 119, 148
ballistic electron emission
    microscopy (BEEM), 658
ballistic electron emission
    spectroscopy, 480
ballistic injection of charge
    carriers, 496
ballistic point contact, 15
ballistic transport, 8, 15, 243,
    495, 655
BCS, 365, 390, 449
Bijl-Feynman
    expression, 41
    theory, 38
biological materials, 473
biological matter, 551
biological systems, 551
biological molecules, 557
biosensors, 552
Block
    function, 119,
    oscillation, 57, 66, 76
    oscillator, 138
Bohr radius, 227, 293, 297, 639

Bohr-Sommerfeld quantization, 28, 259

Boltzmann
distribution, 43
transport equation, 498

Born approximation, 133

Bose particle (Boson), 27, 37, 42, 44, 45, 46, 51, 308, 448

Bose particle representation, 47

Bose particle system, 45, 47

bound states
man-made, 61
of holes, 62

Bragg
reflection, 89
scattering, 66, 439

Brillouin zone, 64, 66, 94, 136, 165, 179

carrier confinement, 60

center of mass (CM) motion, 222

charge density wave (CDW), 447

charge pump, 352

chemical beam epitaxy (CBE), 59

chemical potential, 17

chemical vapor deposition (CVD), 598

chemisorption, 564

Chern-Simons field theory, 46

clusters, 693

coherence length, 405, 435

coherent tunneling, 129

confinement effect, 178

confinement potential, 210

Cooper pairs, 339, 506

Corbino geometry, 13

Coulomb
attraction, 456

blockade, 243, 246, 268, 334, 347
correlations, 37, 293
effects, 218
force, 41, 303, 484
interaction, 42, 182, 227, 303
repulsion, 436
staircase, 336, 347

covalency, 365

cuprate superconductors, 433

cyclotron
diameter, 28, 256, 262
energy, 24
frequency, 37, 228
mass, 150
resonance, 153
splitting, 10

dangling bonds (DB), 518, 640

de Broglie wavelength, 57, 117, 508

density of states (DOS), 12, 15, 58, 104, 366, 514

dielectric function, 172

differential transmission spectrum (DTS), 296

diluted magnetic semiconductors, 102

disordered system, 36

dispersion relations, 125

DNA, 551

do-it-yourself quantum mechanics, 57

double barrier, 57, 275

double barrier resonant tunneling, 62

double barrier resonant tunneling diode, 63

double quantum wells, 130

Drüde model, 655

electroabsorption, 72, 320
electrolyte, 583
electron-electron correlation, 30
electron-electron interactions, 11, 14, 246, 268, 505
electron-energy-loss spectroscopy (EELS), 380, 384
electron focusing, 258, 506
electron-phonon interaction, 495, 506
electron optics, 244
electron waveguide, 244
electro-optical modulator, 147
Eliashberg function, 496
envelope functions approximation, 118
erbium-doped fiber amplifier (EDFA), 683
etch-decoration transmission electron
microscopy (ED-TEM), 605
exchange-correlation potential, 167
excitation gap, 37
exciton binding energy, 297
exciton-exciton interaction, 299, 301

Fang-Howard wavefunction, 281
far-infrared (FIR), 209, 224
far-intrared spectroscopy, 228
Fermi
    liquid, 377, 435, 445
    momentum, 441
    pressure, 227
    statistics, 37
    wavelength, 15, 57, 105
fermion, 45, 51, 367, 433, 448
Feynman diagram, 444

field effect transistors (FET), 552
field ion microscope (FIM), 610, 615
filling factor, 15, 18, 19, 29
fine structure constant, 8
flux quantum, 43
four-wave-mixing (FWM), 301
fractional quantum Hall effect (FQHE), 10, 23, 24, 25, 29, 35, 38, 40, 41, 42, 44, 46, 48, 49, 51, 99, 204, 223, 311
Franz-Keldish effect, 69, 320
friction force microscopy, 486

gauge invariance, 14
Ginsburg criterion, 402
Ginsburg-Landau
    description, 48
    theory, 51, 435
grain boundaries junctions (GBJs), 406
Green's function, 182
group velocity, 15, 65, 251

half-integral filling factor, 12, 13
Hall
    bar geometry, 17
    classical, 14
    conductivity, 35, 36
    effect, 36
    plateaus, 9, 10, 11, 12, 13, 27, 35, 36
    resistance, 12, 17, 18, 19, 265
Hamaker constant, 634
Hamiltonian, 45, 436
Hartree Fock, 373
Hartree potential, 167
heat capacity, 402

Hebel-Slichter peak, 435
Heisenberg model, 438
heteroepitaxy, 59
heterointerfaces, 59
high-electron-mobility field-effect
    transistor (HEMT), 99
highest occupied molecular orbital
    (HOMO)
high-temperature (HiTc)
    superconductivity, 106, 377,
    385
high-temperature
    superconductors, 453, 507
high Tc materials, 431
high Tc oxides, 399
holographic lithography, 211
Hubbard
    band, 381
    gap, 370, 456
    Hamiltonian, 460
    model, 436, 640

inelastic light scattering, 191
inelastic scattering, 13
integer quantum Hall effect
    (IQHE) see quantum Hall
    effect
integral integer filling factor, 13,
    16
Inter-Landau-Level transitions,
    233

jellium, 533
Josephson
    effect, 8
    junctions, 339, 353
    oscillation, 345

Karringa relaxation, 12
Kondo system, 501
Kosterlitz-Thouless point, 48, 50
Kramers degeneracy, 123, 128
Kronig-Penney, 56, 164, 173
Kubo formula, 51

laminar crystal structure, 415
Langmuir-Blodgett, 578, 646
Langmuir monolayers, 567
Langmuir-Schäfer technique, 567
Larmor frequency, 479
Laughlin's
    ground state, 39, 42, 47, 50
    theory, 14, 25
    wave function, 25, 37, 43
Landau level, 10, 12, 16, 37, 39, 40,
    262, 308
Landau level filling factor, 24, 48
Landau quantization, 508
Landauer - Büttiker formalism, 243
lateral quantization, 211
laterally confined
    nanostructures, 140
liquid metal ion source (LMIS), 614
liquid phase epitaxy (LPE), 92
local density approximation
    (LDA), 166, 172, 380, 643
Lorentz
    force, 276, 286
    transformation, 36
lowest unoccupied molecular
    orbital (LUMO), 553
Luttinger
    hamiltonian, 121
    liquid, 377
    matrix 131
    parameters, 286

macromolecules, 542, 549

magnetic semiconductor, 86

magnetic spectroscopy, 103

magneto-absorption, 95, 96, 152

magnetophonon, 39

magneto-phonon oscillations, 95

magnetoplasmon, 39, 228

magneto-tunneling, 62

man-made bound states, 61

Maxwell's equations, 48

mean field theory, 45, 46

mesa etched structures, 211

mesoscopic quantum regime, 57

metallo organic chemical
    vapor deposition (MOCVD),
    59, 84, 92, 117, 164

metal-nitride-oxide-semiconductor
    (MNOS) technology, 597

miniaturization, 467, 551

modulation doped field effect
    transistor (Mod-FET), 313

modulation doped system, 230,
    311

modulation doping, 98

molecular beam epitaxy
    (MBE), 59, 61, 84, 92, 117,
    143, 164,
    277, 668

molecular biology, 573

molecular electronics, 468

Monte Carlo simulation, 41

Mott-Hubbard-Heisenberg, 431

Mott-Hubbard localization, 442

Mott insulator, 371

nanometer-scale objects, 573

nanometer-scale science
    and technology, 466

nano-scale magnets, 600

nano-scale patterns, 581

nanoscopic structures, 551

Néel temperature, 434

negative differential conductance, 72

negative differential resistance
    (NDR), 67, 315, 537

negative magnetoresistance, 261

Neutron scattering, 41, 457

Newtonian mechanics, 55

NMR Knight shift, 367

NMR Knight shift measurements, 385

optical amplifier, 682

optical bistability

optical communications, 147

optical fiber, 677

optoelectronic devices, 293

oscillator strength, 40, 41, 42

pairing interactions, 446

pairing mechanism, 460

pairing state, 446

parabolic confinement
    potential, 214, 229

paramagnetic phase, 443

Pauli principle, 297

peak-to-valley current ratio, 280

penetration depth, 403

phase transitions, 591

phonon
    assisted resonant tunneling, 62
    confinement, 197
    density of states, 459
    spectra, 500

photoelectrochemical technique
    (PEC), 583

photo emission
    spectroscopy, 386, 527, 656

photoluminescence, 71, 72, 176, 210

photoluminescence excitation (PLE), 230

photon emission, 478, 514

photonics, 688

photonic switches, 293

photoreflectance, 101, 176

plasma frequency, 42

plasmon, 42, 202, 226

plasmon dispersion, 229

plasmon excitations, 201

plasmon gap, 47

point-contact spectroscopy, 497

polymer films, 579

polytype heterostructures, 63

polytype superlattice, 63

proteins, 559

pseudopotential, 166

pyramidal apex oxygen, 456

quantized Hall resistance, 8, 10, 13, 14, 15, 260

quantum confined energy states, 211

quantum confinement, 7, 15, 281

quantum dot, 7, 58, 104, 139, 204, 209, 212, 229, 246

quantum electrodynamics (QED), 8, 9

quantum fluctuations, 370

quantum Hall effect (QHE), 7, 8, 10, 15, 16, 18, 20, 24, 28, 162, 249

quantum heterostructure (QHS), 83, 84, 106, 117

quantum interference, 259

quantum point contact, 15, 243, 247

quantum spin liquid excitations (spinons), 448

quantum structure, 56

quantum well (QW), 55, 58, 61, 117, 119, 192, 287, 293

quantum wire, 7, 58, 104, 139, 204, 209, 212, 229, 246

quasi-liquid, 636

quasiparticle, 24, 25, 27, 28, 386, 450

quasiparticle excitation, 44

Rabi frequency, 304

Raman
    measurements, 70
    scattering, 94, 97, 192, 458
    spectroscopy, 210

reflection high-energy electron diffractometer (RHEED), 86

resonant magnetotunneling, 289

resonant tunneling, 28, 56, 61, 62, 93, 104, 129, 145, 268, 275, 311, 647

resonant tunneling magneto-spectroscopy, 276

resonant tunneling transistor, 674

RNA, 551

roton minimum, 37, 38, 41

scanning electron microscopy (SEM), 91, 577, 588

scanning potentiometry, 480

scanning probe, 574

scanning probe methods (SPM), 466

scanning tunneling microscopy (STM), 106, 163, 335, 352, 465, 513, 549, 574, 632

scanning tunneling spectroscopy (STS), 513, 639
Schottky barrier, 658
Schottky contact, 69
Schrieffer's spin bag, 375, 442
Schrödinger equation, 118, 166, 214
self consistent Born approximation (SCBA) 12
self electro-optic devices (SEED), 319, 691
semimagnetic superlattice 151
sensitive magnetometers (SQUID's), 339, 354
Sharvin
    contact, 498
    resistance, 498
short period superlattice, 161
Shubnikov-de Haas
    oscillations, 95, 215, 267, 508
Si-MOSFET, 7, 13, 210
single electron tunneling (SET), 334, 339, 480
single-particle excitation, 37
soliton, 339, 684
spin
    density wave (SDW), 437
    flips, 438
    polaron, 436
    relaxation, 12, 126
    splitting, 10, 12, 15
    wave, 438
spin degeneracy, 15
spin fluctuation, 379, 436
SQUID, 601
Stark
    ladders, 57, 66, 67, 73, 76
    localization, 56, 97
    shifts, 72
    states, 67, 68

STM-derived local probe methods, 466
STM imaging, 472
STM tip, 539
STM vibrational spectroscopy, 543
Stoke's theorem, 49
strained-layer heterostructure, 101
superconducting superlattice, 419
superconductors
    high Tc, 365, 377, 415
    low Tc, 401
    metallic, 433
    organic, 367
    type-II, 50
superfluid helium, 37
superlattice (SL), 55, 56, 57, 101, 137, 143, 163, 193, 204, 294
supramolecular chemistry, 550
surface chemistry, 514, 542
surface plasmon 476
surface topography, 471

tailor-made material, 163
Tamm states, 74
Taylor cone, 592
Tersoff-Hamann theory, 541
thermal profilometry, 480
tip-surface interaction, 635
translation symmetry, 37
transmission coefficient, 60, 517
transmission electron microscopy (TEM), 91
transmission probability, 15
tunnel barrier, 472
tunneling spectroscopy, 290, 535
turnstiles, 351

ultra-high mobility, 42

ultra-small tunnel junctions, 333
umklapp process, 66

van der Waals
    attraction, 633
    force, 482
    interaction, 566
    radii, 566
van Hove
    anomalies, 368
    singularity, 173, 179
Verwey-Mott insulation, 370
vortex, vortices, 404

Wannier-Stark quantization
    (WSQ), 128, 143, 144, 146,
    157

wide-gap semiconductor, 154
wide-gap superlattices, 156
Wigner crystal, 28, 51, 204
WKB approximation, 260, 284
work function, 472

xenon, 602
x-ray absorption spectroscopy
    (XAS), 380, 458
x-ray diffractogram, 417

Zeeman splitting, 501
zero-gap semiconductors, 147
zinc blende lattice, 118